INSTRUMENTATION, CONTROL, AND AUTOMATION IN THE POWER INDUSTRY VOLUME 38

Proceedings of the Thirty-Eighth
Power Instrumentation Symposium
(5th Annual ISA/EPRI Joint Controls and Automation Conference)

June 1995

ISA Program Chairman
Gordon McFarland
Honeywell

EPRI Program Chairman
Mark Perakis
Electric Power Research Institute

1945 • 1995

PUBLICATION POLICY

Technical papers may not be reproduced in any form without written permission from ISA. The Society reserves the exclusive right of publication in its periodicals of all papers presented at the annual ISA Conference, at ISA Symposia, and at meetings co-sponsored by ISA when the Society acts as publisher. Papers not selected for such publications will be released to authors upon request.

In any event, following oral presentation, other publications are urged to publish up to 300-word exerpts of ISA papers, provided credits are given to the author, the meeting, and the Society using the Society's name in full, rather than simply "ISA."

Reprints of articles in this publication are available on a custom basis at reasonable prices in quantities of 100 or more.

Permission to photocopy for internal or personal use, or the internal or personal use of specific clients, is granted by Instrument Society of America for libraries and other users registered with the Copyright Clearance Center (CCC), provided that the fee of $.50 per page is paid directly to CCC, 21 Congress Street, Salem, MA 01970. Payment should include the fee code information at the top of the first page of papers copied which appears in the form of:

0074-056X/95[First Page]-[Last Page]/$0 + .50pp
[ISSN/Year of Publication/Page Numbers of Article/Fee for Copying] ($.50 Per Page)

Special requests should be addressed to:

Training and Publications Services
ISA
P.O. Box 12277
Research Triangle Park, NC 27709

For further information concerning publications policy and reprint quotations, contact:

Training & Publications Services
ISA
PO Box 12277
Research Triangle Park, NC 27709
Phone: (919) 549-8411
FAX: (919) 832-0237

ISSN: 0074-056X

ISBN: 1-55617-549-3

Library of Congress Card Number 63-25575

Fifth International Joint ISA POWID/EPRI Controls and Instrumentation Conference Reviewers

Lewis Adams	TU Electric
Donald Alexander	Detroit Edison
Mike Annon	I&C Engineering Associates
Dan Antonellis	Foxboro Company
Jeffrey Arpin	ABB Combustion Engineering
Laverl Ashcroft	Tucson Electric Power Company
Ronald Battle	Martin Marietta Energy Systems
James Batug	Pennsylvania Power & Light Company
James Bayless III	Black & Veatch
Lester Broeker	PSI Energy
Tom Broyles	Stone & Webster Engineering
Augustine Chen	Black & Veatch
Don Christopher	Houston Lighting & Power
Gary Cohee	Entergy
Ronald Cole	Entergy Operations, Inc.
Danny Crow	TU Electric
Brad DeBow	Lockheed Idaho Technologies Company
Terry Dimmery	Duke Power Company
Don Dziubakowski	Bailey Controls
Dale Evely	Southern Company Services
Dave Farrell	Honeywell
Daniel Franczak	Sargent & Lundy Engineers
Phil Fredricks	Raytheon Engineers & Constructors
Donald Frerichs	Bailey Controls
Don Frey	Computer & Control Consultants
Dwayne Fry	Oak Ridge National Laboratory
William Fryman	Illinois Power Company
Richard Givan	Consultant
Edwin Good	Consultant
William Gordon	Bechtel Savannah River Inc.
Glen Hadley	Iowa Electric
H.M. Hashemian	AMS Corporation
Ronald Hicks	Black & Veatch
Robert Hill	Amtech Services
Wayne Holland	Southern Company Services
Kevin Holmstrom	Northern States Power Company
Robert Hubby	Leeds & Northrup
Roger Hull	Westinghouse Electric
Poong Hyun Seong	Korea Advanced Inst. of Science & Tech
Ronald Johnson	Sargent & Lundy Engineers
Paul Kenny	Forney Corporation
Norman Kerman	Duquesne Light Company

Les King	Hoosier Energy
Kofi Korsah	Oak Ridge National Laboratory
Michael Krueger	Commonwealth Edison
George Kunder	Metropolitan Edison
Don Labbe	Stone & Webster
Michael Leahy	PECO Energy
Daniel Lee	Bailey Controls Company
Roger Lewis	Southern Company Services
William Maxwell	Lower Colorado River Authority
Gordon McFarland	Honeywell
James Nay	Westinghouse Electric
Milton Neher	Central & South West Services
Rudolf Neustadter	Raytheon Engineers & Constructors
Phil Newman	Sorrento Electronics
Don Olson	ABB Lummus Crest
Mohan Pankhania	Raytheon Engineers & Constructors
Ross Papilla	Southern California Edison Company
Robert Pierce	Baltimore Gas & Electric Company
Norbert Pobanz	Bechtel National, Inc.
John Poelma	Nevada Power Company
Clayton Price	EPRI/NMAC
Grant Provins	Bailey Controls
Robert Rand	Bailey Controls
David Roney	Raytheon Engineers & Constructors
Richard Roop	Hoosier Energy
E. Ross Forman	Process Automation Technology
Tom Russell	Honeywell
Ramsay Sawaya	Northern States Power Company
Uma Seshadri	Indira Gandhi Centre for Atomic Research
Frank Short	Houston Lighting & Power Company
Chuck Skidmore	Florida Power & Light
William Sotos	American Electric Power
Fred Stearns	Lower Colorado River Authority
Len Sterle	Allen-Bradley Company
Steve Stultz	Baltimore Gas & Electric Company
Cyrus Taft	EPRI I&C Center
C. Richard Turley, Jr.	Westinghouse Electric
Joseph Vavrek	Sargent & Lundy
Larry Voss	Iowa Electric
John Wallace	Wisconsin Public Service Corporation
Robert Webb	Pacific Gas & Electric
Joe Weiss	EPRI
John Wester	Central Power & Light
Peter Wicyk	Commonwealth Edison
Marjorie Widmeyer	Supply System

Reed Wiegle	Canus Corporation
John Wray	Mollerus Engineering Corporation

FIFTH INTERNATIONAL JOINT ISA POWID/EPRI
CONTROLS AND INSTRUMENTATION CONFERENCE

FOREWORD

The Fifth International Joint ISA POWID/EPRI Controls and Instrumentation Conference coincides with ISA's 50th anniversary. This conference will be following a tradition of timely, innovative and inspiring technical papers and discussions. The keynote luncheon address will address ISA's distinguished history and mission for the future.

During the 1994 conference, industry executives indicated that the power industry is undergoing a revolution. Traditionally highly regulated and monopolistic providers of energy, utilities and facing competition from independent power producers, customer self-generation, and neighboring utilities, and a strong societal sentiment to de-regulate the industry. This leads to our theme for the 1995 conference, "Challenging Competition through Advanced Controls and Instrumentation". In following with our theme, we developed eight informative technical sessions. Plant Information Network Technologies and Expert & Intelligent System Technology provide for improved staff effectiveness through conversion of the copious data into useful information and expert advise, delivered at the right time and the right place. Advanced Control and Advanced Instrumentation Technologies provides a look at innovative applications of emerging technologies that improve plant process performance. Fossil Plant DCS Applications and Nuclear Control Retrofits provide insights to new I&C as enabling technology. And, Nuclear Power Performance & Improvement and Turbine Control Technologies provide an overview of both process and staff performance enhancements. In addition to the technical sessions, there is a panel session comprised of executives representing utilities and control suppliers. The panel will share their perspective on competitive challenges in the power generation industry, how advanced I&C can be used to face the challenges, and where new I&C technology is headed.

These proceedings have been issued in conjunction with the conference so that attendees can follow along with the Authors' presentations as well as leave the conference with the technical papers for reference when returning to the job. To achieve this we had to hold to firm dates for both paper submittal, and issuance to ISA for publication. Unfortunately, a few papers did not make the proceedings publication. We will attempt to include the missing papers in the 1996 proceedings.

We wish to extend our appreciation and sincere thanks to the many people, who largely through their efforts, have made this conference successful: to the Session Developers who structured innovative topics and through their gentle and persistent prodding helped the papers through to completion; to the Authors who provided the enlightening and thought provoking technical papers; to the Executive Panelist who provided their vision of the future and how advanced controls can be utilized in an evolutionary competitive environment. Additionally, we would like to recognize some people for their special support. They include Roger Hull for his coordination, and his team, for review of the papers and provision of guidance to

Authors for enhancements. Kowana Ragland and Carol Ufferman from ISA for coordination of paper submittals and publishing of the proceedings. Lori Adams, EPRI Conference Coordinator, for arranging the site, accommodations, registration, and truly making the conference run.

Gordon McFarland
1995 ISA POWID Co-Program Chairman

Mark Perakis
1995 EPRI Co-Program Chairman

TABLE OF CONTENTS

1994 POWID/EPRI SYMPOSIUM PAPERS

© 0074-056X/95/001-010/$0 + .50pp

PLANT INFORMATION NETWORK
- IMPLEMENTATION GUIDELINES -

AUTHORS

Dr. Nick Hurley, Nehal Maniar

KEYWORDS

Plant, Information, Network, Guideline, Power.

ABSTRACT

This paper presents *practical* advice to computer managers intent on capitalizing on the enormous benefits offered by Plant Information Networks. In a typical Power Utility, these networks tie together the various isolated islands of automation into a global information framework that presents information to anybody in the system having a personal computer. The advantages of both Integration and Distribution of data are unheralded with payback times of *less than four months* frequently being reported. In order to truly make the most of this exciting technology however, there are definite key guidelines that should be followed. This document offers practical pointers. It will cover the evolution of the network from inception and justification to manpower logistics, on to implementation and finally, system administration.

OVERVIEW

A Plant Information Network is broadly described as a system that collects up-to-date process information from a number of sources such as programmable logic controllers (PLCs), distributed control systems (DCSs), remote terminal units (RTUs), Laboratories, etc. and integrates this information, making it available to a large number of plant personnel. The benefits of such networks are substantial and have been widely reported.

This document, however, looks more closely at how such a network is implemented. The diagram on the next page shows the various phases carried out in the implementation of a plant information network. The guidelines that follow should be useful for an engineer (generally referred to as the Champion) about to undertake such a venture. The major phases include:

- Project Approval — How the champion justifies the project,
- Project Launch — The up-front preparation to motivate the project implementation team,
- Planning Phase — The procedures undertaken by the team in the planning phase,
- Implementation — The actual implementation of the project,
- QA — Proving that the system meets the objectives,
- Support — Setting up the mechanisms that will maintain the system.

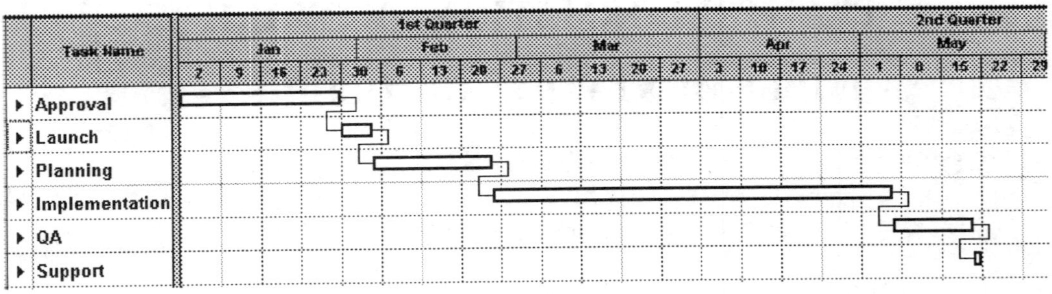

APPROVAL PHASE

The success of this phase is largely dependant on the credibility of the champion and the ability to understand and manoeuver in the administrational and political circles of the plant. Additionally, it often depends on the benefits (both tangible and intangible) that are put forward to management in a project justification document. When discussing the typical tangible benefits, the champion would point to:

- Documented and qualified cases of downtime, disruptions, accidents etc. that clearly would ***not*** have happened if a plant information network had been in place.
- Documented and qualified cases of inefficiencies associated with the tedious collection of information that would be eliminated with a plant information network.
- Estimated improvements in production or efficiencies as a result of advanced applications such as multi-variable control or statistical process control. This type of justification can carry more weight if proven figures from similar sites is obtained.

Intangible benefits are significant and normally tip the balance in forward thinking sites. They include:

- Better plant coordination and visibility through the availability of information.
- Improved quality control and faster reactions to plant upsets as a result of more up-to-date information.
- Better planning based on accurate historical data.

LAUNCHING THE PROJECT

Following the go-ahead for the project, the management first defines a broad mission statement that embodies the essential objectives of the project. A typical mission statement could be similar to the following:
Your mission is to:
Design and implement an information network that:

- *Initially ties together (i) the DCS and PLC in Area A, (ii) the laboratory and QA section, and (iii) the environmental system,*

- *Can be easily expanded to other areas in the plant,*
- *Will provide real-time and historical data in graphic format to (as a minimum) all 27 PC's in the technical, operating and maintenance department,*
- *Link into the corporate wide area network.*

Additionally you must:

- *Achieve a return on investment (ROI) of 6 months (to be measured one year after implementation),*
- *Use technology that complies with the corporate standards,*
- *Show demonstrable functionality within nine months.*

Following the definition of the mission statement, the team leader, (typically the project champion) and team members are selected. It is advisable to keep the number of team members small (below 5) and select the members from divergent areas e.g operations, software, DCS and PLC areas, management and maintenance.

PLANNING PHASE

Under the guidance of the team leader, an implementation strategy with dead-lines, personnel allocation and budgets is put together to meet the requirements of the mission statement. Note that it is advisable that the management (within reason) accept the *team's* strategy. Obviously the Team will be more committed to the strategy they themselves have devised. A typical time line for the planning phase is shown below. It covers a global strategy (where the team members clarify their objectives and set up a needs-

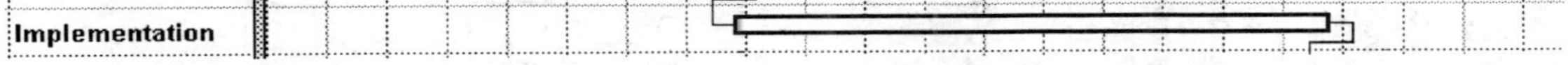

related strategy) followed by a phase where they select the software platform they are to use as the framework for the Plant Information Network. The next section covers the important criteria to bear in mind when selecting the software platform.

SELECTION OF THE SOFTWARE PLATFORM

There are a number of *non-negotiable* features that a plant information package must provide. These are discussed in brief below:

Integration tools - The software must exhibit a clean, structured architecture that enables new data sources and new users to be added simply and quickly without limits. The diagram below shows such a structure. Implicit in the structure is the ability to:

- Provide complex point structures such as PID controls with a simple entity name that covers multiple attributes like Process Variables, Setpoints, Outputs etc.
- Enables the emulation of typical DCS faceplates.
- Provides a common method of addressing ***all*** point structures irrespective of the source address (i.e, a database file should be addressable in the same form as a PLC point.)
- Provide a large number of proven drivers to popular DCS and PLC systems as well as the ability to easily add the more exotic drivers.

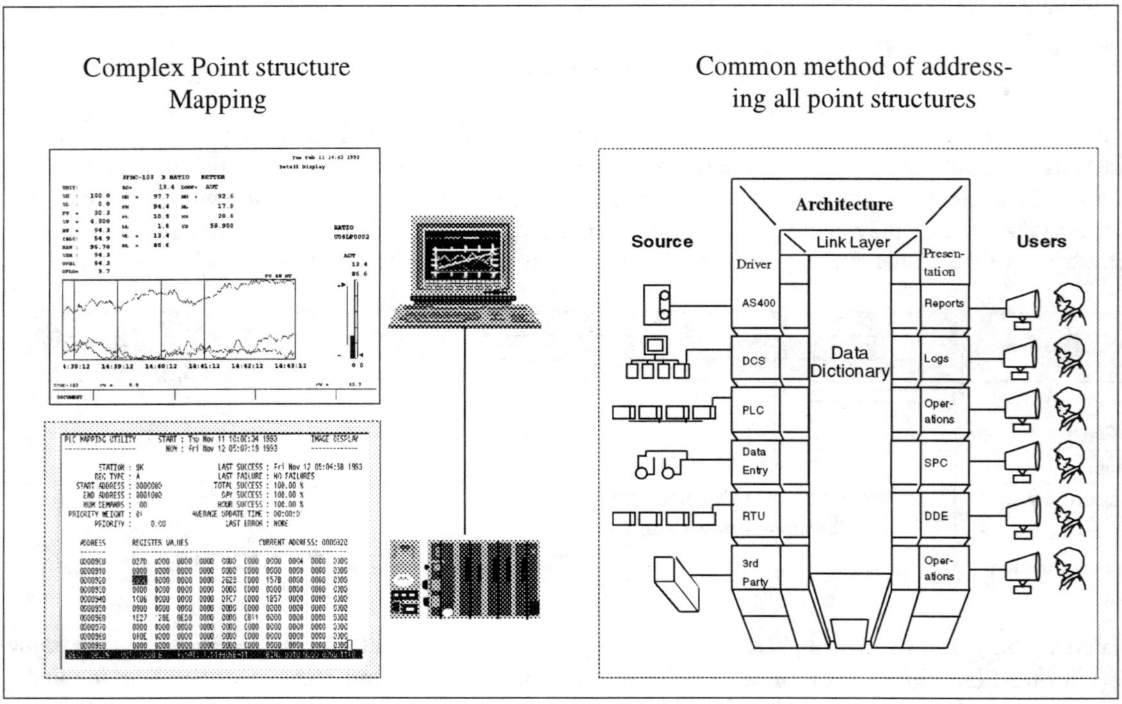

Distribution Tools - The software package must provide the tools that will facilitate and easily administer a network that provides up-to-date, easily understood information directly to the personal computers of key staff. The diagram below shows the elements of such a network. Among the *non-negotiable* features to look for are the following:

- There must be no limit to the number of users that a system supports.
- The system must be able to provide an interactive graphical user interfaces (GUIs) to personal computers and workstations such as IBM compatibles, Mcintoshes, X terminals and RISC workstations.
- The system must support user friendly interface widgets such as pulldown menus, pop-up windows, scrollbars, buttons, browse widgets etc.
- It must be possible to administer the entire network from one central location. Graphics created at one location must be available throughout the organization without tedious down-loading.
- A simple "bullet proof" security system must be provided.

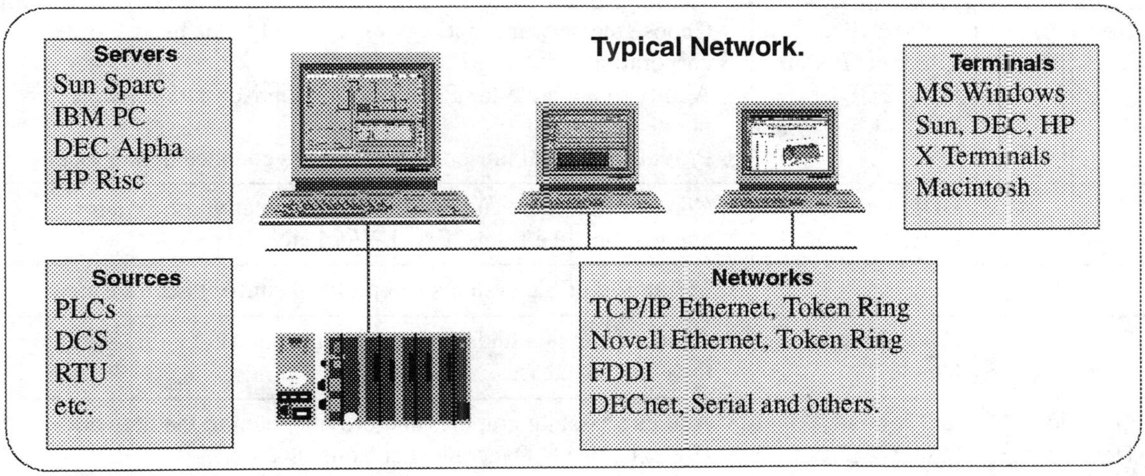

Industry Standards - The software package should support industry standards and should be "open." This means that the team is free to choose the best hardware/software/networking architecture available on the market and can freely "plug" third party packages for specific purposes into the architecture.

This can only be achieved if the team insists that industry standards are conformed to. The standards most important in the implementation of a plant information network that will provide an unconstrained evolution path, free from "brick walls", include those identified in Table 1:

Table 1: Industry Standards Benefits

Relevance	Published Industry Standard	Benefit
Server Hardware	UNIX	Select from any of the major RISC or CISC platforms - e.g, SUN, HP, DEC ALPHA, IBM, Intel Platforms etc.
Terminal Hardware	X Protocol	Choose from any hardware platform such as IBM compatibles, Macintosh, SUN, DEC, X terminals etc.
Networks	TCP/IP, IPX, Ethernet, Token Ring, FDDI, DECnet,	• Choose the network topology most suited to the needs of the corporation. • Ability to integrate the Plant Information Network into the Corporate network. • Provide graphical presentations to users connected via modem.
	DDE and NetDDE	• Connect Real-time and historical data directly into Windows applications like Lotus, Excel, Word etc.
	NFS	• Mount server file systems and perform simple data transfers.
Databases	dBase Format Oracle	• Choose from hundreds of applications that can tie directly into the native databases.
Graphic Development	.dxf format (AutoCAD)	• Animate existing graphics and provide them to the network. • Use existing CAD technical personnel.
Application Interfaces	SQL	• Transfer mission-critical data to and from the corporate database.
	Fortran, C, C++	• Use existing scientific libraries on real-time and historical databases. • Create high speed custom applications like models and simulators.
Window Managers	Windows Motif Open Look	• Integrate applications without having to retrain users.

Tools for Development - The success or failure of the Plant Information Network in terms of ROI will depend largely on the applications that are created specifically for the site. The applications typically start off as graphical presentations of the real-time and historical information, and later have "intelligence" added to them to "close the loop." The selected software package should include as a minimum:

- **An integrated graphics generator.** This will allow engineers to quickly and easily create ergonomic and effective graphic presentations. It should be simple to access the complete spread of information from real-time and historical data from *any* data source in the system. It should also allow the simple addition of widgets such as group displays, charts, browse widgets etc.

- **An integrated scripting language.** This language should be based on a standard, like SQL, but also have extensions that are specifically aimed at the power industry. The language should be usable both in the graphics environment and also as a stand alone, scheduled or triggered program. Applications using this kind of language that are created by **engineers** result in the biggest return of investments.

Typical Graphic Produced using CAD Tools

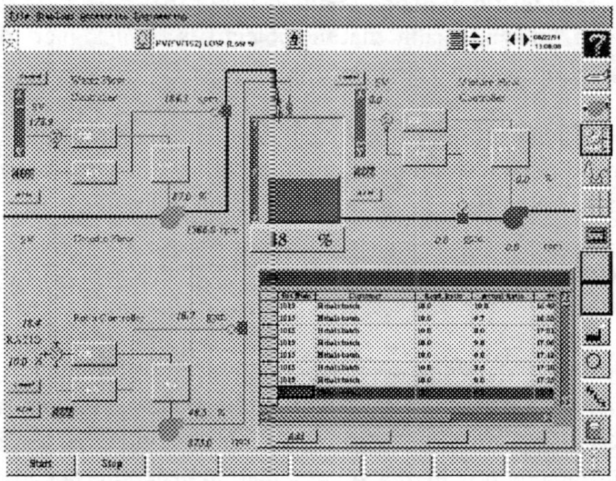

IMPLEMENTATION

The following diagram shows the essential tasks involved in the implementation of a plant information network.

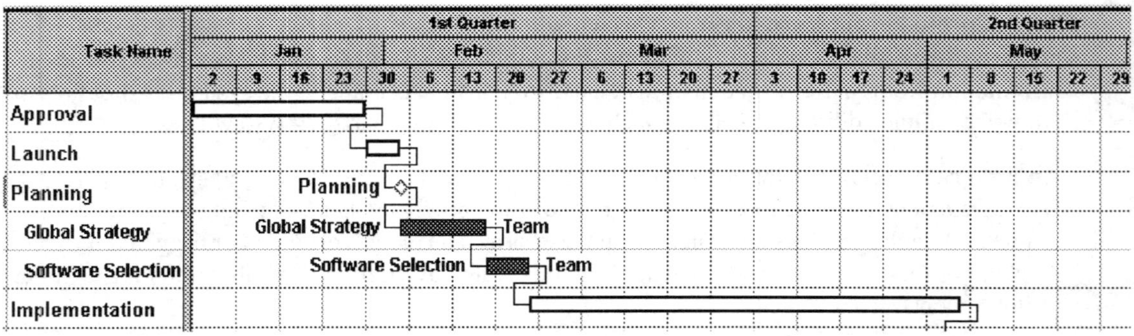

As discussed, the first application should be modest and clearly defined. Later applications may be more ambitious once the engineers have developed hands on, practical experience.

Obviously, the duration of each task will depend on the size of the project. To keep control of the project and to avoid "function bloom," it is advisable that the objectives are designed to keep the complete implementation of the first application within 3 months.

Set up Network - In this task, the MIS department connects the server into the Local Area Network and ensures that all the nominated personal computers on the network can display graphic screens from the server. Also, links to the corporate office from the server through the Wide Area Network are established.

Add Data Sources - This usually involves the integration of at least one DCS or PLC system to collect real-time and historical data.

Configure User Interface - Here the graphic interfaces are designed in the CAD environment and animated to show the real-time and historical data. Additionally, data entry screens for operators and laboratory personnel are designed to capture information that would previously have been written down on paper. The databases are set up and any high level application programs specific to the site are created.

QA Manual - Some time is allocated for a Quality Assurance (QA) manual. the QA manual describes a set of tests that prove the complete workings of the system.

QA AND SUPPORT HANDOVER

QA - Once the system has been set-up, it goes through a series of QA checks (as described in the QA manual) that may necessitate rework.

Support Hand over - Following a successful QA, the system is handed over to a support department that could exist either within the organization or may belong to the software vendor organization. At this stage, the project is considered complete.

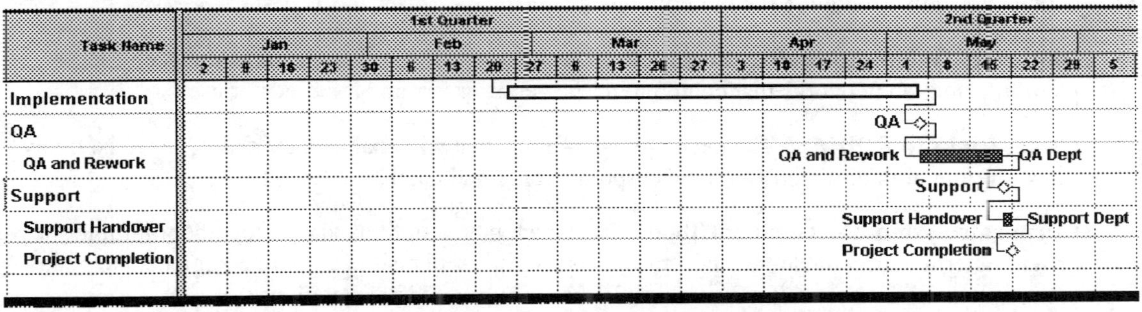

EXPANSION OF THE NETWORK

Once the initial application has been implemented, subsequent applications may be easily added with impressive returns on investment. This is because:

(i) the hardware has been purchased and commissioned,

(ii) the software framework with all the tools has been purchased and

(iii) the engineering staff has practical hands on experience in implementing solutions and a good feel for the kind of results they can create.

MAINTENANCE

The maintenance of a Plant Information Network can essentially be divided into two sections:

System Administration - Typically, the MIS department takes responsibility for the system administration which specifically covers the integrity of the network. The details of password control, file permissions, disk usage, network traffic etc. are covered here.

Application Maintenance - Applications on the network must be clearly documented and easily maintained. It is recommended that the task of support is handed over to support engineers once the project is completed so that project engineers can go on to the next profitable project.

FUTURE TRENDS

The high visibility being accorded to projects such as these has intensified the pursuit for better tools in Plant Information Networks. The benefits attributable to this development will be directly applicable to existing networks. The following development directions are already showing excellent promise.

- **The development of configuration building blocks** - These will be created in an object oriented format, to further simplify the development of applications. Engineers will be able to build applications by simply plugging the building blocks together.
- **Improved System Administration Tools** - These tools will simplify the maintenance of large plant information networks that could include several servers and hundreds of users.
- **Closer Links to RDBMS** - The development of tools that will enable data of all types to be freely transferred in both directions with the corporate relational database.
- **Corporate System Development Tools** - The development of tools that tie together multiple sites providing efficient, reliable transport mechanisms. This will enable corporate engineers to get a composite overview of multiple sites within the same graphic presentations.
- **Multimedia Support Tools** - Tools that integrate video and audio facilities into the network.
- **Integration of new front end technologies** - like the Instrument Bus.
- **Closer integration of specialized applications** - Applications such as expert systems, neural networks, SPC, simulation etc. will be integrated directly into the framework of the system. Currently these applications are started separately and require programmers to implement the data exchange mechanisms.

SUMMARY

Whilst the benefits of implementing plant information networks are attractive, engineers have found the prospect of organizing and managing such a project daunting. It is hoped that the practical suggestions outlined in this document will be useful as guidelines to modernize and improve the efficiency of our operations in the power generation industry.

© 0074-056X/95/011-020/$0 + .50pp

PLANT INFORMATION NETWORK (PIN) EXPERIENCE AT EASTLAKE POWER PLANT

William R. Duge
Jay R. Pomeroy
Sidney J Stone
CENTERIOR SERVICE COMPANY

John Cunningham-Reid
John Turk
THE ROVISYS COMPANY

Mark DeCoster
Russ Pflasterer
ELECTRIC POWER RESEARCH INSTITUTE

ABSTRACT

The controls upgrade project at the Centerior Energy Eastlake Plant has been enhanced by the addition of a Plant Information Network (PIN). The PIN provides an engineered hardware and software system that helps individuals to perform their job functions more efficiently and accurately. As the utility industry downsizes and has increased demand for older fossil units to operate in load following modes the ability to have and use historical information assists in problem and potential problem analysis that effect availability and efficiency.

Centerior has added a research and development effort in this project by participation with the Electric Power Research Institute (EPRI). EPRI has brought the concepts of software protocol standardization into play by providing a standard software protocol template for design and implementation at various utilities around the world. The RoviSys Company provided system integration of the control system, data acquisition and EPRI protocols and designed the red tagging system. This paper presents the experience and results of implementing a Plant Information Network (PIN) along with a controls modernization upgrade project using a Bailey Controls distributed microprocessor based control system for Eastlake Plant Units 3 & 4 and identifies the benefits and advances accomplished.

PROJECT BACKGROUND

Centerior Energy Corporation was created in 1986 and is the holding company for the Cleveland Electric Illuminating Company serving north eastern Ohio and the Toledo Edison Company serving the north western Ohio areas. The generating capacity mix of the Cleveland Electric Illuminating Company is approximately 2500MW fossil and 700MW nuclear. The generating capacity mix for the Toledo Edison Company is approximately 670MW fossil and 700MW nuclear. The fossil fired power plants were constructed primarily in the 1950's and were considered state of the art drum type reheat units at that time. The youngest of the fossil units at Eastlake are supercritical type reheat units and were constructed in the early 1970's. The intended use of the Eastlake units was to provide base load capacity for the rapidly expanding industrial electricity demand of the late 1960's and early 1970's.

The Eastlake Plant is located approximately 22 miles east of Cleveland, Ohio on Lake Erie near the mouth of the Chagrin River. There are six units at Eastlake, of which Units 3 & 4 are affected by this project. Unit #3 is rated at 132MW net and began commercial operation in 1953. The steam generator is a controlled circulation twin furnace boiler with tangential firing using pulverized coal. Main and reheat steam are supplied to a tandem compound, triple flow turbine generator. The original pneumatic controls were state of the art for that time. Unit #4 is rated at 240MW net and began commercial operation in 1956. The steam generator is a controlled circulation boiler with separate (divided) superheat and reheat furnaces. The fuel is pulverized coal. The steam generator supplies steam to a tandem compound triple flow turbine generator. The original pneumatic controls are the same as Unit #3.

Centerior is a member of the Electric Power Research Institute (EPRI) headquartered in Palo Alto, California. EPRI has recently developed a concept called Tailored Collaboration for member utilities to reinvest part of their annual dues to enhance and add value to ongoing projects where the hosting utilities and other utilities may benefit. The Eastlake Units 3 & 4 Control Modernization Project is being used for this purpose and EPRI and Centerior jointly agreed on the scope of work for this Tailored Collaboration Project to meet mutual objectives.

Centerior has implemented a program to inspect all of the fossil fired units within the system to assess the condition of components and anticipated remaining life. This program is known as Unit Career Planning (UCP). UCP establishes remaining unit life, availability and heat rate targets. An inspection of the existing control equipment as part of UCP studies revealed that the pneumatic systems are at the end of their useful life. Hundreds of individual components are obsolete and difficult to replace. The units are operating in a load following mode rather than the original base load design which makes quick accurate response extremely desirable. Many parts were identified as not maintainable over the next five years. In many cases, the original supplier no longer stocks these parts and manufactures them only by special order. Future availability of parts is questionable. In order to maintain unit availability and achieve the corporate goals defined by the Unit Career Plan, the controls are being replaced with a microprocessor-based Distributed Control System (DCS). These systems have proven to be reliable for load following and transient responses encountered in current fossil plant operations.

As the automatic control systems age and degrade, the operator is called upon to regulate more plant systems manually. The operator is then less able to observe, anticipate, and respond to other operating situations as they occur. The risk to plant personnel and equipment increases dramatically. The loss of a single control system component significantly increases the risk of a boiler or precipitator explosion. Even the smallest reheat generating unit should not be run with all control loops in manual for extended periods of time. The amount of information being monitored exceeds the ability of the best trained operators to respond. The controls are degrading just as they are called upon to perform increased duty due to increased load following and reserve shutdown modes of operation. The resultant temperature and pressure excursions subject boiler and turbine parts to high stresses. High temperatures lead to metal creep and temperature swings to metal fatigue. Creep and fatigue are the primary mechanisms of failure for boiler tubes, turbine shells and rotors. Minimizing temperature and pressure excursions will help preserve the boiler and turbine components and meet the corporate mission statements optimizing the use of capital funds. Along with the implementation of a DCS an increase in the quality of process data becomes available. This data is valuable for operating the unit from shift to shift, but it is also valuable for analyses trouble shooting and planning by various groups within the power plant. For this reason significantly new diveranion to unit operation.

SCOPE OF THE PLANT INFORMATION NETWORK (PIN)

At the outset of the project Centerior decided to evaluate the various data historian software available. Centerior conducted a survey of the available software and system integrator and found that there was a well established experience base with Oil Systems' Plant Information (PI) software. Not only was there the extensive experience base but there was also the varied options and potential upward migration of the software that allowed the phased type of implementation that was anticipated to be possible. The availability of experienced third party system integrators further supported the selection of PI.

Many control upgrades on older fossil units such as Eastlake involve only the combustion control and balance of plant modulating control loops. It was decided that all breaker or motor control stop/starts and indications that were previously operated from the benchboard or vertical board, except for functions involved in synchronizing the unit, would be done in the DCS. Centerior committed to extensive Operations training because of the dramatic and rapid change from operating pneumatic hand-auto stations to graphic displays. The control modernization of Eastlake Units 3 & 4 addressed all controls and existing control components and their interactions from the operation and maintenance standpoint. Control algorithm upgrades were implemented on most all of the major equipment associated with the units. Changes in operating philosophy were adopted were also adopted as a result of the advanced controls.

Burner management logic for ignitors, pulverizer feeder, crusher, and fuel trips was developed and implemented into the DCS. All interlocks and permissives found on breaker or motor start/stop logic is shown on CRT graphics to operators to enable trouble-shooting of missing permissives. Combustion controls as well as other advanced controls were implemented to provide turbine follow and full coordinated control capability for the units. Other important items, in addition to device status found on hard wired control panels were the lockout and tag out function, breaker control power, and trip coil status. All of these functions were incorporated into the DCS logic.

The DCS cabinets are dedicated for Modules, Termination Units or Communications. The Unit 3 data highway is contained within a single cabinet for ease in maintenance and trouble shooting. The Unit 4 data highway is in a separate cabinet along with the Computer Interface Units (CIUs) for communications with engineering workstation and the Plant Information Network (PIN). The Unit 3 module cabinets are in the "A" row, while the termination cabinets are in the "B" row. The "C" row holds Bailey cabinets for flame scanner controllers, vibration monitors, acoustic leak systems and Sequence of Events Recorders (SERs). The Unit 4 termination cabinets are arranged next to the module cabinets to which they are connected. The primary reason for choosing this alternate arrangement was to even out the overhead cable tray loading for the field wiring. With 50% more cable (due to more I/O), it was important to ensure that the cable trays fills were balanced. This approach also shortened the lengths of the interconnecting cables.

Software used on the PIN client PC's (i.e. x-windows, Pathworks, visionware, xvision, EXCEL, VAX terminal emulation). Terminal emulation, X-windows emulation, and Microsoft windows 3.1 are well integrated into PI giving the remote user access to complete PI functionality.

EPRI's Plant Monitoring Workstation (PMW) is an integrated collection of programs that monitor, trend, trouble shoot and report plant performance. In addition to providing real time performance indices that are useful to operators, it provides plant performance engineers with interactive access to the performance calculations, their execution, configuration and results. It also gives plant engineers easy access to manually input data for fuel analysis and unburned carbon. This manually input data has a significant effect on overall performance but is either too difficult or too expensive to supply in real time. With the ability to store and retrieve data through the PIN, the PMW provides a means to demonstrate the relative efficiency and availability improvements by comparing unit performance to baseline data. The goal of any performance monitoring system strategy is to minimize operating costs for a unit . The minimum cost comes from the use of on line information that is provided to the operator for corrective action. The accuracy and quality of calculated performance factors will depend on the accuracy of the instrumentation being used in the field. Centerior took the strategy that on line performance monitoring was not to provide the absolute accuracy of the Performance Test Code. On line performance information was to provide relative performance monitoring information that would be used for discerning problems and areas of performance degradation.

The sequence and frequency of sootblowing in a pulverized coal boiler plays a role in unit performance due to the cleanliness of the various heat transfer surfaces. The use and sequencing of sootblowers is typically controlled by the operator who recognizes the need for corrective action in cleaning the heat transfer and he then takes the appropriate action. In the case of Eastlake Units 3 & 4, the Diamond Power Microselect Sootblowing Control System (SCS) was used for the retrofit of the sootblowing system. The SCS has a built in RS-232 serial port for communication with the PIN computer. Monitoring sootblowers sequencing along with the resultant change in temperatures will allow operations to optimize sootblowing, thus reducing erosion and extending tube life.

The Power Monitoring System (PMS) was installed on the 2300 volt switchgear. The PMS utilizes a Square D Programmable Logic Controller (PLC) in the electric house and monitors 18 individual channels for the major pieces of equipment of units 1 through 3. The PMS does not monitor equipment on other units at the Eastlake Plant. The reason for using selected data from the PMS was to enable the plant to establish a predictive maintenance program through the PIN. The PIN provided a central location for information to be obtained by Maintenance Engineers so that run time information is used as a flag to remind an individual to prepare a work order for inspecting a major piece of equipment. In addition, the use of motor amperage can assist in identifying the equipment that has degraded performance and requires immediate attention during a planned weekend shutdown. As a result, a forced outage may be avoided or the ability to bring the unit to full load when required will be possible due to potential problem analysis.

In many power plants there is not sufficient staff to constantly calibrate and tune the controls after a successful controls retrofit or upgrade. As a result, many instruments rarely get calibrated and only then if there is some operating condition that demands there repair. The PIN has been used at the Eastlake Plant as a tool to compare the existing setting of an instrument and compare it to the range that should actually be applied. Once out of range, the PIN is designed to send an alarm to a printer to notify the appropriate group of people that attention is a particular device is required. In reality, the use of a Diagnostic Monitoring system assists in mitigating alarms that the operator will eventually have to deal with over time. The Diagnostic Monitoring system being used for correct instrument calibration represents the first tier of alarm management for the power plant correct instrument calibration. Diagnostic monitoring system monitors and alarm motor run time, transmitters over range. control valve deviation from set point, operating close to alarm limits, and the acoustic leak detector system.

The installation of a CRT - only, "total soft" control system has led to new requirements for implementing Centerior's Lock-Out Tag-Out Procedure. The use of standard OIS RED TAG allows the operator access to the PCU configuration menu. This is not desirable. A unique method for applying a HOLD-OFF: DO NOT OPERATE RED TAG to the soft control stations was developed by Centerior and RoviSys. Every control switch and control station has dedicated control software for "Red Tagging." The software toggles a visible or transparent HOLD-OFF tag like symbol on the control face place, or a miniature HOLD-OFF symbol around the pick point on the DCS graphics. This miniature HOLD-OFF symbol is identical to one that appears on the RED TAG/HOLD-OFF summary screen. This symbol provides the look of tags on hard control switches and stations on a benchboard. The data base for the Paperless Tagging System resides in the PIN but can be accessed on the control room operator's CRT. The software for red tagging is one of the permissive for starting equipment, and for opening/closing motor operated valves. The READY state will not appear on the control station if the red tag software is not reset for that piece of equipment. The red tag software also provides a interlock in logic such that the output is disabled when the red tag is applied. The red tag system has the ability to recognize a discrepancy between the DCS red tag and the PIN red tag and call this to the operator's attention. Multiple red tags can be logged into the PIN as well as the other colored tags being used at the station. The software for red tagging does not disable or hold in place analog control stations and their final control elements such as fan dampers. This action could have been implemented but was not, as the existing tag-out procedures do not require hold in place or disabling of control stations outputs. The stations must be in the manual mode to apply the red tag. The red tag serves as a reminder to the operator not to adjust the control output (like the tag on the benchboard did in the past). Only those stations that have a final element associated with the station have a red tag software.

EPRI Works is a Utility Communications Architectural (UCA)/Out a Access Integration Services (DAIS) compliant software system that allows PI work stations to access data from DCS and PMW. EPRI Works also allows the system for view data from new applications from a common graphical user interface (GUI) without having to import data into the control PI database, or develop specialized software components.

PIN ENGINEERING AND DESIGN

The heart of the PIN system configuration is a VAX cluster (see FIGURE 1). The architecture is considered to be distributed because there is a microVAX 3100-40 dedicated, or distributed, to each of the units and a central microVAX3100-90 which is the central file server. The design of the PIN came out of operational considerations that permit flexibility between operating units and maximizing computing power.

There were two basic reasons that this VAX cluster architecture was developed in the distributed manner. First, the system would be more responsive with three CPU's instead of one. Secondly, the design is such that the loss of any single VAX would not interrupt the system operation. The EL-9 VAX acts as a file server for 9 PC client users, which are the primary users of the PI Data Base Management System (DBMS). The VAX's and PC comprise the PIN Local Area Network (LAN). VAX ties are a combination of thin wire Ethernet and optical fibers. Optical fibers were used because of the interferences from electrical equipment and unit generators. The PC's are connected to the network through existing tested pair wiring using 10-base-T hubs. Using the 10-base-T allows for office environment moves and follows standard CSC practice for using communication wires (twisted pair) for LAN.

The control room used for operation of the power plant is an area that does not lend itself to additional cabinets and computers that will require access and trouble shooting from time to time. For this reason a "computer room" was established below the control room that accommodates the Bailey Controls cabinets as well as the supporting hardware such as Engineering Work Stations (EWS) and the PIN computers. Due to the nature of the controls being a critical aspect of unit operations, the EWS's and the PIN hardware located in a "Security Room" adjacent to the Computer Room so that access to the files could be controlled.

The implementation of a PIN requires a full time project manager to assist in developing a plan and focus and to assure that the implementation of the design is being followed. Important issues need to be identified for the schedule and milestones must be defined and retained as targets. In the case of Eastlake the PIN became the backbone of the paperless red tagging system that was used to assist operators in bringing each of the units back on line safely and successfully. As a result then the PIN had to be delivered prior to start-up to provide this need. The Project Manager also must follow the staffing level to allocate the proper engineers, programmers, system technicians and supporting staff to meet the project objectives and milestones. The Project Manager must follow-up on the preparation of the project documentation as discussed in the next section.

Throughout the PIN there are various End User of the system that with be interfacing will the system and using the information. These individuals vary in their use and need. As a result of the flexibility of the PI system, each end user was given the opportunity to have input into their own custom menu.

The PI menu structure for the Eastlake Plant consists of 10 Custom menus and 100 display pages. The Custom Menus direct the PI user to specific areas of the plant. The display pages allow viewing of 40 displays per page totalling 4000 displays. These displays can be graphics, trends or any other application that PI can execute. For the Eastlake Plant the Custom Menu Layout is shown in Figure II, and the Unit 3 Menu Structure is shown in Figure III.

Each DCS system required approximately 10 months from issue of purchase order to system delivery. The detailed design and drawing preparation of the field element demolition and upgrades took about 10 months and was a concurrent activity to the DCS activities.

PROJECT DOCUMENTATION

A complete documentation package was prepared to install the PIN. The following items are integral parts of the installation documentation.

1. Drawings and Procedures
   Installation procedures were documented, including equipment placement and mounting, network connections and grounding requirements were prepared. Drawings were included for clarity. References to vendor literature was also be referenced as needed.

2. Circuit Schedules and Wiring Diagrams
   Detailed wiring diagrams were provided to show proper wiring to the PIN hardware. Power wiring to the computers and ICI's were also documented.

3. Cable and Conduit Drawings
   Ethernet cables, RS-232 Cables, and custom interface cables were documented on AutoCAD drawings. Conduit runs, cables routing and connections were documented with a set of drawings for each location including the Computer Room, the Control Room, the Electrical Control Room, the Relay Room and the Service Building user office area.

4. Equipment Layout Drawings
   Layout of hardware such as Cabinet Layout for INFI 90 ICI's Equipment Layout of computers in their respective areas, etc.

A detailed schematic of the architecture was also developed. The as-installed schematic will be included as part of the System Manual.

A Plant Information Network System Document was developed to document the detailed architecture, functionality, hardware and configuration of the Eastlake PIN System. The System Design Document defines the scope of work performed by the integrator. As such, it had to undergo revisions during the project as more details became available and as design issues were resolved.

1. Functional Descriptions
   A detailed functional description was prepared for all custom software. This critical step in the design process assisted in the development by providing a stable target for all parties. Functional descriptions were updated with all as delivered options or corrections.

2. Flow Charts
   Flow charts were used to document the design software where appropriate.

3. Source Code
   A copy of all Source code listings for custom software was prepared and organized logically and well commented. Documentation of the vendor information and revisions of compilers used during custom software development were also included.

A System User Manual was developed and serves as a reference for both new and experienced users. It consists of approximately 100 pages of text and diagrams. Navigation through the menus and file structures has been detailed. Customized configuration of database, reports, graphics etc. were documented. The system integrator distributed the manuals prior to training so that the users could become more familiar with the documentation before operating the PIN system.

A System Master Manual was developed and consists of Vendor Manuals, overall system diagrams and operations, and all configuration documentation. Each section was updated to be made consistent with the hardware as installed.

VALIDATION AND ACCEPTANCE TESTING

Checkout of the System will serve to verify the configuration of the systems and software interfaces as the design progresses and before factory acceptance testing. Incremental checkout will speed the overall implementation by catching problems before significant time is lost. A formal Acceptance Test will take place after checkout.

1. Validation Structural Test of PI Tagging Map to DCS
   The PI database will be validated with a point by point review for consistency with the latest available DCS console database. This hard copy database review will validate all the point attributes which are critical for the database. The completion of this validation will ensure an exact match of the database information transferred from the DCS console database to the Plant Information database and therefore design control procedures must be strictly maintained after this step.

2. Validation Structural Test of Custom Software
   The custom software (I/O drivers, MMI, etc.) will be validated with a review of code by a Senior Engineer along with the engineer who writes the actual code. While the drivers will be tested with actual hardware, it is unlikely that all possible combinations of hardware and software conditions can be simulated. Validation at checkout will serve to identify potential problems that may be difficult or impossible to simulate.

3. Mini-Checkout of Physical Network
   The purpose of the Mini-Checkout is to allow confirmation that the network connections between various computers and printers are functional. After the mini-Checkout, cabling and hardware installation can proceed with confidence that the connections will function as desired. This will allow installation to proceed in parallel with design and configuration, improving schedule flexibility relative to manpower available during the outage. accurate configurations will be verified in later tests.

   The system was assembled at the system integrator's facility, including all hardware and software for Unit 3, network cable including spooned lengths to simulate long runs, and any available hardware to be attached to the PIN. Repeaters and the optical legs of the network were tested at this time.

   Available subsystems will be attached to the PIN. Data transfer between physical devices on Network, through interface driver software will be verified for each driver or standard interface. Where multiple applications use the same driver, only one was tested at this stage. Subsets of the tags in each application will be used to verify the base software and configuration of the drivers. Network functions for the EWS computers, backup managers and printing functions will also be tested in this manner.

The purpose of these Acceptance Tests will be to demonstrate that all interfaces between PI and application tasks, user PC's and network utilities are functional.

A key test is the verification that points are linked to the actual I/O described and that the values are reported accurately in terms of scaling, engineering units, post processing etc. It is generally desirable to test as much of the system as is possible before shipment. Testing at the staging area is less costly in terms of engineering time and process downtime, and problems can usually be more easily resolved than they can in the plant. In the case of the PIN system, the hardware necessary to test a complete system with all I/O was not available. Therefore, final testing of the full system had to be completed at Eastlake. Sufficient testing occurred before shipment to offer a very high confidence factor in the functionality and configuration of the system. Checkout and validation will occur before FAT. The method of conducting the Factory Acceptance Test (FAT) can be summarized as follows:

1. Testing
   Tests will be in accordance with written procedures as described in each of the respective FAT sections and shall be conducted in a continuous manner in accordance with the test schedule. All the features of the Plant Information Network (PIN) system should be utilized to make the testing as efficient and expeditious as possible.

2. Test Changes
   When necessary, mutually agreed upon changes to the test procedures were made due to errors or circumstances. Changes are accomplished by marking up the test procedures and having the changes initialed by CSC.

3. <u>Discrepancy Reports</u>
During FAT, if a discrepancy is discovered, a Discrepancy Report (DR) is prepared by the test personnel to document the problem. A log of discrepancy reports will be established and a sequential serial number shall be assigned to each DR. All discrepancies noted by the DR's shall be corrected prior to the final acceptance by CSC.

Site Acceptance Testing (SAT) was conducted as an abbreviated version of the FAT. During DCS system check-out a data file was created using the PIN data base and exported to a PC spread sheet program. Through the use of additional software within the PIN a link to this PC spread sheet was used to perform a point-by-point check out. Points in question could be forced to bad quality and then restored as a validation technique. The PC spread sheet program could be sorted in different ways which served as a check list with comments for check out record keeping.

EPRI OBJECTIVES

The objective of the Integrated Control and Information System (ICIS) is to integrate the unit control systems, diagnostic monitors and expert systems at Eastlake Plant into one comprehension system. The ICIS consists of two distinct parts: the Distributed Control Systems (DCS) and the Plant Information Network (PIN).

The PIN has been the primary focus of this paper. However, the information from the DCS has historical valve beyond the day to day operation of the units. Providing the plant personnel with access to operating data (both current and historical) permits more efficient use of their time on a daily basis as well as for short and long term planning.

The PIN will consist of a Digital Equipment Corporation (DEC) VAX computer running Oil Systems, Inc., Plant Information (PI) System software. The combination of software and hardware, will be capable of collecting data from the DCS and other application software and store or archive it to a removable storage media. The PIN will make this archived data available.

The PIN application network will be a Utility Communications Architecture (UCA) compliant communications network that is an EPRI developed product. UCA is an EPRI standard based on Open Systems Interconnection (OSI) standards developed by the International Standards Organization (ISO). OSI is a suite of open standard communications protocols which allow computers from different vendors to communicate. Selected PIN end user PCs will be equipped with a Data Access Integration Service (DAIS) compliant modules that allows other system database to communicate with one another. DAIS is a component of UCA.

The system architecture at Eastlake Plant is such that each unit has a specific data highway or sub-ring. The sub-rings are in turn connected to the plant super loop to permit open access to the network through an Engineering Work Station (EWS). This provides flexibility for phased Controls Modernization project and a common data highway access when required.

The objective of this portion of the project is to provide a Plant Information Network (PIN) that has the hardware and software capabilities to archive plant data and distribute it to End Users on an applications network. The PIN is the host system for a new paperless Red Tagging made available for use by the operators in the DCS graphic displays. End Users consist of plant personnel who can access the data through personal computers. The PIN will provide End Users access to time stamped data from the following systems:

- DCS - both current and historical valves
- Performance Monitoring Workstation (PMW) - calculation of unit performance
- Power Monitoring System - medium voltage system electrical data
- Sootblowing Control System
- Continuous Emissions Monitoring

In addition, the paperless Red Tagging system resides on the PIN and may be accessed by End Users in a read only query mode.

CONCLUSION

Centerior's experience with Controls retrofit on Eastlake Units 3 & 4 has been successful. Building upon the state of the art Bailey Controls technology the enhancements have accelerated operator acceptance of the system.

The integration of the DCS and the PIN into an Integrated Control and Information System (ICIS) has brought innovative, cutting edge technology to the Eastlake Plant. The PIN's greatest asset is that it handles data collection and storage, as well as presenting that data in a user friendly graphical interface all in one integrated software package. Through these advancements people will be able to work smarter and more efficiently in an increasingly competitive market.

Since the DCS/PIN/PMW was a combined installation, Centerior was able to minimize to overlapping functions of each System, through integrated design. Each component to the Eastlake installation has its own strong and weak points. For example, DCS Systems typically don't allow for long term data archiving; however, that is a PIN strong point. Similarly, most DCS and PIN systems have an option for adding performance calculations, but a PMW system will offer a wider array of performance calculations than the add-on packages from DCS or PIN. By starting from scratch, Centerior was able to save money vs. having to add these various systems to an existing system by taking advantage of each system's strength, and deleting the "weak" options.

The value of the PIN for the Eastlake Plant and Centerior is the easy access to historical operating data. The data permits individuals at the power plant to evaluate data, create reports and solve problems or potential problems before they happen and with planning.

In addition, the PIN became an effective link for providing the safe paperless tagging system to be accessed by the operators without the addition of more CRT's into the control room.

The PIN is being used in its early stages. The Shift Supervisors prepare their shift reports on the custom report format that they designed after their old hand written report of the past.

The operators are using the Paperless Red Tagging system for safety removing or returning equipment to service.

On line performance monitoring graphics are being developed with the operators so the meaningful and controllable information is being presented efficiently.

There are those at the Eastlake Station that are thrilled that the PIN has been made available to them. Others need to get their feet wet and grow into its' use. With any new system there are growing pains and familiarity that come with time.

There is a vision that the PIN could be used some day for the Maintenance Work Order system such that all the repeated paper work could be computerized. Individuals see this as the way to minimize their repeated tasks and also not have to repeat documentation once entered into the system. Less paperwork results in more time to give attention the equipment and devices that make the electricity.

FIGURE 1

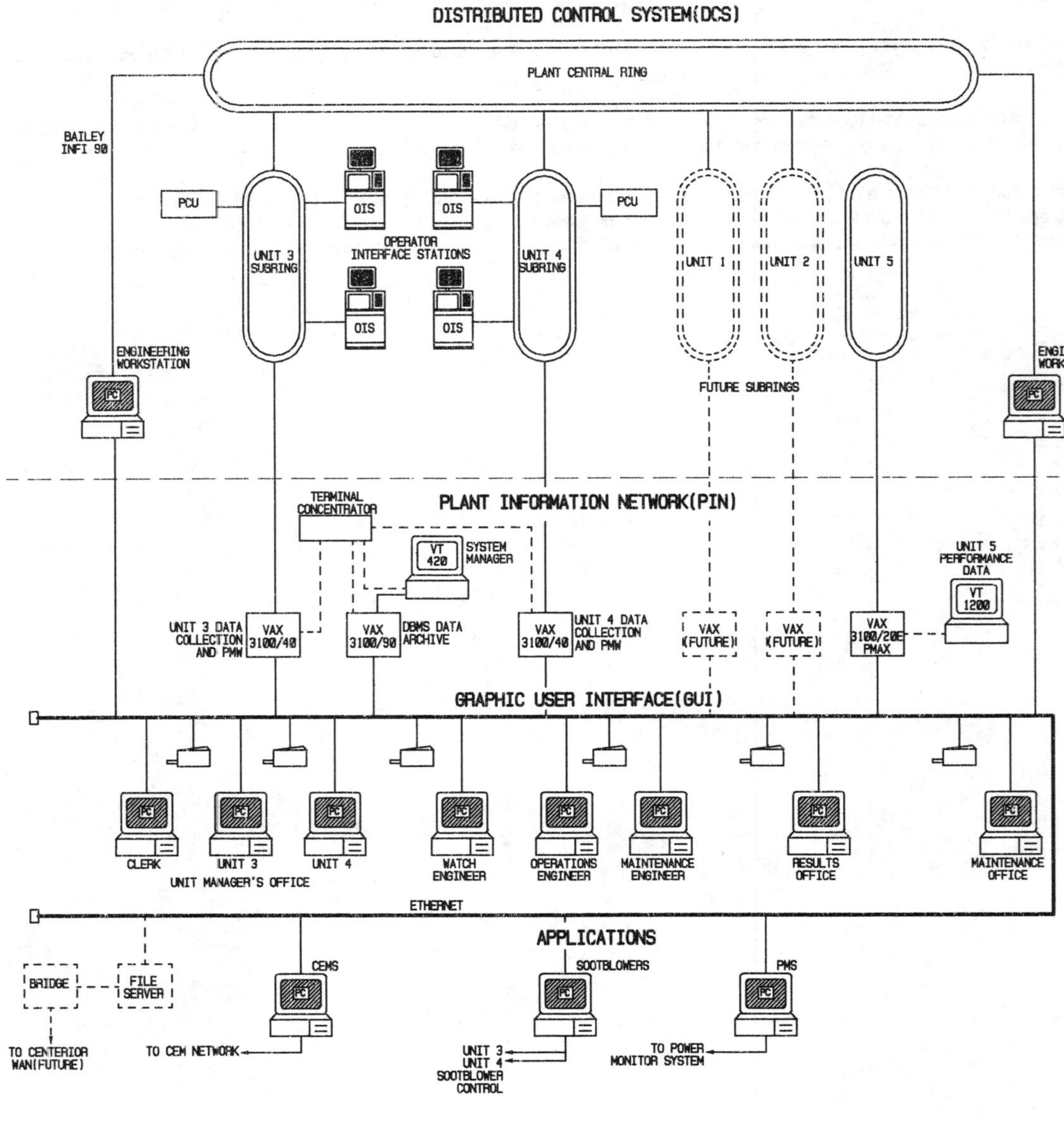

EASTLAKE PLANT INFORMATION NETWORK SYSTEM ARCHITECTURE

Figure II

Custom Menu Layout

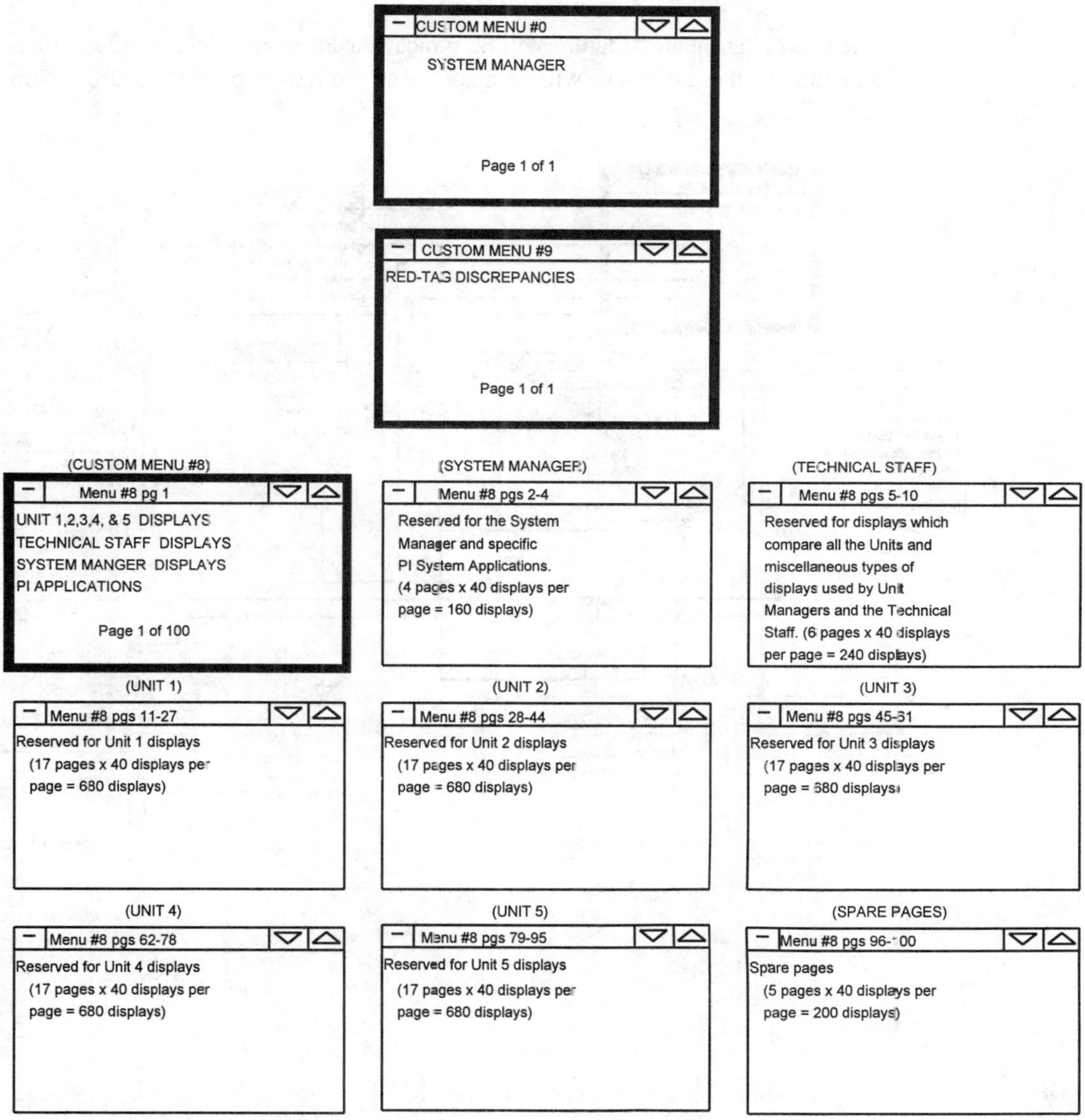

Figure III

PI Menu Structure for Eastlake Plant UNIT 3

The following menu structure will be typical for all Units of the Eastlake Plant. Selection of these displays will be made using the mouse or keyboard function keys

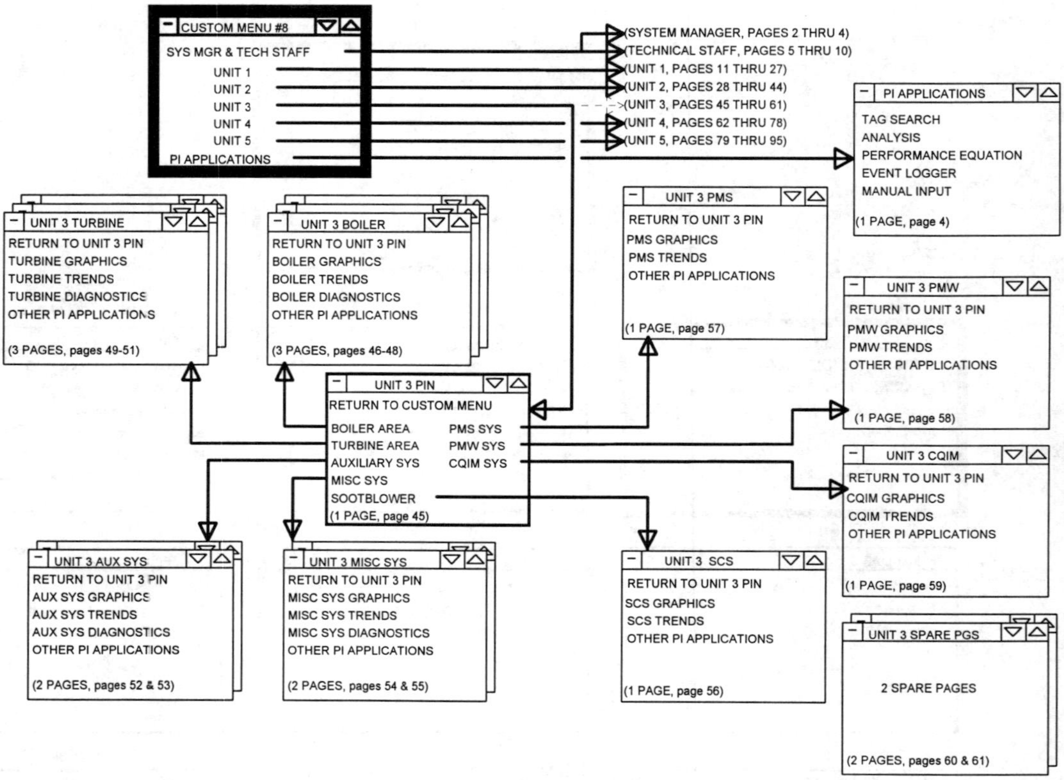

© 0074-056X/95/021-030/$0 + .50pp

APPLIED PATTERN RECOGNITION FOR PLANT MONITORING AND DATA VALIDATION

R. D. Griebenow, P.E.
E. J. Hansen
Performance Consulting Services, Inc.
644 North Fourth Street
Montrose, CO 81402

A. L. Sudduth, P.E.
Engineering Consultant
Duke Power Company
526 South Church Street
Charlotte, NC 28201

KEYWORDS

Calibration, data validation, fault detection

ABSTRACT

In an attempt to improve plant performance, reduce operating and maintenance costs, and meet the requirements of new regulation, many utility companies are implementing automated systems and new technologies for plant monitoring and control. However, the usefulness of these systems is dependent upon the reliability of the instrumentation. Erroneous input data can result in misleading performance assessments, inappropriate operator actions, inefficient plant operation, excessive plant emissions, and a host of other undesirable effects. Duke Power Company is applying advanced pattern recognition (APR) techniques to detect and eliminate faulty input data to on-line plant performance and continuous emission monitoring systems. This paper will present the basis for Duke Power's application, outline the pattern recognition methodology, and discuss its application to continuous emission monitoring systems at Duke's Lincoln Combustion Turbine facility.

INTRODUCTION

Measurement signals obtained from process sensors and associated electronic or pneumatic equipment are the only path by which information concerning process status is conveyed to the operator, the automatic control system, or a computer-based operator support system. It is, therefore, very important that the quality of process information be high. Erroneous information from sensors must be quickly and effectively eliminated from consideration by the operator and by any analysis or control functions performed by the process computers to ensure reliable, efficient operation. The use of erroneous data in an automatic control function could result in unnecessary process shutdowns or in equipment damage. Analysis of invalid sensor signals by an operator or computer-based system may result in misleading or incorrect conclusions and inappropriate responses (e.g., an invalid measurement could be mistaken for a process fault).

Over the years, there have been many techniques employed in an effort to improve data reliability. Prior to the arrival of computer-based control systems, a plant operator may have had a manual switch to select which channel was to be displayed or fed to the control system when multiple measurement channels were available for a particular process state. If the selected channel failed the operator would move the switch to the other channel to avoid an adverse impact on the process. With the new technology available in control rooms today, a much wider range of options has become available.

Computer-based control systems provide a mechanism for automating data validation and sensor selection. While redundant instruments may still be applied to critical process measurements, on-line data analysis can help to ensure signal validity and accurate signal selection. Physical and empirical models can also be run in near real-time on plant computers to provide analytical redundancy of plant measurements for data validation.

Duke Power Company has been applying various technologies for data validation at nuclear and fossil plant sites for years. While traditional validation techniques have provided some benefit, Duke personnel continue to search for improved methods to ensure data accuracy and reliability. To this end, Duke Power has initiated a pilot project to implement data validation and calibration monitoring based on Advanced Pattern Recognition (APR) at its new Lincoln combustion turbine site. While this technology has similarities to some traditional analytical methods for data validation, early studies indicate that it offers improved accuracy and much greater fault tolerance. In addition, the APR approach can provide analytical redundancy for nearly all plant measurements with minimal engineering effort, whereas traditional methods frequently require extensive engineering effort for the development and maintenance of detailed physical models.

TRADITIONAL SIGNAL VALIDATION

Signal validation consists of methodologies for distinguishing measurement failures from process faults and selecting which instrumentation signals to use in control and analysis functions. There are two legacy methods of signal validation which have been used extensively in process monitoring systems - comparisons among multiple values representing the same measurement and the use of reasonableness checking among multiple signals having a quantifiable relationship with the measurement being validated.

Comparison methods are based on the availability of at least two measurements (direct or derived) of a desired process state. These redundant measurements may then be used to make some judgment about the validity of the measurement signals. The simplest comparison methods involve the installation of two sensors at the same location for the same process state. These redundant sensors are compared with each other, and disagreement between them, larger than a threshold (which is typically related to the amount of anticipated measurement noise), is considered indicative of failure of one of the measurements. However, if only two measurements are available, no decision can be made on which to accept. An unambiguous measurement quality determination requires more than two measurements. When at least three measurements are available for comparison, it is possible to make some logical choice of which to accept or reject, and to form a "best estimate" of the true value of the process state. When there are no physically redundant devices, it is possible to use analytically redundant measurements (1). An analytically redundant measurement can be found when there is a process model which can be used to derive a representative value of a directly measured state from measurements of other states. Once three or more representations of a particular measurement are available, there are methods which may be used to discriminate failed measurements and select the most representative "true" value of the measurement.

The other traditional method of signal validation, reasonableness checking, has been implemented much less frequently than direct comparison methods. This is a more qualitative or symptomatic method which is based on a comparison between the measurement value and certain reference values based on a qualitative estimate of what the measurement should be or on values associated with known failure modes of the measurement channel. In reasonableness checking, one takes advantage of the fact that deviations of measurements from expected values do not generally occur in an isolated fashion - that process faults usually result in the deviation of groups of measurements. Thus if a single measurement has an unexpected value or a value characteristic of measurement failure, it is most likely in error.

This form of reasonableness checking may also he applied to the derivative of a measurement. For most measurement situations, it is possible to estimate the upper bound of the derivative based on process knowledge or model studies, and then use that estimate as a limit for the allowable rate of change of the channel. This technique is particularly useful for measurement channels with relatively long time constants, such as temperature detectors in thermowells, or for measurements of process parameters which have long time constants themselves. A common form of applying limits on the derivative is to monitor two redundant measurements, immediately rejecting the one whose derivative assumes a high value. A system based on this method has been applied successfully on a limited number of measurements at Duke Power Company's Oconee Nuclear Station.

The usefulness of the methods described above has been severely limited by the cost and complexity of their implementation. Installing redundant sensors is very expensive and tends to be limited to a small subset of the total number of sensors which might have significant impact on process reliability. Comparison methods also generally compare only two or three measurements to each other, or a measurement to a fixed limit. Consequently, this approach is not very robust and unlikely to cover a large subset of process measurements. In addition, comparison methods cannot be used for diagnosis of any type of process failure other than that of isolated sensors, and every comparison must be enumerated by the designer and appropriate limits assigned in advance.

There is, therefore, motivation for the implementation of more robust methodologies of signal validation. Methodologies which are able to use all measurement information and diagnose all sensors in the process, can be assembled and configured with minimal human intervention, and can adapt to changing process conditions encountered in normal operation, such as operation at various power levels.

PATTERN RECOGNITION BASICS

Pattern recognition treats any process or system as a set of numeric data values. For example, a power plant is viewed as a list of pressure, temperature, and flow values rather than an assemblage of turbines, pipes, and heat exchangers. To understand how pattern recognition works with data from a power plant, consider a hypothetical system that has just two significant measurable parameters, P1 and P2. Data for this simple process can be collected through a series of tests and a graph can be constructed showing one parameter versus the other.

The individual points in Figure 1 represent discrete operating states for a simple process. That is, each point represents a unique characteristic condition of the process. In order to accurately describe a single process state at any given time, the values of all the system parameters must be known (just two in this example -- P1 and P2).

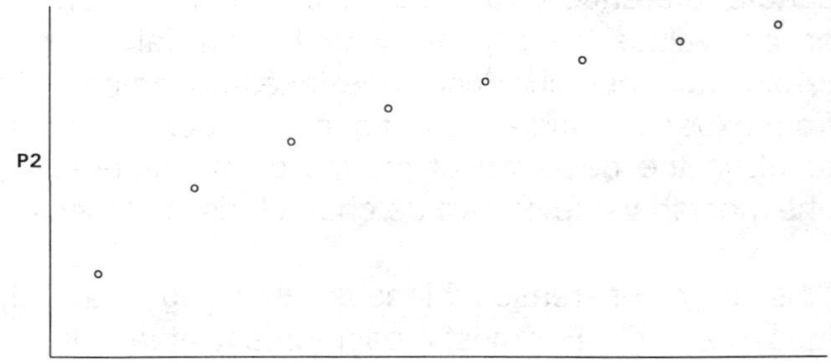

Figure 1. Plotting Process Parameters

The process can then be generalized to estimate behavior at operating states for which no discrete data values have been collected by placing a continuous curve through the plotted points (Figure 2). It should be noted that, when using pattern recognition techniques, these points are considered to be interrelated and are not separated into input and output (independent and dependent) categories.

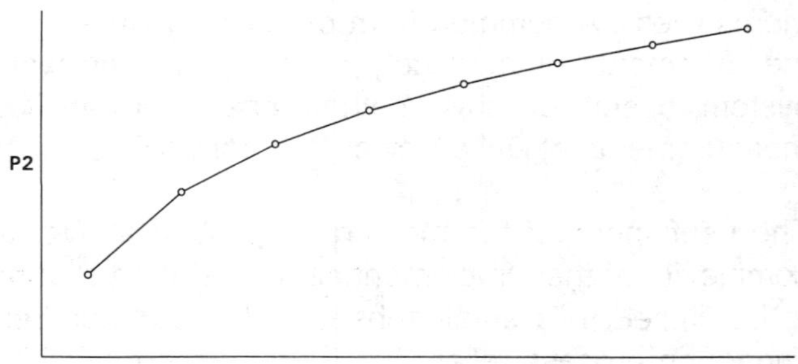

Figure 2. Generalizing Process Behavior

This modest procedure can lead to some very useful results. For example, if at a later time the value of one of the parameters is known, the value for the other can be readily estimated (Figure 3). In addition, the generalized plot could be used to determine whether or not the process is currently operating in a manner consistent with past operation (Figure 4).

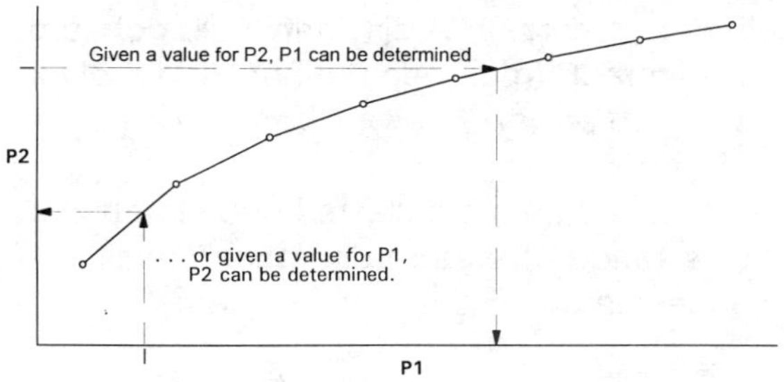

Figure 3. Determining Unknowns

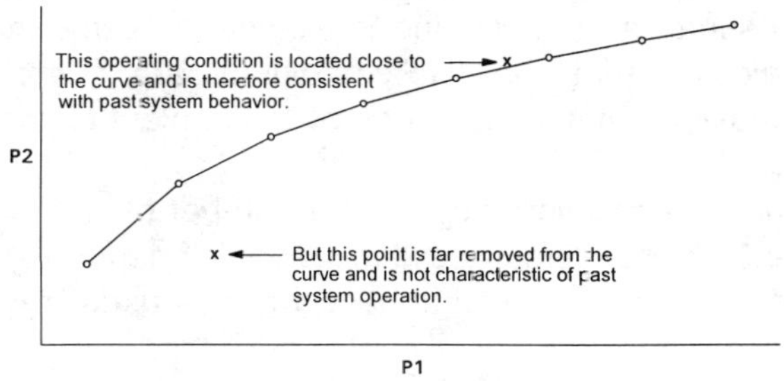

Figure 4. Assessing Process Operation

Although the examples above are presented graphically, similar results can be obtained numerically. However, most processes cannot be characterized sufficiently with just two parameters. For example, a complex process such as a power plant requires the measurement of many different parameters to describe plant conditions completely. Furthermore, most of the parameters associated with a complex process are interrelated and cannot be assessed simply as pairs of variables isolated from the rest of the system. So with most real-world processes it is necessary to examine many different parameters simultaneously rather than pairing them off in sets of two.

Instead of the simple case of P1 vs. P2 for two parameters, we now have a more complicated situation of P1 vs. P2 vs. P3 vs. P4 vs. . . . P*n* for *n* different parameters (or "dimensions") associated with a complex process. Although a graph showing all parameters plotted simultaneously for a complex process is not feasible, numerical methods analogous to the concepts presented graphically above can be employed. In other words, numerical techniques can be applied that utilize historical data values collected from a complex process to form the numerical equivalent of a system performance surface.

ADVANCED PATTERN RECOGNITION METHODOLOGY

Advance Pattern Recognition (APR) works with system data that is captured and arranged in arrays called "data records". A data record is simply a snapshot of the system data for a single instant in time. The individual items which comprise the data record are called "points" and the values associated with these points are called "point values". Table 1 illustrates these concepts for a system consisting of five measured points.

Point ID	Point Description	Point Value
1	Main Steam Temperature	1005.2
2	Main Steam Pressure	1925.5
3	Condenser Pressure	2.05
4	Feedwater Flow	2716.7
5	Generator Power	435.8

Table 1

APR analysis is performed by quantifying the "similarity" between any two data records that are being compared for purposes of creating modeled estimates and other essential functions. Computed similarities are scalar values that range between zero and one. A similarity value of one indicates the two plant conditions being compared are identical (e.g., each temperature, pressure, and flow value is exactly the same in both data records). A similarity value of zero indicates that the two conditions are completely different from each other (e.g., plant conditions at full power compared to plant conditions at zero power).

Prior to analyzing a system, a number of snapshots of plant data are collected and stored in a file. These "reference data records" generally cover a range of system operating conditions and act as a knowledge base which is used to define the characteristics of the system.

After the reference data records have been collected and stored, a new snapshot of system data called the "input data record" is obtained for assessment. The input data record is first compared with each of the reference data records using the APR similarity operation. Several data records are selected from the reference set that have the highest similarity to the input record, while striving to bound each of the input point values. The data records selected from the larger reference set in this manner are referred to as "nearest neighbor" records. Once the nearest neighbor records have been selected, the similarity between all pairs of nearest neighbor records is computed, forming a "recognition matrix" of similarity values. This recognition matrix is then utilized in conjunction with the input data record to compute values for each of the monitored points. That is, a computed point value is calculated for each of the input points. The computed point values are collectively known as the "output data record". Figure 5 summarizes this procedure.

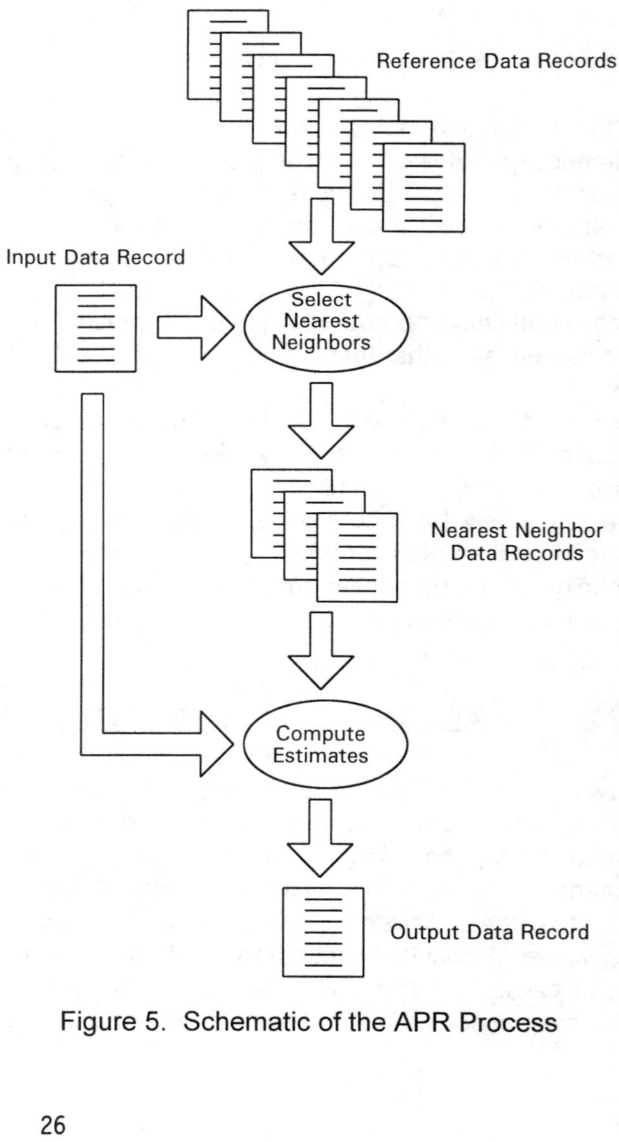

Figure 5. Schematic of the APR Process

The output record is an extremely accurate representation of how the system should be behaving based both on past performance and on current operation. The calculated values are highly fault-tolerant because defective plant parameters in the input record do not markedly bias or affect the accuracy of the computations. In addition, for situations where one or more plant parameters are completely missing from the input record, APR will provide very accurate predictions for these parameters as well.

Once the output record has been computed, it may be compared with the input record for further data manipulations. Again, it should be noted that the APR approach calculate an output point value for each and every input point (Figure 6). The significance of any differences between the input and output values is generally viewed in the context of the specific application (e.g., indication of a signal failure, calibration drift, abnormal system operation, etc.).

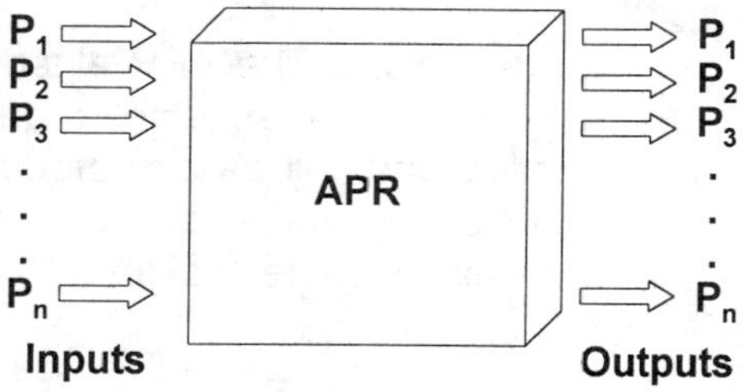

Figure 6. An Output for Every Input

ADVANCED PATTERN RECOGNITION CHARACTERISTICS

The APR technology has several important characteristics:

Fault Tolerance: APR minimizes the effects of incorrect or missing plant measurements. This is due to the fact that it treats all monitored parameters as being interrelated. APR uses all parameters simultaneously to determine all of the predicted values so the impact of individual defects is minimized.

Localized Modeling: APR forms a local model for every new input data record. Because it does not operate using a single set of coefficients, complex non-linear systems can be modeled more accurately and with fewer examples than are required with other techniques. In addition, localized modeling automatically takes into account changes in a system as it progresses through different operating conditions.

No Iterative Training Phase: APR does not require an iterative training phase. Because no iterative training phase is involved, the reference data can be changed as often as desired allowing immediate adaptation to new information.

High Dimensionality: APR is essentially unrestricted in the number of plant parameters that it can handle in a single model. In practical applications, it has been successfully applied to systems containing as many as 400 different parameters.

Models All Parameters: APR is unusual in that it models all variables simultaneously. This is essential for detecting faults in individual plant parameters and for assessing abnormalities in overall system operation. Most other techniques must separate system parameters into dependent and independent sets which precludes the ability to pinpoint specific parameters that are deviating from normal operating behavior.

Repeatable: APR is 100% repeatable. There are no random settings of any initial conditions. If the analysis is executed twice using the same reference data set, it will produce identical results.

Deterministic: The APR algorithm is a well-behaved deterministic relationship. There are no iteration schemes requiring convergence criteria. The algorithm has no failure modes and will always produce an optimal solution given the available information. Further, it is unique in that it enables the determination of uncertainties associated with the final estimated values.

THE DUKE POWER COMPANY APPLICATION

Duke Power Company has initiated a project to apply APR technology for data validation and calibration monitoring at its new Lincoln combustion turbine site. The site will include 16 General Electric 7001EA combustion turbines rated at 80 MW nominal. The first unit was commissioned in December 1994, and commissioning of the remaining 15 units is expected to be complete by mid-1996. It is anticipated that these turbines will be used to meet peak load requirements, operating with an average annual capacity factor of less than ten percent. The units will be controlled by GE Mark V turbine control systems. The entire plant, including balance of plant and auxiliary systems, will be monitored by a Westinghouse WDPF distributed control system which will be used for monitoring and display. The WDPF will also provide the interface to the system operating center from which the units will be remotely operated. Duke Power personnel considered that, given the infrequent operation of these units and the high level of automation, a data validation and calibration monitoring system would be essential to ensure reliable and efficient operation of the combustion turbines.

The data acquisition and archival will be handled by the Oil Systems, Inc. Plant Information (PI) Systems, running on a DEC Alpha processor under the OpenVMS operating system. The APR algorithms will be integrated with the PI data base to validate all monitored data. Data input to the APR system will be performed by using PI Application Program Interface (API) calls to the PI database. The APR system will evaluate the quality of all specified input data and perform the following functions.

1. Estimation of missing or invalid input values
2. Identification of calibration deficiencies
3. Validation of a subset of data
4. Creation of a valid data set in the presence of missing or suspect data

Upper and lower bounds will be computed for each analog input in the engineering units of the input using methods combining standard statistics with APR techniques. Uncertainty values provided by the APR system will be similar to standard deviations of standard statistical methods but differ in the sense that they change with plant operating conditions. The APR system will compute uncertainties for each process variable from the similarity between the input data record and the output data record along with a set of statistically derived weighting factors. Process variables detected outside of their estimated uncertainty bounds will be classified as suspect or faulted (Figure 7). The system will provide a data quality status to the operator display and supply analysis results to the PI database using calls to the PI APIs. Results submitted to the PI database will include:

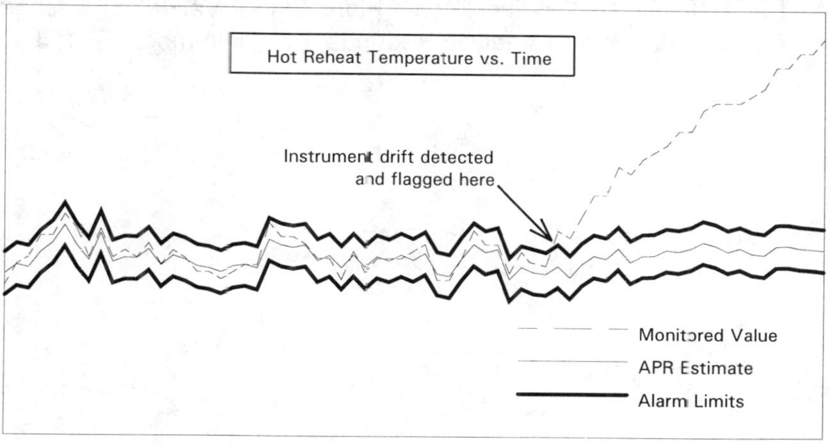

Figure 7. Data Trend Plot

- Actual input value for analog points if good quality
- Actual input state of digital points
- Estimated value based on pattern recognition
- Upper and lower bounds of the estimate for analog points
- Quality status

System should be in full operation by June 1, 1995. Based on the results of the Lincoln Combustion Turbine site installation, Duke Power will extend use of the APR technology into several other areas, including data validation for on-line performance monitoring systems, continuous emission monitoring data validation and emission predictions, and calorimetrics and balance of plant performance data validation in the nuclear plants.

CONCLUSION

The APR technology appears to offer a more accurate and reliable method for validating plant data and detecting calibration drift than traditional data validation techniques. In addition, it does not require complex, high-maintenance process models to relate process parameters. Early analysis indicates that APR-based data validation will result in greatly increased data confidence, providing for improved operating efficiencies. Given the current installation schedule, detailed results should be available for presentation by mid-June, 1995.

REFERENCES

1. EPRI NP-2110, "On-line Power Plant Signal Validation Technique Utilizing Parity Space Representation and Analytic Redundancy", November 1981, J. C. Deckert et. al.

© 0074-056X/95/031-041/$0 + .50pp

THERMINATOR: USING AN EXPERT SYSTEM TO IMPROVE POWER PLANT HEAT RATE

Robert F. Nelson
Georgia Power Company
62 Lake Mirror Road
Forest Park, GA 30050

Joseph Weiss
Electric Power Research Institute
3412 Hillview Avenue
Palo Alto, CA 94303

Mark Perakis
Electrical Power Research Institute
3412 Hillview Avenue
Palo Alto, CA 94303

ABSTRACT

Therminator is an expert system designed to improve power plant heat rate. It builds on the successes of earlier systems that improved heat rate. These included a prototype expert system evaluated on a solar power plant, a heat rate controllable parameters monitor currently in use on a fossil power plant, and heat rate rules developed for generic fossil power plants. The overall approach, architecture, current status, and experience with this expert system are presented.

INTRODUCTION

Improving the heat rate of a power plant has significant benefits in terms of dollars saved. In particular, it is estimated that a 1% improvement in heat rate saves a typical power plant unit over $300,000 a year. The goal of the current project is to provide significant improvement in the heat rate of a coal-fired power plant by building on prior successful work. The system is designed to aid the plant operator's interaction with controllable parameters of the plant, and to specify maintenance and control actions to improve heat rate.

Therminator was developed for Georgia Power's Plant Branch Unit 1 under contract with the Electric Power Research Institute (EPRI). The primary project objectives were to provide a methodology for developing a real-time heat rate expert operating advisor and to discover the benefits to be achieved from its utilization.

The prior work on which the current project is based includes:

- ThermoPlus, an expert system to improve heat rate on a solar power plant.
- Plant Branch Unit No. 3 Heat Rate Monitor (HR Monitor), a program to provide real-time, on-line feedback to the operator on the controllable parameters affecting heat rate.
- HEATXPRT, a generic set of diagnostics for providing guidance to plant operators to improve heat rate.
- Operations, plant process, and performance experience of plant operators and other personnel at Georgia Power's Plant Branch.
- EPRI Human Factors Guidelines for Power Plants.

ThermoPlus

In the early 1980's, most papers on expert systems stated that expert systems were not well-suited for real-time applications. In 1983, a cooperative effort began to create a real-time expert system by The Foxboro Company, E.I. du Pont de Nemours and Company, and the University of Delaware (see references 2 and 3). FALCON (Fault AnaLyzer CONsultant) was designed to provide real-time fault detection and analysis on an adipic acid reactor. The success of FALCON led to a new effort to apply an expert system to a power plant. Georgia Power and The Foxboro Company were both interested in the application of expert systems to process control. Georgia Power Company was willing to test an expert system on their research facility, a combined solar and fossil power plant at Shenandoah. A cooperative effort was formed to test an expert system, in real-time and in a closed-loop situation, to improve the heat rate on the research power plant. Work on this system began in 1988, and several key questions were successfully answered. The results are summarized below and in reference 4.

ThermoPlus demonstrated that it is possible to use an expert system approach to monitor heat rate in real time and on line. It demonstrated that an expert system could make recommendations to improve heat rate in real time. Finally, it demonstrated the advantages of an iterative development approach. The knowledge base was developed using five prototypes, with each prototype furthering the communication between the experts and the knowledge engineers. This helped the experts better define the expert system's needs.

HR Monitor

The Georgia Power Company has demonstrated that plant operators can have a significant impact on power plant efficiency through their constant control over many parameters affecting the overall heat rate. In 1989, Plant Branch personnel placed a controllable parameters heat rate monitor in the control room of Unit No. 3 to facilitate improvement in heat rate. Heat rate deviations from design targets were monitored prior to introduction to the plant operators. After training on the heat rate monitor, the operators were asked to report on heat rate performances during their shifts. A 148 BTU/kWh difference in average heat rate reduction was noted between the control and the operator generated reports.

The heat rate monitor assisted the operator by monitoring the 10 most important parameters to the efficient operation of the unit. The monitor provides them with information about the current net unit heat rate (normalized for comparison between shifts) and shows areas in which improvement can be made. The monitor gives real-time information to the operator, and provides an approximation of the amount of benefit or penalty of each deviation.

In addition to the net heat rate, the heat rate monitor displays the actual value, target value, BTU/kWh deviation, and $/shift cost for each of the following controllable parameters:

- Main steam temperature
- Main steam pressure
- Hot reheat steam temperature
- Reheat spray
- Condenser vacuum
- Excess air
- Gas out temperature
- Station service
- High pressure feedwater heater performance
- Low pressure feedwater heater performance

The current project integrates an expert system with the heat rate monitor and uses the calculated results as input to its knowledge base.

HEATXPRT

Significant progress was made towards creating an on-line, real-time expert system to help improve heat rate on fossil power plants in the development of HEATXPRT by Sargent & Lundy and the University of California at Berkeley (reference 5). This system had several goals. It would validate data provided by the control system. It would assist plant operators in evaluating and diagnosing plant performance. It would analyze and diagnose deviations in heat rate and controllable parameters for fossil fuel power plants. It would provide practical and reliable probable causes of performance deficiencies. It would suggest corrective control and maintenance actions. Therminator builds on this earlier concept and meets many of its objectives.

The earlier system also developed methods for sensor validation and decision/fault trees for fossil power plants (references 6 through 9). The current project combines this sensor validation approach with that developed for ThermoPlus, using the strengths of each. HEATXPRT's methodology was not strictly heuristic but rather employed a probabilistic network approach. Nonetheless, the current project made extensive use of the earlier project's work that defined the relationships between controllable parameters and heat rate (reference 9). Summary sheets relating plant conditions to supported causes of heat rate degradation that had been developed for the earlier system were readily converted into a heuristic format, or pseudo code rules. Finally, the pseudo code rules were converted into a coded knowledge base for use in an expert system shell.

Work at Plant Branch tailored the results from HEATXPRT to Therminator for use on Plant Branch Unit 1. For example, some rules derived from the earlier system concern burner tilt, which does not exist on Unit 1. Other rules required instrumentation not available on Unit 1. Some new rules were needed that were unique to Plant Branch Unit No. 1. For example, when a pulverizer mill is off-line at reduced loads, Unit 1 is much easier to control with main steam pressure at 50 p.s.i.g. below target. Thus, rules were modified to allow for a different target when a mill is down.

Process Operator Experience

The Therminator knowledge base was developed in close cooperation with Plant Branch personnel because of their detailed process expertise. Including the plant operators, engineers, and management at Plant Branch from the beginning of this work was beneficial to the project's progress. Their involvement assured them that they were creating a tool to assist plant operators, not a system to replace them. The expert system had to free the operator from routine problems and offer solutions that augment his knowledge of plant optimization. Furthermore, it was important that the expert system was seen as worthwhile and that management supported the effort.

A key document used during the development of the Therminator knowledge base was "Basics of Power Plant Efficiency" (reference 1). This report was developed by personnel from Plant Branch with cooperation from General Physics Corporation and is used to train operators at Plant Branch. In addition, rules developed from other sources, e.g., HEATXPRT, were reviewed by plant personnel. Plant personnel also summarized rules on specially designed forms and shared the results with those personnel developing the system.

The user interface was also developed with plant involvement. Both hardware and software had to work together in a manner pleasing and informative to the operator. A poorly designed user interface could result in an expert system that would not be used. Since the plant operators were the best and final judges of user interface, full advantage was taken to use their ideas and opinions throughout development of displays and operating methodology.

Human Factors Guidelines for Power Plants

The user interface is a key to determining the success of an expert system. If the user interface is difficult to use, if it does not provide the required functionality, or if it is vastly different from what the user knows, both it and the application behind the user interface will not be used.

Guidelines for user interfaces, especially those based on CRT's, are available (references 10 and 11). For the current project, much work has gone into design issues such as the overall system requirements, the screen structures and content, the screen organization and layout, color coding and other visual coding techniques, menu design, information formats, and input and control devices.

In addition, rapid prototyping and iterative design of the user interface, following the approach of reference 12, was used for the current project. A prototyping tool specifically designed for user interfaces made the task efficient. Showing plant personnel an interactive user interface greatly enhances the feedback received to improve the user interface design. Iteration, or making suggested changes and going back to the plant personnel with an updated prototype, helped solicit further improvements in the user interface.

THE APPROACH TO THERMINATOR

The primary goal of this project is to reduce the heat rate on a commercial, full-scale, fossil-fired power plant. Other goals include:

- Using a real-time expert system on a full-scale power plant;
- Creating an environment for success by including plant personnel from the beginning;
- Building on prior successful work so as not to reinvent what has already been done; and
- Developing an open and portable architecture for this system.

Therminator development was a team effort between EPRI, Sargent and Lundy, The Foxboro Company, Georgia Power Company, Kaman Sciences, General Physics, and several consultants.

Figure 1 shows the logical architecture for the expert system. The power plant is controlled by a distributed control system (DCS), which in turn is interfaced to the expert system (Therminator). The user interface is used not only to control the process, but also to interact with the expert system. Sensor data flows from the process to the control system, and then to the expert system and the user interface. Control suggestions from the expert system are displayed for the operator; with operator permission, the expert system suggestions could be issued to the control system for control action. The plant operator can also issue control commands, and can override the expert system (including preventing the expert system from issuing commands to the control system, and even disabling the expert system entirely). The plant operator can also interact with the expert system to obtain explanations for its recommendations.

Software Architecture

The Plant Branch Unit No. 1 hardware environment contains sensors, field bus modules, control processors, and workstations. An information network interface connects to a SPARC 2, on which the Therminator (and heat rate monitor) reside. An application processor is used to control a hard disk drive as well as perform other computational functions for the control system. Other hardware environments can be accommodated because of the modular design of the expert system

Figure 2 shows the software architecture. The control system interfaces to the expert system through the Application Interface Software (AIS), a library of C language functions that call data in and out of the control system into third party software. The modules of the expert system include data gathering, data validation (built on HEATXPRT), heat rate calculations (HR Monitor), and rules which make heat rate improvement recommendations to the operator (or control actions if operator permission is granted).

Data gathering starts with AIS passing current process variables to the expert system. These variables are then smoothed and accumulated until ready for validation. Data validation uses many of the approaches used by HEATXPRT (see references 6 and 8). This frees the expert system from the chore of avoiding inconsistent, bad, and noisy data. Heat rate calculations are done by HR Monitor, similar to that already in use at Plant Branch Unit No. 3. The monitor was implemented in a library of C functions that contain equations for control parameter targets, heat rate impacts, and the ASME Heat Loss Method for calculating boiler efficiencies. The monitor was embedded with the expert system operating concurrently.

The expert system analyzes data and sends recommendations and actions to the control system via the Application Interface Software. It saves its recommendations and the current data in a file for later analysis. Therminator is implemented in CLIPS, a non-proprietary expert system language developed at NASA. CLIPS can run (in different versions) on a workstation or on an IBM-compatible personal computer. Segregating the expert system from the control system enhances its portability to other control system platforms.

User Interface

Considerable effort has gone into developing the user interface. After the task and users were defined, a prototype user interface was created. Existing user interface guidelines were consulted and used (references 10 and 11). The prototype was created on a PC-based prototyping tool designed specifically for user interfaces, following the general approach described in reference 12. Figure 3 shows the heat rate monitor screen from the current version of the user interface.

The PC-based prototype of the expert system's user interface was then shown to the engineers and operators at Plant Branch. Three cycles of reviews and modifications occurred before the user interface was ready for implementation on the control system for Plant Branch Unit 1. Initially, the user interface will be on its own CRT, always available for operator use. Later on, the configuration may be totally incorporated into the control system operator interface, and available upon call-up from any operator station.

Knowledge Base

The structure of the knowledge base is shown in figure 4. The main modules are sensor validation, plant operating conditions checks, analysis, and operator advice or closed loop control only with the process operator's explicit permission. Most of the sensor validation is done using methods from HEATXPRT that consist of statistically based comparisons of parameter values against target ranges. The parameter targets are modeled from several plant parameters such as generation load and mill output. The knowledge base then determines whether there is an adequate set of validated data for further analysis.

Checking the plant operating conditions is essential. The expert system determines whether the process is in a state for which it can do a proper analysis. For example, the expert system is not designed for use during start up or shut down, nor can it operate well in transient conditions. The expert system does not make any recommendations, and will alert the operator, if the plant is not in a condition for which it can do an analysis.

The analysis section develops recommendations for improving plant heat rate. It analyzes plant conditions for steady-state operation, and no changes will be recommended during significant transients. It determines whether control loops are in remote (the computer can control the set points). Control is physically impossible unless these conditions are met. The system makes certain that loops which may require control actions are responsive, primarily by ensuring that the controlled variables are within a small range around the set points. If deviations occur, it determines whether steady-state operation has resumed within reasonable time periods. After analyzing plant conditions and looking for control or maintenance actions that will reduce the heat rate, all results are passed to rules for interacting with the process operator.

Finally, Therminator will act on its decisions. It will pass information to the user interface informing the operator whether or not expert system analysis is possible, if any unusual conditions are encountered, and any recommendations for control or maintenance actions, and why. The expert system can recommend control actions provided all necessary plant conditions are met, provided operator permission for closed loop control has been specifically granted by the operator, and provided control actions are recommended.

THE CURRENT STATUS AND EXPERIENCE WITH THERMINATOR

The Therminator Expert System is still being evaluated at Plant Branch Unit No 1. It was first put in service during May 1994 to collect data only (over 432 input points are analyzed by the system). The collection of heat rate data continued from May 1994 until the unit's outage in November 1994. Data collection resumed in January 1995 when the unit came back in service and is continuing at present. During July 1994 heat rate tests were conducted to establish a base line for heat rate performance and to calibrate the heat rate monitor portion of the system.

During August 1994, the expert was operated for the first time with a subset of the Therminator rule base. The rules selected for testing were those that performed diagnostics on the feedwater heaters. It was believed that the feedwater heaters had the greatest potential for heat rate improvement and that the feedwater heaters needed the most work. When first evaluated, the feedwater heater rules indicated that several heaters were operating with low condensate levels. Sight glasses on the units, however, indicated that the condensate levels were normal. Analysis of the feedwater heater data revealed high drain water temperatures and normal to low terminal temperature differences (TTD) for several heaters. This analysis indicated that the feedwater heaters acted as though they had low condensate levels. A test was performed on August 18, 1994 to test the low condensate level hypothesis. When Feedwater Heater No.16's condensate level was raised 4 inches, there was no change in the heater performance. However, when the condensate level was raised an additional 3 inches, the drain temperature dropped 6 degrees F without significant change in TTD.

Rules continued to be added throughout September 1994 and operation with the expert system's complete rule base began in October 1994. During the first week in November 1994, just prior to a planned outage, the expert system was indicating a fouled condenser condition in addition to the chronically low feedwater heater condensate levels. This unit had not had previous problems with fouling. During the outage, extensive fouling was found in the Unit 1 condenser. (Unit 1 had recently had its condenser tubes replaces with stainless steel. The previous condenser tubes were a copper alloy that were believed to have suppressed the "slime" build up found during the outage.) The heat rate penalty was 187 BTUs/kWh in November. When the unit came back in service in January 1995, the expert system stopped indicating condenser fouling and the heat rate penalty from condenser back pressure dropped below 50 BTUs/kWh.

In January 1995, when the unit was back in service, the issue of low feedwater heater condensate levels remained unresolved. The drain water from Feedwater Heater No. 18 dumps directly into the hotwell. The drain water temperature is usually from 6 to 10 degrees F above target values (a drain cooler approach of 10 degrees F is indicated by design). Although the waste heat content of the hot drain water is small, i.e. 6 to 10 BTUs/kWh, its impact on condenser back pressure is uncertain. It should be noted that a 10 degree F difference in temperature in a closed vessel results in a saturation pressure difference of about .5 inches of Hg. Testing was inconclusive due to control problems on the feedwater heater drain valve. Higher condensate levels were not able to be established and repairs were not possible while the unit was operating.

The current schedule includes the introduction of the expert system into the control room in February 1995 and continued testing and evaluation through September 1995. Heat rate data will be compared with the baseline at that time.

SUMMARY AND CONCLUSIONS

Therminator was built on prior experience. It was developed in close consultation with plant operators and other experts at the target plant. It uses a heat rate monitor currently in use on a power plant. It builds on a generic set of diagnostics designed to provide guidance to plant operators in order to improve heat rate.

This expert system was developed in an iterative approach. The software, hardware, user interface, and knowledge base went through a repeated cycle of design and review, homing in on a prototype for testing on Georgia Power's Plant Branch Unit 1. Base line heat rate measurements were made during 1994 using the expert system before being introduced into the control room. Continued testing is planned after January, 1995 when Therminator is planned to be installed in the control room for operator use.

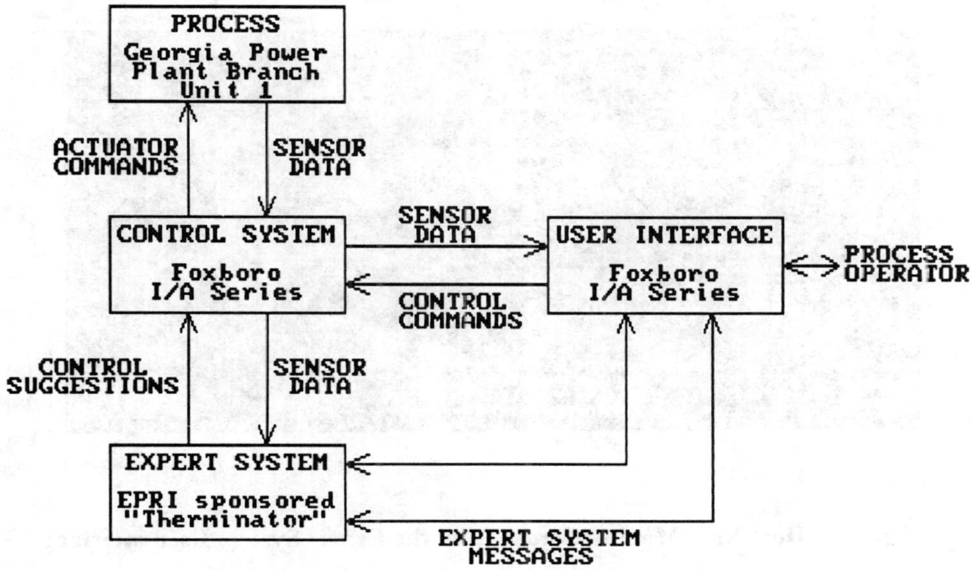

Figure 1: The Expert System Logical Architecture

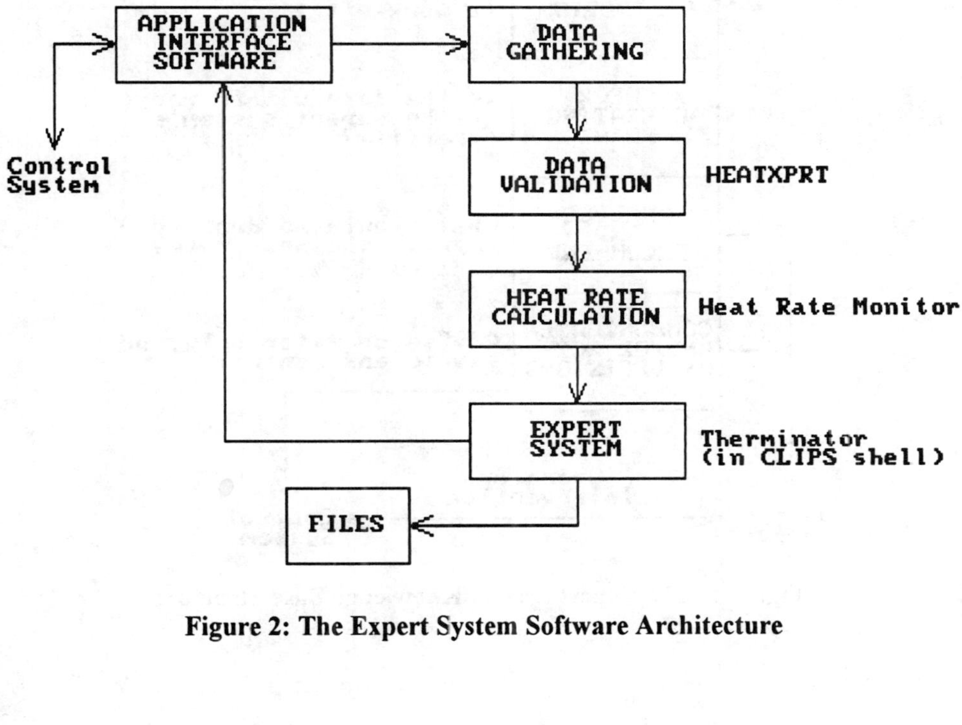

Figure 2: The Expert System Software Architecture

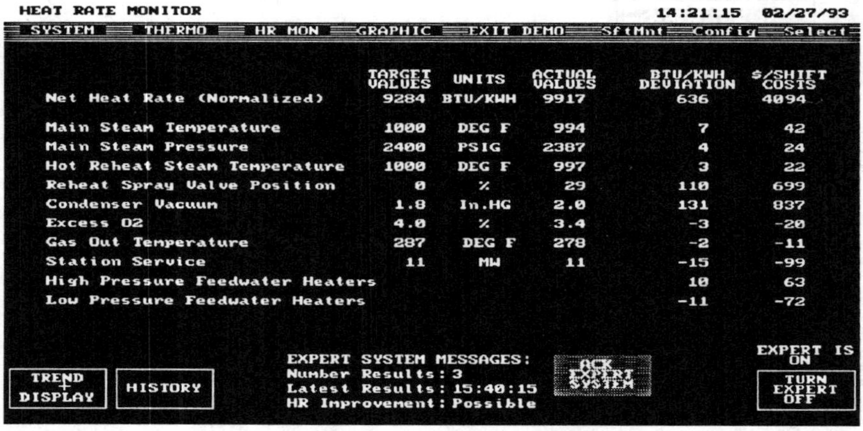

Figure 3: Heat Rate Monitor Screen from the Expert System User Interface

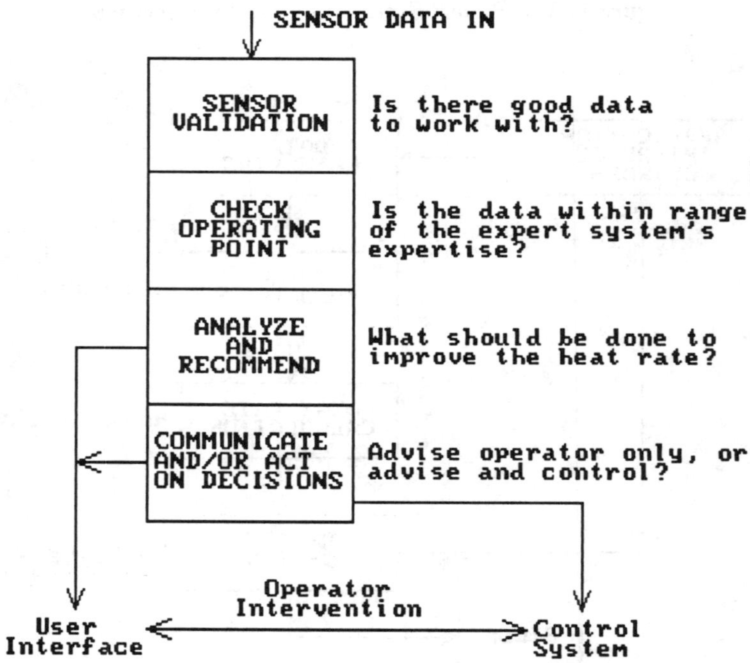

Figure 4: The Expert System Knowledge Base Structure

REFERENCES

1. Banger, John W., Jerry E. Brown, John A. Morse, and John S. Weaver, Jr., "Basics of Power Plant Efficiency," internal Georgia Power Company report, TIS Number 12146856, Revision 4, November 3, 1989.

2. Dhurjati, P., D.E. Lamb, and D.L. Chester, "Experience in the Development of an Expert System for Fault Diagnosis in a Commercial Scale Chemical Process," Foundations of Computer Aided Process Operations, pp 589-625, CACHE Publications, Elsevier Science Publishers, New York, 1987.

3. Shirley, Richard S., "Falcon: Final Report on a Real Time Diagnostic Expert System", EPRI Conference on Power Plant Controls and Automation, Miami, FL, February, 1989.

4. Shirley, Richard S., Harry Forbes, and Robert F. Nelson, "A Supervisory Control Expert System to Improve Power Plant Heat Rate," Proceedings of the Canadian Chemical Engineering Conference and Exhibition, Dalhousie University, Halifax, Nova Scotia, Canada, July, 1990.

5. Henry, R.E., D. Sopocy, J.A. Montanus, M.A. Blanco, K.W. Bleakley, and S.A. Perry, "Helping Operators Improve Plant Performance: An On-Line Expert System," Proceedings of the American Power Conference, 54th Annual Meeting, Volume 54-I, pp 671-685, Chicago, 1992.

6. Agogino, Alice M., Young-jin Kim, and William H Wood III, "Final Report for Sensor Validation for HEATXPRT," Department of Mechanical Engineering, University of California at Berkeley, Berkeley, CA 94720, April, 1992.

7. Frogner, B., A.E. Tome, and S.M. Divakaruni, "EPRI Demonstration of Decision Tree Approach to Heat Rate Diagnostics for Fossil Power Plants," presented at a seminar on expert system applications in power plants, Boston, MA, available from EPRI, 3412 Hillview Ave., Palo Alto, CA 94303, May, 1987.

8. Kim, Young-jin, and Alice M. Agogino, "Signal Validation for Expert System Development," Department of Mechanical Engineering, University of California at Berkeley, Berkeley, CA, 94720, 1991.

9. Sopocy, D.M., R.E. Henry, J.A. Montanus, S.G. Warren, B. Frogner, A Agogino, Y. Kim, "HEATXPRT, knowledge Base Development, Second Review Cycle," EPRI Research Project 2923-5, EPRI, 3412 Hillview Ave. Palo Alto, CA 94303, Revision 1, Nov. 12, 1990.

10. Beare. A.N., C.D. Gaddy, and J.C. Taylor, "Human Factors Guidelines for Fossil Power Plant Control Rooms and Remote Control Stations." EPRI Research Project 2710-20, EPRI, 3412 Hillview Ave., Palo Alto, CA 94303, final report, December, 1991.

11. Shirley, Richard S., Computer Graphics for Industrial Applications, Prentice Hall, Englewood Cliffs, NJ, ISBN 0-13-041732-7, January, 1992.

12. Shirley, Richard S., "Portable Human Interface Prototypes On A Personal Computer (With Three Process Control Examples)", Proceedings of ISA/88, Houston, TX, October, 1988.

©0074-056X/95/043-056/$0 + .50pp

ENTERPRISE COORDINATION At Entergy's Lewis Creek Units 1 and 2 and Nelson Unit 4

Gary A. Cohee
Senior Engineer
ENTERGY
P.O. Box 2951
Beaumont, Texas 77704

Dave Broske
Project Manager
Electric Power Research Institute
P.O. Box 10412
Palo Alto, California 94303

Robert N. Hubby
MAX Control Systems, Inc.
Applications Specialist
North Wales, Pennsylvania 19454

ABSTRACT

Today's focus on the need for improvement of controls and monitoring for cyclic operation stems from a power industry requirement to assume an ever larger share of the system load change requirements. The source of this need is the significant growth of power sources that provide no assistance in load regulation such as cogenerators and independent power producers.

Management of today's fossil fired power plants must recognize two broader constraints in addition to simply meeting the load demand: first — the need for reliable operation with minimum expenditure of equipment life; and second — the need to conform to environmental regulations including NO_X and other environmental emission limits.

The authors will discuss two approaches currently being implemented at Entergy's Lewis Creek and Nelson Stations to enable power plants to provide the best performance achievable with equipment in service with minimum expenditure of equipment life within the requisite environmental constraints.

Implementation will be accomplished through the following task development:

1. The Plant Automation Coordinator which provides expert best operating practice for plant start-up and shutdown supervision with optional equipment automation available for critical plant operations.

2. The Process Constraint Coordinator to provide the easy incorporation of plant operating constraints, plant life expenditure, environmental and regulatory constraints in to the real-time operating strategy of boiler control. By the dynamic on-line calculation of maximum allowable unit rate of change and maximum and minimum operating limits based on available equipment in service and life expenditure and environmental restrictions, intelligent automatic unit dispatch is readily achievable.

Operating experience with these new strategies will be shared as available.

INTRODUCTION AND BACKGROUND

The Implementation of Automation Design Concepts is part of a joint development program between the Electric Power Research Institute (EPRI) and Leeds & Northrup Company. EPRI, in order to commercialize the results of advanced technology in the utility industry has sought out a number of joint development projects with major suppliers to the utility industry under tailored collaboration contracts both with the utility and the supplier. It is the result of such a tailored collaboration contract that will be shared with you today.

In the conclusion to the paper *Automation Strategies for Cyclic Operation* (reference 1) the author concludes "while advanced boiler control strategies have provided responsive unit performance over a wide range limited only by the auxiliaries being controlled, system extensions to extend the concept of "unit coordination" are essential to support cycling operation." Additionally the concept of "extended coordination" is needed to provide the best performance available with equipment in service with minimum expenditure of equipment life within the requisite environmental constraints.

A further important consideration has come forward as these automation design concepts have been implemented with a number of utility users, that is that the knowledge base for implementing advanced automation strategies lies with the utility user's plant such as operating procedures charts and plant operating practice. So an ideal advanced automation strategy would be one that was very flexible, that could easily accommodate and encode current plant operating practice and then could be easily expanded by appropriate utility engineering/plant personnel as future automation needs are developed. This need is even more significant today as utilities look for ways to reduce manpower in order to compete as a supplier of generation capacity in a competitive market.

TASK 1 PLANT AUTOMATION COORDINATOR

Based on these needs, the Plant Automation Coordinator has been designed to accommodate a full range of automation strategies from the simple automation of a single auxiliary device to a single pushbutton start of an entire plant. This flexibility is particularly important since the level of plant instrumentation and system input/outputs for the installed distributed control system is already determined at ENTERGY's Nelson 4 and Lewis Creek 1 and 2 Stations. A second consideration is to be able to apply automation strategies to assist with the operation of a limited number of auxiliaries which may be causing operational problems in the plant.

With this range of requirements for plant start-up and shutdown in mind three implementation levels are provided for the plant automation coordinator:

Level 1	Algorithm for control of a single device (device core logic for motor or valve)
Level 2	Algorithm for control of a group of devices associated with a major auxiliary (FD, ID Fan, BF Pump etc.)
Level 3	Algorithm to implement operator guidance messages to start-up, run and shutdown the plant

Level 1 and level 2 algorithms are resident in the distributed control systems (DCS) distributed processing units (DPU) and are accomplished by modifying the existing DPU control strategy recipe configuration. Level 3 is implemented in the applications processor which is part of the information management function of the workstation for the DCS.

Beginning with the plant supervisory level 3, as shown in Figures 1-4, the operator will follow event level operator guidance messages which are simply the required actions to be accomplished from the plant start-up/shutdown procedures with checkpoints to confirm that actions have been completed. Event level operator guidance messages are organized into thirty groups of twenty-four messages each, to provide flexible access to the relevant plant operating procedure.

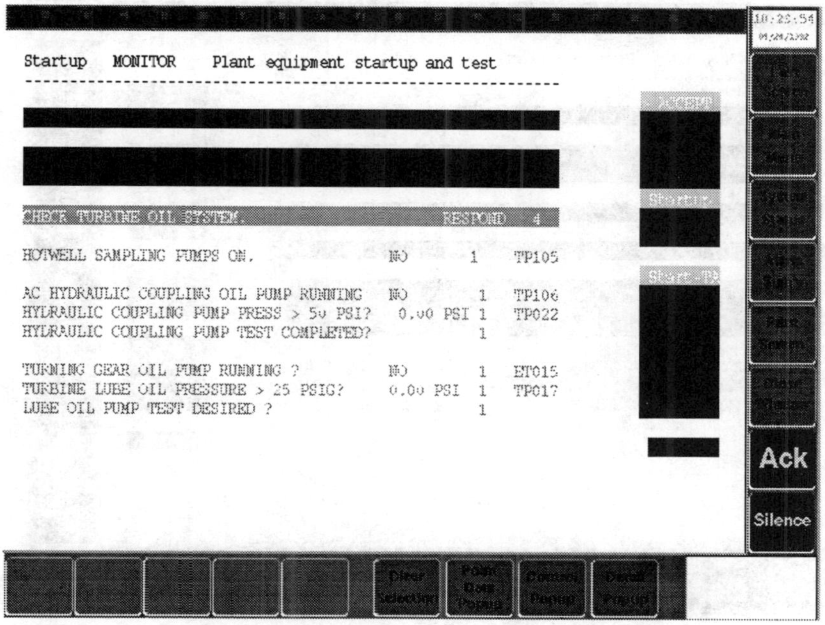

Figure 1

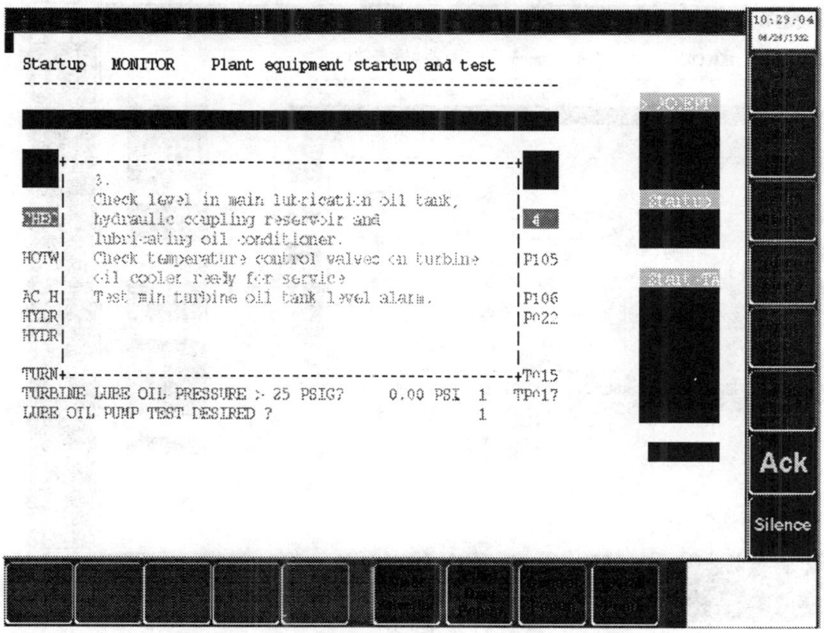

Figure 2

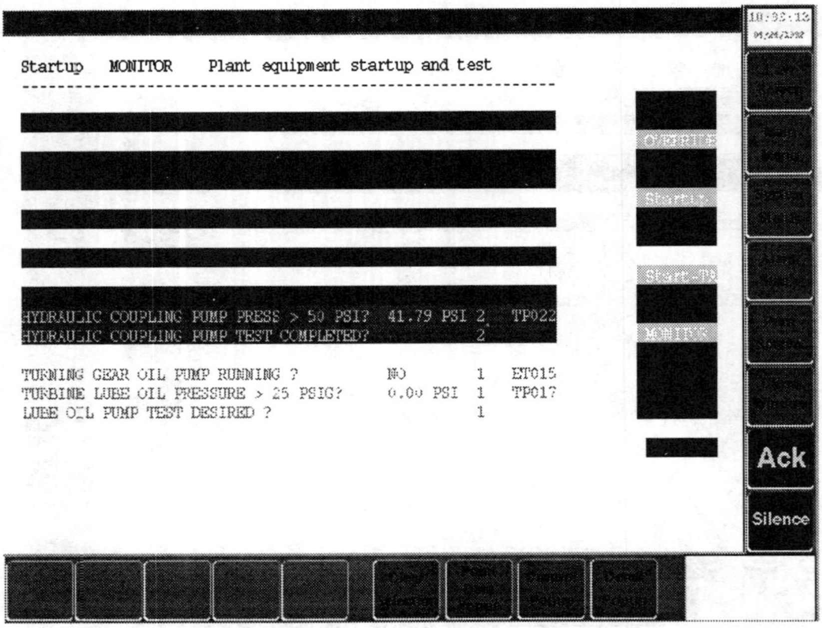

Figure 3

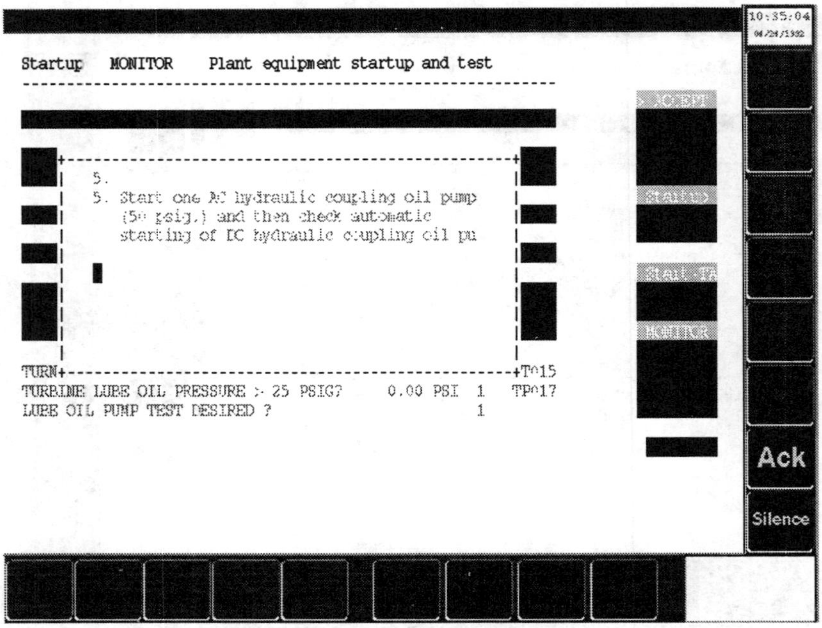

Figure 4

Status of actions can be shown as follows (please note that color selection can be made in accordance with user color conventions):

- Red - Actions to be completed
- Blue - Completed actions
- Yellow - Overrides taken to complete actions following physical verification of acceptable plant status/conditions
- Green - Future activities to be completed beyond those currently being processed by the operator which are shown in red

Explanatory operator guidance HELP messages including any special advice or cautions as to how to proceed to complete a given step can be provided as a HELP pop up on the bottom of the CRT screen, and are available by depressing the HELP key. These HELP messages are simply for operator guidance and do not replace any of the normal operator response to alarm activity. Completion of each event action is logged in the DCS event log, to provide a complete time based record of the start-up.

In Figure 1 the event being checked is the status of the turbine oil system. The help message of Figure 2 includes details of what is to be checked. Similarly in Figure 3 the event is the test of the ac hydraulic coupling oil pump. The Figure 4 help explains the details of this test.

Two operating modes are available from level 3: (1) a "monitor" or "supervisory" mode where all start/stop actions are completed manually by the operator in accordance with the plant procedures, (2) a "control" or "automatic" mode where required actions are implemented automatically. Provision for monitor mode/control mode is individually selectable for each step in the start-up/shutdown procedure.

The ability to override a specified step will allow the operator to continue with the start-up in the event of a malfunction of the associated interlocking circuitry. All override actions will be logged in the DCS event log.

At Lewis Creek the operator guidance messages are organized as follows:

Group #	Group Title	Number of Steps
1	Boiler Start-up Procedure	24
2	Main Turbine Start-up (prior to Roll on Steam)	11
3	Main Turbine Start-up (rolling on steam)	2
4	Main Turbine Pre Start	11
5	Main Turbine Cold Start Rolling Procedure	21
6	Main Turbine Hot Start Rolling Procedure	9
7	Unit Synchronization	11
	TOTAL	89

For Nelson 4 the operator guidance messages are organized as follows:

Group #	Group Title	Number of Steps
1	Boiler Cooling Water Start-up	24
2	Circulating Water Start-up	10
3	Circulating Water System Fill	8
4	Circulating Water Pump Start	10
5	Main Condensate System Start-up	24
6	Fill Main Boiler - Cond Booster Pump Fill	10
7	Fill Main Boiler - Aux BFP Fill	9
8	Fill Main Boiler - Main BFP Fill	24
9	Vent Valves on 8th Elevation	5
10	Purge Furnace	16
11	Establish Air Flow	15

Group #	Group Title	Number of Steps
12	Light Off Gas Burners	19
13	Establish Flash Tank Pressure	15
14	MAX 1000 Start-up Guide - Ramping	16
15	Roll and Synchronize Turbine	19
16	Shutdown Steps 1-15 to 250 MW	15
17	Shutdown Steps 16-26 to 170 MW	11
18	Prepare to Start Deramp	21
19	Reserve Shutdown	6
20	Boiler Leak Cool Down	7
21	Boiler Off-line/Tuning Gear Tagout	10
22	Break Vacuum	8
23	Shutdown - Condensate Pumps	6
24	Shutdown - Bearing Cooling Water	6
25	Shutdown - Circulating Water Pumps	7
26	Shutdown - Inservice Status Procedure - Stator Cooling Water System - EHC System	2
	TOTAL	324

Supporting the level 3 operator guidance messages will be level 2 or level 1 algorithms as shown in Figures 5 and 6 within the distributed processing unit, which can be triggered from the level 3 program when the associated sequence steps required to start a major auxiliary. The current status and progress to completion for the level 2 algorithm is shown on a "pop up" display that can be accessed from the associated system overview display - airflow/furnace pressure; feedwater; etc. Similar to level 3, each of the sequence steps can be placed on "automatic" or "supervisory" mode at the discretion of the operator. Message background colors (similar to level 3) for each of the steps to be completed by the level 2 sequence are used to show the current status of actions.

Figure 5

Level 1 algorithms within the distributed processing unit are configured to start/stop, open/close-shut any device in the plant. When on "control" or "automatic" mode, level 1 algorithms can be triggered from level 2 or level 3 programs as appropriate. The current operating status/condition of the device is shown on a "pop up" display that can be accessed from any sequence control display as shown in Figure 6.

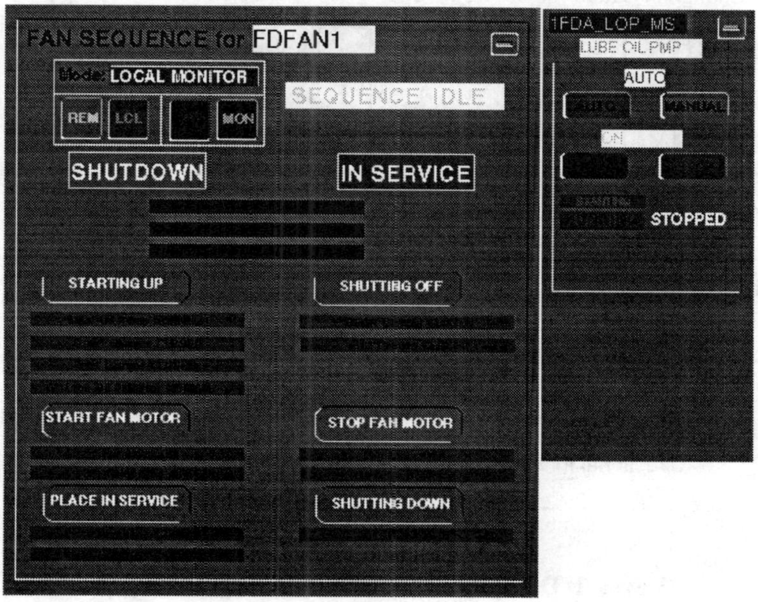

Figure 6

The heart of the level 1 algorithm is the device core logic subroutine as shown in Figure 7 which is used to start/stop, open/close-shut start/trip, insert/retract any plant device. This standard logic template which is a custom algorithm includes the common logic features for any plant binary device. It also includes user configurable options such as momentary or maintained start/stop etc., commands or stop an intermediate position.

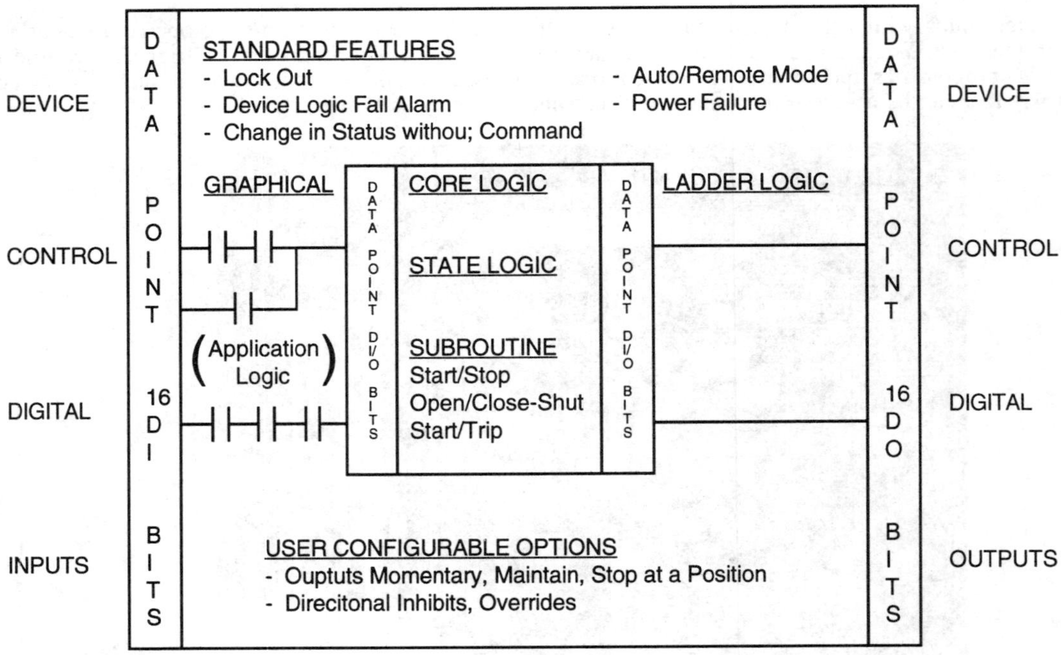

Functional View
Level 1 Device Control - Core Logic

Figure 7

Additional optional logic features include: directional inhibits to block an operation, override to achieve a specified status upon command.

Standard core logic features include: lockout capability; device logic fail alarm in the event of not achieving command status within a configured time, a change in status without command, a conflict in status feedback, or power failure without lockout; reset logic to the existing state of the binary device (part of alarm acknowledge); auto (remote) mode which allows the device to be operated from higher level logic; power fail to block a start/open, etc., operation.

In Figure 7 only abbreviated application logic is shown. More typical examples of application logic and customized logic to specific device requirements is easily accomplished by the addition of graphically configured application logic, this logic can be easily added to the level 1 device core logic as shown in Figures 8 and 9. The graphically configured custom logic supports a full range of troubleshooting and self diagnostic ladder logic features such as power flow indication through and/or gates, timers, set reset functions, input/output status forcing functions. Alternately standard nongraphical custom logic is also available for customization if desired. With these standard flexible tools, users of the *Plant Automation Coordinator* can easily expand and evolve their plant logic systems as may be dictated by future needs.

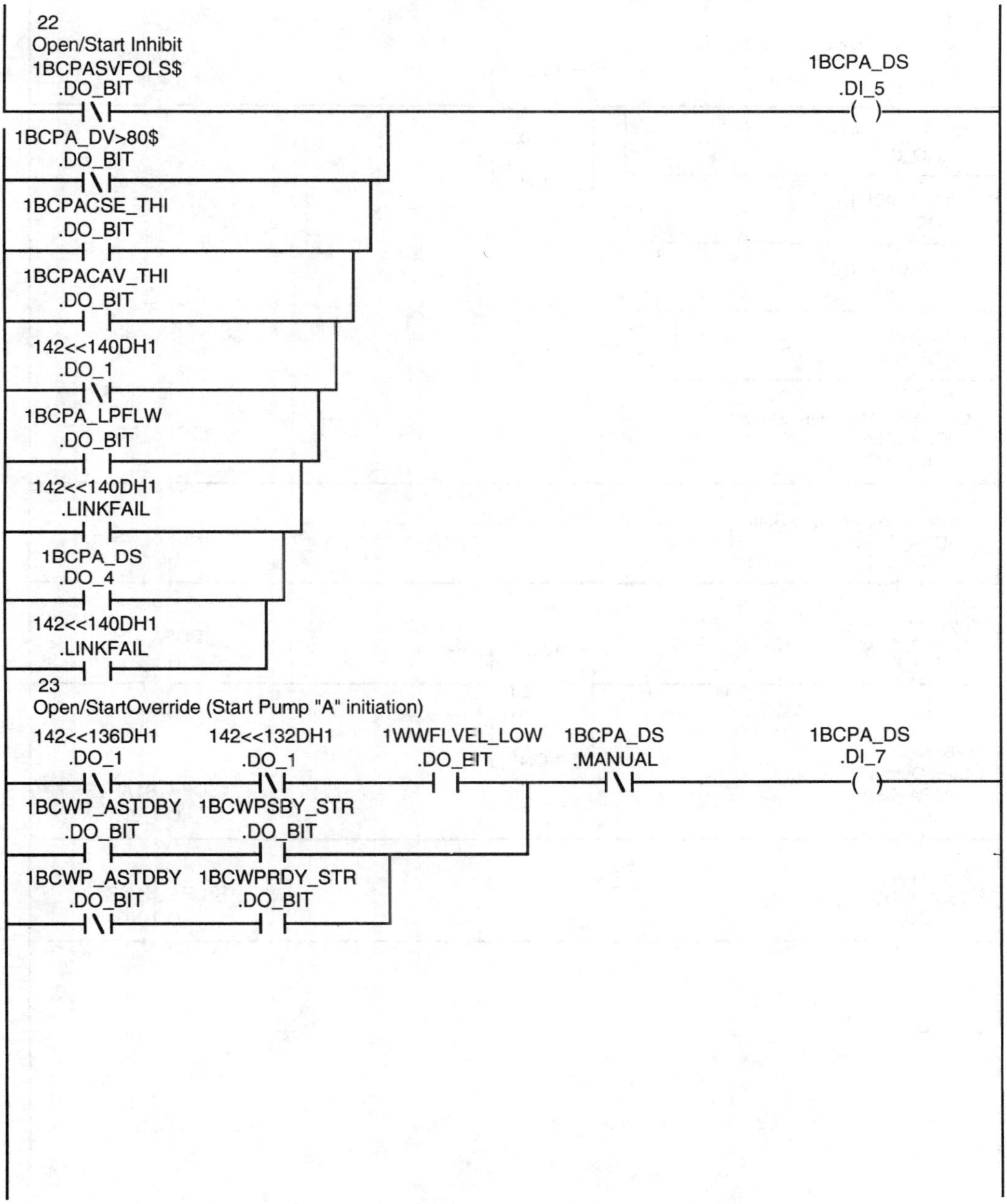

Application Logic
Boiler Circulation Water Pump
Open/Start Inhibit; Open Start Override

Figure 8

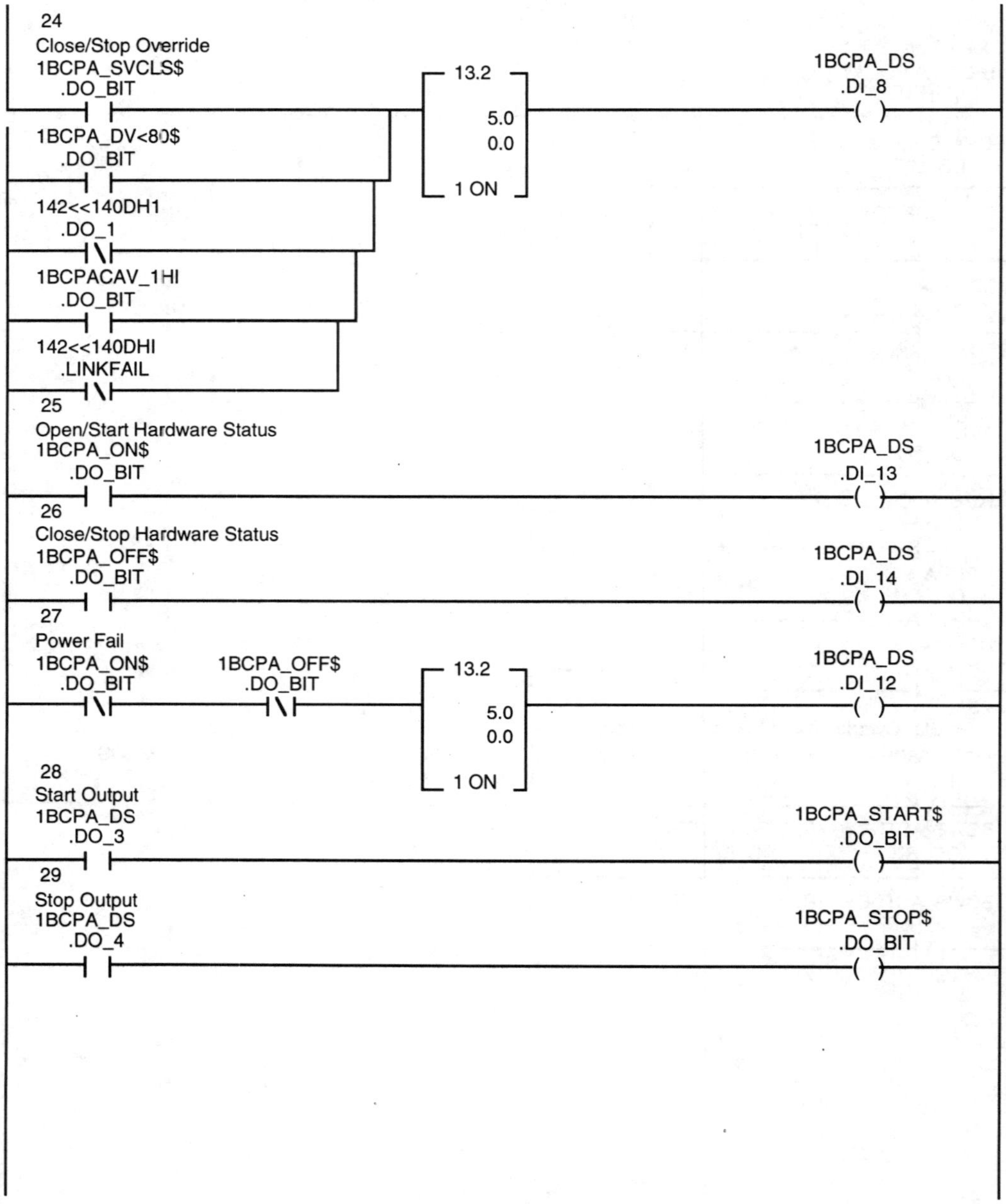

Application Logic
Boiler Circulation Water Pump
Stop/Close Override; Open/Start, Close Stop Hardware Status
Power Fail, Start Output, Stop Output

Figure 9

The level 1 core logic supports a full range of CRT operator display features within standard customer specified control pop ups including: two pushbuttons for directing device status; two mode select pushbuttons for device auto (remote/manual (local) mode selection; one pushbutton to reset logic in alarm condition to the existing state of the binary device; an optional pushbutton to STOP the device at an intermediate position.

Full device status reporting for CRT display pop ups is available from the level 1 core logic to show: device mode (auto (remote)/manual (local), lockout); device auto mode inhibit; the status of the device based on field inputs; the status of outputs commanded by the operating logic; the status of the device operating logic (lockout, reset, (directional) in travel, (directional) inhibit or override); or device power failure - thus the operator will be fully informed of the exact status of the binary device in question as well as what is preventing the attainment of the desired status.

Within the "pop up" windows, specific displays can be developed in accordance with the user's graphic standards and conventions.

The initial implementation of the Task 1 *Plant Automation Coordinator* at Nelson 4 and Lewis Creek Stations shall include the following Level 1 and Level 2 Automation.

Quantity	Auxiliaries	Level 1 Device Control	Level 2 Group Control
2	FD Fans	2	1
2	Circulating Water Pumps	2	1
1	Main Boiler Feedpump	1	1
1	Start-up Boiler Feedpump	1	
2	Condensate Pumps	2	1

TASK 2 PROCESS CONSTRAINT COORDINATOR

This algorithm provides for the application of multiple constraints to set the maximum/minimum generating limits on a plant as well as the maximum allowable rate (increase or decrease) of load change that can be accommodated with the current equipment in service

The sources of constraint conditions include:

- Equipment Life expenditure constraints
- Control System transient constraints
- Equipment Limitation constraints
- Error constraints

Significant advances have been made in the ability to analyze and diagnose the well being of a power plant on-line as has been demonstrated at EPRI's maintenance and diagnostic center at Philadelphia Electric's Eddystone Power Station (Reference 4). The Process Constraint Coordinator now allows the orderly introduction of constraints from maintenance and diagnostic programs on unit operations as appropriate to support best unit availability with minimum expenditure of equipment life (see reference 3).

Beginning with the unit master station (UMS) display as shown in Figure 10 which will normally be used to operate the unit, existence of a constraint condition will be shown by changing the background color of maximum rate of change setter. By accessing a "pop up" display from the UMS rate of change setter the major area of the constraining conditions is shown on the Process Constraint Coordinator graphic (Figure 11). Depending on the number of sources of constraints further access to detail displays may be required to show the exact source of the limiting conditions.

Constraint variables can be: direct process values; process rate of change values; process value deviation from setpoints or demands; process deviations from other process values.

The constraint function establishes the allowable unit rate of change in each direction as a function of the constraint variable. A constraint is active when its allowable unit rate of change is less than that which is set by the operator. The process constraint coordinator limits the allowable unit rate of change to whichever constraint calls for the lowest rate of change.

To fully inform the operator of constraining functions affecting the ability to change load, a bar chart display is provided for each constraining function (event limit display functions are noted in brackets "()" since there is no constraining variable).

Display information includes the constraint variable: long title; value (constraint time remaining); high/low alarm limits (not applicable); point alarm condition; the allowable rate of change for each direction and their relationship to the operator set unit rate of change; the constraint function point alarm condition when the constraint is active; and the constraint function point auto/manual status with hot spots for operator selection.

The constraint coordinator display thus shows the allowable limit, and the auto/manual status. The operator can then make an informed decision about the operating constraint consistent with achieving best unit availability with minimum expenditure of equipment life. Trend for the limiting stress condition including tag and variable identification may optionally be provided.

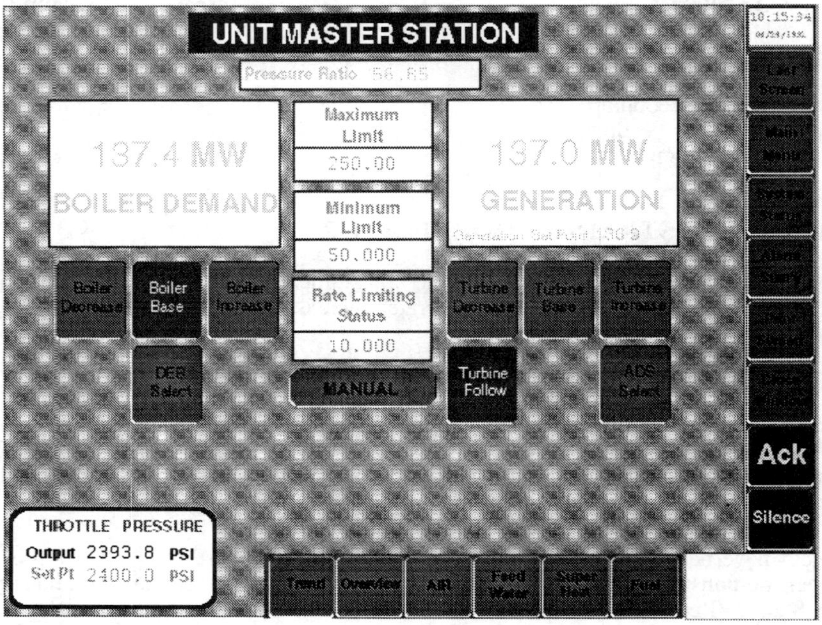

Figure 10

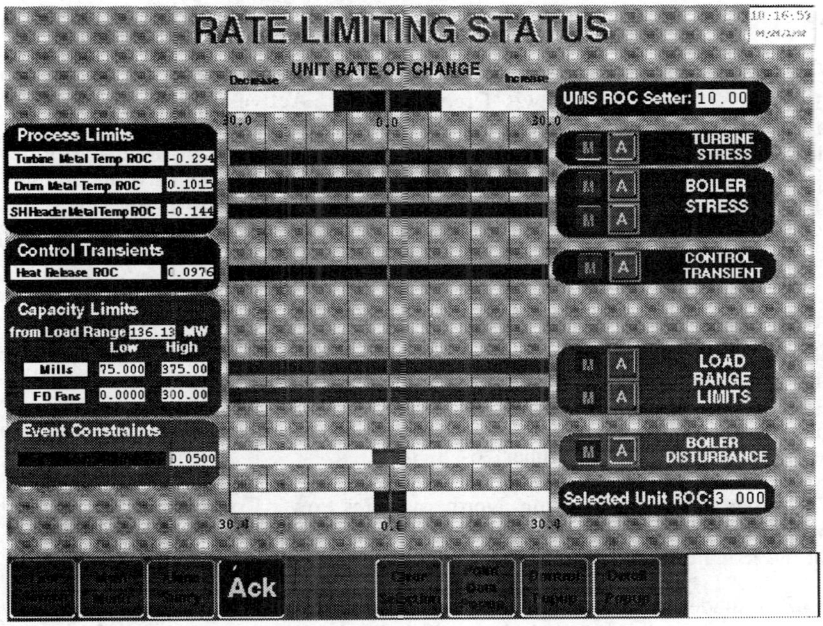

Figure 11

The initial implementation of the Task 2 Process Constraint Coordinator at Nelson 4 and Lewis Creek Stations will include the following constraints:

Constraint Type		Constraint Description
Equipment Life Expenditure	-	Turbine Metal Temperature Rate of Change
	-	Boiler Header Temperature Rate of Change
	-	Boiler Drum Temperature Rate of Change*
Control Transient	-	Load Demand Error
	-	Superheat Temperature Error
	-	Superheat Temperature Approaching Saturation*
	-	Reheat Temperature Error
	-	Reheat Temperature Approaching Saturation*
	-	Throttle Pressure Excessive Error
	-	Fuel Control Excessive Error
Equipment Limits	-	Fuel Regulating Equipment at a maximum/minimum limit (function of # burners in service and burner pressure)
Event Constraint	-	Boiler Water Purity
	-	NOx Greater than Allowable Limit
*Lewis Creek only	-	Sliding Pressure Mode*

A major benefit associated with the use of the Process Constraint Coordinator is the ability to calculate on-line how far a unit can go at any moment in time and how fast it can get there. Use of the Process Constraint Coordinator in this way supports the possibility of intelligent unit dispatch since the real unit performance parameters are applied to the dispatch functions.

CONCLUSIONS

The application of automation design concepts presented today is scheduled for completion within the next 6 months at ENTERGY's Nelson 4 and Lewis Creek Stations. Active discussions are proceeding with ENTERGY to establish the technical details of the scope to be implemented at each plant. The standardized approaches to plant automation and plant/process coordination resulting from these two joint development tasks will be a major step forward in providing more responsive and more responsive individual boiler turbine units for the management of system load. The Authors look forward to reporting upon the results achieved at a future technical conference including the cost benefits attained and the resultant return on investment.

REFERENCES

1. R. N. Hubby *Automation Strategies for Cyclic Operation*, presented at EPRI Conference on Fossil Plant Cycling, Washington DC, December 4-6, 1990.

2. R. N. Hubby *Performance Monitoring for Economic Dispatch*, EPRI Conference on Heat Rate Improvement, Knoxville, TN, September 26-28, 1989.

3. Richard W. Radtke and Conrad Colson, Northern States Power Co., *Coordination of Unit Equipment Capabilities Into Load Management Systems*, Instrumentation in the Power Industry Volume 31, presented at the ISA Power Symposium, St. Petersburg, FL, May 23-25, 1988.

4. W. J. Leonard, Philadelphia Electric Co., *Improved Performance from the Eddystone Unit No. 1 and No. 2*, Instrumentation in the Power Industry Volume 31, presented at the ISA Power Symposium - Ste. Petersburg, FL, May 23-25, 1988.

5. R. N. Hubby *Draft Functional Specification EPRI RP2710-24 Task 1-4*, Revision date 1/15/92.

6. R. N. Hubby *Advances in the Design and Implementation of Automation Design Concepts (Task 1 and 2)* presented at Israel Section of ISA Technical Conference, Herzilia Israel January 23, 1992.

7. Joseph M. Weiss EPRI, Richard J. Colsher EPRI, and James J. Hlebik L&N *Interconnectivity of Diagnostic Technologies at the EPRI M&D Center at Eddystone*, presented at the Israel Section of ISA Conference, Tel Aviv, Israel, January 26-28, 1993.

8. Gary A. Cohee ENTERGY and Dave Broske EPRI *Implementation of Automation Design Concepts at Entergy's Lewis Creek Units 1 and 2 and Nelson Unit 4*, presented at the EPRI Fossil Plant Cycling Conference, New Orleans, Louisiana, September 14-16, 1994.

© 0074-056X/95/057-065/$0 + .50pp

PREDICTIVE EMISSION MONITORING SYSTEMS: LOW-COST CONTINUOUS EMISSION MONITORING

Paul Reinermann, III
Director, Environmental and Regulatory Affairs
Pavilion Technologies, Inc.
12112 Technology Boulevard
Austin, Texas 78727

James Keeler
Chief Technical Officer
Pavilion Technologies, Inc.
12112 Technology Boulevard
Austin, Texas 78727

KEYWORDS

Air Quality, Computers, Environment, Gas Analyzers, Input/Output Systems, Life Cycle Cost, Multivariable, Neural Networks, On-Line Analyzers, Regulatory Compliance, Software

ABSTRACT

Predictive emission monitoring systems (PEMS) can be used as an alternative to continuous emissions monitoring systems (CEMS) to satisfy monitoring requirements established under the Federal Clean Air Act Amendments of 1990. PEMS offer the advantages of lower cost, lower maintenance, higher reliability and better emissions control than traditional hardware CEMS. This paper discusses the regulatory history of PEMS, addresses the various types of PEMS, highlights high order nonlinear regression model based PEMS and provides case studies on PEMS installations.

INTRODUCTION

The proposed Enhanced Monitoring Program, published on October 22, 1993 in the Federal Register, allowed parameter monitoring systems such as predictive emission monitoring systems (PEMS) but the language had some restrictions that made these low-cost alternatives to traditional continuous emission monitoring systems (CEMS) not as favorable as envisioned by PEMS manufacturers and users. Under the proposed regulation, the proposal and establishment of a PEMS as an enhanced monitoring protocol was arduous. Since the proposal of the Enhanced Monitoring Program and in consideration of the over 5,000 comments received, the EPA has indicated that the final Enhanced Monitoring Program will be more favorable to PEMS. Industry has proposed specific language and protocols that EPA has indicated will be reflected in the final Enhanced Monitoring Program and appear in the first issue of the Enhanced Monitoring Reference Document.

PROPOSAL OF A PEMS

In the proposed Enhanced Monitoring Program, if the best established monitoring was not proposed, then the owner or operator "shall identify all technologically feasible monitoring methodologies for a particular emissions unit in order to select as a proposed enhanced monitoring protocol the best other monitoring methodology for providing sufficiently representative, accurate, precise, reliable, frequent and timely data to satisfy the requirements of this part at the particular emissions unit." Cost was not a factor in determining whether or not a proposed enhanced monitoring protocol was the best other monitoring methodology.

For example, a traditional CEM would be considered the best established monitoring for nitrogen oxide emissions from an industrial boiler. In order for the owner or operator to use a PEMS, the owner or operator would have to identify all technologically feasible monitoring and justify why the selected monitoring technology is the best for the situation without cost being a factor.

At a November 9, 1994, Enhanced Monitoring meeting, the EPA released a Staff Working Draft of the Enhanced Monitoring Program that will change the enhanced monitoring proposal process. When proposing the best monitoring for a particular emission unit, the owner or operator does not have to propose the best technologically feasible monitoring methodology. The owner or operator must provide a list of all available established monitoring for a particular emission unit (in lieu of all technological feasible monitoring) and other monitoring considered. The relative cost of different monitoring methodologies can be considered by the owner or operator, however, cost does not have to be considered by the permitting authority.

A December 28, 1994, *Federal Register* notice, solicited comments on five aspects of the Enhanced Monitoring, one being whether or not cost should be a more significant factor. The final rule may direct the permitting authority to consider cost when approving a proposed enhanced monitoring protocol.

ESTABLISHMENT OF A PEMS

The proposed Enhanced Monitoring Program required that potentially a large number of reference test runs would need to be utilized to demonstrate that a PEMS meets relative accuracy requirements. For each parameter which is known to affect emissions, a minimum of three test runs would need to be performed at representative points in addition to performing three test runs at three process operating loads. For a PEMS which utilizes eight process sensors, twenty-four test runs would have to be performed for the eight process sensors and then three additional test runs at three process operating loads for a minimum of thirty-three test runs.

Based upon comments received and information provided by the workgroup formed by Pavilion, the EPA indicated on November 9, 1994 that they will reduce the total number of test runs to nine. Essentially, three test runs at low, normal and high emission rates must be conducted.

At the November 9, 1994 meeting referenced above, the EPA indicted that PEMS Enhanced Monitoring Protocols will be included in the first issue of the Enhanced Monitoring Reference. Since that time, EPA has transmitted five PEMS Enhanced Monitoring Protocols (EMP's) to the 10 EPA Regional Offices and selected State agencies for their review prior to being incorporated into the Enhanced Monitoring Reference Document. The five EMP's were based upon information provided by current users of PEMS to the EPA, Emission Measurement Center on October 26, 1994. The information demonstrated that the five PEMS installations were equivalent to traditional CEMS in all of the applications. The five installations were:

1. Monitoring of NOx emissions from a 221 mmBTU/hr heat input natural gas fired boiler subject to 40 CFR Part 60, Subpart Db;
2. Monitoring of NOx and CO emissions from an industrial furnace;
3. Monitoring of NOx and CO emissions from a natural gas-fired turbine with dry-low NOx;
4. Monitoring of NOx and CO emissions from a natural gas-fired turbine with steam injection; and
5. Monitoring of NOx emissions from a natural gas-fired reciprocating internal combustion engine.

PROCESS MODELING - HEART AND SOUL OF A PEMS

Process modeling methods generally fall into two broad classes: first- and regression modeling methods. Models developed by these methods have been used in PEMS. Both methods have advantages and disadvantages. This section compares and contrasts these two methods, concentrating on the high order, multivariate, nonlinear regression modeling.

FIRST-PRINCIPLES MODELS

First-principles modeling techniques describe the process dynamics via the physical equations designed to approximate the system. Examples include kinetics, thermodynamics, and fluid-flow equations. These models are typically expressed in the form of a set of algebraic expressions and/or coupled partial differential equations with coefficients fitted to the data of the system. First-principle models have been used to create PEMS for natural gas fired turbines and reciprocating internal combustion.

LINEAR REGRESSION MODELS

The most widely known form of regression is linear regression-fitting a straight line or a plane to a data set. The simplest form of linear regression minimizes the sum squared error of the model output versus the plant output. That is, given a set of output values $y = \{y1(t), y2(t), yn(t)\}$ and input data values $x = \{x1(t), x2(t), xm(t)\}$ sampled at times $t = \{1,2,N\}$, linear regression minimizes the sum-squared error, E, of the predicted values p(x) and the output values y, where:

$$E = \sum_{t=1}^{N} (y - p(x))^2 \qquad \text{Equation 1}$$

For a linear system, the predicted outputs, p(x), are a linear sum of a matrix of coefficients A times the inputs:

$$p_i = \sum_{j=1}^{m} A_{ij} x_j + b_i$$

or, in matrix notation,

$$\mathbf{p} = \mathbf{Ax} + \mathbf{b}$$

Assuming a linear system, Equation 1 can be solved analytically via matrix inversion.

Linear regression can only model linear processes (or only the linear component of nonlinear processes). However, most plants exhibit some degree of nonlinear behavior, and, therefore, most linear models are not as accurate as nonlinear regression models.

NONLINEAR REGRESSION MODELS

Nonlinear regression techniques model the output as a nonlinear function of the input variables:

$$\mathbf{p} = \mathbf{f}(\mathbf{x})$$

where **f** is a nonlinear function. Nonlinear regression is a well-studied discipline, with polynomial regression perhaps being the most common traditional form. Polynomial regression, however, has many disadvantages including "overfitting" the data, that is, the data used in model development may be fit perfectly, but any new data point may yield an output very far from the true value.

The newest and most sophisticated modeling software allows the use of both first-principles information and advanced nonlinear regression techniques. The key to the modeling methodology is based on major advances in nonlinear regression that have occurred over the past decade which facilitate the building of high-dimensional nonlinear models without the problems of overfitting. These new regression methods are computationally intensive, and their development was facilitated by the widespread availability of powerful computers. The underlying mathematics of these regression models and several different regression algorithms have been reviewed in hundreds of papers and books on the subject [References 1-7], and several authors [References 8-9] have proven that these regression structures are capable of approximating any continuous multivariate function with very weak restrictions to arbitrary accuracy, a property called "universal approximation".

NONLINEAR REGRESSION MODEL BASED PEMS

The general procedure for building, installing and certifying a nonlinear regression model based PEMS is as follows:

1. Collect emissions data with a highly-accurate reference method or CEM to the stack, as depicted in Figure 1. Emissions (NO_x, CO, O_2, CO_2, SO_2, VOC, HAPS, etc.)
2. Operate the process and/or pollution control device over the range of anticipated operations for 3 to 5 days and concurrently collect process and/or pollution control device data with the emissions data.
3. When the data gathering period is finished, use a modeling tool such as *Process Insights* to build emissions and sensor validation models as a function of the process variables.
4. Install the PEMS in the computer which receives the process data.
5. Perform a relative accuracy test audit (RATA) of the PEMS to demonstrate conformance with accuracy requirements.

BENEFITS OF NONLINEAR REGRESSION MODEL BASED PEMS

Regardless of the modeling method employed, achieving the EPA 95% up-time with a PEMS for demonstration of continuous compliance requires a stringent quality assurance program. Since PEMS are computer software programs which rely on sturdy process sensors, achieving the required up-time is almost assured. In order to assure that the up-time requirement is met, the ability to detect sensor failures, and, if possible, compensate for those failures is necessary. Sensor validation is a crucial step to ensure accuracy and reliability of the emissions model. If sensor drifts or failures go undetected, the process could be out of compliance without detection, or the system could determine emissions that are higher than the proper values. Either case is unacceptable for an accurate, reliable system. When a nonlinear regression model based PEMS is created, a Sensor Validation System can be also be created.

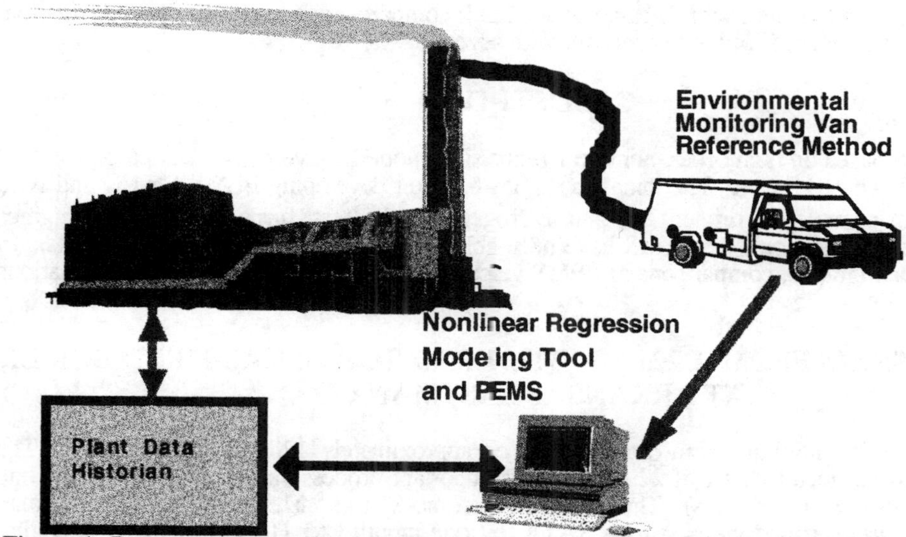

Figure 1: Process and emissions data are collected simultaneously from the process data historian and by a hardware CEMS or a reference method monitoring van, respectively The emissions model is built with a modeling tool such as Process Insights by correlating these two sets of data. Then the PEMS is installed on the plant computer which receives the process data and the PEMS runs real-time, continuously.

A Sensor Validation System is based on an accurate models of all of the sensors used in the emissions model as a function of other sensors in the plant. These sensor models enable the accurate prediction of process data so that in the event of a process sensor failure or drift, the PEMS can continue to determine emissions accurately. This validation and data reconciliation function ensures that the PEMS will continuously provide accurate monitoring values in a reliable and robust fashion.

Sensor validation is based on solid principles that are commonly used in all measurements, essentially, checking one sensor measurement against an independent measurement. This is exactly how calibration of a sensor is done; a sensor is compared against an independent reference value and if it is off, it is adjusted or replaced. This is the same concept behind a Sensor Validation System. It is based on independent checks of each sensor, but the reference in the case of the sensor validation system is a calculation based on data from the other sensors.

Consider the following: Sixteen temperature transducers located at various positions on a process. The temperature measurements of each transducer is correlated and interdependent. If, for example, these transducers are located on a pipe and one sees that transducer A reads 134 degrees , C reads 135.2 degrees, but B reads 57 degrees, then if B is located between A and C on the pipe, it is evident that B is wrong. In fact, B should be measuring around 134.6 degrees. Note that with the information provided here, it is unclear as to whether B has failed or whether A or C have failed. Based on common sense, since A and C agree, the assumption is that B has failed. The same is true in calibration of any instrument. If you compare an instrument to a reference, and you see that there is a significant deviation, it could be that the instrument is off, or the reference is off or both. To check which one is off, another instrument to validate our system. If two instruments agree, but one is off, then it is highly probable that the one that is off is bad. If we wanted to be very sure, we would compare it once again to another instrument. If three agree and one is off, then we are quite confident that the one that deviates from the other three is in error.

A Sensor Validation System uses this same principle. It compares each sensor to peer information. In fact, it compares not just with one peer, but with several

CASE STUDIES

To date, PEMS based on high order, nonlinear regression models have been installed on over 30 processes in the United States. The initial focus has been on developing PEMS for CO and NO_x emissions from natural gas, refinery gas and hydrogen fired heaters, furnaces, turbines, engines and boilers. In all instances, the PEMS have been able to pass all required certification tests. In the following paragraphs, comparisons of PEMS versus CEMS are provided for three installations are provided.

NOx EMISSIONS FROM A 221 MMBTU/HR NATURAL GAS-FIRED BOILER AT ARKANSAS EASTMAN CO.

This PEMS was initially built from data collected on approximately 120 process variables. The final model was reduced to a set of 22 of the most important process variables as inputs and one output (NOx lb/mmBTU readings). The NOx emission model was installed at Arkansas Eastman in May, 1993, was certified via passing a RATA test one month later [11], and is now operating continuously under an operating permit with the Arkansas Department of Pollution Control and Ecology. This was the first PEMS installed on Subpart Db boiler and done with the approval of EPA, Region VI in conjunction with EPA Headquarters.

Note the accuracy of this PEMS in Figure 2 below. Even in the fast dynamics regions the model tracks the measured NOx emissions. This system has been in operation for almost two years and has passed all RATA tests. It has achieved better than 99% up-time with 100% pollution compliance.

Figure 2. NOx Emissions from a 221 MMBTU Boiler at Arkansas Eastman, Batesville, Arkansas

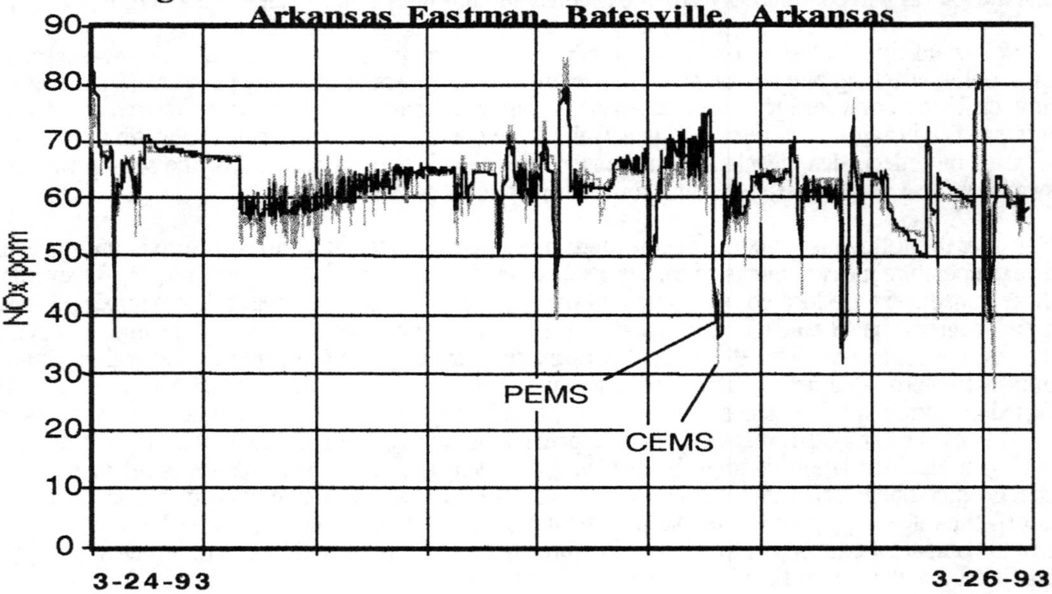

NOx AND CO EMISSIONS FROM A 100 MW COGENERATION GAS TURBINE AT DOW CHEMICAL, FREEPORT, TEXAS

The PEMS in this application was built from data collected on 184 process variables. The final model was reduced to 17 important inputs and 3 outputs (NO_x ppm, CO ppm and O_2% readings). The NOx emission model achieved an R^2 correlation of 0.997 (accuracy of 99.7%). Figures 3 and 4 display the accuracy for the NO_x and CO emission test sets. The original RATA on this turbine for NO_x was 2%, 3% and 4% at the low, medium and high ranges respectively compared to a requirement of 20%.

This PEMS has been in operation for approximately 1.5 years. It has passed all RATA tests (5 total to date), with better than 98% up-time. Eight other units at this facility have had PEMS installed and all have demonstrated the same exceptional performance. A total of fifteen will eventually be installed. The savings of these 15 PEMS versus 15 hardware CEMS is estimated at over $3 million for installation and over $150,000 in annual operation and maintenance costs.

Figure 3. NO_x Emissions from a 100 MW Gas Turbine

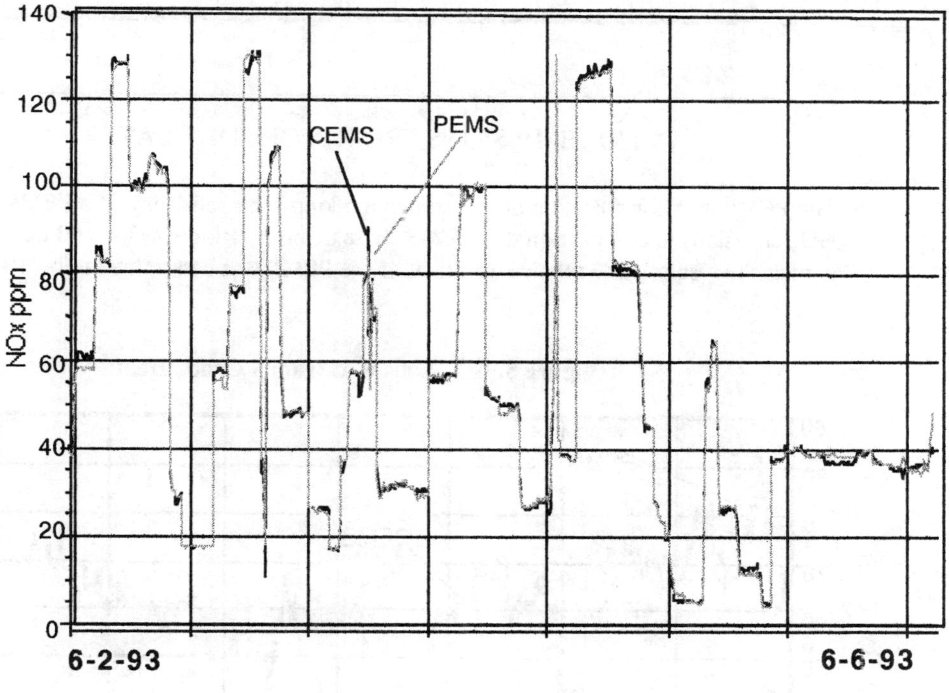

Figure 4. CO Emissions from a 100 MW Gas Turbine

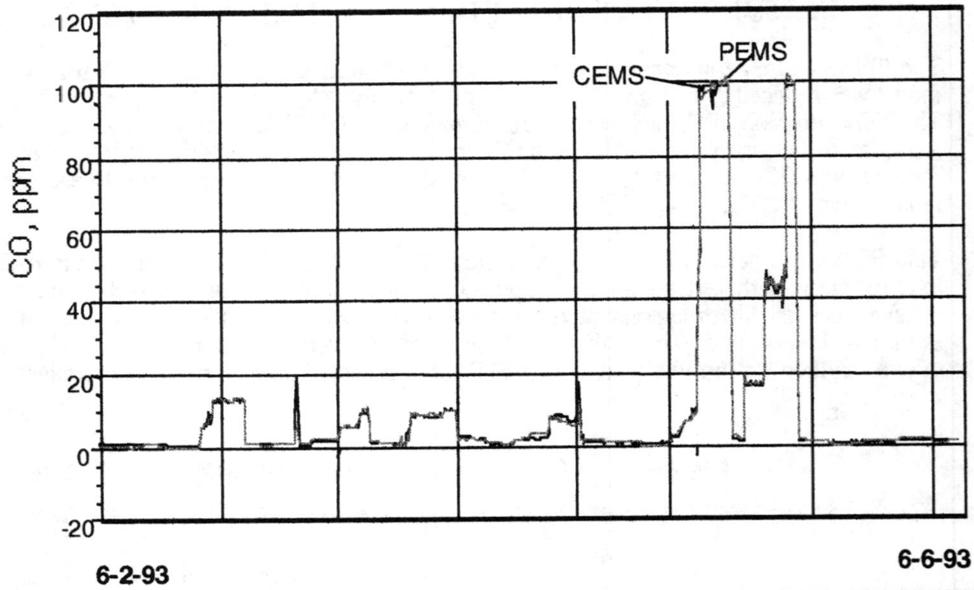

NO_x EMISSIONS FROM A CRUDE HEATER

The PEMS in this application demonstrated the long-term reliability of a PEMS to determine NO_x emissions in comparison to a CEMS. The demonstration was for 100 days and Figure 5 depicts the results. The relative accuracy of the PEMS was determined to be 3%.

Figure 5. NO_x Emissions from a Crude Heater

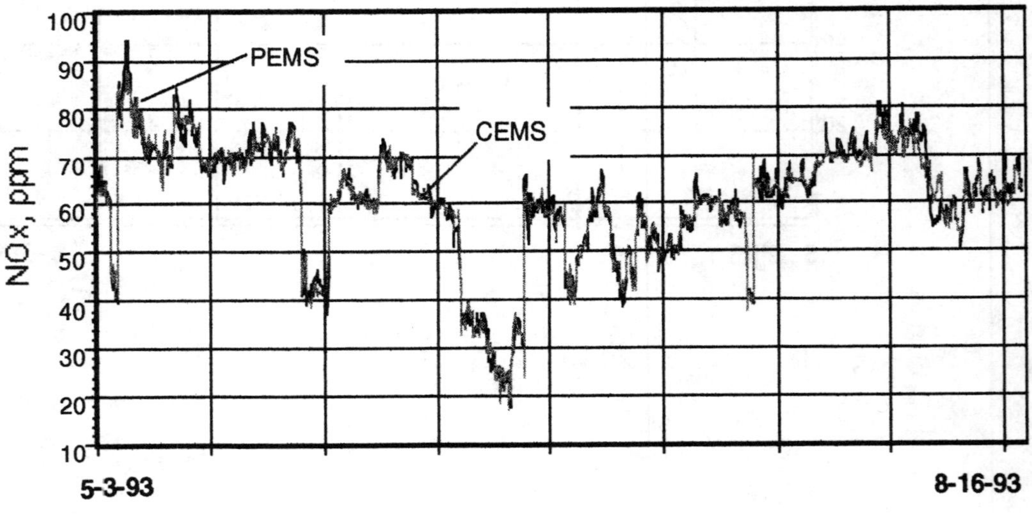

CONCLUSIONS

The above case studies demonstrate that high order, nonlinear modeling tools such as *Process Insights* can build accurate emission models for a variety of unit operations that require emissions monitoring. These models, when coupled with Sensor Validation System to create a PEMS, provide highly accurate and reliable emissions monitoring. A PEMS has several advantages including:

1. The initial cost of installation and certification of PEMS is 1/3 to 1/2 the cost of a CEMS. The installation costs through certification costs of a PEMS is $70,000-$125,000 in comparison to $150,000-$300,000 or more for a hardware CEM.
2. PEMS has proven to be as accurate as the best available established monitoring.
3. The maintenance costs of PEMS are small in comparison to those of a CEMS. The maintenance costs for a PEMS is $10,000-$20,000 in comparison to $50,000 to $80,000 for a hardware CEM.
4. The Sensor Validation System provides PEMS with minute-by-minute quality assurance checks and superior reliability in comparison to a CEM.
5. PEMS satisfies the regulations of 117.580(c) recently passed in Texas for PEMS.
6. Perhaps most importantly, nonlinear regression modeling tools can provide very valuable information on the plant operations. CEMS can only provide information on whether the emissions are within the regulations or not. In contrast, PEMS can provide information on how to operate the process and optimize the process to achieve maximum output and minimum emissions.

PEMS provide for increased profits and a cleaner environment.

References

[1] D. Rumelhart and J. McClelland, 1986. Parallel Distributed Processing, Cambridge, MA: The MIT Press.

[2] Advances in Neural Information Processing Systems 1-5, 1989-1993. San Mateo, CA: Morgan Kaufmann Publishers.

[3] J. Hertz, A. Krogh, and R. Palmer, 1991. Introduction to the Theory of Neural Computation. Addison-Wesley.

[4] V. Rao Vemuri, 1992. Artificial Neural Networks: Concepts and Control Applications. Los Alamitos, CA: IEEE Computer Society Press.

[5] W. Miller, R. Sutton, and P. Werbos, 1990. Neural Networks for Control. Cambridge, MA: The MIT Press.

[6] E. Hartman and J. D. Keeler, 1991, "Predicting the Future: Advantages of Semi-local Units," Neural Computation 3, 566-579.

[7] J. D. Keeler, 1992. "Vision of Neural Networks & Fuzzy Logic for Prediction and Optimization of Manufacturing Processes", SPIE Conference Proceedings, Vol. 1709, pg. 447-456, SPIE Publishing, Bellingham, Washington.

[8] M. Hornik, M. Stinchcombe, and H. White, 1989. Multilayer feedforward networks are universal approximators. Neural Networks 2, 359-368

[9] E. Hartman, J. D. Keeler, and J. Kowalski, 1990. Layered Neural Networks with Gaussian Hidden Units as Universal Approximations. Neural Computation 2, 210-215.

[10] P. J. Werbos, 1974. Beyond regression: New tools for prediction and analysis in the behavioral sciences, Ph.D. Thesis, Harvard University, Cambridge, MA.

[11] M. Collins, W. Jackson, K. Terhune, J. Havener, 1993, "NOx Emissions using *Process Insights* and the *Software CEM*" Pavilion technical report (to be published).

0074-056X/95/067-075/$0 + .50pp

INTEGRATED PROGNOSTICS, MAINTENANCE AND LIFE EXTENDING CONTROL OF POWER PLANTS†

by

Asok Ray
Chen-Kuo Weng
Mechanical Engineering Department
The Pennsylvania State University
University Park, PA 16802
Tel: (814)865-6377; FAX: (814) 863-4848; e_mail: A2R@ecl.psu.edu

KEYWORDS

Life Prediction, Failure Prognosis, Condition-based Maintenance, and Robust control

ABSTRACT

This paper presents the concept of a methodology for synthesis of life extending control systems along with failure prognostics and condition-based maintenance of fossil power plants. The proposed control synthesis methodology relies on augmentation of the current system-theoretic techniques for synthesis of optimal decision and control policies with governing equations and inequality constraints that model the structural properties of materials for the purpose of damage representation. This decision and control system has a hierarchical structure and is required to be robust, i.e., tolerant of a variety of errors resulting from parametric and non-parametric uncertainties in modeling of plant and damage dynamics, and continuously varying or discrete-event disturbances. It also allows prediction of impending failures well in advance based on an analytical model of the material damage even though the available sensors may not be able to detect any change in the plant behavior. The power plant availability is envisioned to be significantly enhanced by making maintenance decisions based on the sensor data and analytical prediction of the remaining service life especially if the plant is subjected to load following operations and repeated start-up and shutdown.

INTRODUCTION

The cost of generating electricity, which covers capital investment, plant operation and maintenance, has become extremely important in all stages of power generation. Nearly 70% of the fossil power plants that represent more than 40% of U.S. electric power generation capacity will be over 30 years old by the year 2000. Since most of these plants are designed for a service life of 40 years, the utility companies are faced with the problem of operating the aging power plants with continued reliable service to the customers while simultaneously matching varying load demands, satisfying the pollutant emission requirements, and achieving high plant performance to meet increased competition for lower cost per kwh. Therefore, enhancement of structural integrity, service life prediction of critical components, and prognosis of impending failures can yield significant improvement in operations planning and maintenance scheduling of power plants. Somewhat similar problems exist in production systems (e.g., petrochemical and processing industries and manufacturing plants) and mechanical systems (e.g., aircraft and rail transport). These economical and technological issues have spawned a strong interest in failure prognostics, condition-based maintenance, and life extending control.

---

† The work reported here is supported by National Science Foundation under Grant No. ECS-9216386 and Electric Power Research Institute under Contract RP8030-5.

FAILURE PROGNOSIS AND DECISION-MAKING FOR MAINTENANCE

Failure prognosis and risk analysis are envisioned to be an augmentation of failure diagnosis to predict, locate, and reconfigure impending failures of critical plant components. In addition to the modeling errors of damage dynamics, prediction of both the damage rate $\dot{v}$ and hence the accumulated damage v are affected by plant disturbances and sensor noise. Furthermore, the failure prognosis system must be cognizant of the plant operational requirements, mission objectives, and the limitations and inaccuracies of the predicted damage information. The damage calculations are based on the current plant state x, damage state v, and any additional available information such as history of measurements of the local strain and temperature.

Traditionally, the risk index and residual service life [Bolotin, 1989] of mechanical structures and machine components are calculated off-line on the basis of statistical models of the mechanical properties of the material, operating conditions, major disruptions (e.g., emergency shutdown of the plant from full load), and accidents (e.g., seismic and large atmospheric disturbances). The service life is finite as the damage accumulation is monotonic with time. Furthermore, the plant operations and maintenance are scheduled based on the fact that damage accumulation is likely to be accelerated in the event of accidents and major disruptions. Therefore, on-line monitoring of the accumulated damage will allow refinements of the risk index and residual service life prediction [Bogdonoff and Kozin, 1985; Bolotin, 1989] as time progresses. Specifically, prediction of the accumulated damage and the current damage rate will assist the plant engineer in making dynamic decisions regarding the safety procedures, mission accomplishments, and the time interval between major maintenance actions. The concept of service life prediction based on the current plant and damage states is outlined below.

Let $\varphi_j(t)$, $j=0,1,2,\cdots,m$ represent predicted occurrences of impending accidents or major malfunctions in plant operations after the current instant of time, t, such that the set φ_o indicates the absence of any accidents or disruptions, and the event sets φ_j, $j=0,1,2,\cdots$, m are exhaustive and mutually exclusive. Then, in the absence of any accidents or disruptions, the service life θ_o at time t is obtained in the probabilistic setting in terms of the plant state x(t), damage state v(t) and desired plant output profile $Y_d(t) = \{y_d(\tau): \tau \geq t\}$ $Y_d(t) = \{y_d(\tau): \tau \geq t\}$as:

$$\theta_o\big(t, x(t), v(t), Y_d(t)\big) = \mathrm{Sup}\big\{\tau \in [t,\infty)\text{: } \Pr\{v(\tau) \in \Omega \mid \varphi_o(t)\} \geq \varepsilon\big\} \tag{1}$$

where Ω is the set of the admissible damage vectors for the plant to remain in service; and $\varepsilon \in (0,1)$ is the threshold probability that the damage vector v(t) remains within Ω during the service period, i.e., while the plant is in operation. Eq. (1) implies that θ_o is the largest time for which the damage vector remains within the admissible range with probability greater than or equal to ε.

The service life of the plant is reduced if an accident or disruption occurs, i.e., $\theta_j \leq \theta_o$ for $j=1,2,\cdots,m$. Hence, similar to Eq. (1), the service life θ_j based on the occurrence of the event $\varphi_j(t)$ is obtained in the probabilistic setting:

$$\theta_j\big(t, x(t), v(t), Y_d(t)\big) = \mathrm{Sup}\,\big\{\tau \in [t,\theta_0]\text{: } \Pr\{v(\tau) \in \Omega \mid \varphi_j(t)\} \geq \varepsilon\big\},\quad j=1,2,\cdots, m \tag{2}$$

Then, the expected service life at time, t, can be obtained as:

$$\Theta\big(t, x(t), v(t), Y_d(t)\big) = \sum_{j=0}^{m} \theta_j\big(t, x(t), v(t), Y_d(t)\big)\, \pi_j(t) \tag{3}$$

where $\pi_j(t)$, $j=0,1,2,\cdots,m$ are the *a priori* probabilities of occurrence of the events φ_j over the interval $[t,\theta_0]$.

In the absence of any accident or major disruption after the current time t, the service life θ_o in Eq. (1) can be obtained from the stochastic differential equations of plant and damage dynamics. However, computation of θ_j for $j=1,2,\cdots,m$ in Eq. (2) is more complex because the damage accumulation depends not only on the transient states of the plant and damage dynamics as a result of the accident or disruption, but also on the instant of its occurrence. Advanced statistical analysis [Bogdonoff and Kozin, 1985; Sobczyk and Spencer, 1992] is necessary for predicting the service life from Eq. (2) to make maintenance decisions. A possible approach is to formulate the distribution functions of the damage process by solving the Kolmogorov diffusion equations whenever the maintenance decisions can be made off-line with available computational resources. An alternative approach proposed here is to identify the type of distribution (e.g., lognormal) from experimental data, and then generate the requisite statistical information from the first two moments of the plant and damage states. Although this information may not provide exact statistics, it is sufficiently accurate for generating early warnings for impending failures [Ray, 1990] as well as for making maintenance decisions.

The above information will also be inputs to a hybrid (i.e., combined continuous-time discrete-event) decision system for condition-based maintenance. This decision system will incorporate nonlinear stochastic models of the plant and damage dynamics in a continuous setting coupled with a discrete-event model representing accidents and disruptions as uncontrollable events and maintenance actions as controllable events. Since solutions of the nonlinear stochastic differential equations excited by randomly occurring discrete events may not be achievable except possibly by Monte Carlo simulation [Rubinstein, 1986], approximate solutions will be obtained from simplified models of the plant dynamics supported by the available sensor data history. These results, in turn, will be verified by simulation.

The major implications of the proposed service life prediction and risk analysis are: (i) generation of early warnings resulting in diminished risk of loss of human life and plant equipment; (ii) enhancement of plant availability by avoiding an emergency shutdown; and (iii) reduction of the plant operational cost by dynamic rescheduling of the maintenance plan via on-line assessment of damage in the critical components.

INTEGRATED DECISION AND CONTROL SYSTEM ARCHITECTURE

The integrated decision and control system for prognostics, condition-based maintenance and life extension is built upon the principle of hybrid supervisory control [Ramadge and Wonham, 1987; Garcia, 1993]. This concept of supervisory control offers a systematic and rigorous approach to hierarchical control of hybrid processes where the continuous process (e.g., power or processing plant) is controlled at a lower level of hierarchy, and the decisions for operation and maintenance are made at upper levels. The rationale for this approach is that the inputs to the upper levels are a combination of continuous signals and discrete events and that the resulting decisions are implemented as discrete events. Fig. 1 shows the general concept of the integrated decision and control system which is hierarchically structured in three levels of operation, namely, coordination, supervisory, and execution [Garcia, 1993; Garcia et al., 1994]. As their names suggest, these three levels perform different tasks and are partially ordered according to the principle of increasing intelligence with decreasing precision [Saridis, 1989]. While the lowest level (i.e., the execution level) is generally of high precision and low intelligence, the higher levels are of increasing intelligence and diminishing precision. There is exactly one coordinator for control of the entire plant (e.g., a power plant or a processing plant or a large rolling mill) but there may be one or more supervisor(s) over which the coordinator exercises control. Each subsystem, representing a major plant component (e.g., the furnace-boiler or the turbine-generator in a fossil power plant) whose performance and service life are to be controlled belongs to the supervisory level. Each supervisor, in turn, controls its own executor which communicates with the continuously varying process of the plant subsystem. The operating principle of the supervisory control system is briefly described below.

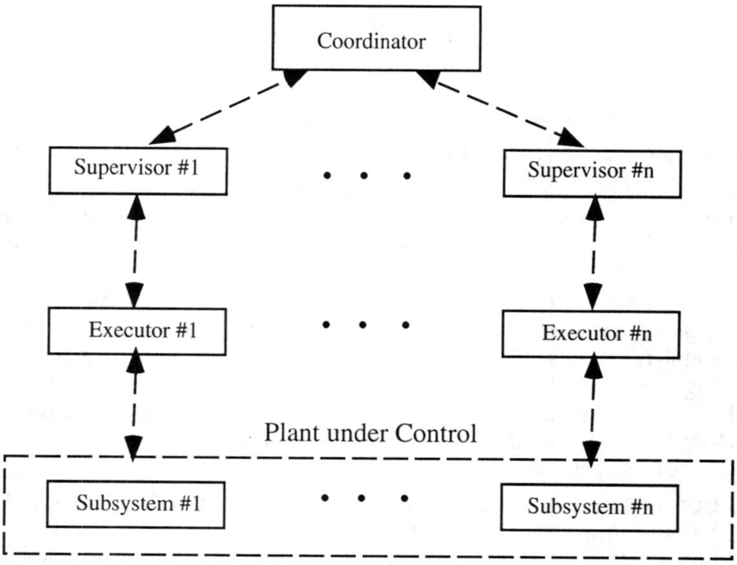

Fig. 1 Concept of the Integrated Decision and Control System

The architecture of the proposed integrated supervisory control system is presented in Fig. 2 in which only one supervisor and one executor are shown for clarity. It consists of two loops, namely, the outer loop and the inner loop. The decisions on plant operation and maintenance are made at the coordination level in the outer loop. This information is communicated up to the human users and down to the supervisor(s). Life extending control of a given subsystem via damage mitigation is the sole function of a supervisory level controller which has its own executor for communicating with the continuously varying control system within the inner loop. The role of the executor is to modify some of the control signals within the inner loop so that the detrimental effects of the uncontrollable environmental events (e.g., failures, accidents, or operational disruptions) acting upon the plant are mitigated. It is also responsible for selecting a feedforward and feedback control policy based on the control strategy received from the supervisor.

The inner loop consists of two major functional modules: sensor signal validation and information processing; and feedforward/feedback control. The sensor signal validation and information processing module feeds validated sensor signals to the control system module within the inner loop, and plant and damage information to the interface and life extension controller modules in the outer loop as seen in Fig. 2. The discrete-event interface at the supervisory level receives both continuously varying signals of validated sensor data and predicted damage, and discrete information of plant operation status such as past incidents of equipment failures and unscheduled plant outage. The interface generates discrete events which are, in turn, inputs to both the supervisory system and the decision process interface at the coordination level.

A sequence of discrete commands, called strategy patterns, which are controllable events and provide the information regarding the desired plant performance and service life are issued to each individual supervisor by the coordinator. Strategy patterns represent the coordination level decision for plant operation and maintenance actions to mitigate detrimental effects of actual or anticipated failures and changes in the operation strategy. Similarly, a sequence of discrete commands, called control patterns, which are controllable events and represent the supervisory level actions in response to the strategy patterns from the coordinator and the events of subsystem operation generated by the discrete-event interface, are issued to the execution level by each supervisor. The control patterns are interpreted at the execution level to generate continuous signals to modify the control signals in the inner loop.

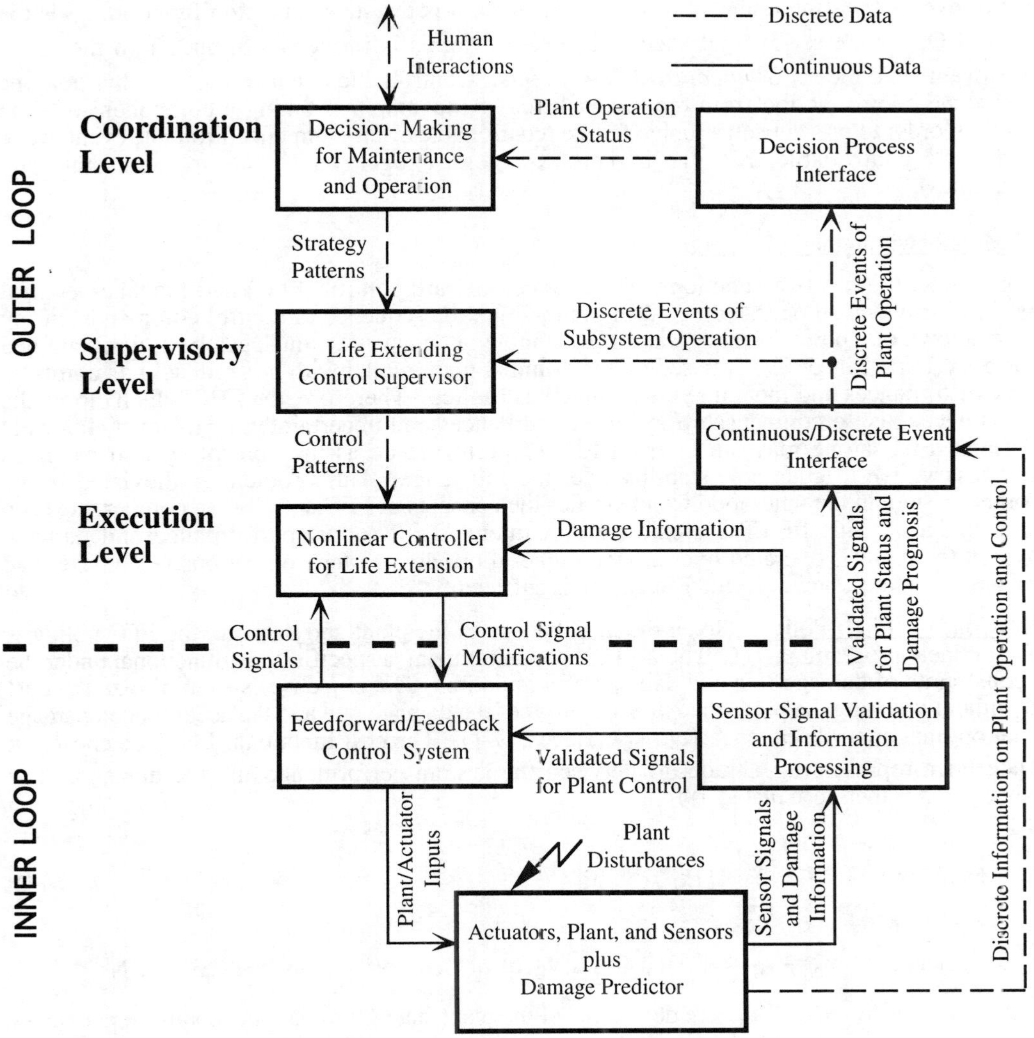

Fig. 2 Architecture of the Integrated Decision and Control System

The functions of the coordination, supervisory and execution levels in the outer loop are realized as minimization of maintenance cost and the probability of subsystem failures that may result in loss of plant availability. To accomplish these tasks, the coordinator exercises control over the supervisors, and each supervisor on its executor by enabling or disabling events through discrete choices of strategy patterns and control patterns, respectively. Discrete-event phenomena at these three levels are modeled as interacting *finite-state automata* following the recent work in this field [Ramadge and Wonham, 1987; Cislak et al., 1988; Ozveren and Willsky, 1989; Garcia, 1993]. In this setting, for example, the states of an automaton reflect status (e.g., the material degradation condition with respect to the need for maintenance, repair, or replacement) of a plant subsystem; and the events interconnecting the states correspond to discrete and abrupt changes in the equipment state, or gradual degradation of the material condition as it crosses a predetermined threshold of the predicted remaining life. The automaton for each of the interacting process at three levels is represented by a quintuple, $G = (Q, \Sigma, \delta, q_0, Q_m)$, where Q is the finite set of states, Σ is the finite set of events (i.e., the alphabet of the automaton), δ is the state transition function, q_0 is the initial state, and Q_m is the set of final states. The set of events is further partitioned into the set of *controllable* events and the set of *uncontrollable* events. Controllable events are events that may be prevented by some control action from the higher level. An example is taking a component off-line (control action) in order to prevent catastrophic failure (controllable event). An uncontrollable event is one that is not directly controllable from the higher level (e.g., exogeneous inputs to the system and unpredicted failures).

The Control Module within the Inner Loop

The control module has two functions, namely, feedforward control (FFC) and feedback control (FBC). The FFC provides an open loop control policy (e.g., the sequence of control commands during start-up and shutdown of a power plant or take-off or landing of an aircraft) which can be preprogrammed into the control system. Since FFC is based on the nominal plant model with no feedback, its ability to overcome any disturbances and modeling uncertainties is limited. Therefore, the FBC which closes the control loop is necessary to compensate for these disturbances and uncertainties. However, for wide range operations, FBC alone may suffer from loss of performance as the control system becomes excessively conservative to guarantee stability over the full range. This problem is alleviated in the proposed control system in the inner loop which unifies the concepts of FFC and FBC to compensate each other's disadvantages. This FF/FBC architecture is intended for robust performance and damage mitigation when the plant is operated over a wide range. For narrow range operations (e.g., base load operation of a power plant), however, the FBC alone is sufficient.

Feedforward Control (FFC) Policy: Given the initial states of the plant and damage, the FFC policy is obtained via nonlinear programming [Gill et al., 1981] by minimizing a specified cost functional under the prescribed constraints of damage rate and damage accumulation. The objective is to minimize the cost functional J without violating the prescribed upper bounds of the damage rate and the accumulated damage to identify an optimal control sequence $\{u_k: k=0,1,2,\cdots, N-1\}$. The cost functional J is to be chosen in an appropriate form representing a trade-off between the system performance and the damage. The optimization problem is then formulated as follows:

$$\text{Minimize:} \quad J = \sum_{k=0}^{N-1} J_k(\tilde{x}_{k+1}, \dot{v}_k, \tilde{u}_k) \tag{4}$$

$$\text{Subject to:} \quad (v_N - v_0) < \Gamma \text{ and } 0 \le h(v_k, q(x_k), k) < \beta(k) \quad \text{for } k=1,2,3,\dots,N \tag{5}$$

where $\tilde{x}_k = x_k - x_{ss}$ and $\tilde{u}_k = u_k - u_{ss}$ are deviations of the plant state vector and the control input vector from the respective final steady state values of x_{ss} and u_{ss}; and $\beta(k) \in \mathbf{R}^r$ and $\Gamma \in \mathbf{R}^r$ are specified tolerances for the damage rate and accumulated damage, respectively. Optimization in Eqs. (4) and (5) are to be carried out off-line for several maneuvers, and the feedforward block for a given maneuver is shown in Fig. 3. The feedforward control policy is selected based on the control patterns received by the executor from its supervisor.

Feedback Control (FBC) Policy: The structure of the life extending control system is shown in Fig. 3 in which the feedforward and linear feedback control (FF/FBC) are exercised in the inner loop to ensure robust performance; the nonlinear controller for life extension in the outer loop is identical to the execution level controller for life extension. In FF/FBC, u^{ff} and y^{ff} represent the sequences of the plant input and plant output, respectively, generated as an optimal solution of the open-loop control problem via nonlinear programming. As mentioned above, a feedback controller is necessary to compensate for the plant modeling uncertainties and disturbances such that the trajectory of the actual plant output, y, should be close to that of the desired plant output y^{ff} which serves as the reference trajectory. The resulting error in the plant output is an input to the feedback controller and the signal Δu modifies the plant/actuator input in an attempt to compensate for errors resulting from plant disturbance and noise.

The linear robust feedback control policy is formulated by using one or more techniques of robust multi-input multi-output (MIMO) control synthesis, which rely on approximation of the plant dynamics by a linear time-invariant model (e.g., H_2-based LQG/LTR [Stein and Athans, 1987], H_2/H_∞ optimization [Doyle et al., 1989], μ-synthesis [Packard and Doyle, 1993; Stein and Doyle, 1991; Skogestad et al., 1988] or ℓ_1-synthesis [Diaz-Bobillo and Dahleh ,1993]). In this case, the FBC is synthesized based on a plant model which is linearized at a desired point on the plant trajectory, and the (slowly) time-varying plant parameters, if any, are averaged over the time horizon of interest. Selection of the frequency-dependent weighting parameters in the FBC synthesis procedure, in turn, depends on the region over which the accuracy of the linearized model is acceptable. Since the damage rate in the closed-loop control system may violate the specified constraints due to the additional compensation generated by the feedback control effort Δu, the nonlinear controller is incorporated into the system to reduce the damage rate and accumulation. This is accomplished by control signal modifications y^{cor} in the reference trajectory and u^{cor} in the control effort. The rationale for modifying the reference trajectory $\{y^{ff}\}$ is that, due to plant disturbances and sensor noise, it may not be possible to follow this trajectory without violating the damage constraints. The additional control effort u^{cor} is intended to provide a fast corrective action to the control input, u, whenever necessary.

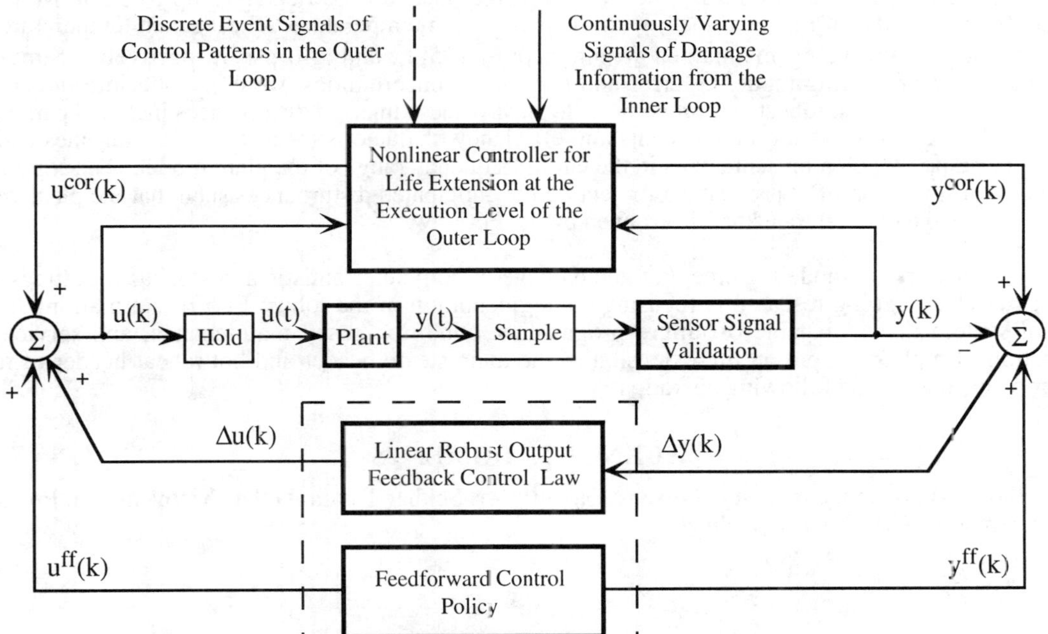

Fig. 3 Structure of the Life Extending Control System

The control signal modifications, y^{cor} and u^{cor}, from the outer loop can be considered as exogenous inputs to the linear robust control system in the inner loop. Since the plant is nonlinear and the feedback control system is synthesized based on a model linearized at an operating point, y^{cor} and u^{cor} must be norm-bounded such that the postulated bounds of plant modeling uncertainty are not exceeded. This would assure the stability and performance robustness of the control system. A possible approach to synthesis of the nonlinear control system in Fig. 3 is to postulate a control structure and then optimize the control parameters based on a cost functional which depends on the control patterns from supervisory level, the control signals from FFC/FBC and damage information from the sensor signal validation and information processing module.

SUMMARY AND CURRENT STATUS OF THE RESEARCH PROJECT

The concept of life extending control systems synthesis, presented in this paper, will satisfy multiple objectives of fossil power plant operations which include robust maneuvering of rapid load changes, failure prognostics of critical plant components, prediction of the remaining service life, condition-based maintenance, and service life extension . In addition, other objectives like emission reduction can be also included in this control framework. This research work is in progress and the current status is summarized below.

The concept of robust feedforward/feedback control (FF/FBC) has been realized, and its efficacy has been investigated by simulation experiments [Weng, 1994]. The synthesized FF/FBC law has been tested via real-time simulation to maneuver the load of the a 525 MWe fossil power plant from 100% down to 40% load in six minutes. Similarly, the plant load has been increased from 40% up to 100% load in six minutes. The results of these simulation experiments showed that, in the absence of plant modeling uncertainties, the FF/FBC system is able to operate the plant for load changes up to 52.50 MWe per minute (which is 10% of the rated load) with the main steam temperature, reheat steam temperature and main steam pressure being maintained within $\pm 10^{o}$F, $\pm 15^{o}$F, and ±40 psi, respectively. Simulated disturbances, representing modeling error and parametric uncertainties, were injected into the control system to investigate the robustness properties. Typically, these injected disturbances included parametric errors such as control valve time constants and efficiency characteristics of the steam turbines and the feedwater pump, which represent errors in the entire frequency range of the plant model. The simulation results suggest that the FBC is capable of rejecting the anticipated disturbances such that the plant would closely follow the optimal trajectory determined by the FFC law.

The research work on modeling creep/fatigue damage in both deterministic and stochastic settings is in progress. This ongoing research is focusing on augmentation of the robust FF/FBC into an intelligent control system that is capable of failure prognostics, condition-based maintenance, and service life prediction. Emphasis is put on life extension of the main steam header and hot reheat header of fossil power plants under load following operations.

ACKNOWLEDGMENTS

The authors acknowledge benefits of discussion with Mr. Sekhar Tangirala, Dr. Xiaowen Dai, Professor Marc Carpino and Mr. Michael Holmes

REFERENCES

J.L. Bogdonoff and F. Kozin, 1985, *Probabilistic Models of Cumulative Damage,* Wiley Interscience, New York.

V.V. Bolotin, 1989, *Prediction of Service Life for Machines and Structures,* ASME Press.

R. Cieslak, C. Desclaux, A.S. Fawaz and P. Varaiya, 1988, "Supervisory Control of Discrete-Event Processes with Partial Observations," *IEEE Transactions on Automatic Control,* Vol. 33, No. 3, March, pp. 249-260.

I.J. Diaz-Bobillo and M.A. Dahleh, 1993, "Minimization of the Maximum Peak-to-Peak Gain: The General Multiblock Problem," *IEEE Transactions on Automatic Control,* Vol. 38, No. 10, October, pp. 1459-1482.

J.C. Doyle, K. Glover, P. Khargonekar, P. and B. Francis, 1989, "State-Space Solutions to Standard H_2 and H_∞ Control Problems," *IEEE Transactions on Automatic Control,* Vol. 34, No. 8, August, pp. 831-847.

H.E. Garcia, 1993, *A Reconfigurable Hybrid Supervisory Control System*, Ph.D. Dissertation, Department of Electrical Engineering, The Pennsylvania State University, University Park, PA 16802.

H.E. Garcia, A. Ray and R.M. Edwards, 1994, "A Reconfigurable Hybrid System and Its Application to Power Plant Control," *IEEE Trans. Control Systems Technology*, in press.

C.M. Ozveren and A.S. Willsky, 1989, *Aggregation and Multi-Level Control in Discrete Event Dynamic Systems,* Massachusetts Institute of Technology, Technical Report No. LIDS-P-1902.

A. Packard and J. Doyle, 1993, "The Complex Structured Singular Value," Automatica, Vol. 29, No. 1, pp. 71-109.

P.J. Ramadge andW.M. Wonham, 1987, "Supervisory Control of a Class of Discrete Event Processes," SIAM J. Control and Optimization, Vo.l. 25, No. 1, January, pp. 206-230.

A. Ray, 1990, "Multi-level Hypotheses Testing for Signal Validation in Continuous-Time Processes," *ASME Journal of Dynamic Systems, Measurement and Control*, December, pp. 787-790.

R.Y. Rubinstein, 1986, *Monte Carlo Optimization, Simulation and Sensitivity of Queueing Networks*, Wiley, New York.

G.N. Saridis, 1989, "Analytic Formulation of the Principle of Increasing Precision with Decreasing Intelligence for Intelligent Machines," Automatica, Vol. 25, pp. 461-467.

S. Skogestad, M. Morari, and J.C. Doyle, 1988, "Robust Control Of Ill-Conditioned Plants: High Purity Distillation," *IEEE Transactions on Automatic Control*, Vol. 33, No. 12, December, pp. 1092-1105.

K. Sobczyk, K. and B.F. Spencer, Jr., 1992, *Random Fatigue: Data to Theory*, Academic Press, Boston, MA.

G. Stein, G. and M.A. Athans, 1987, "The LQG/LTR Procedure for Multivariable Feedback Control Design," *IEEE Transactions on Automatic Control*, Vol. AC-32, No. 2, February, pp. 105-114.

G. Stein and J.C. Doyle, 1991, "Beyond Singular Values and Loop Shapes," *IAIAA Journal of Guidance, Control and Dynamics,* Vol. 14, No. 2, February, pp. 105-114.

C-K. Weng, 1994, *Robust Wide Range Control of Electric Power Plants*, Ph.D. Dissertation, Department of Mechanical Engineering, The Pennsylvania State University, University Park, PA 16802.

©0074-056X/95/077-085/$0 + .50pp

ADVANCED FEEDWATER HEATER LEVEL CONTROL

Scott W. Robinson
Systems Specialist
Commonwealth Edison Co.
Downers Grove, IL

Donald K. Frerichs
Chief Application Engineer
Bailey Controls Co.
Wickliffe, OH

KEYWORDS

Feedwater Heater, Advanced, Optimize, Level, Heat Rate

ABSTRACT

The economical significance of poor feedwater heater performance can be realized by examining its negative affect on unit heat rate. EPRI reports detail and categorize the leading heater failure mechanisms [1] [2] [3] [4]. This paper discusses an aggressive heater level control strategy designed to alleviate heater damage attributed to inadequate condensate level. Emphasis is placed on simplified instrumentation, along with performance driven advanced Distributed Control System (DCS) solutions.

INTRODUCTION

Commonwealth Edison's Powerton Generating Station is located in mid-state Illinois, two miles South of Pekin, Illinois. Units #5 and #6 were placed into commercial service in July, 1972 and January, 1976, respectively. The turbine-generators are 818 MW General Electric 18 stage tandem-compound reheat design with four-flow low pressure exhaust, presently rerated to 750 MW while burning low sulfur western fuels. A unique unit design feature has each turbine supplied by two half-size Babcock & Wilcox Universal Pressure once-thru, cyclone coal fired, balanced draft, reheat boiler steam generators with individual start-up bypass systems. The regenerative cycle incorporates seven horizontal heat exchangers. There are three low pressure, one deaerating, and three close coupled high pressure heaters installed in two parallel circuits. Each high pressure heater circuit supplies feedwater to the economizer of one half-size boiler. Individual heater circuit strings were provided with full feedwater bypass capability. Original high pressure heaters were supplied with carbon steel materials. Figure 1 illustrates the close coupled H.P. heater circuit arrangement.

Documented high pressure heater tube failures were experienced within a few years of plant operation. Initial failures were incidental, but started increasing by the late 1970's and early 1980's. A mid 1980's company thrust channeled station efforts from base-loaded to cycling operations. Station performance objectives were changed to focus on optimizing unit generation ramp rates, load-following ability and low load turndown capability. This new operating strategy introduced potential heater level concerns that traditional self-contained level controllers could not resolve. As cycling operations progressed, heater maintenance and repair costs continued to escalate and station heat rate suffered. The late 1980's saw Unit #5 high pressure heater circuits permanently bypassed with Unit #6 heater circuits permanently bypassed by the early 1990's. Tube failures were prevalent at the 5 to 7 o'clock tube sheet positions. The heater manufacturer stated that failures at these lower locations are indicative of tube erosion from the sandblasting effect of the high velocity steam-water droplet mixture flowing through the drain subcooling zone; i.e., improper liquid level control to seal the drain subcooler. Figure 2 illustrates the cross section of a typical horizontal three zone feedwater heater.

In Spring of 1992, the Unit #5 analog control system was replaced with a Bailey INFI 90 Distributed Control System (DCS). In the Spring of 1994, an identical upgrade was installed on Unit #6. With all four boiler close coupled high pressure heater circuits (three heaters per boiler) slated for replacement starting Fall of 1994, the decision was made to integrate the heater level controls and turbine water induction protection schemes into the newly installed distributed systems. This way all the reliability, flexibility, performance and monitoring features provided with a digital system could be used. Additionally, the DCS software configurations would provide redundancy features designed to optimize heater operating hours, thermal performance, and turbine water induction prevention.

THE CAUSES

Four unique phenomena combine to cause either heater tube failures or poor thermal performance. The following sections describe the four individual phenomena uncovered by our investigation.

Level Sensor Inaccuracy and Reliability - Each heater was equipped with two self-contained pneumatic level controllers piped to the side of the heater. The proportional plus reset Normal Level Controller positioned the heater drain valve to maintain level at a fixed setpoint. The proportional only Emergency Level Controller primary function was to quickly open the emergency drain valve to prevent water induction to the turbine. The emergency controller performed actual level control on the #5 heater whenever low unit load conditions supplied insufficient pressure to lift the cascading drain flow to the higher elevation deaerator heater. During low load operation, the #5 heater level was maintained at the higher emergency controller setpoint, covering more tubes and further degrading heater performance. In addition, the controller's highly sensitive 10% proportional band provided oscillatory level control.

Instrument sensing tap pipe sizing and location contributed to level sensing errors. As carbon steel components eroded/corroded inside the heater shell, a powdery sediment plugged the lower sensing line piping from the multiple heater penetrations running to each instrument. Even the larger 2" pneumatic controller sensing lines experienced partial pluggage. Sensing taps were located in the turbulent vicinity of the single bottom-mounted drain nozzle. The violent fluid action near the nozzle played havoc with the torque tube design of the pneumatic controllers. Normal level controller sensitivity was decreased to compensate for the process anomaly, resulting in sluggish controller response.

Physical instrument location and high process temperatures proved detrimental to pneumatic controller performance. All instruments had been physically installed between the parallel close coupled heater circuits exposing the pneumatics to ambient operating temperatures exceeding 140°F.

Controller pneumatic signal amplifiers required constant maintenance attention as internal diaphragm materials routinely turned brittle and inoperable.

The high liquid level float switch, the primary water induction protection device, also had failings. Calibration routines often found the mechanical movement rusty or non-repeatable, mercury switch bottles broken, and housing covers discarded or misplaced. Ironically, no low liquid level switch was originally supplied with the heaters.

Sight glasses were a constant source of steam and water leaks, and were found valved out more often than not. With sight glasses inoperable, technicians were required to "rough-in" heater levels by adjusting the setpoint, then calculating the Drain Cooler Approach Temperature (DCA) from information provided by the station Data Acquisition System (DAS). Setpoint adjustments were repeated until actual DCA temperature (drain subcooler outlet minus feedwater inlet temperature) was in close approximation to the heater manufacturer's load vs. DCA design criteria. Any significant load variations, however, prompted additional level setpoint adjustments.

The self-contained level controllers could not compensate for extreme changes in fluid density introduced by the wide range of heater shell pressures. Level sensors normally calibrated and pressure corrected for one operating pressure were inherently inaccurate at all loads except the original calibration point. Depending on the operating pressure, level errors approaching 2 inches could be experienced. This degree of error could have easily allowed steam to enter the drain subcooler or submerge tubes even though the level appeared at setpoint.

Drain Valve Sensitivity Changes - Just as the wide range of operating pressures causes level sensor inaccuracies, it also subjects the associated drain valves to a wide range of differential pressures. This in turn causes the sensitivity of the valves to change as a function of load. For example, at low loads with low pressure differential across a drain valve, 10% valve motion causes a small change in drain flow. At high loads, this same 10% valve motion causes a large change in drain flow due to the larger pressure drop. If the level controller is well tuned for low load conditions, it will be unstable at high loads. The only practical tuning solution for self-contained level controllers was to tune them well for high loads and accept very sluggish performance at low loads.

Rapid Load Changes - As controls personnel focused on station performance, rates of unit load change grew increasingly faster and sluggish performance of the traditional level controls became even more unacceptable. Due to slow pneumatic controller output response and process tuning practices, it is highly probable that intermittent steam was entering the drain subcooler during load reductions and tubes were flooded excessively during load increases.

Liquid Level Slope - Perhaps the least understood cause of the heater level problem is the phenomenon of liquid level slope. It may come as a shock to many, but the level in a horizontal feedwater heater is not horizontal! Actually, heater level has a definite downward slope toward the inlet to the drain subcooler. The longer the heater, the greater the slope. Also, the degree of slope is not constant and changes as a function of heating steam flow (extraction flow plus cascaded drain flow). Figure 3 illustrates the liquid level slope phenomenon. This condition exists on all horizontal feedwater heater designs. Thus even if a traditional controller could maintain the level at setpoint, steam might still enter the drain subcooler due to the liquid level slope effect.

ADVANCED CONTROL SOLUTION

Each of the above causes of feedwater heater liquid level problems has been addressed with an advanced control strategy implemented at ComEd's Powerton Station as discussed below.

Level Sensor Inaccuracy and Reliability - Two vertical motion spring supported displacement style electronic level transmitters with remote electronics and one magnetic coupled float/indicator follower style local level indicator were specified for each heater. All three level devices are zero referenced to the heater inner diameter invert and sense a 32" range.

All instruments are installed on a 6" standpipe harness located at the U-tube end of the heater away from the normal drain, emergency drain and extraction steam piping penetrations. The instrument standpipe harness requires only one upper and lower heater penetration, positions all instruments to the outside of the parallel heater circuits, and includes a flanged inspection tee for visual borascopic access.

Remote level transmitter electronics and remaining instruments are consolidated onto prefabricated local panels positioned in close heater proximity, but away from high process and ambient temperatures.

Innovative software design combines level transmitter functions of control, interlocking, and alarming. The strategy emphasizes transmitter redundancy, deviation detection and alarming, and on-line maintenance/calibration features. Two HI/HI level signal monitor outputs, one per level transmitter, provide redundant water induction circuit actuation. HI/HI level output relays are partitioned and driven by separate slave modules for added system integrity. Figure 4 illustrates the simplified feedwater heater level control scheme.

The extended 32" level transmitter range provides additional design benefits. Instruments will no longer be physically relocated to compensate for heater installation or operating irregularities determined after start-up. Level or signal monitor setpoints will be adjusted through the DCS to meet dynamic operating requirements. The added trend and monitoring range will help analyze the extent of level transients experienced during any major heater upset.

Heater extraction steam inlet pressure is used to compensate level measurements for the effect of pressure on fluid density [5]. The advanced control strategy uses pressure compensation on all heaters, including the deaerating heater, assuring accurate level measurements at all loads.

Drain Valve Sensitivity Changes - The heart of each heater level control resides in a single Advanced PID controller used to develop the split-range position demand to both normal and emergency drain valves. This single controller action is then modified to meet the functional requirements of both linear trim drain valves.

Two function generators condition the Advanced PID controller output signal to provide individual drain valve demands. The demand curves establish an overlap that starts opening the emergency drain valve just prior to the full open demand to the normal drain valve. This demand overlap allows both valves to smoothly control a single level setpoint, but places emphasis on the normal drain valve for level control [5].

Today's DCS designs include adaptive tuning capabilities that can change controller tuning constants as a function of one or more process variables. Heater shell pressure inputs, already used for level measurement compensation, are also used to develop controller adaptive gain parameters for the normal drain valve. The pressure drop across the drain valve was approximated by subtracting the downstream heater shell pressure from that of the upstream heater shell pressure. The repeatable process relationship is scaled to modify the controller gain to correct for normal drain valve sensitivity over the entire load range. In like fashion, a high level digital input adapts the controller integral rate to provide quick emergency drain valve response during a high liquid level occurrence.

Rapid Load Changes - Traditional level control strategies simply use the PID control algorithm. This algorithm has the disadvantage of waiting for an error in level before it takes any corrective action to move the drain valve. During rapid load changes, this response is inadequate. Conversely, the feedforward control method begins to move the final element (drain valve) in response to a function of an independent index (typically unit load, feedwater flow, or even heater shell pressure). The feedforward feature assists the controller in quickly stabilizing level during load transients, an obvious benefit that fosters continued improvements in unit generation and load-following proficiency without compromising heater integrity.

The traditional method of calibrating the feedforward signals involves taking data at numerous loads and then plotting a function generator curve. For this project, the data acquisition process has been automated using the DCS to take data during normal operation as part of a real time regression strategy. The result of the data regression is a set of tuning constants for a third order polynomial equation used as the feedforward. The regression algorithm is constantly in service and advises plant staff if the latest curve fit is significantly different than prior curve fits.

Some heaters adversely suffer from swell and shrink effects, particularly the swell effect that occurs upon turbine runbacks or turbine trips without an accompanying boiler trip. For these situations, swell and shrink compensation can be added to the heater level controls. This compensation temporarily biases the level sensor signal as a multivariable function of rate of change of level, shell pressure, and feedwater flow. This method is common practice in drum level control applications for both utility and large industrial boilers and will be incorporated into the Powerton strategy if significant swell and shrink effects are detected during operation upsets.

Liquid Level Slope - By incorporating the preceding three improvements (reliable accurate level measurement, adaptive PID tuning, and a well tuned feedforward controller), the level control strategy accurately maintains the heater liquid inventory over all operating conditions. But given the phenomenon of liquid level slope, the question becomes what is the correct setpoint? A patented algorithm, Mdc2000 by Neundorfer, Inc., was developed with feedback from utilities and feedwater heater consultants. The sophisticated algorithm calculates optimal level setpoints based on heater thermal performance measurements and correlates these with boiler load.

The Mdc2000 algorithm automates the data gathering and analysis method necessary to measure the onset of steam carryover into the drains. Heater drain cooler approach (DCA) temperature can be used to detect when steam begins to enter the drain subcooler [6]. The operator simply initiates the calibration process for each heater at low, medium, and high loads. Mdc2000 then plots the empirical data at each load, and regresses the data from the three load points to develop a curve of minimum level setpoint versus load. This curve is biased by a tuneable safety factor, and then transmitted as the setpoint to the feedforward level control strategy described earlier. For added integrity, all calculated level setpoint versus load curve values are stored in retrievable non-volatile memory addresses.

The Mdc2000 algorithm also provides a constant monitoring of each heater's performance. Using the data collected during the calibration process, Mdc2000 compares the current DCA with that expected for the current load. If there is a significant difference, the algorithm will sound an alarm and correct the setpoint in a safe direction to protect the heater. Typical examples of items that might cause such a situation are heater tube leaks, drain valve malfunctions or drifted DCA thermocouples/level sensors.

IMPLEMENTATION

The entire advanced feedwater heater control strategy has been implemented in a Bailey INFI 90 Distributed Control System. Standard software configuration tools were used to develop logic for the self-calibrating feedforward level control using adaptive tuning and pressure compensated level sensors. The Neundorfer, Inc., Mdc2000 algorithm was executed as a "C" language program in the same DCS microprocessor that accomplished the I/O processing and level control for the three close coupled high pressure feedwater heaters on each boiler.

CONCLUSIONS

ComEd's Powerton Generating Station has experienced premature feedwater heater failures and inadequate heater performance. These problems have significantly raised operating costs and limited unit capacity. ComEd and Bailey Controls Company have co-developed a state-of-the-art level control strategy that uses the unique capabilities of the DCS to enhance heater level control, performance, and monitoring; an insurance policy that guarantees maximum financial return on the capital invested to install twelve stainless steel tube high pressure heaters.

REFERENCES

1 "Failure Cause Analysis - Feedwater Heaters," EPRI Report CS-1776, April 1981.

2. "Failure Cause Analysis - Condenser and Associated Systems," EPRI Report CS-2378, June 1982.

3. "Proceedings: Symposium on State-of-the-Art Feedwater Heater Technology," EPRI Report CS/NP-3743, October 1984.

4. "Proceedings: 1992 Feedwater Heater Technology Symposium," EPRI Report TR-102923, September 1993.

5. Kugler, J.; "Computerized Heater Level Controls and Interlocks - A First Attempt"; 1990 American Power Conference.

6. Thompson, N., Franklin, M. W. Andrew, J. M., & Baker, J. L.; "Performance Driven Level Control of Feedwater Heaters"; 1992 EPRI Heat Rate Improvement Conference.

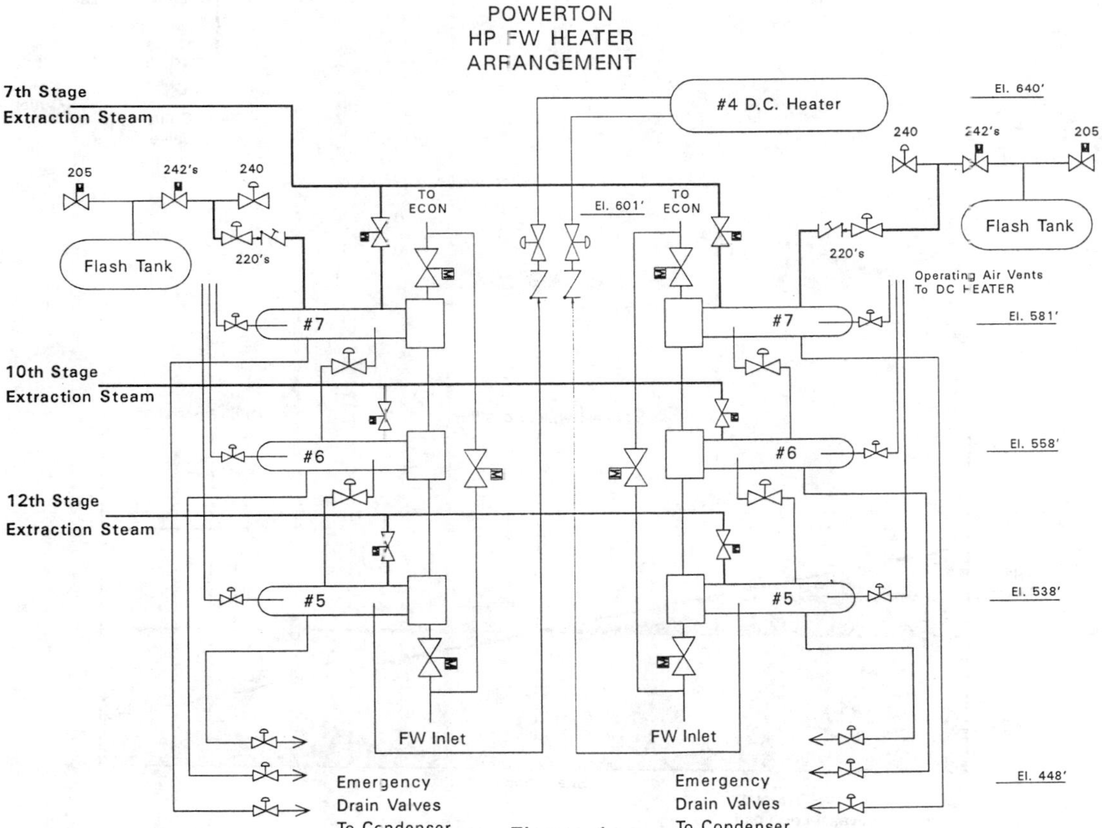

Figure 1

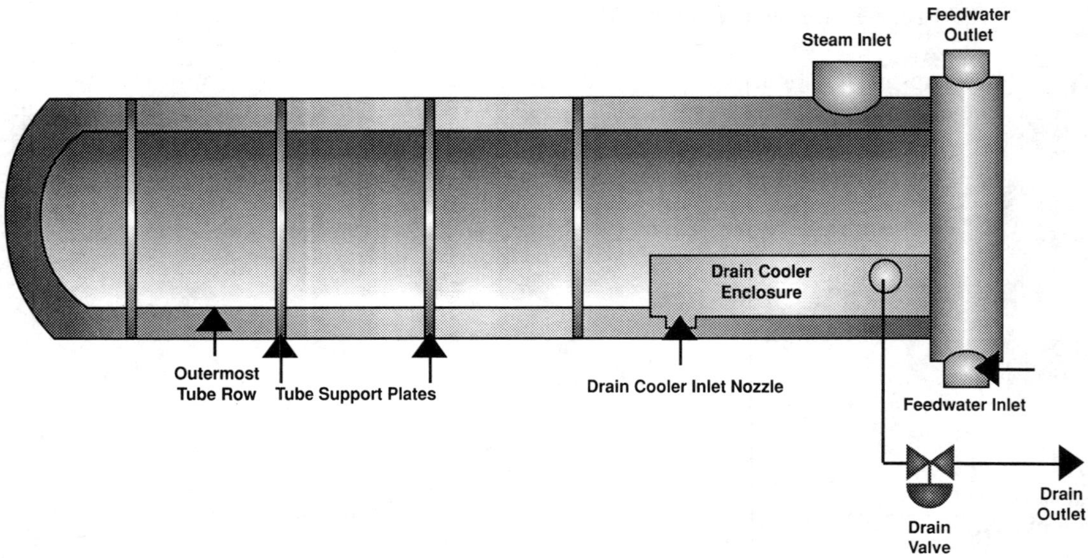

FIGURE 2 – Typical Horizontal Feedwater Heater

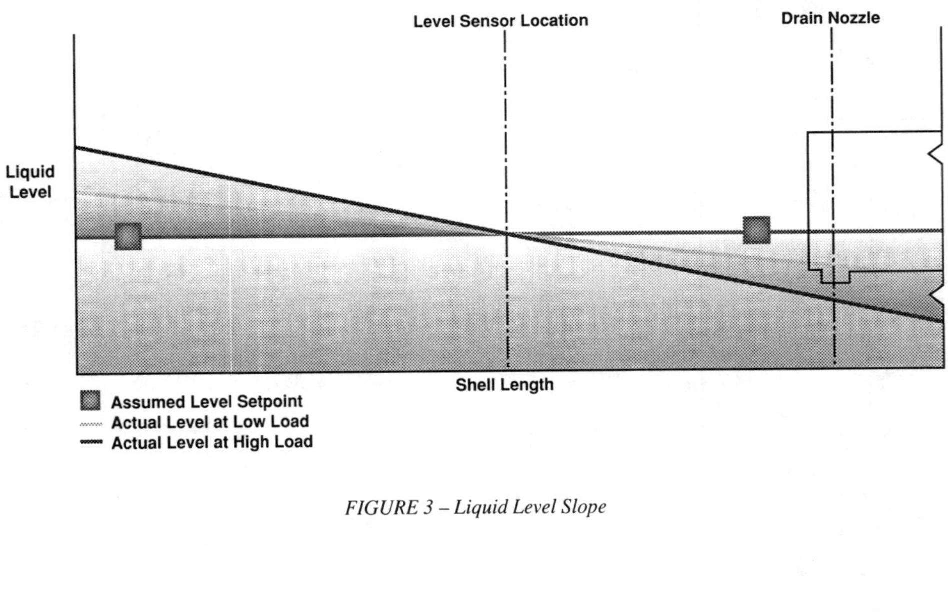

FIGURE 3 – Liquid Level Slope

FEEDWATER HEATER LEVEL CONTROL FUNCTIONAL DIAGRAM

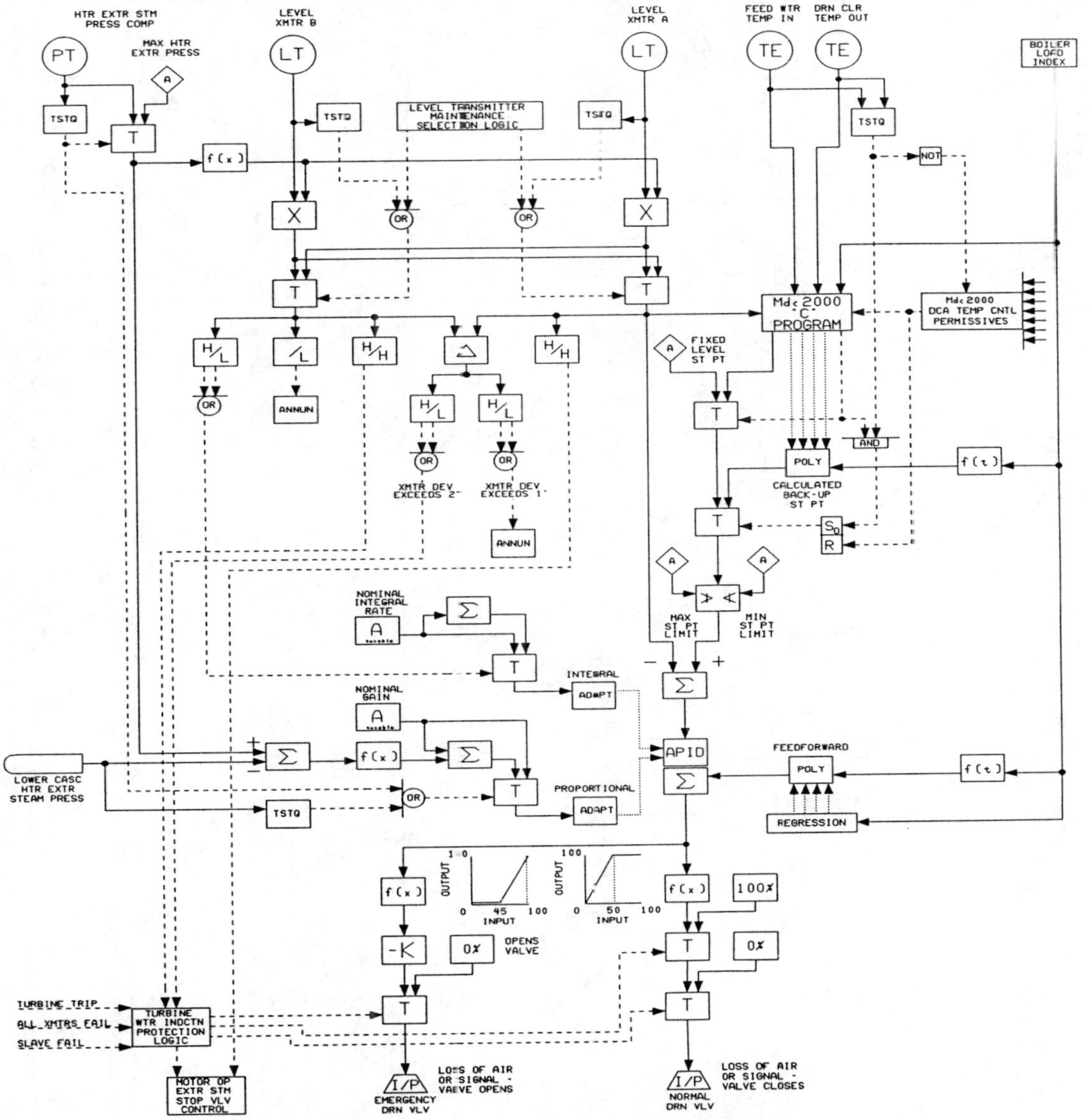

Simplified Infi 90 Control and Mdc2000 Logic

FIGURE 4

0074-056X/95/087-096/$0 + .50pp

OPTIMUM COMBUSTION
Through Preemptive Constraint Control of an Oil/Gas Fired 300MW Electric Generating Unit

Paul C. Schultz
Principal Engineer

Florida Power & Light Company
P.O. Box 14000
Juno Beach, FL 33408-0420

Albert J. Waltz, PE
Consultant

PCC Technologies
2008 N. Memorial Ct.
Pasadena, TX 77502

KEYWORDS

Optimum Combustion, Constraint Control, Model Predictive Control, Combustion Control

ABSTRACT

A combustion optimal air to fuel ratio exists because too much air, mixed with fuel, wastes heat and too little air wastes fuel. A model predictive constraint controller adjusts the air to fuel ratio to control stack gas opacity and combustibles at optimal setpoints. The setpoints can be changed to either maximize combustion efficiency or maximize megawatt generation at peak demand. NO_x emissions are reduced by operating with less oxygen.

Preemptive Constraint Control uses internal models plus clear logic to control oxygen, opacity and combustibles in the stack gas. The multivariable control yields a synergism that provides stable responsive control even when subject to large disturbances.

Control performance and challenges encountered during implementation of advanced controls, on an oil/gas fired 300MW Electric Generating Unit at the Florida Power & Light Company Riviera Station, are discussed.

INTRODUCTION

Optimum combustion is achieved in boilers and furnaces by adjusting the air to fuel ratio to burn almost all of the fuel without adding too much air. Air is heated from ambient to stack temperature; so, adding too much air wastes heat. Adding too little air allows unburnt fuel, in the form of carbon monoxide (CO) and hydrogen, to exit the stack. These two opposing potential energy losses provide the opportunity to optimize combustion by adjusting the air to fuel ratio for minimum heat loss.

Measurement of oxygen and combustibles in the stack gas indicates combustion efficiency. Using these measurements as feedback to an advanced control system enables optimum combustion control. The oxygen or combustibles measurements by themselves do not provide good control. Combining both measurements, in an advanced control system, generates a synergism where the advantages of each offset the disadvantages of the other.

Florida Power & Light Company faced the challenge of improving operation of a 300 megawatt oil/gas fired electric generating station in a highly visible residential and tourist area on the south east coast of Florida. The unit is predominately oil fired and can smoke severely if the air to fuel ratio goes too low. A very important operating objective is to prevent high opacity operation while providing either peak power or maximum efficiency.

The FPL Riviera Plant consists of two identical units placed in service in 1962/1963. The General Electric turbine generator unit is a tandem compound, 18 stage, triple flow, impulse-reaction, seven point extraction, condensing, reheat type. Nameplate rating is 260MW normal with 300MW peak operation at higher inlet pressure. The boiler is Foster Wheeler 1,950,000 pound of steam per hour at 1000°F/1000°F and 2000PSIG conditions. They burn number six (bunker C) high sulfur fuel oil and are alternately fired with natural gas. With use of 100PSIG over pressure, removing the seventh stage high pressure feed water heater and using reheater sprays the boiler can supply the energy for peaking operation.

Flue gas exits the boiler and enters two separate air pre-heaters via separate east and west ducts. One Ametek/Thermox WAG-INC flue gas analyzer is provided for each duct. Extractive samples are analyzed for oxygen and combustibles. The combustibles measurement includes hydrogen as well as carbon monoxide (CO) and therefore gives a parts per million reading significantly higher than from in-situ infrared type instruments that only measure CO.

The boiler controls have been upgraded to a Bailey Infi-90 system. FPL designed and installed this new system in 1992/1993. It included steam temperature, combustion, draft, drum level, deaerator and assorted miscellaneous control loops. The operator interface selected was conventional HARD hand auto stations for back-up safety and reliability. A CRT operator interface was also installed to provide the single point operator interface, trending tools and an engineering interface to change or add loops.

Before the 1992 upgrade, the analyzers were not used for automatic control and had been allowed to fall in a state of disrepair from time to time. As a result the automatic excess air controls failed to work well and consequently were left in manual. A new infrared CO analyzer and four in-situ oxygen probes, two for each duct, are to be installed with the objective of improving instrument reliability.

The need for a robust, automatic excess air control system was targeted as a likely application to introduce a new higher level of control. A survey of the industry controllers, (as opposed to the state of the art), revealed that neuro, fuzzy and artificial intelligence controllers are not mature enough for the utility environment. The best thing going seems to be the multivariable, model based controllers. Based on some good reports, PCC Technologies was selected to start a proof of concept project, at the Riviera Plant, using Preemptive Constraint Control® (PCC). The PCC multivariable controller was applied to provide optimum combustion using oxygen and combustibles measurements from the original analyzers.

CONTROLLING AT THE OPTIMUM

Combustion is optimal at the minimum of the sum of the combustibles and excess air heat loss curves. (See Figure 1) Excess air heat loss decreases linearly with decreasing excess air while combustibles increases nonlinearly. The shaded area represents the sum of the two heat loss curves.

The economic optimal combustibles target is the value on the combustibles curve directly under the minimum heat loss sum. Optimum combustion is achieved at about 130 PPM CO. The combustibles measurement will be somewhat higher at the optimum since it is a measure of both hydrogen and CO.

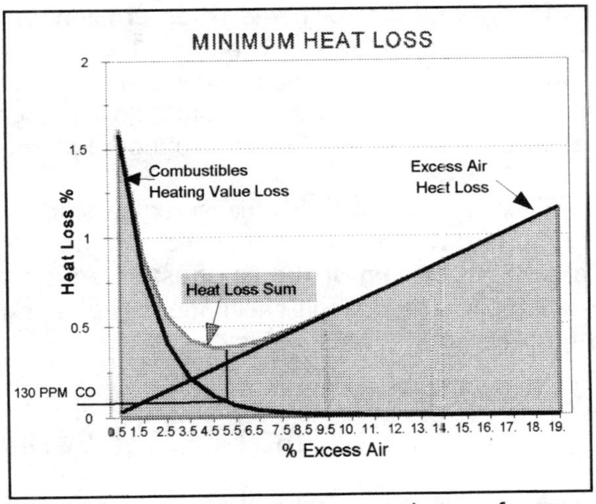

Figure 1 Individual curves and sum of excess air and combustibles heating value loss.

Total heat loss changes little between 100 PPM to 300 PPM CO as evidenced by the broad valley in the sum of the heat loss curves.

Many units operate with 17 percent or greater excess air, far to the right side of Figure 1 where the combustibles measurement is nearly zero. Increasing combustibles, by reducing excess air from 17 percent to bring CO above 100 PPM, produces a large decrease in excess air heat loss. The excess air reduction yields a 0.4 percent reduction in total fuel cost when the stack gas temperature is 300°F.

The optimum concentration of combustibles in the stack is independent of firing load and varies slightly with stack gas temperature. The excess air heat loss lines converge near the optimum on the left side of Figure 2. Stack gas temperature has a small effect on the combustibles optimum concentration.

Higher stack gas temperature causes a significant increase in heat loss when operating with high excess air at the right side of Figure 2. The potential heat loss savings increase significantly with temperature.

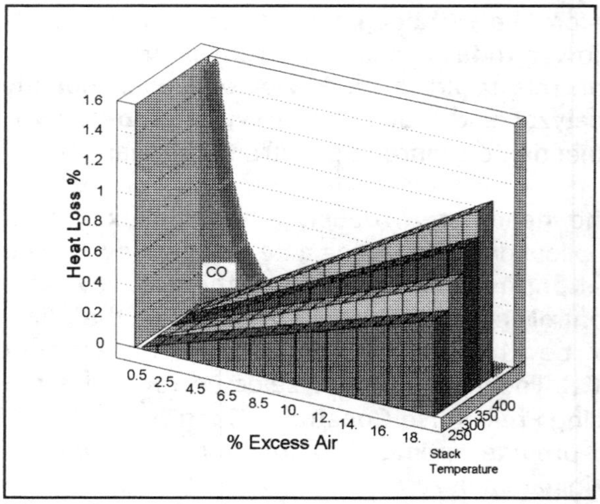

Figure 2 Stack temperature effect on heat loss.

The two oxygen measurements in Riviera Unit 4 correlate well with changes in air to fuel ratio when adjusted for transport lag dead time. Changing oxygen measurements reflect changes in combustion; however, oxygen is an indirect measure of combustion efficiency. There is no single oxygen value that yields maximum combustion efficiency. The oxygen measurement, for optimum combustion, changes with furnace load and other factors such as burner condition, air temperature, moisture content and air leakage down stream of the combustion zone. Some other measurement is needed to control combustion quality.

The Thermox combustibles measurements provide a direct measure of the stack gas fuel loss and the quality of combustion. The combustibles measurement is minimally affected by air leaks down stream of the combustion zone since there are negligible combustibles in the entering air. The difficult challenge is regulating excess air to maintain combustibles at the optimal target.

FIRST LEVEL REGULATORY CONTROL

Advanced controls can compensate for dead time and long process time constants; however, it is always better to reduce or eliminate the problems at the first level of regulation, rather than compensate for them. Flow and pressure controllers provide the first level of regulation. Higher level controls cascading to the first level, can respond no more rapidly than the slowest first level loop.

It is desirable to have a good response to changes in power demand. Unit 4 fuel oil flow responds rapidly; but, is necessarily cross limited by air flow to be sure that the addition of air keeps up with any increase in fuel. An air to fuel ratio controller cascades to air demand (air flow setpoint) through cross limiting control logic.

Responsive air flow is a challenge. Large pneumatic positioners manipulate jack shafts to adjust louvers on the forced and induced draft fans. The rather long response of air flow to a step increase in air demand is six seconds pure dead time followed by a moderate increase then a gradually approach to the air demand. (See Figure 3)

The slow response of air flow causes a correspondingly slow response of fuel flow to load increases, when cross limited by air. Improvements in air flow control improve overall response to unit load changes.

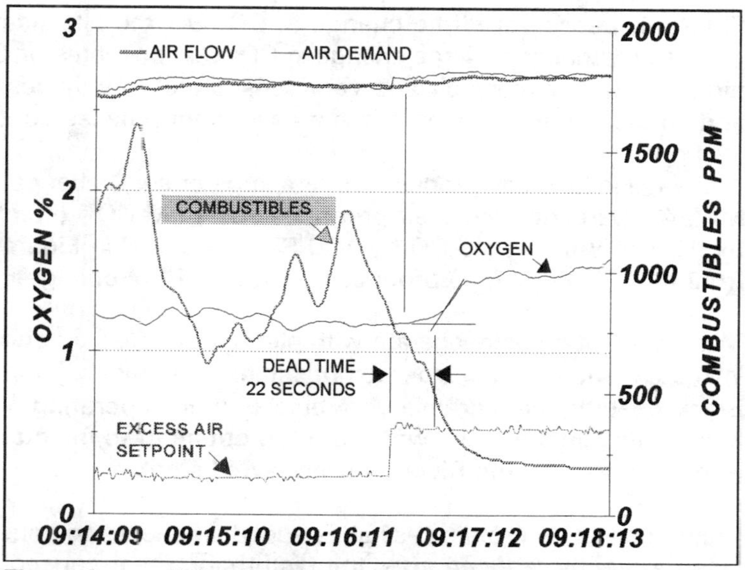

Figure 3 Riviera Unit 4 open loop step test at 190 MW 8/20/94.

THE EXCESS AIR CONTROL CHALLENGE

Industry experience has revealed a number of problems controlling excess air based on measurements of either oxygen or combustibles. No one stack gas oxygen value indicates optimum combustion. Even when corrected for changes in load, the optimum oxygen value is affected by many other variables.

Differences in individual burner operation produce different concentrations of oxygen and combustibles in the products of combustion. Stratification and swirling of the gas cause measurements to vary as the various stratified compositions pass over the measuring probes. Measurements could be averaged; however, averaging slows control response and requires operating further from limits to prevent individual measurement limit violations.

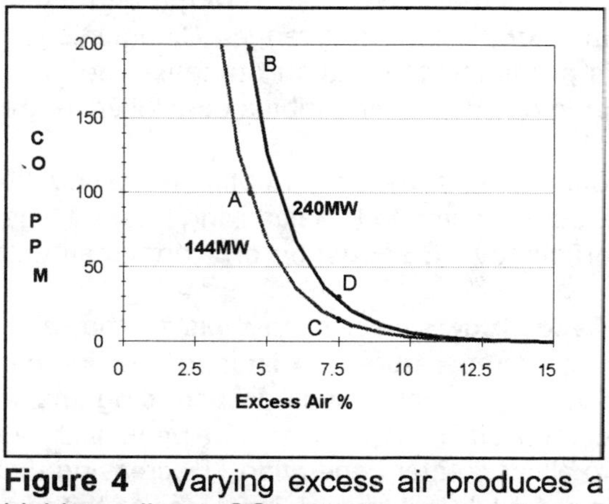

Figure 4 Varying excess air produces a highly nonlinear CO response depending on the initial excess air and unit load.

Combustibles in the stack gas provide an excellent measure of combustion efficiency; but, exhibit a highly nonlinear change with excess air. Large changes in air to fuel ratio, when operating at the right hand side of

Figure 4, cause very little change in CO. At the left side of Figure 4, a small change in excess air produces a large change in CO. Combustibles or CO control is very insensitive at high values of excess air and very sensitive at low values. This highly nonlinear response and sudden large changes in the measurement makes control tuning difficult.

A change in load can produce either a large or small change in combustibles or CO. At 144 MW, 4.5 percent excess air produces 100 PPM CO. (Point A in Figure 4) Increasing load by 100 MW increases CO by 100 PPM. (Point B in Figure 4) Changing load by 100 MW from point C to point D produces only a 14 PPM change in CO.

Many generating units operate with plenty of excess air and few combustibles in the stack. Without responsive control, combustibles can quickly jump from reasonable to high levels. Stack opacity can increase very quickly while operating oil fired boilers at low excess air. Most units operate well away from the optimum to maintain a generous comfort factor. It is imperative that the Riviera boilers avoid smoking.

Under manual control, excess air is added whenever combustibles or opacity increases; but, less frequently reduced after the disturbance that caused high combustibles is past. The result is uneconomic operation at high levels of excess air.

By themselves, neither combustibles nor oxygen provides a good measurement for optimum combustion control. Optimum combustion is achieved by using measurements of oxygen, combustibles and opacity in an advanced control system.

ADVANCED CONTROL

Maximum combustion efficiency is achieved by controlling combustibles at an optimal setpoint provided oxygen remains within an acceptable range and opacity is at an acceptable level. A maximum oxygen target and a minimum oxygen target are set in PCC to establish the oxygen operating range. Combustibles are controlled at the optimum as long as oxygen is predicted to remain in range. The maximum or minimum oxygen constraint controllers preempt the combustibles controller to prevent oxygen from going out of range.

Controlling at the target optimum combustibles, maximizes combustion efficiency. At peak electric demand, maximizing electric generation is more important than combustion efficiency. Combustion efficiency can be sacrificed to generate maximum power.

As load increases, hotter gas to the reheat section increases reheat steam temperature. High temperature can limit further increases in load. More power can be generated by reducing excess air and sacrificing the loss of combustibles in the stack. Raising the combustibles setpoint, to a value well above the combustion optimum, minimizes excess air to allow greater generation. Decreasing excess air reduces reheat temperature. Excess air is reduced until opacity increases to the maximum opacity target. Controlling at the opacity maximum acceptable target increases generating capacity.

Optimum control is maintained through PCC model predictive control. Models are used to predict future limit violations. Logic is applied to select the proper combustibles, oxygen or opacity limit controller so that the control system responds only to meaningful measurement variations.

Internal process models are used to compensate for nonlinearity, estimating the effect of air to fuel ratio adjustments on flue gas combustibles, oxygen, and opacity. Dynamic models compensate for dead time providing stable responsive control. The models are based on open loop step test data and are in matrix form for rapid execution.

Dead times are associated with each measurement based on the measurement location along the gas path. The models adapt to changes in transport lag dead time with load. Dead time compensation enables responsive tuning settings that would cause instability in uncompensated controllers. This is particularly significant in light of the 22 second dead time which is quite long relative to an 11 second process time constant. (See Figure 3)

Examining control response provides some insight to the PCC advanced control logic. Combustibles can drop to low values upon a decrease in load. If CO was 100 PPM at 240 MW, Figure 4 shows that CO would drop to 50 PPM for a 100 MW decrease in load. The PCC combustibles controller decreases the air to fuel ratio in response to this drop. Meanwhile, if an internal model predicts that the oxygen will drop below the low limit, the PCC minimum oxygen controller preempts the combustibles controller, taking control of the air to fuel ratio before the oxygen reaches the minimum target. A similar scenario takes place upon sudden large increases in combustibles.

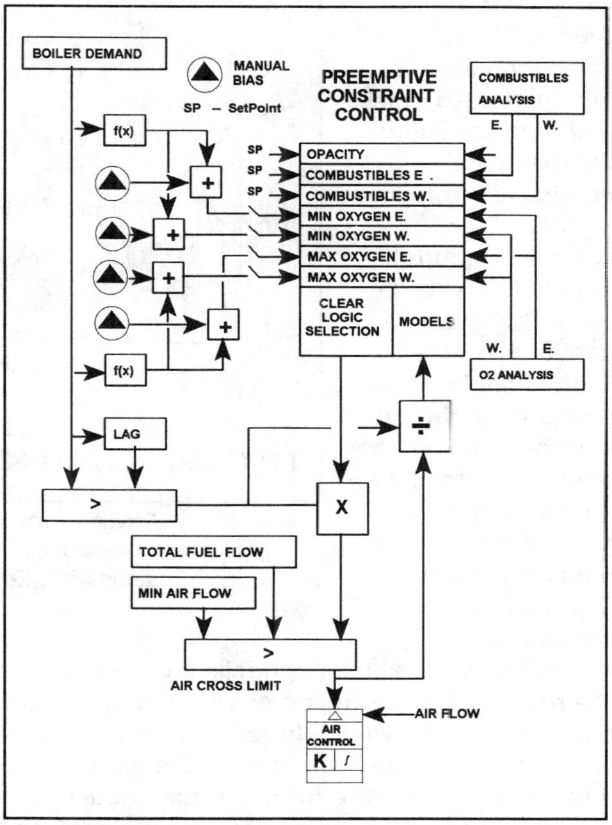

Figure 5. Optimum Combustion Control Block Diagram

Maximum and minimum oxygen constraint controllers monitor and moderate the opacity and combustibles controllers enabling very responsive controller tuning. High gains provide responsive correction of low magnitude disturbances. Excessive responses to high magnitude disturbances are preempted by the oxygen constraint controllers. Individual constraint tuning provides for a wide range of stable control performance tailored to each individual operating unit.

Maximum and minimum oxygen targets are changed as a function of load and may be biased by operating personnel. (See Figure 5) This defines a changing range of acceptable oxygen values. The range is increased at minimum load because less stable combustion conditions occasionally produce high spikes of combustibles. A higher maximum oxygen target enables control of the combustibles spikes. The opacity controller preempts the combustibles controllers to preclude opacity from exceeding a high constraint.

IMPLEMENTATION OPERATING EXPERIENCE

Optimum Combustion Control (OCC) was implemented in the FPL Riviera Unit 4 in the fall of 1994. Open loop step tests provided dead times and time constants for the model predictive controllers in PCC. Each constraint controller was tuned with the other PCC module constraints off. With multiple constraints on, the control response of one controller will be modified by a second controller as a measurement approaches a constraint.

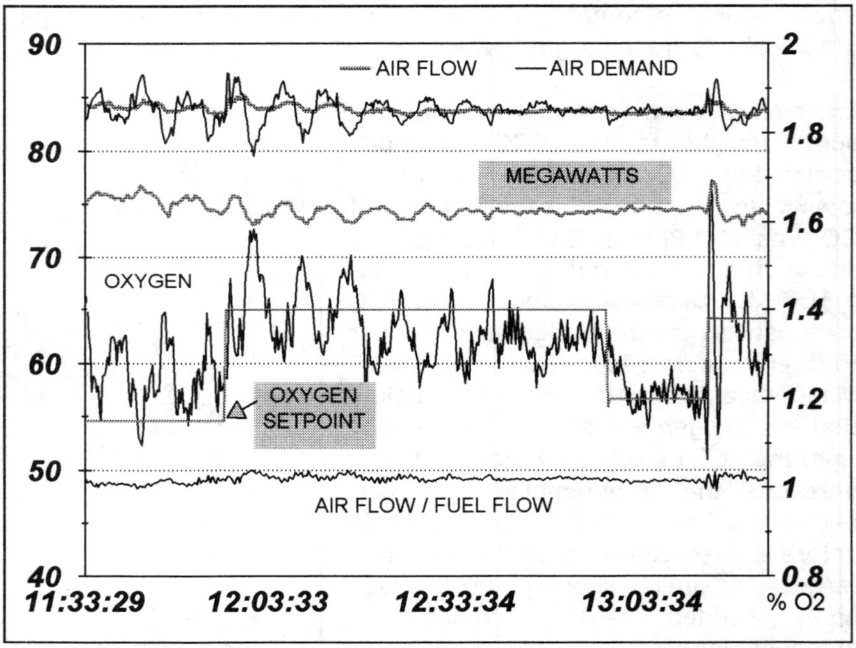

Figure 6 Oxygen constraint initial tuning closed loop step test response. Riviera Unit 4 8/30/94. Gain was set high then reduced for the steps after 12:33.

The individual constraint controllers were tuned just like any PID controller. Figure 6 shows the results of initial tuning of an oxygen constraint when gain was set high then reduced. Reset time was eventually reduced to decrease offset. The oxygen constraint controllers were tuned to be less responsive and the combustibles or opacity controllers more responsive so that the former would moderate any over responsive control by the latter.

The more detailed view of the right side of Figure 6, provided by Figure 7, shows why combustion control is a very challenging task. The first step down in oxygen setpoint evokes minimal response from the unit. The subsequent increase evokes a very strong response. Process response is not only nonlinear but discontinuous.

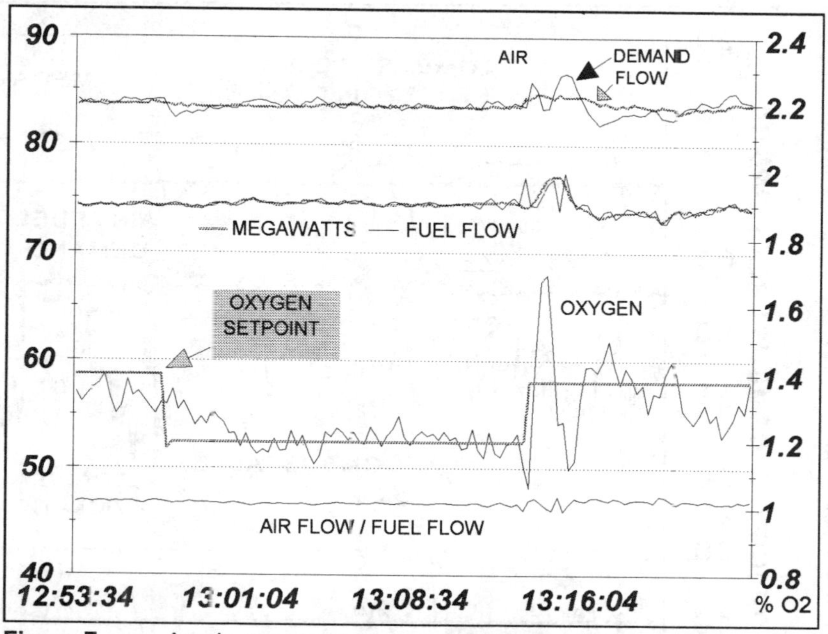

Figure 7 An inverse response disturbance from a fuel control interaction is rapidly damped by PCC model predictive control. Riviera Unit 4 8/30/94.

The increase in air flow drops flame temperature causing an immediate drop in steam pressure. Conventional feed forward from pressure, kicks fuel flow up when the pressure drop exceeds a dead band. This is done to overcome the thermal inertia in the boiler to load changes. In this case the much faster increase in fuel causes the oxygen to drop with the increase in air flow producing the inverse response of oxygen. When the fuel flow drops back down the oxygen overshoots the setpoint severely because of the increased air flow. Although the PCC model predictive control makes some significant changes in air demand, it makes smart changes that promote a rapid return to smooth control.

At the end of control commissioning, FPL wanted to test the robustness of OCC with a severe load change. Unit 4 was cleared to hold megawatts constant at 195 MW then drop load by 30 MW. OCC was set to opacity control by raising the combustibles setpoint to 700 PPM while combustibles remained mostly below 200 PPM. The opacity control lined out at ten. It's important to note that an environmental report must be filed if opacity exceeds 20 for more than six minutes. Visual observation of the stack shows little change until opacity exceeds 15.

After a short run at 195 MW the load was dropped 30 MW as rapidly as possible. As the load dropped, opacity climbed steeply and OCC increased the air to fuel ratio in response. The increase in opacity was stopped just shy of 20 and brought back down to 10. (See Figure 8) Once it was apparent the OCC had handled the load change well, the load was raised again and the unit was returned to dispatch.

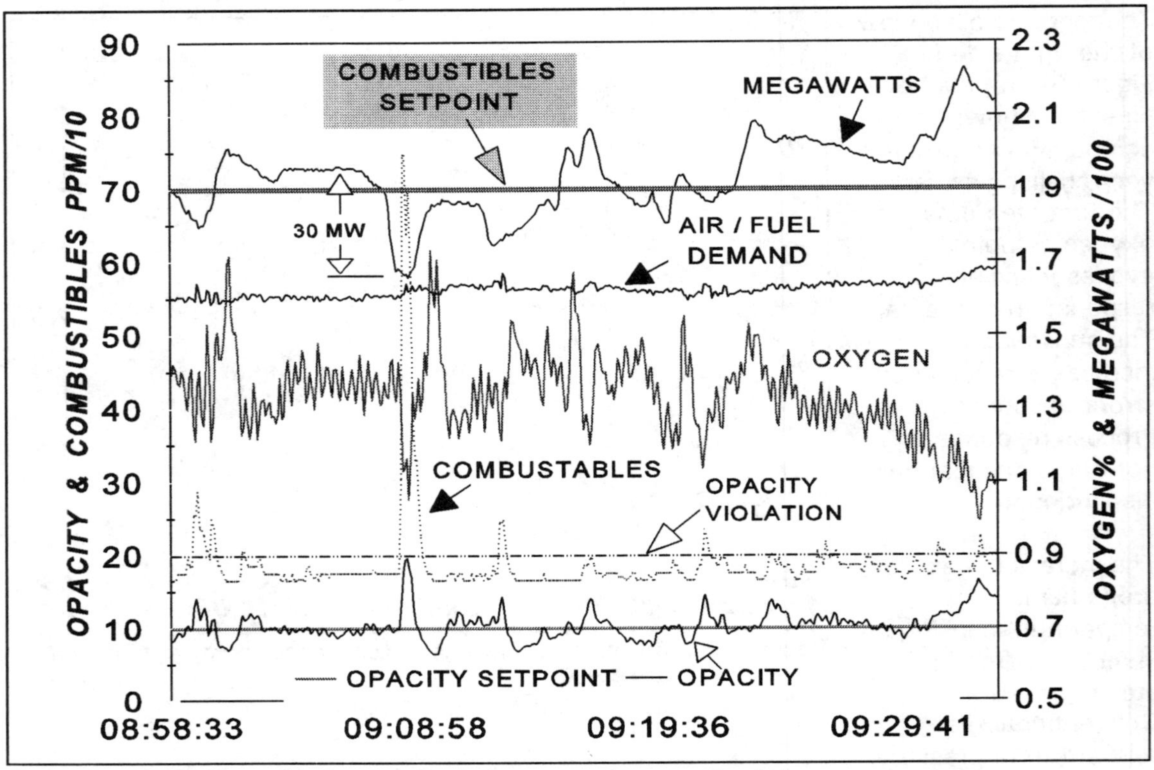

Figure 8 Optimum Combustion Control response to severe load change prevents an opacity violation. Riviera Unit 4 10/5/94.

CONCLUSIONS

Measurement of combustibles in the stack gas is a direct measure of fuel loss. Regulating combustibles at an optimum value maintains combustion at the optimum trade off between fuel loss and excess air heat loss.

Responsive control is the most important element in achieving optimum combustion control. Multivariable controllers, applied to combustion control, must address the response requirements of each controlled measurement by providing adaptive dead time compensation and compensating for control interactions. Incorporating model predictive control in the multivariable controller, provides the needed response for robust control. The multivariable controller should include adaptive model predictive algorithms that provide exceptional control for the trivial case of controlling a single measurement.

Utilizing measurements of oxygen, combustibles and opacity, in a multivariable control of excess air, provides a synergism where the resulting control is better than can be expected using the measurements individually or in a cascade control structure.

0074-056X/95/097-106/$0 + .50pp

Real Time Robust Adaptive Control of a Steam Generation Plant

James Broderick, Gordon Pellegrinetti, Joseph Bentsman† and Ross Blauwkamp
Department of Mechanical & Industrial Engineering
University of Illinois at Urbana-Champaign
1206 West Green Street, Urbana, IL 61801
217.244.1076 *e-mail:* gordon@joe.me.uiuc.edu

KEYWORDS

Receding Horizon Control, Boiler Control, Robust Control, Real-Time Implementation

ABSTRACT

The recently introduced self-tuning Minimax Predictive Controller (MPC) is implemented in real-time on a high fidelity simulation of a fossil fuel heating-cogeneration boiler. The development of the MPC algorithm is outlined in its full multi-input multi-output (MIMO) form. A multi-loop structure of MPC controllers is then applied to a nonlinear real time boiler simulator, and evaluated with respect to standard control schemes.

1. INTRODUCTION

MPC is a control scheme in which the controller input-output relationship (control law) is recomputed at each sampling instant. The control law is selected to minimize the peaks of the spectra of the future predicted tracking error and the incremental control signal over a finite horizon. Their new control law is then used to compute the next incremental control. Since this occurs at each sampling instant, MPC is a receding horizon control strategy.

The plant to be considered is a nonlinear boiler simulator including time delays, load disturbances and measurement noise. The plant is controlled by using three single-input single-output (SISO) two-degree-of-freedom MPC controllers with Extended Least Squares (ELS) identification. The performance capabilities of the multi-loop MPC control scheme is demonstrated through simulation on the real-time boiler model through a variety of setpoint variations and load changes. A hardware and software framework for implementation of a multi-processing extension is proposed.

The paper is organized as follows. Section 2 outlines the MIMO form of the MPC control scheme. The development and features of the nonlinear boiler model is reviewed in Section 3. The implementation of the controller on the model is presented in Section 4 with details of the necessary code, and the results of the simulation. Finally, conclusions are presented in Section 5.

2. MINIMAX PREDICTIVE CONTROL

Predictive control methods have been used effectively to address constraints and predicted reference sequences. Traditional predictive methods, such as the GPC law, minimize the integral (on the unit circle) of the finite sum of the error and control spectra [3]. The corresponding (mixed H_2/H_∞) minimax control

* This work was supported by the EPRI Contract EPRI RP-8010-19, the NSF Grant MSS-8957198, and by U. S. Army CERL under Contract DACA 88-88-D-0024-48.
† Author to whom correspondence should be addressed

law's design philosophy would be to penalize, instead, the peak of this finite sum of the disturbance spectrum to achieve good stability robustness and disturbance rejection, while minimizing the reference spectrum in the usual minimum variance sense to obtain good tracking properties. Such a (mixed H_2/H_∞) minimax predictive multivariable control law (MPC) is introduced in [6]. The MPC algorithm is a receding horizon control law and has an organization similar to GPC.

Since both prediction and optimization are carried out in the minimax (H_∞) sense, the MPC is expected to be inherently robust in the face of plant uncertainty and disturbances. The receding horizon strategy is chosen with the view of subsequently incorporating input constraints into the controller synthesis, a practically important task much easier handled by receding horizon designs (e.g. the GPC) than the standard state space H_∞ methods of Rotstein and Sideris [5].

2.1. Discrete Multivariable Model Description

The discrete time m–input and r–output multivariable plant is assumed to be described by a Controlled Auto-Regressive Integrated Moving Average (CARIMA) model:

$$A(z^{-1})y(t) = B(z^{-1})u(t) + C(z^{-1})\Delta^{-1}\varepsilon(t) \tag{2.1}$$

where $\Delta = 1 - z^{-1}$ and $\varepsilon(t)$ is a white noise source. Expression (3.1) can be rewritten as:

$$y(t) = W(z^{-1})\Delta u(t) + W_d(z^{-1})\varepsilon(t) \tag{2.2}$$

where $W(z^{-1}) \triangleq \left(\Delta A(z^{-1})\right)^{-1} B(z^{-1})$, and

$$W_d(z^{-1}) \triangleq \left(\Delta A(z^{-1})\right)^{-1} C(z^{-1}),. \tag{2.3}$$

Polynomial matrices $A(z^{-1})$ and $B(z^{-1})$ are not necessarily coprime so long as the plant does not contain any unstable hidden modes, and $C(z^{-1})$ is assumed to be strictly Hurwitz polynomial matrix. Also, the poles of the system (i.e. poles of $W(z^{-1})$) can be either stable or unstable. The overall multivariable feedback system is shown in Fig. 2–1 with $\tilde{A}(z^{-1}) \triangleq \Delta A(z^{-1})$. The plant may also be represented by the right co-

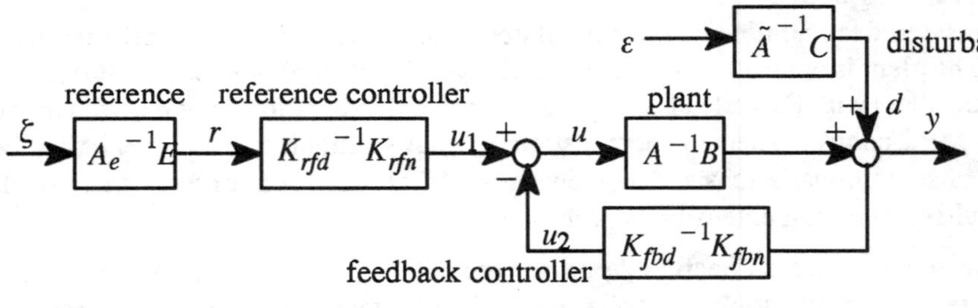

Fig. 2–1 Two degree of freedom system

prime fraction decomposition

$$\tilde{A}(z^{-1})^{-1}B(z^{-1}) = B_1(z^{-1})\tilde{A}_1(z^{-1})^{-1}.$$

In the subsequent development, the reference signal, $r(t)$, is assumed to be generated by a white noise source, $\zeta(t)$. The reference system can be expressed as

$$r(t) = W_r(z^{-1})\zeta(t), \tag{2.4}$$

where

$$W_r(z^{-1}) \triangleq A_e(z^{-1})^{-1}E(z^{-1}). \tag{2.5}$$

The reference system is asymptotically stable and causal. Also, the white noise sources $\{\varepsilon(t)\}$ and $\{\zeta(t)\}$ are assumed to be zero mean and mutually statistically independent. Without loss of generality, the variances of $\varepsilon(t)$ and $\zeta(t)$ are taken to be identity.

The left and right coprime factorization of the feedback controller and reference controller polynomial matrices are respectively:

$$K_{fb}(z^{-1}) \triangleq K_{fbd}(z^{-1})^{-1}K_{fbn}(z^{-1}) = K_{1fbn}(z^{-1})K_{1fbd}(z^{-1})^{-1} \quad \text{and}$$
$$K_{ref}(z^{-1}) \triangleq K_{refd}(z^{-1})^{-1}K_{refn}(z^{-1}) = K_{1refn}(z^{-1})K_{1refd}(z^{-1})^{-1} \tag{2.6}$$

2.2. The Minimax Predictive Multivariable Control

2.2.1 . The Wiener K-Step-Ahead Predictor

Given the set of measurements $\mathcal{Z} = \{y(\tau_1), u(\tau_2); \tau_1 \le t, \tau_2 \le t + k - d\}$, let $\hat{y}(t + k|t)$ be the output of the k-step-ahead predictor of feedback system shown in Figure 2-1. Then the k-step-ahead predictor [2] that minimizes the estimation error

$$\tilde{e}(t + k|t) \triangleq y(t + k) - \hat{y}(t + k|t)$$

under the cost criterion

$$J = \mathcal{E}\left[\tilde{e}(t + k|t)^T\tilde{e}(t + k|t)|\mathcal{Z}\right] = tr\mathcal{E}\left[\tilde{e}(t + k|t)\tilde{e}(t + k|t)^T|\mathcal{Z}\right] \tag{2.7}$$

is given as

$$H_{Pk} = H_{1nk}C_1^{-1}, \tag{2.8}$$

where the $\mathcal{E}[\cdot]$ is the mathematical expectation and H_{1nk} and C_1 satisfies the following right coprime decomposition

$$\left[H_{1nk}, C_1\right]\tilde{A}_1^{-1} = \left[\tilde{A}^{-1}H_{nk}, \tilde{A}^{-1}C\right], \quad C_1(0) = I. \tag{2.9}$$

The polynomial matrix $H_{nk}(z^{-1})$ is computed from the solution (H_{nk}, F_k) minimum degree with respect to $F_k(z^{-1})$ of the following polynomial matrix Diophantine equation

$$\tilde{A}F_k + H_{nk}C * z^{-g_{pk}} = CC * z^{-g_{pk}+k} \tag{2.10}$$

where g_{pk} is the smallest positive integer which ensures the above polynomial matrices to be in the negative power of z.

Remark 1. The optimal k-step-ahead predictor can be written as

$$\hat{y}(t + k|t) = H_{Pk}d(t) + W\Delta u(t + k). \tag{2.11}$$

2.2.2 Cost function Definition

The MIMO MPC problem is to minimize over all stabilizing controllers, K_{fb} and K_{ref}, the performance index:

$$J^{MPC} \triangleq \|X_1(z^{-1})\|_{L_\infty} + \frac{1}{2\pi j}\oint_{|z|=1} trX_2(z^{-1})\frac{dz}{z}, \tag{2.12}$$

where $X_1(z^{-1})$ and $X_2(z^{-1})$ are defined as:

$$X_1(z^{-1}) = \sum_{k=1}^{N}\left[(I_r - WM)H_{Pk}W_d(I_r - WM)H_{Pk}W_d) *\right] + \sum_{k=1}^{N}\lambda I\left[MH_{Pk}W_d(MH_{Pk}W_d) *\right] \tag{2.13}$$

and

$$X_2(z^{-1}) = \sum_{k=1}^{N}\left[(I_m - WS_mK_{ref})W_r\left((I_m - WS_mK_{ref})W_r\right)*\right] + \sum_{k=1}^{N}\lambda\left[S_mK_{ref}W_r\left(S_mK_{ref}W_r\right)*\right] \quad (2.14)$$

and defining the input sensitivity:

$$S_m \triangleq (I_m + K_{fb}W)^{-1}, \quad (2.15)$$

and the control sensitivity:

$$M \triangleq K_{fb}S_r = S_mK_{fb} \quad (2.16)$$

2.2.3 The MIMO MPC Control Law

After computing all k-step-ahead predictors by (2.11), the MIMO MPC control law for the system in (2.1) and the cost function in (2.12) is computed by first calculating the following spectral factors:

$$D_fD_f* \triangleq \sum_{k=N_1}^{N_2} H_{Pk}CC * H_{Pk}*, \quad (2.17)$$

$$D_rD_r* \triangleq \sum_{k=N_1}^{N_2} EE*, \quad (2.18)$$

$$D_c * D_c \triangleq B_1 * B_1 + \lambda\tilde{A}_1 * \tilde{A}_1, \quad (2.19)$$

$$D_{0c}D_{0c} * = D_c * D_c, \quad (2.20)$$

where D_f, D_r, D_c, and D_{oc} are strictly Hurwitz spectral factors. The prediction horizon is from N_1 to N_2 and the control horizon is from 1 to N_2. Second, compute the nonminimum phase and strictly Hurwitz spectral factor, Λ, and equalizing cost, γ, which satisfy

$$\Lambda * \Lambda = \gamma^2 I_r - D_f * \tilde{A} *^{-1}\left(I_r - D_c *^{-1} B_1 * B_1 D_c^{-1}\right)\tilde{A}^{-1} D_f \quad (2.21)$$

Third, compute the solutions (G_M, H_M, F_M) of minimum degree with respect to F_M of the regulation matrix Diophantine equations:

$$D_c * G_M z^{-g_m} + D_{0c}F_M\tilde{A}_2' = B_1 * D_{2f}' z^{-g_m} \quad (2.22)$$

and $$D_c * H_M z^{-g_m} - D_{0c}F_M B_2' = \lambda\tilde{A} *_1 D_{3f}' z^{-g_m}, \quad (2.23)$$

where $\tilde{A}_2'$, B_2', D_{2f}', and D_{3f}' satisfy the following right coprime decompositions:

$$\tilde{A}_2' D_{2f}'^{-1} = F_{Ms}^{-1}\Lambda D_f^{-1}\tilde{A}$$

and $$B_2' D_{3f}'^{-1} = F_{Ms}^{-1}\Lambda D_f^{-1}B,$$

and F_{Ms} is the stable spectral factor of F_M (i.e. $F_{Ms} * F_{Ms} = F_M * F_M$).

Fourth, also compute the solutions (M_M, N_M) of minimum degree with respect to N_M of the reference matrix Diophantine equation:

$$D_c * M_M z^{-g_r} + N_M A_{3r} = B_1 * D_{2r} z^{-g_r-(N_2-N_1)} \quad (2.24)$$

where A_{3r} and D_{2r} satisfy the following right coprime factorization

$$A_{3r}D_{2r}^{-1} = D_r^{-1}A_e,$$

and the integers g_m and g_r are the smallest positive numbers which make the above polynomials matrices functions of z^{-1}.

Then the optimal feedback MPC controller is given by

$$K_{fb}^{MPC} = \left(H_M D_{3f}'^{-1}\right)^{-1} G_M D_{2f}'^{-1}, \tag{2.25}$$

and the characteristic polynomial (implicit equation) of the closed–loop frozen–time system has the form

$$G_M D_{2f}'^{-1} B_1 + H_M D_{3f}'^{-1} \tilde{A}_1 = D_c. \tag{2.26}$$

Since both D_f and D_c are stable spectral factors, the closed–loop system,

$$\left(I_r + WK_{fb}^{MPC}\right)^{-1} = \left[I_r + B_1 \tilde{A}_1^{-1}\left(H_M D_{3f}'^{-1}\right)^{-1} G_M D_{2f}'^{-1}\right]^{-1} \tag{2.27}$$

is stable.

The optimal reference MPC controller is given by:

$$K_{ref}^{MPC} = D_{1c}^{-1} M_M z^{(N_2-N_1)} D_{2r}^{-1} D_{0f} F_{Ms} D_C H_{1M}^{-1}. \tag{2.28}$$

where H_{1M}, G_{1M}, and D_{0f} satisfy the following left coprime decomposition:

$$\left[G_M D_{2f}^{-1}, H_M D_{3f}^{-1}\right] = D_{0f}^{-1}\left[G_{1M}, H_{1M}\right]. \tag{2.29}$$

and D_{1c} satisfies $D_c \tilde{A}_1^{-1} = \tilde{A}^{-1} D_{1c}$.

The optimal control law is:

$$\Delta u(t+k) = K_{ref}^{MPC} r(t+k) - K_{fb}^{MPC} \hat{y}(t+k), \tag{2.30}$$

Finally, the minimum cost achieved by the controller is:

$$J_{\min}^{MPC} = \gamma_{H_\infty}^2 I_r + \frac{1}{2\pi j} \oint_{|z|=1} tr\left(D_c{*}^{-1} N_M N_M{*} D_c^{-1} + D_r{*} A_e{*}^{-1}\left(I_r - D_c{*}^{-1} B_1{*} B_1 D_c^{-1}\right) A_e^{-1} D_r\right) \frac{dz}{z}. \tag{2.31}$$

3. BOILER MODELING

3.1. Features of the Boiler Model and Plant

We consider boiler No. 2 at Abbott Power Plant in Champaign, Illinois, which is used for both heating and electric power generation (cogeneration) and is rated at 175,000 lb/hr of steam at 325 PSI. The boiler is gas– or oil–fired, with two essential components: a combustion chamber and a drum that contains a fluid mixture of water and steam at high temperature. The measurable outputs are the header pressure, the excess oxygen in the stack, and the water level. The inputs available for control consist of the rates of fuel and air flows into the combustion chamber and water flow into the drum. A full nonlinear, control–oriented model was developed by the researchers to capture the significant dynamics of the boiler process [4], utilizing previous work by Åström[1], physical laws, identification experiments, and heuristic knowledge of boiler dynamics. A nonlinear combustion equation was included to model the excess oxygen in the stack and the stoichiometric fuel to air ratio for complete combustion. Dynamic models of the plant measurement noise and load disturbance were obtained. The time delays τ_i of the system include measurement instrumentation delays, process delays, and transport delays, that result in a complex nonlinear relation for τ_i which depends on travel distances, gas flow rates, flow resistance, and gas buoyancy relationships. The details of the model development are presented in [4].

The model has the form:

$$\dot{x}_1(t) = c_{11} d_{ld}(t) x_1^{\frac{9}{8}}(t) + c_{12} u_1(t-\tau_1) - c_{13} u_3(t-\tau_3) \tag{3.1}$$

$$\dot{x}_2(t) = c_{21}x_2(t) + \frac{c_{22}u_2(t-\tau_2) - c_{23}u_1(t-\tau_1) - c_{24}u_1(t-\tau_1)x_2(t)}{c_{25}u_2(t-\tau_2) - c_{26}u_1(t-\tau_1)} \tag{3.2}$$

$$\dot{x}_3(t) = -c_{31}x_1(t) - c_{32}d_{ld}(t)x_1(t) + c_{33}u_3(t-\tau_3) \tag{3.3}$$

$$\dot{d}_{ld}(t) = -c_{41}d_{ld}(t) + c_{42}u_1(t-\tau_1) + c_{43} + n_5(t) \tag{3.4}$$

$$y_1(t+\tau_4) = c_{51}x_1(t) + n_1(t) \tag{3.5}$$

$$y_2(t+\tau_5) = c_{61}x_2(t) + n_2(t) \tag{3.6}$$

$$y_3(t+\tau_6) = c_{70}x_1(t) + c_{71}x_3(t) + c_{72}d_{ld}(t)x_1(t) + c_{73}u_3(t) + c_{74}u_1(t-\tau_1) + \frac{(c_{75}x_1(t) + c_{76})(1 - c_{77}x_3(t))}{x_3(t)(x_1(t) + c_{78})} + c_{79} + n_3(t) \tag{3.7}$$

$$y_4(t+\tau_7) = (c_{81}d_{ld}(t) + c_{82})x_1(t) + n_4(t) \tag{3.8}$$

where $\mathbf{x}_1$ is the drum pressure state (kgf/cm^2); $\mathbf{y}_1$ is the measured drum pressure (PSI); $\mathbf{y}_2$ and $\mathbf{x}_2$ are the measured excess oxygen level and its state, respectively (percent); $\mathbf{x}_3$ is the system fluid density (kg/m^3); $\mathbf{y}_3$ is the drum water level (in.); $\mathbf{y}_4$ is the steam flow rate (kg/s); $\mathbf{u}_1$, $\mathbf{u}_2$, $\mathbf{u}_3$ are respectively the fuel, air, and feed water flow rate inputs which take values between 0–1; $\mathbf{d}_{ld}$ is the load disturbance intensity (0–1); and the variables $\mathbf{n}_i$ are the outputs of first–order colored noise models driven by zero mean, unit variance white noise. A detailed description of the derivations and physical meanings of the constants is discussed in [4].

3.2. Control Objectives

A properly functioning boiler must satisfy the following basic requirements: (1) a desired steam pressure must be maintained at the outlet of the drum despite variations in the quantity of steam demanded by users; (2) the water in the drum must be maintained at the desired level to prevent overheating of drum components or flooding of steam lines; (3) the mixture of fuel and air in the combustion chamber must meet standards for safety and efficiency, which is usually accomplished by maintaining a desired percentage of oxygen in the stack in excess of that required for a perfect stoichiometric combustion. Control of the system is complicated by coupling effects between each of the boiler subsystems.

4. APPLICATION AND CONTROL IMPLEMENTATION

4.1. Test Setup

The full simulation is run on three 486 computers, which run the boiler model, the controller,and the user interface respectively. (See Fig. 4–1)

The user interface produces real time graphs of the plant outputs, optionally logs data to disk, and on command displays the current model of a single loop. Through the user interface it is possible to choose between different control schemes (Open Loop, PID, GPC, MPC), enable and disable model identification, and set the reference signal to each loop. Communication between the user interface and the control computer is conducted using the NetBIOS protocol on a local Ethernet connection. Typically, a loop is run in PID control for a few hundred samples to allow for model convergence, and then the GPC or MPC controller is selected.

The boiler model includes the equations presented above, as well as actuator dynamics and colored measurement noise. Since the simulation runs in real time using only analog inputs and outputs, it can be used as a prototyping tool for a real control system.

The control computer implements the control scheme shown in Fig. 4–2. The outer loops and feedforward blocks control the plant outputs by computing the desired actuator outputs. Thus, the outputs of the outer loop controllers and feedforward blocks become the setpoints to the inner loops. The outputs of the inner loop controllers are the command signals sent to the actuators. This inner/outer scheme is designed to provide good actuator response and reduce the effects of actuator dynamics on identification. The feedforward

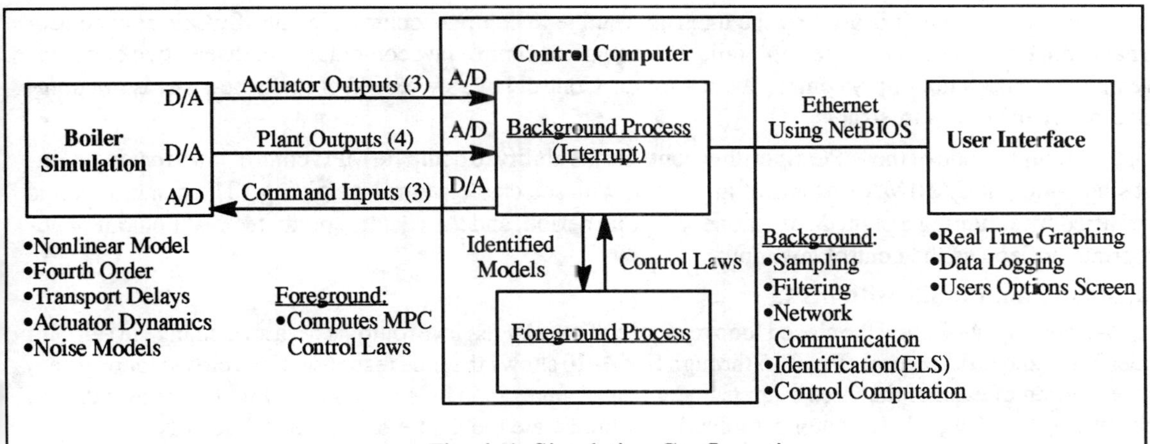

Fig. 4–1 Simulation Configuration

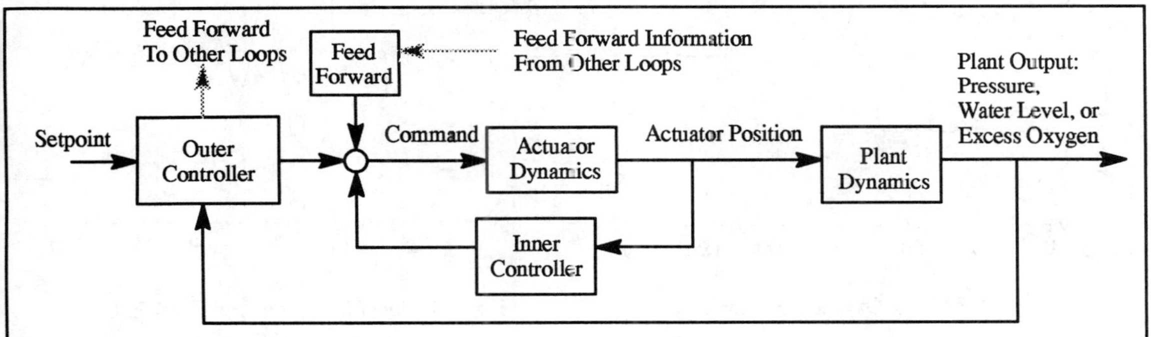

Fig. 4–2 Control Law Implementation on a typical loop

blocks are included to minimize the coupling between the outer loops and are implemented as a simple gain with a delay.

4.2. Implementation Concerns

The control computer uses an analog to digital conversion board with built in sample pacing to sample each plant and actuator output at 5 Hz. This hardware uses direct memory access to move the samples to the computer's memory, and signals the computer that a complete set of samples is taken by generating an interrupt. This scheme completely eliminates input channel polling from the list of tasks performed by the CPU. The interrupt handler routine applies a digital filter to each input to reduce noise, and, if appropriate, updates the CARIMA model of a loop, and computes a new control signal. The inner (actuator) control signals are computed every interrupt and the outer (plant output) loops run at lower speeds since the plant dynamics are comparatively slow. Any new control signal is delayed until the beginning of the next interrupt in order to produce consistency in computational delay.

Since an MPC control law computation is costly in terms of computer time, it cannot be guaranteed to be completed in the interrupt routine before another interrupt occurs. This necessitates the use of a foreground (non interrupt) routine that computes new control laws while interrupts are occurring. When a new control-law becomes available, it is used by the interrupt routine to compute control signals. Since a new control signal is not computed every interrupt, failure to complete control law computation between two consecutive interrupts does not imply control law re-usage. Control laws are reused only if a new law is not computed between control interrupts

The C Language code of the MPC algorithm contains the ability to compute MPC control laws for single-input single-output CARIMA systems of any order, with selectable control weighting. The computation is simplified by assuming a plant delay of one sampling period, and then using a predicted plant output of appropriate horizon as the control law input.

4.3. Simulation Results with MPC

Fig. 4–3 and Fig. 4–4 show the closed loop response of the oxygen level outputs for a step change in setpoint under MPC and GPC control. Fig. 4–5 through Fig. 4–10 shows the time response of the closed loop system for regulation of each of the three outputs under step changes in load input signals. Fig. 4–8 shows the the control command signals (actuator setpoints) to demonstrate the nature of the control activity.

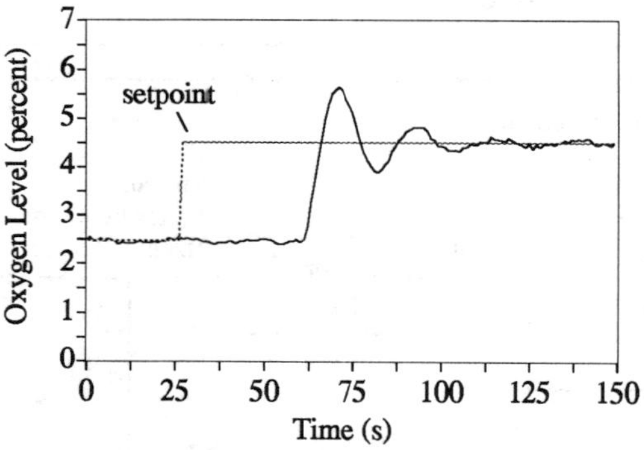

Fig. 4–3 Plot of Oxygen Level for MPC

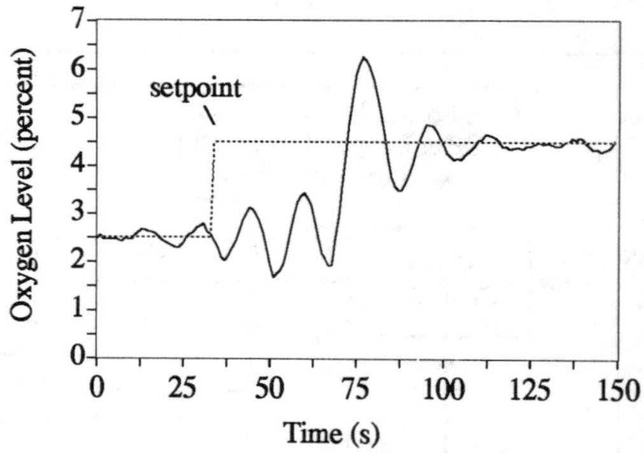

Fig. 4–4 Plot of Excess Oxygen Level for GPC

4.4. Future Improvements

The MPC control scheme can benefit greatly from long range parameter estimation as compared to simple RLS schemes. This is unfortunately, a computationally intensive task that cannot be implemented in real time on the existing setup. An increase in computer power is one solution, but the presence of the network allows the addition of more computers of equal computing power to the task. The control computer already transmits the measured signals to the user interface via the network, thus it is a relatively easy task to add another computer to listen to the measured signals and report updated models.

5. CONCLUSIONS

The configuration of the control implementation is effective in controlling the real-time boiler simulator. Under this configuration, the Minimax Predictive Control algorithm provides a robust solution to the boiler control problem. Fig. 4–3 and Fig. 4–4 show the robustness properties of the algorithm, compared with the traditional GPC. The change of setpoint represents a corresponding change in model parameters, resulting in degraded performance for the less robust GPC. Fig. 4–5 through Fig. 4–10 displays the time response of the closed loop system and the respective reference signals for each of the three outputs under regulation. The control command signals are shown, which display the nature of the control activity, which remains within the feasible range.

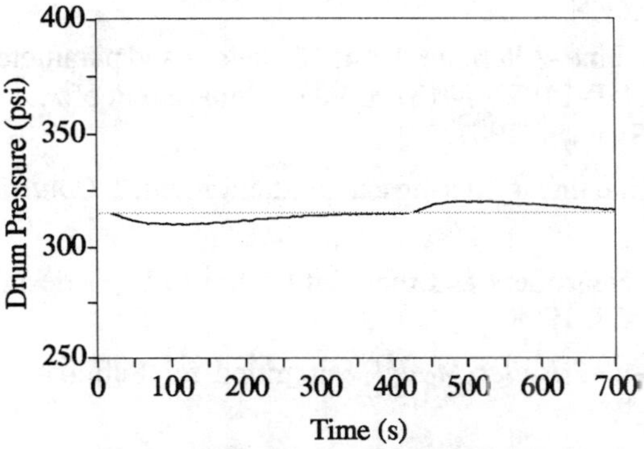

Fig. 4-5 Plot of Drum Pressure and Setpoint

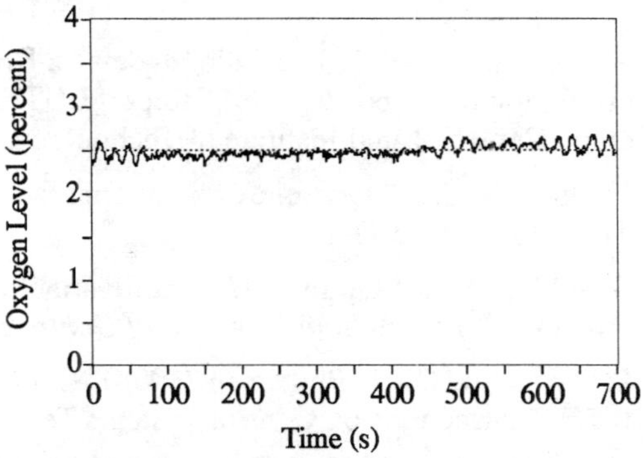

Fig. 4-6 Plot of Excess Oxygen Level and Setpoint

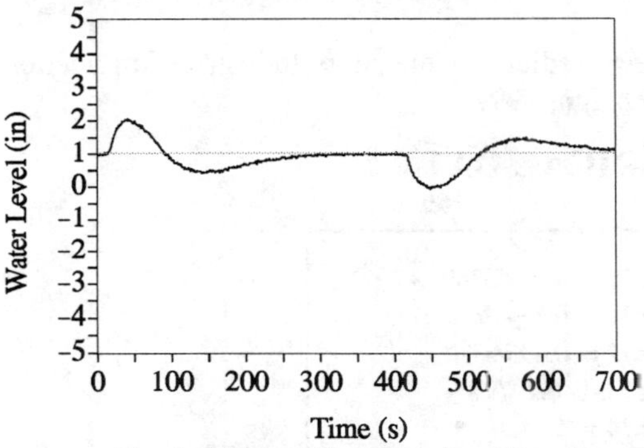

Fig. 4-7 Plot of Water Level and Setpoint

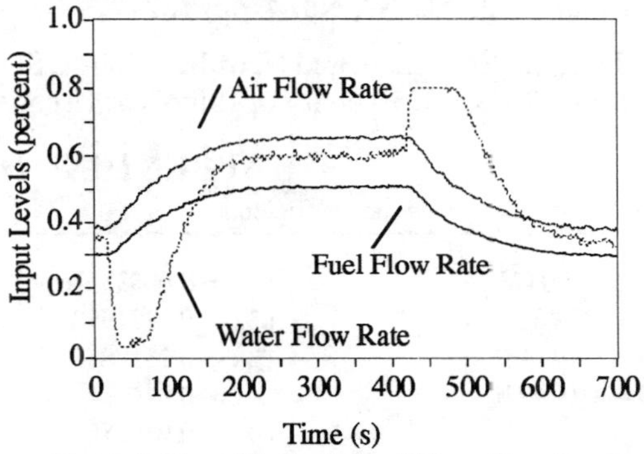

Fig. 4-8 Plot of Fuel, Air, and Water Flow Rates

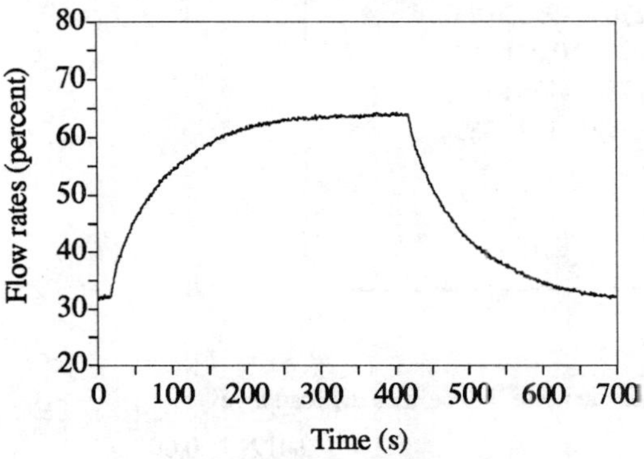

Fig. 4-9 Plot of Steam Flow

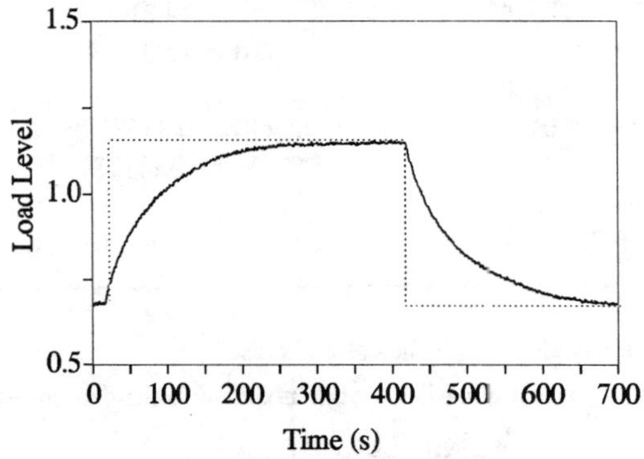

Fig. 4-10 Plot of Load Level (X_4)

REFERENCES

[1] Åström, K., R. Bell: "Dynamic Models for Boiler–Turbine–Alternator Units: Data logs and parameter estimation for a 160 MW unit," Report LUTFD2/ (TFRT-3192)/ 1-137/ (1987), Department of Automatic Control, Lund Institute of Technology, Lund, Sweden, 1987.

[2] Grimble, M. J. . Polynomial systems approach to optimal linear filtering and prediction. *Int. J. Control*, **41**, 1545–1546, 1985.

[3] Grimble, M. J., "Optimal H_∞ multivariable robust controllers and the relationship to LQG design problems," *International Journal of Control*, **48**, 33–58, 1988

[4] Pellegrinetti, G., J. Bentsman (1994). *Control-Oriented Boiler Model.* Submitted for Publication, IEEE Transactions on Control Systems Technology, 1994.

[5] Rotstein, H., A. Sideris,: "H_∞ Optimization with Time-Domain Constraints," *IEEE Trans. Automat. Contr.*, vol. AC-39, No. 4, pp, 762–779, 1994.

[6] Tse, J., J. Bentsman, and N. Miller. Robust Self-Tuning Predictive Control for Industrial Applications. *Ph.D. Thesis,* University of Illinois at Urbana-Champaign, 1992

APPENDIX: NUMERICAL DATA

Table 1. Nonlinear equation coefficients:

c11 = -0.00478	c31 = 0.00533176	c70 = -0.1048569
c12 = 0.280	c32 = 0.0251950	c71 = 0.15479
c13 = 0.01348	c34 = 0.7317058	c72 = 0.4954961
c14 = 0.02493	c41 = 0.04	c73 = -0.20797
	c42 = 0.029988	c74 = 1.2720
c21 = 0.1540357	c43 = 0.018088	c75 = -324212.7805
c22 = 103.5462	c51 = 14.214	c76 = -99556.24778
c23 = 107.4835	c61 = 1.00	c77 = 0.0011850
c24 = 1.95150		c78 = -1704.50476
c25 = 29.04	c81 = 0.85663	c79 = -103.7351
c26 = 1.824	c82 = -0.18128	

$\tau_1 = 2,\ \tau_2 = 2,\ \tau_3 = 3,\ \tau_4 = 3,\ \tau_5 = 4,\ \tau_6 = 10,\ \tau_7 = 2.$

Noise models: n_i is the colored noise

w_i is unit variance white noise, w_d is the deterministic part of the noise that defines the load level

$$n_1 = \frac{0.75s + 0.1}{s + 0.001} w_1, \qquad n_2 = \frac{0.019s + 0.001}{s + 0.024} w_2,$$

$$n_3 = \frac{0.105s + 0.038}{s + 0.010} w_3, \qquad n_4 = \frac{0.01s + 0.0001}{s + 0.001} w_4,$$

$$n_5 = \frac{0.003s + 0.003}{s + 0.0075} w_5 + w_d .$$

Nonlinear Model Operating point information:

$$x_0 = \begin{bmatrix}22.5 & 2.5 & 621.17 & 0.6941\end{bmatrix},$$

$$y_0 = \begin{bmatrix}320 & 2.5 & 0.0 & 9.3053\end{bmatrix},$$

$$u_0 = \begin{bmatrix}0.32270 & 0.39503 & 0.37404\end{bmatrix}.$$

0074-056X/95/107-114/$0 + .50pp

APPROACHABLE MULTI-VARIABLE CONTROL

Gerald M. Bruscia
Principal Power Specialist
Foxboro Company
33 Commercial St.
Foxboro, Mass. 02035

KEYWORDS

Multi-Variable, Advanced Control, Superheat Temperature, Fuzzy Logic, Neural Nets

ABSTRACT

This paper introduces a simplified approach to Advanced Multi-Variable Control as it applies to large utility plant operations.

Many advanced control applications have addressed the problem from the "top down", by applying extensive application programs which require sophisticated modeling to work accurately. These programs, for the most part, must run in an application processor that is separated from the actual control platform. The new approach mentioned in this paper, uses standard control algorithms, which have been enhanced to include automatic self tuning with extension blocks for automatic feedback tuning and automatic feedforward calibration. The extension blocks work in conjunction with a new highly sophisticated block that includes PID or Dead Time capabilities, along with fuzzy logic and neural networks. The combination allows for powerful advanced control to take place in the controller, where all other control is performed.

This paper explores how this approach can be applied to improve the controllability of typically difficult control loops, such as superheat and reheat temperature control, and non linear feedwater regulation. It suggests operational and maintenance improvements due in large part to interface simplicity and consistency, along with a very robust controller.

INTRODUCTION

There have been many different approaches utilized to improve control for large utility boilers. This process has traditionally been difficult to control because of the multitude of variables that directly and indirectly affect it. Some of the disturbances, like air flow and fuel flow are directly measurable. Others, such as fuel BTU changes, are not. To complicate the matter, some of the control loops such as superheat temperature, have variable dead times and long process residence times. This makes them extremely difficult to control close to set point.

Traditionally feedforward control has been implemented in many control loops by obtaining a relationship among the controlled, manipulated and measured load variables such that the manipulated variables can be changed to counteract the measured load variables before an error is detected. In the case of superheat temperature control, measured load, typically in the form of air flow or steam flow, is used to determine the set point to the secondary superheater outlet controller so as to maintain final temperature at design. Experience has shown that load and spray demand are non-linear and also that the time dynamics vary from one load to another. This causes a dynamic imbalance in the response of the loop, which has typically been compensated for by using dynamic elements such as lags, leads, and/or dead times.

The traditional feedforward models do not automatically compensate for either unmeasured load variables or process non-linearity. Moreover, the dynamic imbalance between the measured load variables and the process throughput changes. Due to these affects, no "static" feedforward model can be perfect. What is needed is a control strategy that will automatically re-adjust its PID settings to compensate for unmeasured variables that effect the measurement and also be capable of analyzing multiple related variables to determine their relationship to the manipulated variable, in order to keep the feedforward model correct under all conditions.

Furthermore, many of the current technologies which attempt to compensate for the changing conditions require extensive testing, engineering and commissioning work to obtain a functioning system. Of key importance, and often neglected during the design phase, is the requirement for ongoing maintenance and calibration of the control system elements which compensate for interaction among the variables. In most cases these systems do not receive the maintenance they require to obtain optimal (or even satisfactory) performance. This is due in large part because they are very complicated , model based and require extensive testing and tuning. The result is that many of these systems fall into disuse or are decommissioned completely, leaving the operator with less than optimal control.

The control approach discussed in this paper enables automatic calibration of the feedforward compensation elements, and most importantly, dynamic compensation when a significant shift in process dynamics is detected.

THE ALGORITHMS

The advanced control technique presented here will improve control in three distinct areas: feedback tuning; feedforward calibration and dynamics; and standard PID with the option of selecting a pure dead time function and interactive or non interactive reset.

The feedback element has a pre-tune function that automatically learns the inter-relationships of the process and calculates the appropriate tuning settings and gains. After the pre-tune period, these settings will subsequently be presented to the engineer for manual input into the controller, or they can be automatically updated. This feature can be used as a stand alone feature to speed start up and tuning. The feedback pre-tuner looks for pattern features outside the norm, such as peak values and response timing in the control-error signal. When connected to a specific controller, and run in the self-tune mode, process disturbances, which are outside the operator set noise band, are used to update the controllers PID parameters to automatically adjust the controller to match a predetermined response pattern. Alternatively, when an oscillatory error response starts from the quiet state, fuzzy logic is used to interpolate a stored data table to determine the controller tuning parameters. When the response is not oscillatory, expert rules are used to tighten the tuning until a nearly symmetric load response is achieved.

The feedforward self-tuner operates independently from the feedback tuner and needs no pre-tuner. It uses a projection method (1) to find the smallest changes in the transfer function model parameters consistent with the last set of signal expansion coefficients. In other words, the feedforward tuner can automatically determine the compensations for up to four measured load signals, one of which may be absolute. If the dynamic relationship between any of the feedforward signals and the manipulated variable changes, the feedforward tuner will automatically compensate. Like the feedback tuner, the feedforward tuner uses stored compensatory parameters to establish base settings.

FIGURE 1 illustrates functionally how the controller blocks might be connected to achieve feedback and feedforward multi-variable control for a generic application.

Superheat temperature control was chosen as one example since it is an application that can greatly benefit by this type of control. The fluid path is made up of a primary, secondary, and final superheater, the output of which supplies steam to the high pressure stage of the turbine. See FIGURE 2 for details. The standard control strategy is to control the de-superheater outlet temperature by spraying water upstream of the secondary superheater tubes. The de-superheater outlet temperature set point must anticipate the temperature rise in the secondary stage such that the output is at the proper temperature for delivery to the turbine. It is imperative that outlet temperature be maintained extremely close to setpoint. As mentioned previously, many outside parameters can act to disturb the final superheat temperature, some measurable and some not.

FIGURE 3 illustrates how the multi-variable self tuning blocks can be connected in a simple single stage cascade control strategy. The primary loop controls final superheater outlet temperature, and the secondary controls de-superheater outlet temperature. Traditional control algorithms are limited in the number of feedforward signals that can be accommodated. For this reason, standard approaches have been marginal in their ability to satisfactorily control temperature. Furthermore, unmeasured disturbances are not accounted for in sufficient time to stop disturbances in the final superheat temperature. Using the advanced multi-variable feedforward tune block, three other dependent variables can be used in the control strategy to better account for upsets. Auto-tune will determine the relative gain relationships between the various measurements and the de-superheater outlet temperature, drastically minimizing process upsets to the final temperature. Feedback-tune will compensate for unmeasured disturbances and re-tune the controller, whenever the measured variable deviates outside the noise band.. The dead time feature will help to optimize the primary controllers normal settings, so tighter set point control can be realized. Superheat and reheat actions can be linked with the feedforward tuning acting to decouple the two processes, for example on applications where bypass dampers are used to control reheat but also effect superheat.

BENEFITS

Other multi-variable programs offer modeling capabilities with the ability of relating various process measurements to the manipulated variable. However most that are available today, have large portions of their algorithm in an application processor or are completely resident in it, and are therefore somewhat removed from the direct control process. This makes the integration of control cumbersome. In addition, the control engineer must learn one language or procedure for the basic control strategy and a completely different one , usually mathematically intense, for the advanced control portion. Often, this requires the talents of more than one person.

The control strategy presented in this paper runs in the same control processor that is assigned to perform regulatory control. This offers marked benefits for the user. First and foremost is ease of use, only one control language and one interface must be learned. This reduces training costs and allows advanced control principles to be adapted almost immediately in a project, so that savings can be realized right away. Secondly, the operator interface is the same as for all other control loops. This reduces errors and aids the operator in easy recognition of problems. This is crucial for the acceptability of any new control approach. Understanding being the key to maintainability, the more understood a control philosophy is the more readily it will be maintained and therefore kept in peak operation. Third is real time processing, for advanced control techniques to be effect on real time processes, they must run in the control processor.

Multi-variable control can be achieved today by imbedding sophisticated algorithms into pre-configured control blocks. This brings advanced control capabilities into the average control engineers realm by providing blocks that are easily connected to form complex control strategies.

Advanced control does not have to be "black magic". In fact, it can be very approachable to the plant engineer and operator. This paper offers one such way to achieve it.

References

1. E.H. Bristol and P.D Hanson, "Moment Projection Feedforward Control Adaptation" Research and Development, The Foxboro Company.

2. Peter D. Hanson, "New Adaptive Multivariable Controller" Advanced Systems Development, The Foxboro Company

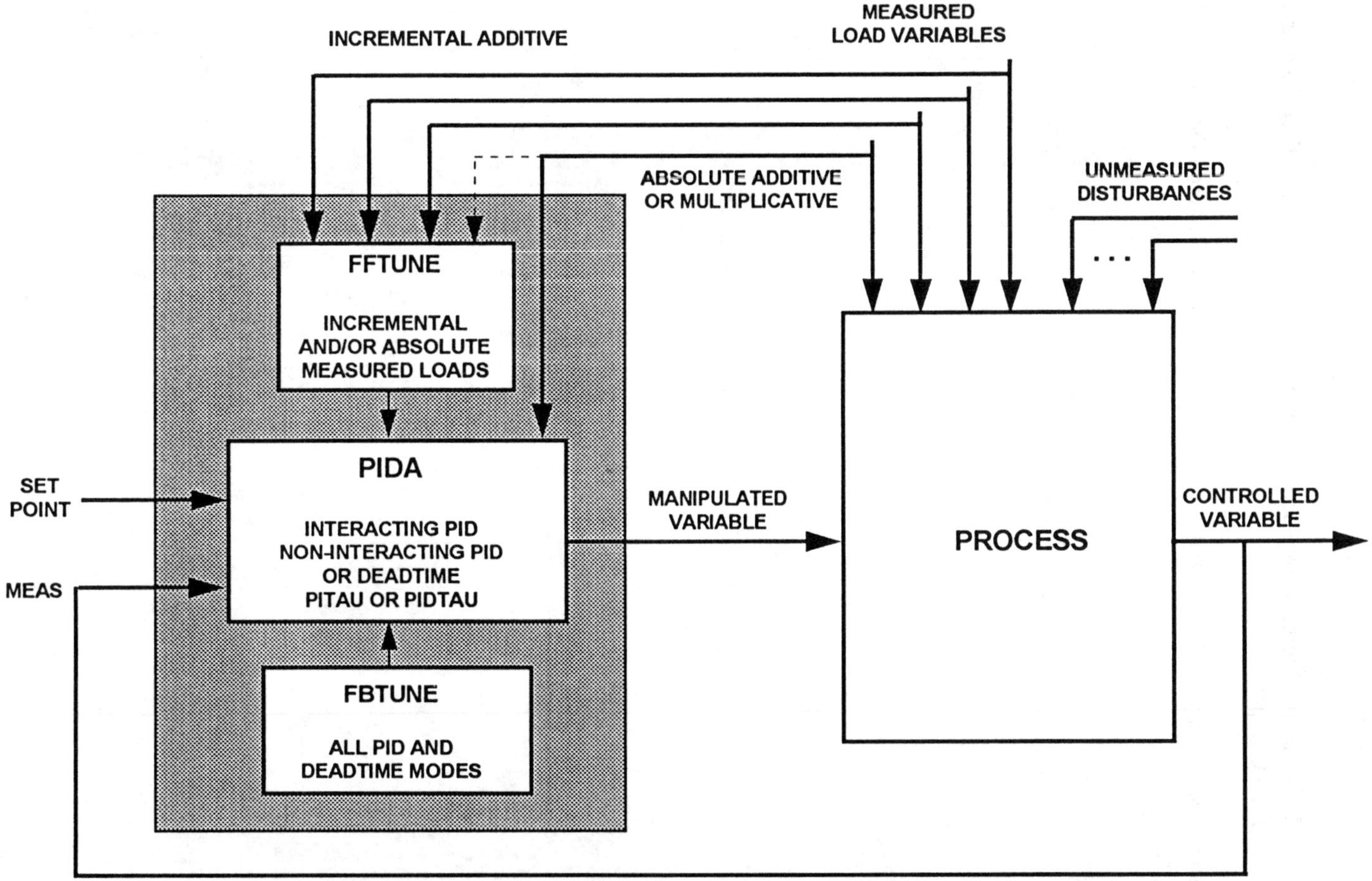

PIDA ADAPTIVE FEEDBACK AND FEEDFORWARD CONTROLLER

FIGURE 1

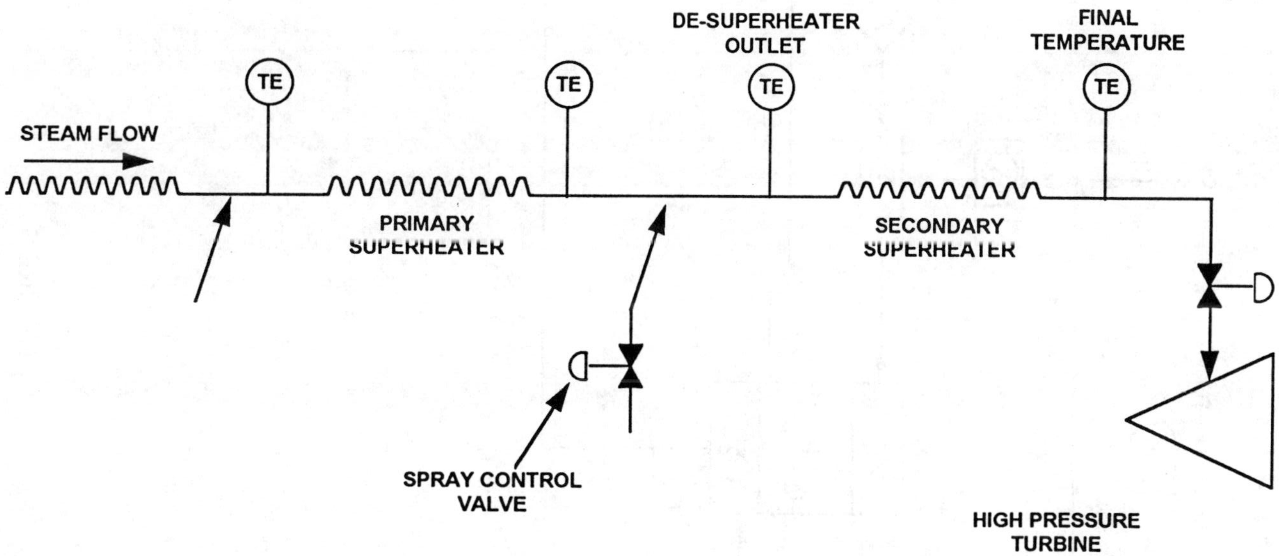
STEAM FLOW
TE
TE
DE-SUPERHEATER
OUTLET
TE
FINAL
TEMPERATURE
TE
PRIMARY
SUPERHEATER
SECONDARY
SUPERHEATER
SPRAY CONTROL
VALVE
HIGH PRESSURE
TURBINE

FIGURE 2

SUPERHEAT STEAM TEMPERATURE FEEDFORWARD LOAD MODELING (SIMPLIFIED DIAGRAM)

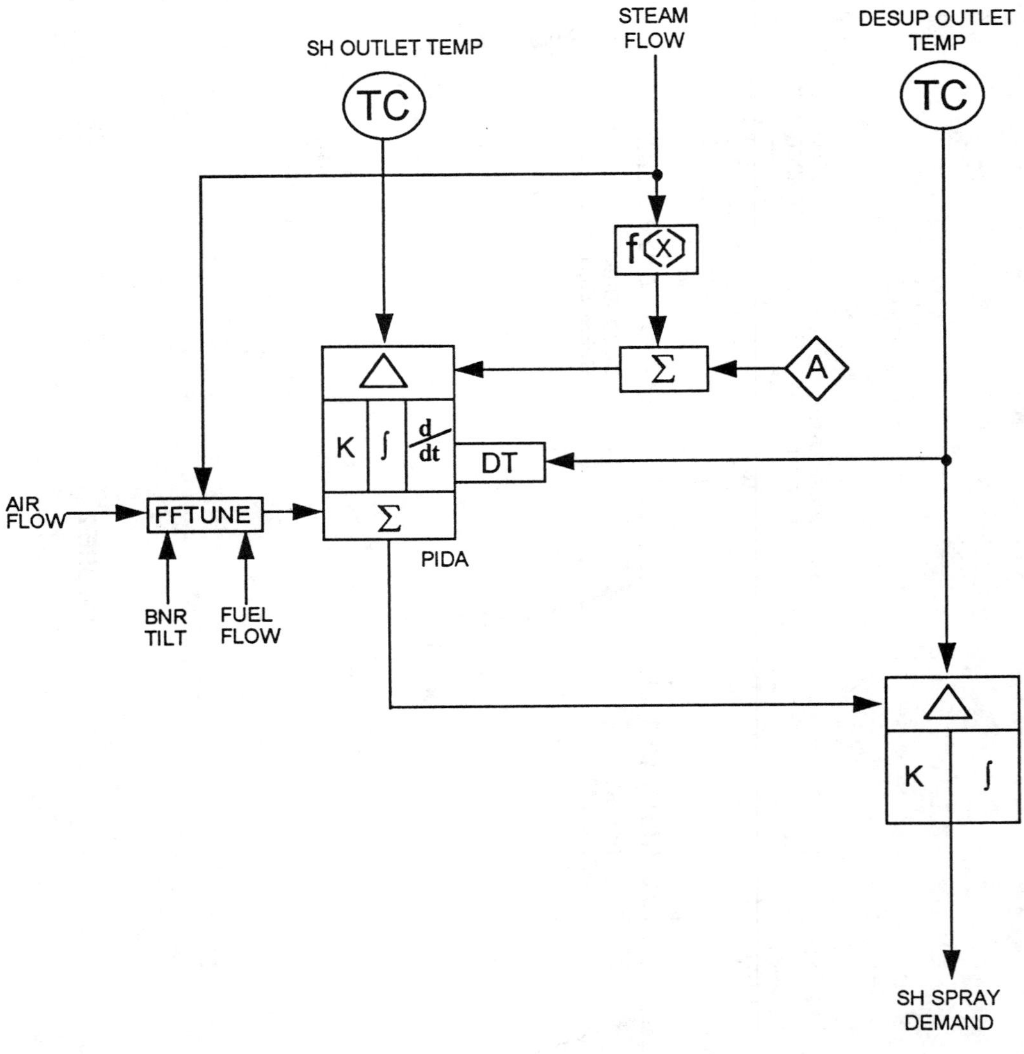

FIGURE 3

0074-056X/95/115-127/$0 + .50pp

CONSOLIDATION OF INDEPENDENT CONTROL SYSTEMS TO A STATION-WIDE DISTRIBUTED CONTROL SYSTEM

Alan Robertson
Senior Purchasing Specialist
Commonwealth Edison Company

Michele Dybel, Bill Pott
Project Engineers
Doyen & Associates, Incorporated

KEYWORDS

Consolidation
Methodology
Studies
Location
Conceptual
Feasibility

ABSTRACT

This paper discusses the methodology used to implement a feasibility study including cost estimates for incorporating independent control systems into a station-wide DCS for operation from a central control location. Systems are either incorporated directly into the DCS or datalinked into the DCS plant wide communication loop.

The systems requiring consolidation include combustion and burner management control, turbine and generator control, pump control, fire protection, precipitator, sootblowing, demineralizing, fly ash handling and other miscellaneous areas.

Alternate consolidation scenarios were developed from which recommended modifications to existing control room and logic rooms were made or new locations identified, based on technical feasibility and installation costs. Estimates included DCS equipment and vendor services, field devices, HVAC and UPS requirements, demolition, new equipment installation and engineering services.

Final recommendations were developed and options presented for consideration. Project implementation schedules were developed including engineering, vendor fabrication, pre-outage and outage construction periods.

INTRODUCTION

In 1993 Commonwealth Edison Company (ComEd) authorized studies for upgrading and consolidating various independent control systems at some of their coal-fired stations. The existing systems included a myriad of DCS-based, PLC-based, single loop, pneumatic and hardwire controls at various locations throughout the stations. The scope of the project involved feasibility studies, conceptual designs and installed cost estimates for upgrading and consolidating the controls to a DCS system utilizing a plant-wide communication network operated from a central location.

Using these studies ComEd conducted economic analyses, (discounted cash flows) and determined that consolidation of controls would allow for department reorganizations eventually resulting in personnel cutbacks. ComEd determined payback periods averaging 8-10 years based on manpower reductions.

During initial station reviews, ComEd targeted specific control systems and station goals for consolidation, discussed operating procedures, and identified potential control/logic room locations. In general, the following control systems required review for consolidation considerations:

- Combustion control (ICC) (existing DCS)
- Burner management (BM) (existing DCS)
- Miscellaneous boiler (controls not part of ICC/BM)
- Pumps
- Turbine
- Generator
- Switchyard
- Sootblowing
- Demineralizing
- Precipitating
- Fly ash handling
- Fire protection
- Coal handling

Subsequent station walkdowns and review of drawings established input/output (I/O) requirements which formed the design basis for the DCS hardware and software requirements. Strategies were developed which targeted control/logic room locations, required field work and equipment necessary for system integration. The work identified in the studies included instrumentation mounted at control panels and at field locations/devices, demolition in control/logic rooms, control/logic room modifications and/or new constructions, and any new DCS equipment power (UPS) and room conditioning (HVAC) requirements.

The conceptual designs presented in the studies formed the basis of the installed cost estimates which detailed the material and labor costs associated with the DCS system installation. The estimates also included all DCS system training, start-up and vendor engineering design. The total installed cost estimates reflected a 20% margin of accuracy. Final recommendations for control systems to be included in the DCS consolidation and for control room and logic room locations were based on installed costs, technical feasibility and, often, station preferences. Based on these recommendations, project schedules were developed spanning the design phase through system start-up.

BACKGROUND

The stations included various boiler and turbine manufacturers with nominal unit ratings from 150 MW to 550 MW. The control rooms were equipped with a variety of controls and instrumentation which required operator interface i.e., screens/keyboards, hand manual/automatic stations, hand control switches/pushbuttons, process indicators, recorders, annunciators and miscellaneous monitoring and diagnostic systems from various vendors. The variety of systems and interfaces made operations and maintenance complex for the following reasons:

- Control locations geographically separated from operators causing communication difficulties.
- Devices and systems with various operating, readout and maintenance procedures.
- Scarcity of spare parts and technicians for servicing older systems.
- Physical space constraints in existing control/logic rooms.

To alleviate the above problems the need was identified to investigate updating and consolidating systems **to increase unit availability, increase overall productivity and identify possible manpower reductions.**

The benefits system integration would provide were identified as follows:

- Operator access to remotely located systems' information.
- Operator access to diagnostics.
- Intelligence on system performance.
- Real time graphic display updates.
- Immediate access to alarm and trip information.
- Management access to station operating information.
- Optimized unit loading changes.
- Increased equipment life span.
- Improved ergonomics through common operator interface and alarming systems.
- Ease of system configuration changes.
- Minimized unit startup time.
- Decreased possibility of unit trips.
- Heat rate improvement.
- Tighter control of process.
- Tighter emission control.
- Lower maintenance expenditures.

METHOD OF INVESTIGATION

The studies approached the system integration on the basis of the following key elements, always keeping in mind the most **cost-effective solutions**:

- Identify the systems for consolidation.
- Establish design criteria and goals.
- Develop the conceptual designs including options.
- Develop installed costs based on the conceptual designs.
- Recommend the optimal conceptual design.
- Establish project schedule.

IDENTIFY SYSTEMS

Station systems to be included in the DCS consolidation were identified in discussions with ComEd engineering and further defined through detailed discussions with station personnel. At all stations, the ICC/BM controls currently existed in a DCS network. ComEd established that the consolidation of all remaining hardwire and non-DCS controls would be into the same DCS vendor's system that operated the ICC/BM controls. Table 1.0 presents targeted control systems and the proposed consolidation strategies for a typical 4-unit station.

ESTABLISH STRATEGY CRITERIA AND GOALS

Station-specific design criteria was reviewed which would impact the conceptual designs (i.e. ongoing or future plans for controls upgrades, new vendor-specific control package preferences to include in consolidation, etc.). Project goals were established with deliverables defined to be a detailed engineering summary report for each station which would include an installed cost estimate and a project schedule.

DEVELOP THE CONCEPTUAL DESIGNS

The conceptual designs were based on the previously referenced targeted systems and any station preferences for new control system installations. The conceptual designs were developed as described in the following phases of design: field investigations, DCS hardware requirements and potential DCS hardware locations.

Field Investigations

The first step in the conceptual design effort was a site walkdown to establish the station physical layout and analyze the following details:

- Input/Output (I/O) requirements for each system.
- Operator interface requirements.
- Location of consoles/logic cabinets.
- Control/logic room space availability.
- Control/logic room HVAC requirements.

- Control and alarm functions.
- Data acquisition requirements.
- Primary sensor/signal converter locations.
- UPS system capacity.
- Hardwire versus Datalink options.
- Demolition considerations.
- Ease of construction considerations.
- Future expansion capabilities.

The key document that was generated during the project was each station's I/O list. The I/O list was based on the systems currently being controlled and monitored and was used to establish the DCS equipment requirements. The I/O list classified the controls as follows:

- Analog and digital control functions
- Analog and digital monitoring functions
- Pulse signals
- Alarms

The I/O count, new instrumentation requirements, and conceptual designs were developed based on the following criteria as established during the walkdowns:

- The devices presently on the control boards(i.e. switches, manual/automatic stations, gauges, etc) will be emulated on the operator consoles.
- Existing control boards and their devices will be retired in place (i.e. termination points maintained and used as field wiring junction locations) unless space requirements for consolidation require their removal.
- Demolition work will be limited to removal of wiring from control board devices to terminal blocks and removal of tubing from gauges and indicators, to nearest shutoff valve.
- New transmitters will be located at the process connections.
- Signal converters will be located at the field devices (i.e. valve, drive unit).
- Thermocouple signals will be wired direct to the DCS cabinets via new thermocouple extension wiring.
- All DCS hardwire switch/ pushbutton operations will be consolidated as soft control with 4 kV operation and greater maintaining parallel existing hardwire controls for emergency operation.
- Electrical/Switchyard controls consolidation will be evaluated for three approaches: install a SCADA system and datalink, install a PLC system and datalink, or consolidate directly to the DCS system.

DCS Hardware Requirements

Based on the I/O count, the number of logic and termination cabinets (collectively termed DCS cabinets) was determined with the station's DCS vendor. In establishing the cabinet count, the layout accounted for station requirements to segregate controls by system (i.e. pump, turbine) and unit (i.e., Units 1 & 2, Units 3 & 4). The cabinet layouts included 20% spare I/O capacity and 20% spare cabinet space. The cabinet layouts also incorporated the controls of auxiliary systems which had a small I/O count that did not require a complete cabinet. Some of the auxiliary systems involved a new or existing control system which would datalink to the DCS highway via one of these cabinets (i.e. sootblowing PLC datalinked into pump controls cabinet).

The remaining DCS hardware to establish involved the number of new operator consoles which followed ComEd's practice of two operator stations and two slave stations per unit. These new consoles were in addition to existing ICC/BM consoles (3-4 per unit).

DCS Hardware Location Options

In addition to DCS hardware, the conceptual design must identify wiring requirements from existing termination points at the control boards to new termination points in the cabinets. Cable routings for signal wiring and data highway cables were dependent on the geographic locations of the cabinets and operator consoles. The optimal design should locate the termination cabinets as close as possible to existing termination points on control boards to minimize new wiring lengths and subsequent costs. Minimal wiring costs were realized when termination cabinets were located in the control rooms containing their respective hardwire control boards (i.e., locate pump DCS cabinets in the pump control room). For this reason, the designs first established potential control room locations to consider for station-wide consolidated control. Since these control rooms house the operator consoles, completing the console layout identifies the remaining space available for DCS cabinets.

The location of operator consoles for consolidation was typically investigated for three locations at each station. The first and most obvious location was to re-use the existing boiler control room(s) since the ICC/BM consoles were there. Existing bench/gaugeboards could be retired in place if sufficient room existed for new consoles or could be removed to provide more space. (This location identified as Option A)

The second location considered was the electrical/generator control room since this area had adequate space for new consoles without the need for demolition. (This location identified as Option B)

The last option considered was the construction of a new control room. This option was possible only at stations that had available space (usually on the turbine deck) and was mainly considered when existing control rooms were too small to accommodate the additional number of consoles and operators required for consolidated station control.

With the Option A and B control rooms established, DCS cabinets could then be located in these control rooms, if space allowed. Many stations had available space for the new cabinets in nearby existing logic rooms which housed the existing ICC/BM controls. Another cabinet location option involved converting existing control rooms to logic rooms (i.e. pump control room became pump logic room). If control room space constraints and existing logic room space constraints prevented use of the above locations, the designs included new logic room constructions.

Depending on the number of cabinets, the new cooling requirements could exceed current HVAC system capabilities for some or all of the potential locations. Heat load calculations were completed for each potential cabinet location and compared to existing HVAC equipment capabilities. A conceptual design using a location with insufficient HVAC involved supplementing or replacing the existing HVAC system to meet the new cabinet conditioning requirements. The conceptual designs incorporated HVAC redundancy for cabinet locations to collaborate with existing ComEd logic room design philosophy which was employed at all stations.

In addition to cabinet HVAC needs, power requirements for the new cabinets were compared to available power to determine what modifications were required to the UPS at the various locations. Table 2.0 presents an example of potential DCS cabinet locations for a typical 4-unit station considering both Options A and B operator console locations.

The costs associated with the new cabling, new/modified HVAC systems and additional power requirements were the three major factors weighed together for each possible combination of potential cabinet locations. The designs were evaluated on this basis to arrive at the most cost-effective logic cabinet locations for each system based on both operator console locations (Options A and B). The lowest cost combination of cabinet locations became the recommended designs (one for Option A, one for Option B) for which the subsequent installed cost estimates were developed.

Initially, certain designs that appeared most cost-effective actually were not when all three of the above cost factors were considered. For example, a given design whose cabinet location involved minimal cabling requirements due to their close proximity to existing control boards would seem to involve the lowest cost. However, it could actually involve extensive HVAC modifications which would make that design more cost intensive than others.

Exhibits 1 and 2 show the two DCS architecture layouts (Options A and B) for a 4 unit ComEd station (Unit 1 and 2 are shown, Units 3 and 4 are similar). Each unit has its own network loop which connects all the consoles, module and termination cabinets for that unit. This allows each unit to run totally independent of the other units. each unit's loop is then connected to a central network loop to allow for plant wide control and communication capability.

Exhibit 1 shows operator control in separate control rooms (Unit 1 and 2 shown, Unit 3 and 4 similar). In this approach the electric/generator control has also been incorporated.

Exhibit 2 shows a central operator control room located in the existing electric/generator room. For this approach, electric/generator control was not incorporated into the DCS, however, the capability still exists.

DEVELOP INSTALLED COST ESTIMATES

Detailed installed costs were developed for Options A and B based on the above recommended conceptual designs. The costs consisted of three (3) major categories of work:

DCS: Costs include hardware, vendor engineering (programming/graphics development), and field start-up services.

System Installation: Costs include purchasing and installation of all cable, conduits, raceways, termination boxes, field sensing devices, tubing, equipment supports, HVAC, UPS, demolition and any miscellaneous hardware external to the DCS. Labor manhours and material costs were obtained from the Means and Richardsons estimating guidelines for project installations during 1995.

Engineering Design Hours: Hours include the A/E project management, engineering, design/drafting and administrative support required for project implementation. (ComEd engineering effort was not included in the A/E engineering design hour estimates. ComEd's engineering effort was estimated independently by them as a fraction of the total A/E hours and was included in the final economic analyses which determined the payback periods for each station's consolidation effort).

The estimates were developed on an independent system basis such that the total installed cost for each system could be singled out. This allowed flexibility to select/reject systems being considered for consolidation and adjust total costs accordingly. The estimates further itemized the work required for each system by discipline (i.e. I&C, mechanical, electrical and structural) and then divided the discipline costs into labor and material costs.

RECOMMENDED CONCEPTUAL DESIGNS

A recommendation was made after determining all technically feasible design options and weighing their costs. Referring to the Table 2.0 example, total installed costs for a typical 4-unit station for the systems identified were in the following ranges:

- Option A estimated cost: $8-9 Million
- Option B estimated cost: $9-10 Million

An estimated 50,000-55,000 A/E engineering hours was projected for the design effort for either option.

Assuming the conceptual designs (Options A and B) were both technically feasible, cost became the overriding factor for a final design recommendation. If the cost differential between alternatives was small, the station preferences was considered in the final recommendation.

Based on the above total cost comparisons for the Option A versus Option B designs, typically no technical advantage was identified which warranted an additional $1.0 Million expenditure for the Option B design. Thus, the recommended conceptual design was the Option A consolidation of operators/operator consoles to the Units 1 & 2 and Units 3 & 4 Boiler Control Rooms with DCS cabinet locations to be as identified in Table 2.0 for the Option A design.

Project Schedule

The project schedule identified the project scope activities and major milestones as follows:

- A/E engineering design
- DCS equipment fabrication
- DCS vendor engineering design
- Demolition of existing equipment
- Installation of field instrumentation and all cabling, cable tray, conduit, tubing, etc.
- Installation of new control/logic rooms, HVAC and UPS systems
- Installation of DCS equipment
- DCS Start-up and testing

Typically, a 24 month project duration was anticipated for a station's consolidation effort which included an 18 month period for A/E engineering.

SUMMARY

The studies of all ComEd stations were completed during a seven (7) month period and the costs developed have projected economic paybacks which were deemed acceptable by ComEd. The consolidation efforts were approved as projects in 1994 for all the stations and are currently in the A/E design phases. The overall success of the studies can be credited to the following key factors:

- Continuous communication between ComEd, A/E and DCS vendor.
- A/E strategic planning sessions assigning key team personnel for all studies at all stations.
- Participation of station personnel in walkdowns and project meetings.
- Review of preliminary findings with ComEd engineering and station personnel before issuing final report recommendations.
- Thorough understanding of ComEd's requirements resulting in cost estimates being evaluated as reasonable for use in their economic analyses.

SUMMARY OF PROPOSED CONSOLIDATION STRATEGIES FOR A TYPICAL 4-UNIT STATION

SYSTEM	EXISTING CONTROLS	EXISTING LOCATION	NEW CONTROLS
Unit 1 & 2 ICC/BM	DCS	Units 1&2 Boiler Control Room	None
Unit 3 ICC	DCS	Units 3&4 Boiler Control Room	None
Unit 3 BM/Feedwater	PLC	Units 3&4 Boiler Control Room	DCS
Unit 4 - ICC/BM	DCS	Units 3&4 Boiler Control Room	None
Units 1&2 Misc. Boiler Controls * (Air heaters, fans, valves, etc.)	Hardwire	Units 1&2 Boiler Control Room	DCS (eliminate PLC)
Unit 1&2 Pump *	Hardwire	Units 1&2 Pump Control Room	DCS
Unit 1&2 Turbine *	Hardwire	Turbine Deck	DCS
Units 1 thru 4 Generator	Hardwire	Electrical/Generator Control Room	DCS
Switchward (12kv, 34kv, 138kv)	SCADA	Electrical/Generator Control Room	Datalink to DCS
Unit 1&2 Sootblowing *	PLC	Units 1 & 2 Boiler Control Room	Datalink to DCS
Units 1 - 4 Demineralizing	PLC & Hardwire	At Equipment	Incorp. hardwire into existing PLC & datalink to DCS
Units 1 - 4 Precipitating	Hardwire	At Equipment	New PLC & datalink to DCS
Units 1 & 2 Fly ash handling *	Hardwire	Units 1 & 2 Boiler Control Room	New PLC & datalink to DCS
Fire Protection	Hardwire vendor-specific panel	Electrical/Generator Control Room	None-Relocate panel only if consolidate to Boiler Control Rooms (cannot datalink)
Units 1 - 4 Coal handling	A. Hardware B. Manual	A. Breaker House B. All locations	New PLC & datalink to DCS Add equipment to automate New PLC & datalink to DCS

Note: For systems denoted *, the existing and proposed new controls for Units 3 & 4 systems are the same as the Units 1 & 2 systems above (i.e. Unit 3 & 4 pump controls are in Unit 3 & 4 Pump Control Room).

SUMMARY OF POTENTIAL CABINET LOCATIONS FOR BOTH CONTROL ROOM LOCATIONS

Option A - Locate Operator Consoles in Unit 1 & 2 and Unit 3 & 4 Boiler Control Rooms for Consolidated Control
Option B - Locate Operator Consoles in Electrical/Generator Control Room for consolidated control

				POTENTIAL CABINET LOCATION OPTIONS			
SYSTEM CONTROLS	TYPICAL I/O TOTAL	NO. OF DCS CABINETS		OPTION A		OPTION B	
		Units 1 & 2	Units 3 & 4	Units 1 & 2	Units 3 & 4	Units 1 & 2	Units 3 & 4
Pump	1700	5	7	• Units 1&2 Pump Control Room	• Units 3 & 4 Control Room	• Units 1&2 Pump Control Room	• Units 3 & 4 Pump Control Room
Turbine	1300	4	5	• Units 1&2 Pump Control Room • Units 1&2 Logic Room	• Units 3 & 4 Pump Control Room • Units 3 & 4 Logic Room	• Units 1&2 Pump Control Room • Units 1&2 Logic Room • Units 1&2 Boiler Control Room	• Units 3 & 4 Pump Control Room • Units 3 & 4 Logic Room • Units 3 & 4 Boiler Control Room
Misc. Boiler	800	3	3	• No Cabinets - Maintain hardwire	• No Cabinets - Maintain hardwire	• Units 1&2 Boiler Control Room	• Units 3 & 4 Boiler Control Room
Electrical	1300	13 (Units 1 - 4)		• Units 1 - 4 Electrical/Generator Control Room		• Units 1 - 4 Electrical/Generator Control Room	

- OPTION A - ESTIMATED COST $8 - 9 MILLION
- OPTION B - ESTIMATED COST $9 - 10 MILLION
- ESTIMATED A/E HOURS OPTION A OR B 50,000 - 55,000

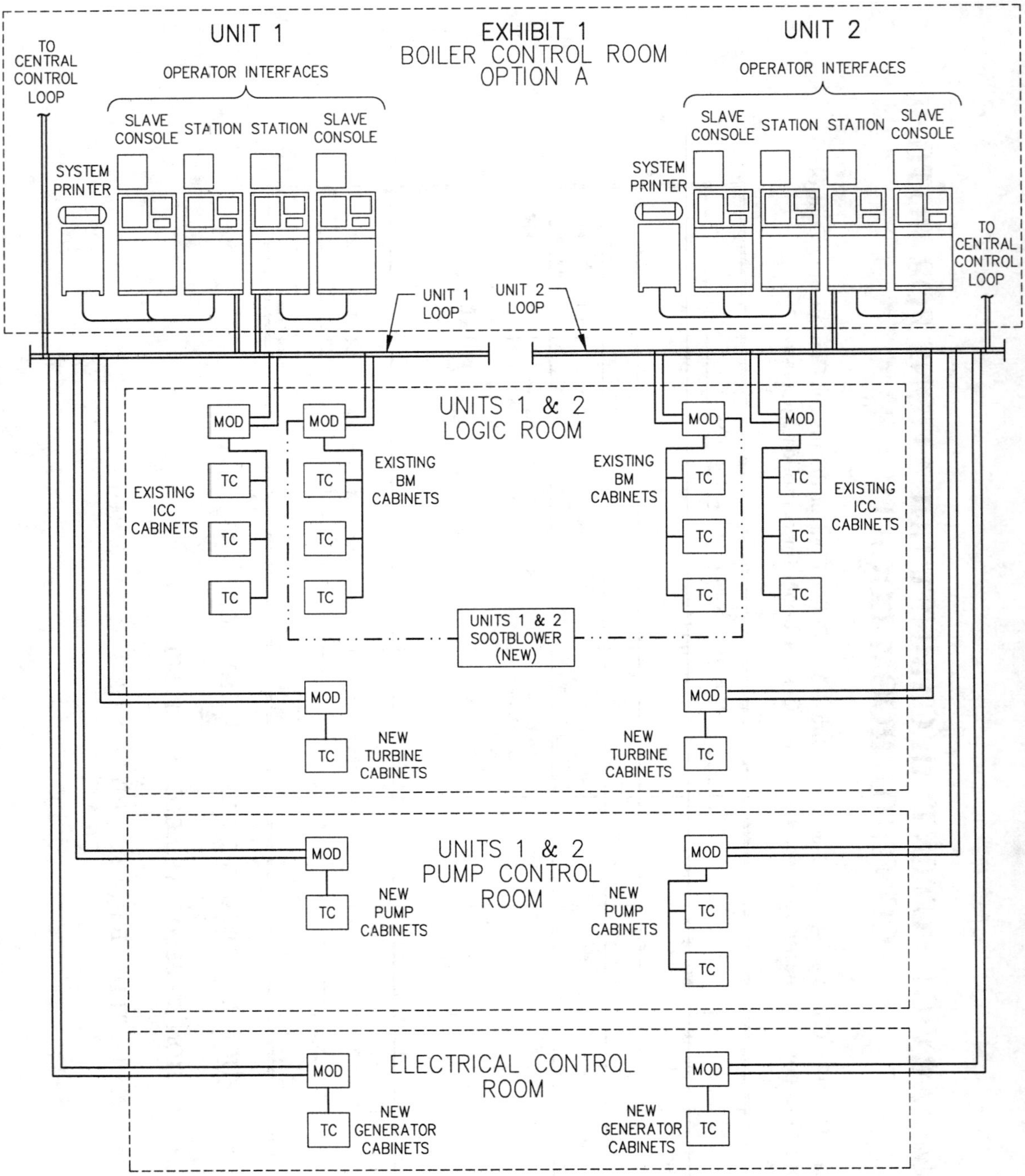

LEGEND

MOD = MODULE/PROCESSOR CABINETS
TC = TERMINATION CABINET
══ = DATA HIGHWAY (REDUNDANT)
– – – – = EQUIPMENT LOCATION AREAS
– · · – = PLC DATA LINK

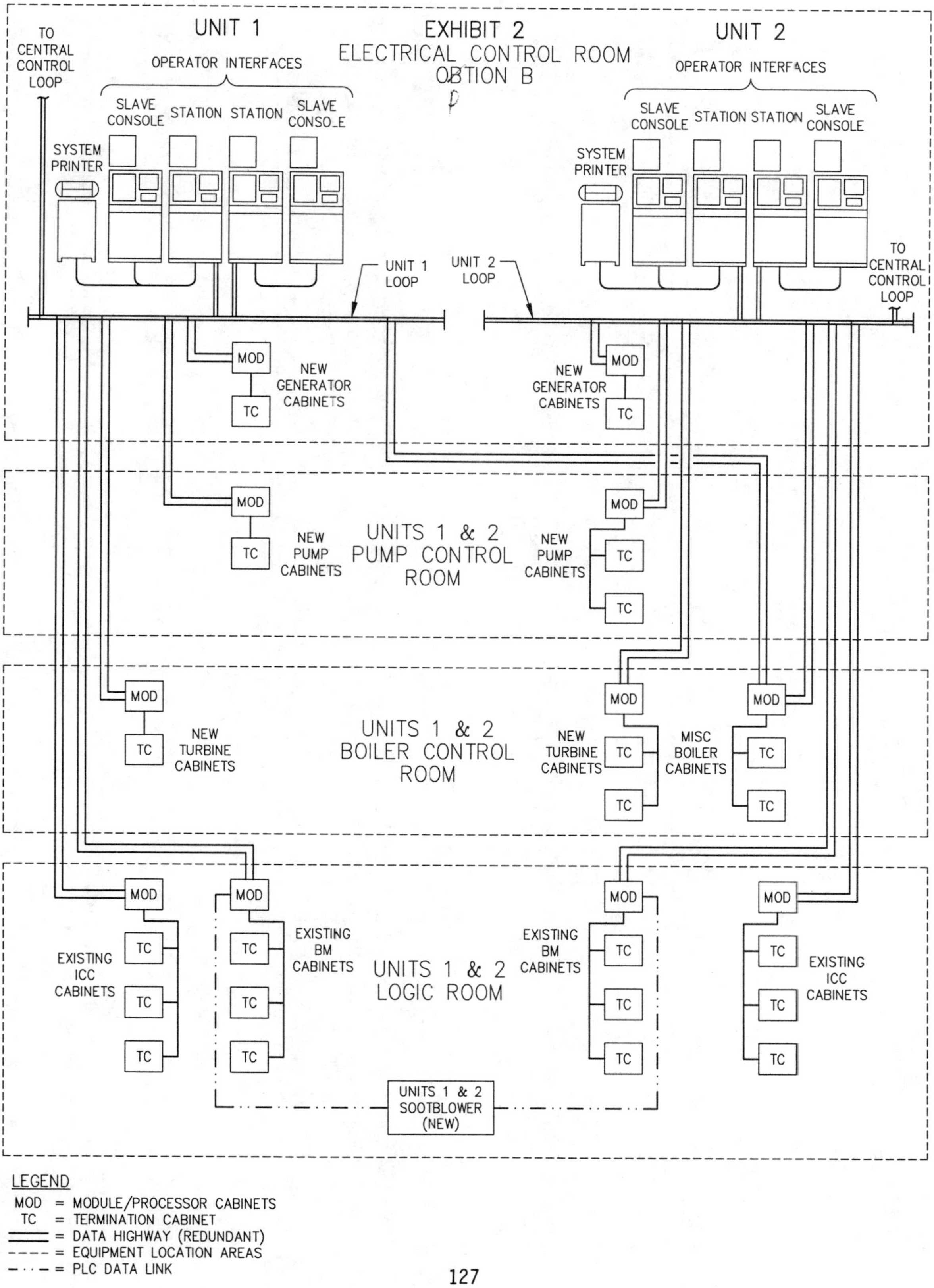
EXHIBIT 2
ELECTRICAL CONTROL ROOM
OPTION B
UNIT 1
OPERATOR INTERFACES
SLAVE CONSOLE
STATION
STATION
SLAVE CONSOLE
SYSTEM PRINTER
TO CENTRAL CONTROL LOOP
UNIT 1 LOOP
UNIT 2 LOOP
UNIT 2
OPERATOR INTERFACES
SLAVE CONSOLE
STATION
STATION
SLAVE CONSOLE
SYSTEM PRINTER
TO CENTRAL CONTROL LOOP
MOD
TC
NEW GENERATOR CABINETS
UNITS 1 & 2
PUMP CONTROL ROOM
NEW PUMP CABINETS
UNITS 1 & 2
BOILER CONTROL ROOM
NEW TURBINE CABINETS
MISC BOILER CABINETS
UNITS 1 & 2
LOGIC ROOM
EXISTING ICC CABINETS
EXISTING BM CABINETS
UNITS 1 & 2 SOOTBLOWER (NEW)
LEGEND
MOD = MODULE/PROCESSOR CABINETS
TC = TERMINATION CABINET
= DATA HIGHWAY (REDUNDANT)
= EQUIPMENT LOCATION AREAS
= PLC DATA LINK

©0074-056X/95/129-137/$0 + .50pp

CONTROL SYSTEMS UPGRADE AT ANCLOTE STATION UNITS 1 & 2

DAVID T. TRUDEL, P.E.
Control Systems Project Engineer
Gilbert/Commonwealth, Inc.
Reading, Pennsylvania

ABSTRACT

This paper discusses the design process associated with the replacement of a plant computer, boiler controls, and burner management systems at a two unit oil fired power plant. Specific items discussed are system architecture, graphics development, and design automation.

INTRODUCTION

Florida Power Corporation was driven to perform a control systems upgrade at its oil fired Anclote station for a number of reasons, primarily an obsolete and unreliable plant computer and analog boiler controls. Problems included limited space in the control room for new Distributed Control System (DCS) equipment and the need to have both units down to replace the computer which was common to both units. The installation was designed to be done in two phases, the first being the replacement of the plant computer while in the second phase, the boiler controls will be replaced in conjunction with a planned gas conversion and Burner Management System (BMS) installation. The plant computer provided an Automatic Turbine Startup (ATS) system which was implemented in the new DCS.

The decision was made to retain the existing control board and all hardwired motor control switches because of budget constraints. Operator consoles (CRTs) were installed in the control board to replace the old computer terminals and additional ones will be installed in the future to replace the hard M/A stations of the existing boiler controls and the BMS inserts. The system when fully implemented will be provided with eight CRT screens per unit.

The Anclote Station consists of two 525 MW oil fired units designed for continuous service located in Holiday, Florida on the gulf coast. Each unit is equipped with a Combustion Engineering tangentially fired drum boiler with a General Electric steam turbine.

Each boiler is a corner fired, forced circulation, pressurized furnace, drum type, utilizing a total of 20 No. 6 fuel oil guns arranged on five elevations and four No. 2 light oil warm ups guns on the lower elevation.

Units one and two were placed in service in 1974 and 1978 respectively.

PROJECT SCOPE

The main project objective was to replace the obsolete and unreliable plant computer with a new DCS that would perform all of the existing plant computer functions, provide enhanced data gathering capabilities, and be easily expandable to incorporate future boiler controls and burner management upgrades. The requirements for the new system were:

- Provide an interface for the I/O of the existing computer.
- Replacement of the automatic turbine startup program.
- Installation of a turbine rotor stress monitor.
- Installation of generator autosychronization capability.
- Interface with a PC based performance monitoring workstation.
- Intelligent alarm management and display.
- Expanded trending capability.
- Connection to the plant Local Area Network (LAN) and corporate Wide Area Network (WAN).
- Four X-Windows based CRTs per unit.
- Collection of historical data in an easily retrievable format through the LAN/WAN interface.
- Interface with the existing sequence of events recorder (SOE) through a serial link.
- Interface with the cooling tower performance monitoring I/O through a serial link.
- Installation of new Uninterruptable Power Supplies.

The expansion of the system to include a boiler controls upgrade and burner management replacement is planned to be performed in conjunction with a conversion to dual fuel firing of No. 6 fuel oil and natural gas. These new systems will be provided with a total soft interface and require the installation of four additional CRTs per unit into the Boiler and Turbine Generator (BTG) Benchboard. The boiler controls upgrade will include a number of improvements including:

- A coordinated control system for faster ramp rates.
- An interface through the system WAN to Energy Load Dispatch signals.
- Hybrid Variable Pressure Control.
- Individual signals for each fuel/air and auxiliary air dampers to provide for individual burner air biasing.
- Improved steam temperature and feedwater control logic.
- Dual fuel firing capability.

The potential modifications of the system to incorporate boiler controls and burner management had to be considered in the design of all equipment layouts, power requirements, operator interfaces, and displays for the computer replacement.

PROJECT SCHEDULE

The computer replacement phase of this project had a total duration from start of engineering to on line operation of 21 months. The concurrent Unit 1 and 2 outages were two weeks and four weeks respectively.

The DCS specification and purchase cycle required six months from project kickoff to placement of a purchase order while another 12 months was required for the DCS implementation and shipment. There was one factory acceptance test for both systems and they were delivered together.

A six week construction preoutage permitted placement of the new module cabinets and network testing prior to the start of the two unit outage. The outage durations allowed for the installation of the DCS termination equipment and I/O termination and testing. The ATS and turbine autosychronization functional testing will be completed during future outages.

SYSTEM CONFIGURATION

A Bailey Controls Company DCS using its Infi 90 I/O hardware and Digital Equipment Corporation VAX based display system with X-window capability was selected for the Distributed Control and Information System (DCIS). The network consists of three loops, one for each unit and one for common equipment (see Figure 1).

Each unit's network had five processor and display driver cabinets, a remote I/O interface for condenser performance monitoring, and connections to custom I/O termination assemblies located in the existing computer I/O termination cabinets. The common loop has the printer interface, engineering workstations, performance monitoring workstations, historian, SOE interface, and cooling tower performance monitoring interface. An interface to the plant LAN and corporate WAN was established through ethernet to connect the display equipment to the process equipment.

This system provides security and reliability by segregating each unit's I/O and displays onto separate data highway loops and ethernet segments, but provides flexibility by allowing all displays to be accessible to all displays with the use of the proper password. It also allows operator displays and historical data to be viewed on PCs connected throughout the Florida Power LAN/WAN.

PROJECT DESIGN

A number of specific design philosophies for interfacing with the existing equipment and field wiring were established to provide an economical and trouble free installation requiring minimum downtime. These requirements included:

- Reuse of all existing field cables without trimming or the installation of new cable lugs since much of the existing cabling is asbestos jacketed.

- Minimize alterations to the BTG for CRT installation.

- Retention of all hardwired annunciators.

- Maximum use of prefabricated cables for field interconnection of DCS equipment.

- One for one replacement of the Boiler Controls cabinets.

- Maximum amount of preoutage construction work to minimize downtime.

I/O termination assemblies were installed in the existing computer analog I/O cabinets on custom frames and connected to the I/O slaves using standard prefabricated cables. The digital I/O cabinets could not be approached in the same manner because of space limitations.

Therefore, custom terminal block assemblies were designed and fabricated for the termination of the existing digital I/O field cables. Custom prefabricated cables were then provided by the vendor to interface the DCS equipment with this I/O. All these I/O interfaces and cables were fully staged and tested during the factory acceptance test.

The DCS vendor was required to layout the I/O on his equipment and also to functionally partition the I/O between the data acquisition and ATS subsystems. Therefore, some compromises to the partitioning were required in order to ensure that new field cables or splices would not be required.

The planned installation of a total of five CRTs into the face of each units BTG board presented a number of problems including minimizing the impact on remaining controls, interference problems behind the board, the retention of hardwired indicators, and the requirement to perform the installation in phases. The design was accomplished by working closely with the operations department to determine their requirements, minor modifications to the vendors equipment, and some selective relocation and rewiring of the BTG board.

The design was facilitated by the preparation of a comprehensive database, which was used to coordinate all aspects of the design. It was used to capture and present information for most of the design functions of the project including interfacing with the DCS vendor, preparation of circuit and raceway schedules, CRT display design, cross reference information for plant and DCS vendor design drawings, and physical termination drawings of the DCS vendor's equipment.

The database included fields for the following:

- I/O data.
- Existing termination information.
- Circuit data.
- Calculation logic.
- Specific locations on CRT displays.
- Alarm data.
- Log data.

The database was maintained on a network server and accessible to all project personnel allowing for a more timely dissemination of information. Revisions were made as data became available and revision levels were maintained in the database. Transfer of the database to the DCS vendor or FPC was by electronic media with reports being minimized as much as practical. The DCS vendor transmitted system design data on electronic media which was incorporated into the master database and used for the preparation of the installation design.

A master report containing all database fields was not maintained. Reports were designed and printed as needed for facilitating the design, transferring data, and printing reports for construction and testing.

AUTO TURBINE STARTUP AND AUTOSYNCHRONIZATION

The existing Automatic Turbine Startup (ATS) logic was replaced with new logic which provided:

- Speed control loop.
- Megawatt control loop.
- Full arc to partial arc transfer logic.
- Improved rotor stress monitor.
- Allowable load rate calculations.
- Generator Autosynchronization.

New turbine stop and governor valve position indicating LVDTs and turbine speed transducers were installed to provide the necessary resolution and accuracy required for precision positioning and reliable position feedback.

CRT GRAPHICS AND ALARMS

CRT graphics and display development began with the establishment of guidelines in a graphics configuration standard developed specifically for the Anclote Plant. This standard covered system configuration philosophy, color schemes, device symbols, alarming, trends, symbol logic, hierarchy of displays and specific graphics to develop. After the first round of comments on the standard, the graphics for the first system were developed by the vendor and reviewed by the plant operators on an active system. Further comments were developed from this review and were incorporated into the final graphics standard. The standard was used as the basis for all graphics and display development including the numerous displays for the operation of the ATS system therefore the extra effort that went into developing a good comprehensive standard was justified.

Graphics development proceeded from this point with each system having four types of displays; graphics, faceplate, status, and trend. All types of screens were as identical as possible to the corresponding screen in the other unit to provide a consistent operator interface. Approximately 370 displays of all types were developed by the project team for each unit.

The following guidelines were used for each type of screen:

I. Graphics

a. The flow diagrams were used as the basis for the graphics since the operators would be familiar with the layout.

b. Care was taken not to clutter the graphics with redundant or impertinent information.

c. Overview displays were developed for each unit and one for the overall plant.

d. Subgraphics were developed where additional details were required.

e. Popup displays were used for all control devices

II. Faceplates

a. Standard faceplates were used

b. Faceplates were grouped to enhance operability

III. Status

a. Alarms were grouped by system

b. Annunciator type templates were used

IV. Trends

a. Trends were grouped by pieces of equipment

b. Fast and slow trends were not grouped together

System alarms were established by using the existing plant computer alarms and SOE alarms as a basis. All active existing alarms were retained because many plant procedures are initiated by a specific alarm. Again, wherever possible the alarms were identical between units.

A hierarchy was established to make the 'navigation' of the system easier for the operators. It provided a consistent 'roadmap' through each process system and was provided with numerous quick keys to minimize the number of keystrokes that were required to access any individual display.

A review was held at the plant for the graphics and alarms with FPC engineering and plant personnel. The plant personnel included operators, shift supervisors, and management which provide different views of the plant. Their comments helped develop a comprehensive set of control screens that should require a minimum number of changes. Also, a cooperative attitude was established by opening up this phase of the project to all levels in the plant.

CONCLUSIONS

This project was fairly normal for the replacement of this equipment, however the extensive use of databases to automate the design process greatly facilitated the project by providing a cost effective means for processing and presenting information to the client and DCS vendor. The initial effort required to create the database was more than recovered by:

- Permitting changes to be integrated cost effectively.
- Providing a means of checking the design without checking every piece of data.
- Providing a base to prepare documents.

REFERENCES

1. Gebhart, C. & Sensenig, R., "Interfacing a Modern Plant Control System with a Corporate Information Network," presented at the 1994 ISA Power Conference.

ANCLOTE UNITS 1 & 2
DCS NETWORK

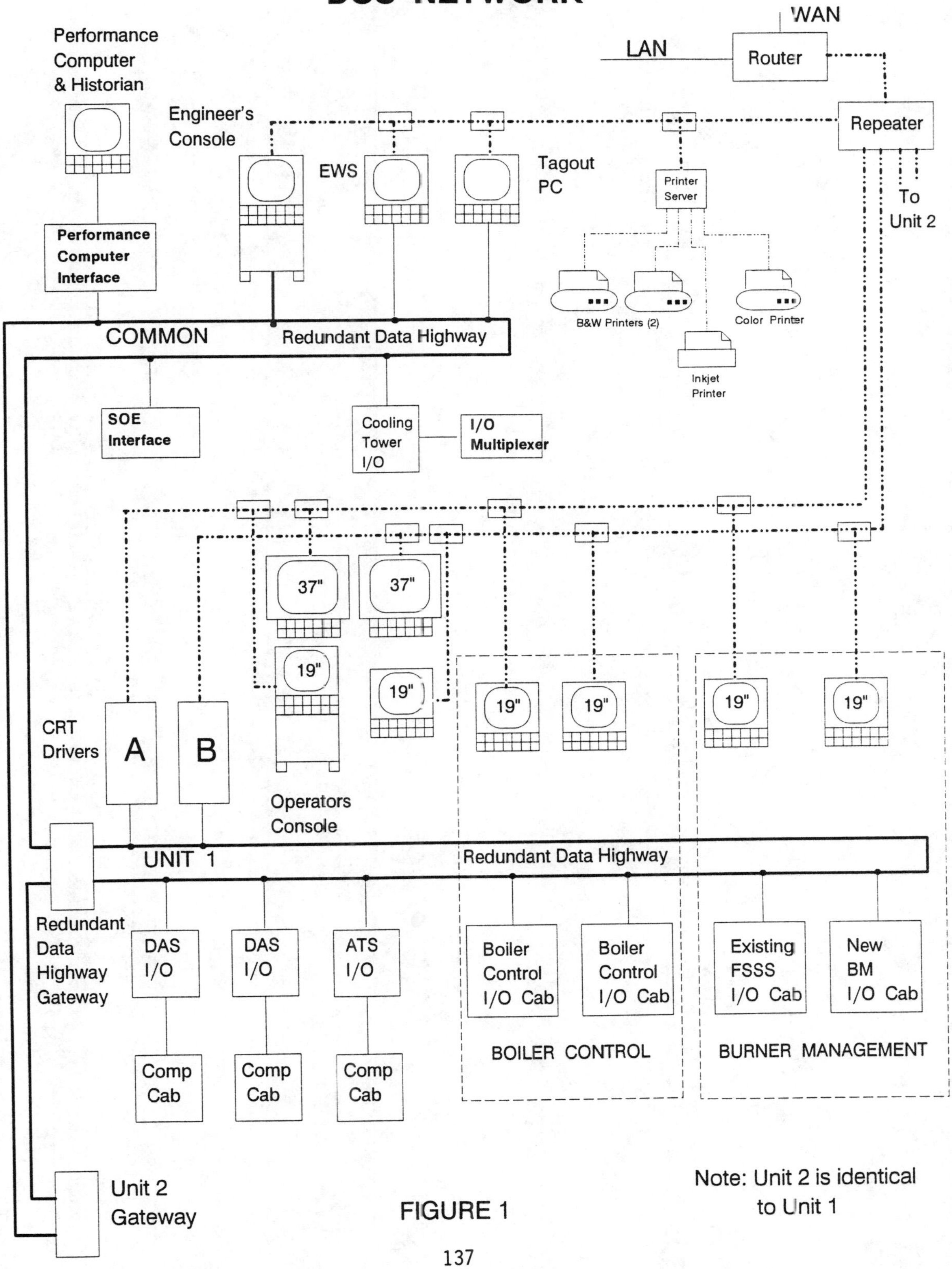

FIGURE 1

)074-056X/95/139-150/$0 + .50pp

SUCCESSFUL DCS UPGRADE EMPHASIZING OPERATOR INTERFACE AND SIMULATOR TRAINING - 1995 ISA POWID/EPRI CONTROLS CONFERENCE -

Robert C. Szczerbicki, PE
Project Engineering Manager
Gilbert/Commonwealth, Inc.
Reading, Pennsylvania

Robert J. Martin
Project Control Systems Engineer
Gilbert/Commonwealth, Inc.
Reading, Pennsylvania

Richard A. Brill, III
Control Systems Engineer
Gilbert/Commonwealth, Inc.
Reading, Pennsylvania

Walter C. Lamm
I&C Project Engineer
Pennsylvania Power & Light
Allentown, Pennsylvania

KEYWORDS

Control Upgrade, Low NO_X Burner, Operator Interface, Human Factors, Graphics, Simulator, Operator Training

ABSTRACT

The paper covers the control system replacement project at PP&L's Brunner Island Unit 3. Although field equipment replacement and Low NO_X burner and air system upgrades were significant aspects of this project as well, the paper deals primarily with the evaluation of operator interface alternatives, human factors evaluation of the control room and CRT console, graphics design, simulator development, and pre-outage operator training on the simulator. Key elements integral to the success of the project are identified.

PROJECT BACKGROUND

Brunner Island Steam Electric Station Unit 3, located in York Haven, Pa, is a Combustion Engineering, once-through, supercritical, pressurized, tangential-fired divided furnace boiler with combined circulation. It is rated at 5,700,000 PPH, 3,830 psig and 1,010^{0} F main steam and 1,010^{0} reheat steam. Six coal mills are Combustion Engineering No. 843, Raymond RPS, with a capacity of 125,000 lb/hr. Each is fed by a Stock Equipment Company rotary volumetric feeder having a capacity of 70 TPH max., 5.2 TPH min.

The turbine is a General Electric, tandem compound type, rated at 730,000 KW (summer) and 740,000 (winter). Steam temperature rating is 1,000^{0} F and pressure rating is 3,500 psig. General Electric also supplied the generator which is rated at 832,000 kVA, 95% P.F., 60 Hz, 3,600 rpm. Terminal voltage is 24,000 volts, with hydrogen cooled rotor and water cooled stator.

Two boiler feed pumps are Worthington Corporation 14-WNC 156 six stage, barrel casing type, with speeds of 5,500 rpm max., and 4,850 rpm min. Their maximum capacity is 6,750 gpm at 10,800 ft. TDH. Each is driven by a General Electric DRVM, 6 impulse stage, condensing turbine rated at 19,550 hp.

The unit is equipped with two Westinghouse (Sturtevant) double well, double inlet, 780,000 cfm forced draft fans and two axial flow 1,400,000 cfm induced draft booster fans, supplied by TLT-Babcock, Inc.

The two air heaters are rotating, regenerative type VI Lungstrom, plus two Aerofin Corp. Preheaters.

Brunner Island Unit 3 was put in service in 1972 with an electronic analog boiler control system, a solid state burner management system, hardwired relay logic for discrete control, and a computer based data acquisition system. A central control room with all operator interface through hardwired control boards was provided.

The intent of this controls replacement project was to replace a Leeds & Northrup (L&N) Direct Energy Balance (DEB) coordinated control system with a microprocessor-based distributed control system. The replacement system was to feature state-of-the-art equipment and a CRT-based man to machine interface (MMI). A control system training simulator was to be included in the scope of the work. Attention was to be given to revising or altering the existing control system functionality to meet applicable codes and standards. Also, dynamic rate limiting functions were to be added as an enhancement. The project was justified due to the fact that the existing control system was no longer manufactured, nor was it supported by the original vendor. It had reached the end of its useful life. A few spare parts were available from other users as they retired their systems, but these parts were aged and had a limited life-span.

In addition, the existing control system did not meet its functional intent. Operator interface with this system was provided primarily through control board mounted Manual/Automatic (M/A) stations. A number of control loops were slow to respond to control commands, did not perform within their design tolerance or became unstable and over-reacted to upset conditions when in automatic mode. Most of the field mounted sensors were obsolete, with output signals which are no longer used in the instrument industry; 0-4 maDC, 1-5maDC, 10-50maDC, etc. These had to be replaced to be compatible with a replacement control system. The new distributed control system was to provide increased functionality, responsiveness and operating efficiency. New cables were planned for new equipment because of new equipment locations, new signal requirements, added signals and the deteriorated condition of existing cables.

The original Burner Management System (BMS) was a high density solid state electronic system supplied as part of the boiler contract. Operator interface was provided through BTG control board mounted switches and indicating lights. This system was replaced as part of the Low NO_x Burner (LNB) project during the same outage that the analog boiler control system was replaced. The new BMS was implemented in DCS hardware in order to provide full integration with the boiler control and data acquisition systems. This allowed control function interface between the boiler control system and the new BMS to utilize the system's redundant communications highway instead of hard-wired signals. It was also possible to integrate control functionality on common CRT graphic displays.

The remaining discrete control consisted of remote control interface with various switchgear and Motor Control Centers (MCCs). Process interlocking was provided through hardwired relay logic. Operator interface was provided through control board mounted switches, indicating lights and ammeters.

Operator interface with the turbine control system was provided through hardwired control elements on the BTG control board. Control was provided for the stop valve bypass valve, governor control, load limit governor, turning gear, oil pumps and extraction valves.

Similarly, hardwired control elements were provided for operator interface with the generator, exciter, main transformer, and auxiliary electric systems. Control of the generator and exciter breakers, exciter voltage and generator synchronization, as well as monitoring of system current, voltage and VARs was provided.

Several other control boards in the control room provided interface with plant auxiliary systems. The environmental panel provided remote control and monitoring of the precipitators and monitoring of flue gas SO_2, CO_2 and opacity. The boiler auxiliaries panel provided control and monitoring of the boiler sootblowing system, precipitator fly ash handling system, and mill pyrites removal system.

The existing alarm system utilized a combination of hardwired annunciators and CRT alarm summary displays provided via the data acquisition system. The BTG control board had five annunciator boxes for the boiler, feedwater, turbine, generator, and miscellaneous alarms. The boiler auxiliaries panel had an annunciator for the precipitator fly ash system and the auxiliary electrical board had an annunciator for associated electrical system alarms.

Initial planning for the control system replacement began in 1992, when the Bailey 1055 data acquisition system was replaced with a micro-processor based distributed data acquisition system. This system paved the way for a distributed system as opposed to a mainframe type of architecture. Operator interface with the new system was provided through two CRTs and one keyboard mounted in the second extension of the Boiler-Turbine-Generator (BTG) control board. One of these CRTs is utilized for alarm summary display and the other displays the operator's current interest. Also, one CRT and keyboard are provided on the operator's desk.

Westinghouse Electric Corporation, Process Control Division, was chosen to provide the control system. They had supplied the Unit 3 distributed data acquisition system, and had, in turn, supplied the replacement boiler control systems for both Units 1 & 2 at Brunner Island. It had been a corporate decision to retain the control board manual/auto stations for all control systems upgrades in order to have an extra margin of safety in the unlikely event that the distributed control system should fail completely.

In 1992, PP&L began a massive program to bring all fossil burning units into compliance with the clean air act by installing Low NO_x burners and burner management control systems.

Gilbert/Commonwealth was contracted to assist in the Brunner Island SES Unit 3 Controls Replacement and Low No_x projects. Staffing for the project began in 1993 in a satellite office in downtown Allentown, just one half block away from the PP&L engineering office. The A/E was responsible for engineering, design, drafting, project management, factory test, construction and start-up support. This work covered both the burner management and boiler control systems which were to be installed during the same fall of 1994 outage.

OPERATOR INTERFACE STUDY

Since the analog combustion control system, the burner management system, and the existing data acquisition system would be integrated into a common distributed control system, PP&L authorized a study of operator interface options. The objective of the operator interface study was to identify the most cost-effective, efficient, functional, and utilitarian operator interface design basis utilizing currently available operator interface technology.

While traditional Manual/Automatic (M/A) stations, switches, pushbuttons and indicating lights could have been retained, there were inherent advantages to providing operator interface exclusively through the CRTs. Future additions and revisions to the system would be less costly because no hard control elements would need to be installed and wired in the control boards. Revision to or addition of CRT graphics would involve software changes only. Calibration and maintenance of control board gauges and recorders would be eliminated reducing maintenance costs. If a CRT console component failed, the failed module or CRT could be replaced reducing repair time. Also, graphic screens could integrate information from various subsystems to provide a more effective operator interface.

The dominant factor in considering revisions to the existing operator interface was available space. Minimal space was available to add new control elements to the existing BTG control board. The existing control room layout left no usable room for additional control boards without relocating the existing control boards. Three options were defined and evaluated for operator interface with the DCS equipment:

1. Hard Panel Operator Interface - Existing M/A stations and burner control insert panel would be replaced with new hard panel equipment. All other existing hard panel equipment would be retained with minor modifications required to some elements such as the annunciators. No changes to the existing CRTs and keyboards would be provided.

2. Hybrid Operator Interface - Some of the existing hard panel control elements would be eliminated and those control functions provided exclusively through new CRTs installed in the existing BTG control board. The eliminated hard panel control elements would include fuel oil system controls, M/A stations not directly associated with a field actuator, and many of the existing recorders. The existing data acquisition system CRTs and electronics would be replaced with newer high resolution CRTs and workstation windows environment. Two additional workstations with four CRTs and two keyboards would be installed in the main section of the BTG control board.

3. CRT Operator Interface - The existing BTG control board including its extensions would be eliminated except for the rear section behind the walk-through which contains electrical protective relaying, electrical metering and generator synchronization. All control and monitoring functions currently provided on the BTG control board front panels would be provided exclusively through a new CRT console installed in front of the remaining rear section of the BTG control board. This console would include eight CRTs and four keyboards, and four relocated annunciator boxes. In order to fit the new CRT console into the control room given the arrangement of the remaining control boards, the environmental monitoring panel and boiler auxiliaries would also be eliminated. The control and monitoring functions provided on these control panels would also be provided exclusively through the new CRT console.

Option three was recommended. The future of operator interface is CRTs. It would be much less costly to incorporate the transition to a highly effective CRT operator interface simultaneously with the replacement of the boiler control and burner management systems than at some point in the future. With the inclusion of the environmental and boiler auxiliaries panels, the control room could be configured to provide the most effective CRT operator interface configuration.

Utilizing CRTs exclusively for operator interface had the additional advantage of reducing the cost of future system additions, reducing maintenance costs because of the elimination of the many control elements on the existing hard panels, and providing information integration enabling the operator to view all related data in a single graphic screen. A further recommendation of the operator interface study was to perform a detailed human factors evaluation of possible control room configurations and CRT console configurations.

After a great deal of discussion and analysis, the decision was made to change to CRT control and monitoring. The only exceptions were the electrical auxiliary control board and the turbine synchronization control.

HUMAN FACTORS EVALUATION

The human factors evaluation provided guidelines for the operator's console and desk design, control room layout and environment, and graphic screen building. The study was performed by reviewing existing operating procedures, observing control room activities, and interviewing plant operating personnel. Ergonomic guidelines were incorporated and final recommendations made.

The equipment specification requirement for eight CRT monitors in the control room was the critical design element of the operator's console. An arched "matrix" of two high and four across was recommended based on primary single-person operation with accommodations for additional operators during critical periods. See Figure 1. To aid in accommodating more than one operator and additional operator work positions, full swivel and tilt adjustability were recommended for all screens. Trackball devices were recommended over both touchscreen and keyboard. The extensive use of windows eliminated touchscreen due to the possibility of a "lingering digit" activating through one or more windows. Graphics were designed to allow access to all keyboard functions allowing the keyboards to be relegated to a drawer freeing up desk space. Dimensional recommendations were made for monitor heights, separation and viewing distance. The additional task requirements of the operating personnel (permit and tagging activities and non-DCS instrumentation monitoring) were also considered in determining the console and desk area configuration. The operator desk work surface was recommended as an extension of the monitor console. The desk and monitor console were designed as an integral unit with minimal underside obstructions to allow the seated operators to move freely from one location to another.

The recommended control room arrangement was based on operator activities and projected traffic patterns. See Figure 2. The only limiting factors of the control room arrangement were the existing BTG termination board and the Auxiliary Electric Board (AEB). Operator interface between the new control monitors and the remaining hard control panel (AEB) needed to be considered in the final control room arrangement. The layout attempted to restrict traffic flow into the main operator area. The arrangement provided more lay-down areas for operator permit and tagging activities while keeping the operator close to the control console monitors while performing these activities.

Human factors design guidelines such as ANSI/HFS 100-1988, NUREG 0-700, and MIL-STD-1472C, combined with observational data and interview results made up the basis for the control room environmental and aesthetic recommendations. Replacement of existing lighting fixtures with an adjustable lighting control system was recommended. This would supply an adequate illumination level at key areas of the control room and would allow operator adjustment of area brightness. Reduction of glare on the monitors and panel mounted instruments was also a consideration. It was recommended that noise levels within the control room be attenuated through the use of wall, ceiling and floor insulations that have known acoustical attenuation properties. Flooring insulations were recommended which would also attenuate vibration as much as possible. Wall and floor covering color and surface finish recommendations were general in nature allowing the control room architect to select final colors and treatments.

Operator task requirements, personnel interviews and design guidelines (ANSI/HFS 100-1988) were used as a basis for the graphic displays recommendations. Limited color coding and retention of existing PP&L color coding were recommended. The requirement that colors be reliably discriminated led to the suggestion that between 7 and 10 different colors be used. Sizes for alphanumerics, graphic lines, and symbols were recommended based on viewing distance. The study recommended the use of iconic symbols over either abstract symbol types (simple geometric shapes) or pictorial symbol types (increased level of detail). The study made several recommendations concerning the design of screen control elements (faceplates and M/A stations). The study advised building standardized control elements and being consistent when highlighting key information areas and in arranging control element components. Recommendations concerning size, orientation, functionality and opening/closing of control elements were made and implemented.

GRAPHIC DISPLAYS

The ability of the operator to obtain information and interact with the system to perform control functions is a critical feature of a DCS man-machine interface. The development of the graphic displays affects this important interface issue. The graphic display building effort was a three step process. First a design basis document was developed which outlined the general graphic building rules and philosophies. The document was reviewed and approved by PP&L. Second, the graphics were laid out and programmed. Periodic reviews with operations personnel were conducted and comments incorporated. Finally, the graphics were tested during factory acceptance testing and fine tuned and revised as needed during the plant start-up.

The purpose of the graphic building design basis document was to detail the general philosophies and rules to be used in an effort to retain consistency between all graphics. This document was modified as needed at the end of the project so it could be used as a guide by plant personnel to build graphic displays in the future which would be consistent with the displays built previously. The design basis document covered such items as: type of graphics to be built, organization, color coding, symbols, alphanumeric sizing and conditionals, faceplate design, and manual/auto station (M/A) design.

It was decided to use process type (P&ID based) graphic displays as the main means of displaying information for operator information and action. This decision was based upon the operators extensive knowledge of the plants physical configuration and the plant equipment functions. Also it was agreed that these types of graphics would allow the operator to extract information quicker by placing analog values in or near their actual process measurement location, thus reducing the need for text description. Other graphic display types were used to supplement the process type graphics including performance graphics, annunciator graphics, and windows for control and help functions.

The graphic hierarchy consisted of five major process groups plus six auxiliary system graphics. Refer to Figure 3 for the details. An overview graphic was developed for each major process group. Organized within each major system was a set of detailed graphics displaying information pertaining to individual components and/or systems. The hierarchy layout was built into a menu type graphic from which the operator could access any process graphic display. See Figure 3 for actual graphic. There were a total of 87 process graphics.

Each process graphic was built from the same standard skeleton. Thus the following features were typical of all process graphics:

- size and shape
- title in upper left corner (title acted as poke target to access the hierarchy/menu graphic)
- direct graphic accessing poke targets in upper right hand corner
- black background
- menu of critical plant parameters and alarm acknowledge and silence pushbuttons in lower right
- blank space on right side for control window pop-up

Color coding for the process graphics was based on existing color coding used by PP&L, since the company uses operators who may work in more than one control room. The color coding is as follows:

SYMBOLS, TEXT, VALUES

COLOR	USAGE
GREY	Static symbol
GREEN	alarms
MAGENTA	bad quality/hardware failure
CYAN	device in auto
RED	on, open, or breaker closed
GOLD	off, closed, or breaker open
OFF WHITE	device in manual
BLACK on WHITE	dynamic values

PROCESS LINES

COLOR	USAGE
LIGHT BLUE	steam
BLUE	water
ORANGE	oil
LIGHT ORANGE	gas
DARK GREY	coal
WHITE	air

The use of industry accepted symbols was employed so that they could be easily recognized by the operators. The symbols were dynamic in color only to indicate the operating status of the device. Discrete devices used color coding to differentiate between on/off (for pumps, fans, mills, etc.) and open/close/mid-position status (valves, and dampers). Analog devices used color coding to differentiate between auto/manual status. Pump, fan, and mill bases and valve actuator symbols were used to indicate if a device was in alarm. Since the use of blinking was discouraged within the vendor's system, a filled symbol versus an unfilled symbol was used to differentiate between acknowledged and unacknowledged alarms.

Analog values and text alarms were shown on a green highlighted background if in alarm and in reverse video once the alarm was acknowledged. A magenta background was used to show a point which was reading bad quality or hardware failure. Text alarms would not appear on the display unless they were in alarm or had bad quality. Use of static text as labels and process identifiers was limited to places where it was needed to add clarity and was built without a highlighted background.

Discrete faceplates were built to accommodate 2, 3, or 4 control device pushbuttons used to operate the equipment or place it into a certain operational mode (auto, jog, standby, etc.). The pushbuttons were backlit so that operators could easily assess the current status of each device. Start, stop, open, and close pushbuttons each contained a "ready" indication which would inform the operator that all permissives and interlocks were satisfied to perform that operation on the device. The border around each pushbutton would illuminate to indicate that the poke target had been successfully activated by the operator and a signal had been output from the system. A text string at the bottom of the face plate was used to identify active alarms associated with that piece of equipment. Typical digital logic for plant switchgear, motor control center devices, and motor operated valves was developed in conjunction with the discrete faceplates. This approach assured that all points required to drive the faceplates were developed within the logic. Also, this standardized approach was beneficial to the programming effort since only a few typical faceplates needed to be built. The poke targets used to call-up the faceplates were modified for each piece of equipment.

This same standardized approach was used in the development of the manual/auto (M/A) stations. Several typical M/A stations were developed consisting of both sliding bars and digital indication to show process variable, setpoint, demand, feedback, and/or bias values. In addition, auto and manual select pushbuttons and increase and decrease pushbuttons for setpoint and demand control were included. Green color coding of elements within the M/A station was used to indicate manual reject, priority raise/lower, and raise/lower inhibit states.

Help windows were developed for all discrete devices. These windows listed all the permissive and interlocks associated with a particular device. The text, listing the permissives and interlocks, was backlit to indicate if each item was true/on or false/off. When an operator attempted to operate a device from its faceplate and a "ready" indication (as discussed above) was not active, he could call up the device help window to see which permissive or interlock was not met.

Annunciator light box (ALB) graphics were developed to eliminate the hardwired annunciators from the control room. Operations personnel assigned priorities of 0, 1, or 2 (0 = lowest and 2 = highest) to all the existing alarms on the hardwired annunciators. Priority zero alarms were designated to the standard alarm summary screen only. A custom alarm package was developed for the higher two priority levels of alarms. An alarm status bar consisting of six icons indicated if a high priority alarm was present in any one of six plant areas (boiler, turbine, generator, feedwater, fuel or miscellaneous). The operator then selected that icon and the appropriate ALB (annunciator lampbox mimic) was displayed. It would show what alarm(s) in the area were active. By clicking on the illuminated box, the appropriate process graphic was displayed so the operator could further assess the situation.

OPERATOR TRAINING SIMULATOR

PP&L has had an operator training simulator for Brunner Island Unit 3 and its sister units, Montour Units 1 and 2, for many years. The simulator used CRTs to emulate the BTG control board. With a CRT based control room decided upon, it made the design of the training simulator easier. A control board mockup would not be necessary. An array of CRTs driven by standard control system vendor hardware and stimulated by several Personal Computers would be used. A separate training room to house the simulator was needed. A new simulator/training room was constructed adjacent to the control room.

Since the new controls would be based on a CRT operator interface and common DCS, it was decided to utilize the DCS hardware and software for the simulator control system and operator interface. The DCS vendor had previously developed an interface with the simulation software vendor running on personal computers. PP&L contracted configuration of the simulator model to the DCS vendor.

All control system input and output points were simulated with derivation either from the plant model or from structured simulator logic reacting to DCS commands or Instructor overrides. All DCS graphics were provided with all dynamic data updated by the model or simulator logic. All control functions not provided through the DCS CRT graphics that are an integral part of the unit startup, shutdown, or normal operating procedures were provided in the simulator model. Model calculations and logic execute a minimum of once a second.

Operator interface with the auxiliary electric control board was provided by CRT hard panel mimics configured as part of the simulator. These include generator field breaker control, generator synchronization, generator main breaker control, and auxiliary electric systems controls. The CRT mimics were created by scanning pictures of the existing control boards into software and adding poke fields for initiating operator control actions.

Instructor functions included on-line programmable malfunctions, freeze/continue, snapshot, snapshot download and parameter listing. The Instructor can download any previously saved snapshot condition, which will initialize the model and control system to the operating condition at the time the snapshot was taken.

The simulator allows the Instructor to initiate system and process disturbances. The following types of disturbances were included:

- Equipment Trip/Failure to Operate
- Transmitter/Sensor Failure
- Process Variable Deviation/Excursion
- Loss of Drive Unit Motive Power
- Failure of a DCS Data Station
- Failure of a DCS Operator CRT Station
- Fuel BTU Deviation
- Equipment Malfunction
- Load Dispatch Requirements
- Mill Fire
- Boiler Tube leak (Minimal or Severe)

See Figure 4 for the Simulator/Operator Training Room Layout

CONCLUSION

Although the initial start-up was hampered by several mechanical problems (boiler tube leaks, condenser leaks, valve packing leaks, etc.), the control and monitoring system performed as expected. After initial tuning, calibration, and ranging, all modulating control loops are in Automatic and the unit responds to Load Dispatch requests in a Coordinated Control Mode. The Start-up Feedwater Transfer Ramp is executed automatically for the first time. Future activities are focused on implementing operator suggested enhancements to the graphics and upgrading the simulator (both model and control system) based on actual plant and control system performance data.

The project was completed within the projected schedule, but slightly over budget. There were many changes throughout the life of the project which improved the unit's performance and operability. Many of the changes were recommended by a joint effort of G/C and PP&L operations personnel. The extensive modifications associated with the Low No_x Burner installation have precluded a detailed analytical evaluation of unit performance on a before and after basis.

REFERENCES

[1] Shirley, R. S., A Fog Index for CRT Displays, ISA Power Symposium, 1986.

[2] Szczerbicki, R. C., Martin, R. J., Keiderling, R., and Miller, R., Successful DCS Upgrade Emphasizing Operator CRT Interface Design and Control Room Modifications, ISA/EPRI Control Conference, 1993.

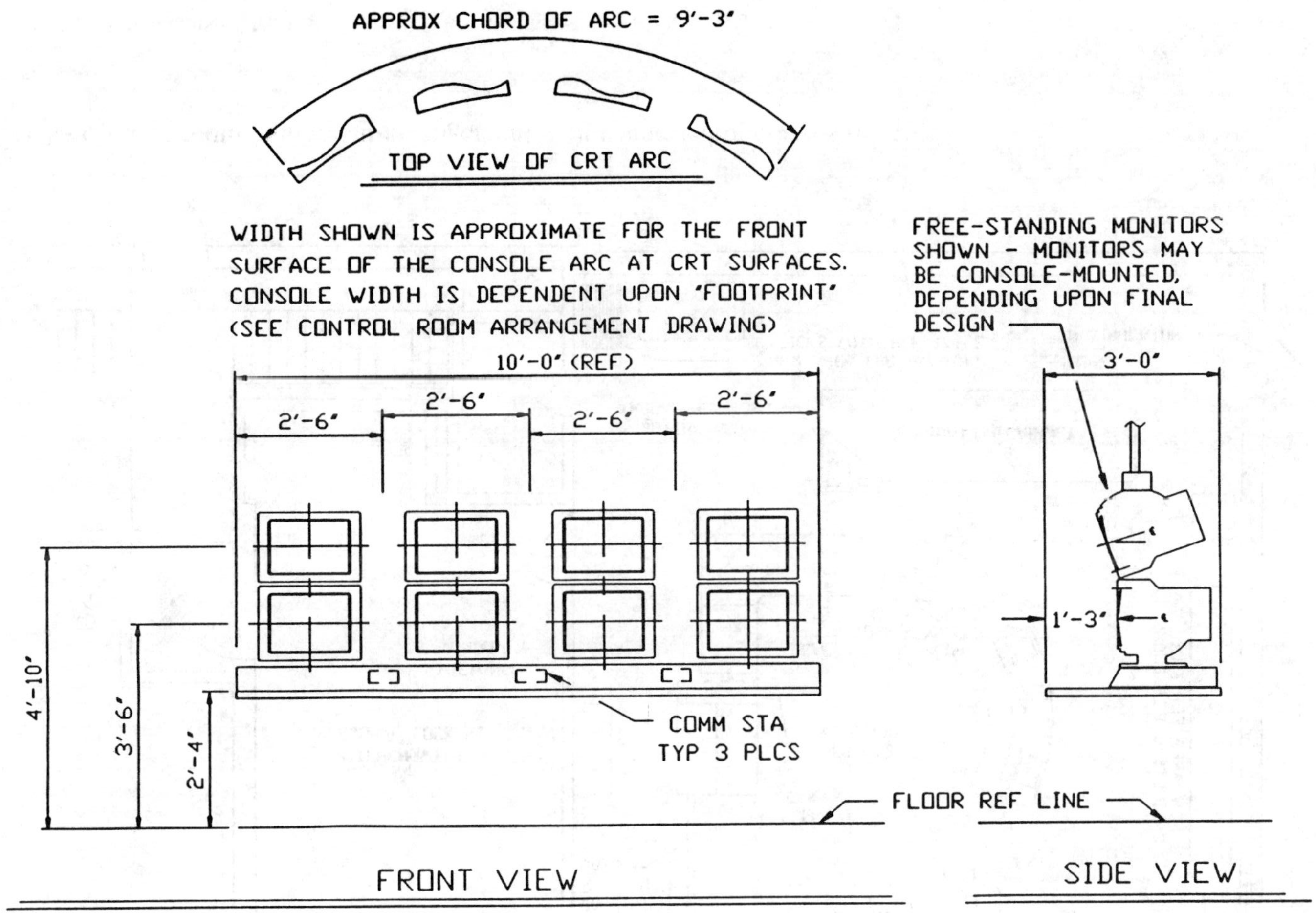

Figure 1 - Monitor arrangement for the DCS control console.

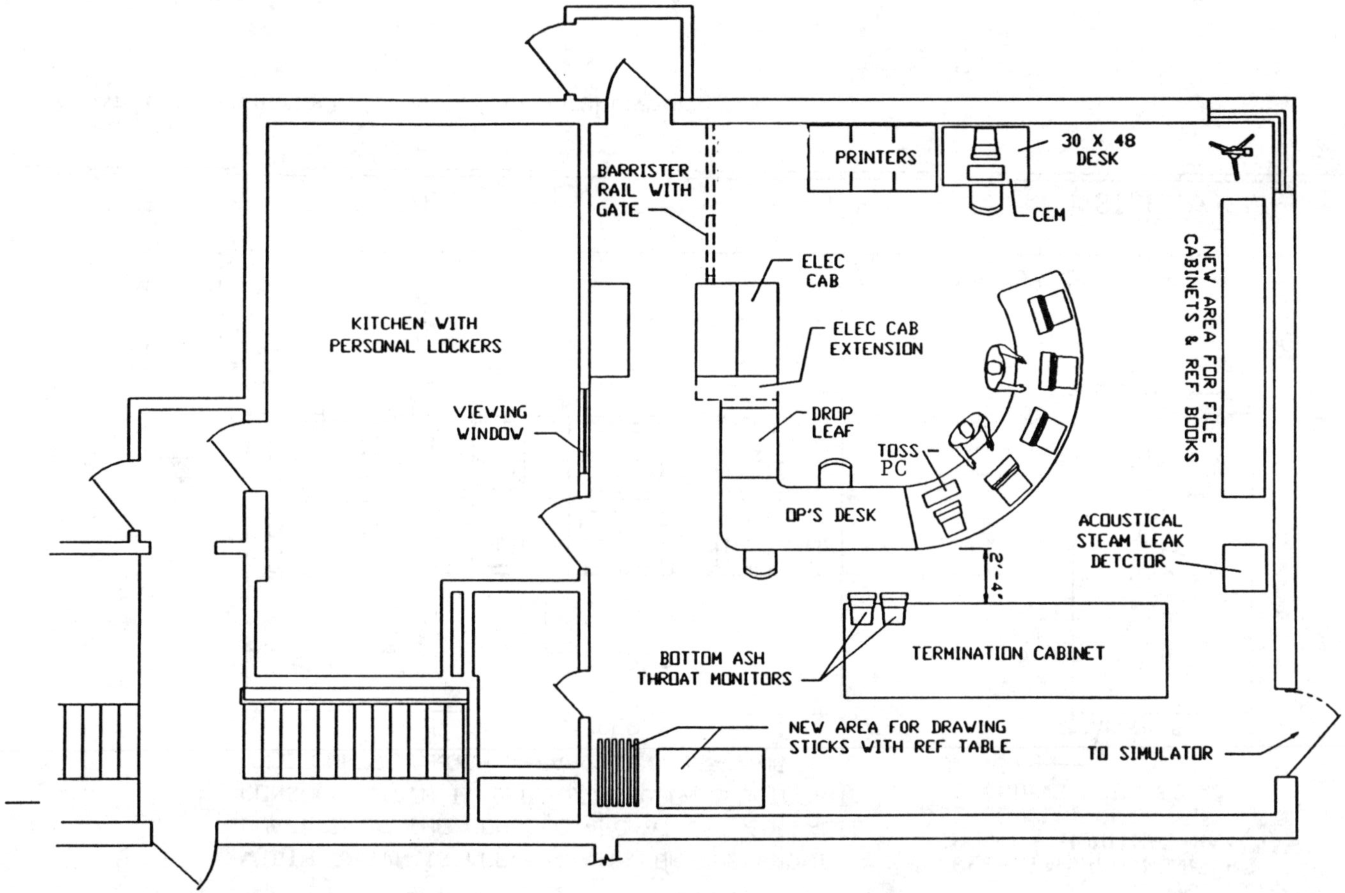

Figure 2 - Control room equipment arrangement with connected operator's desk.

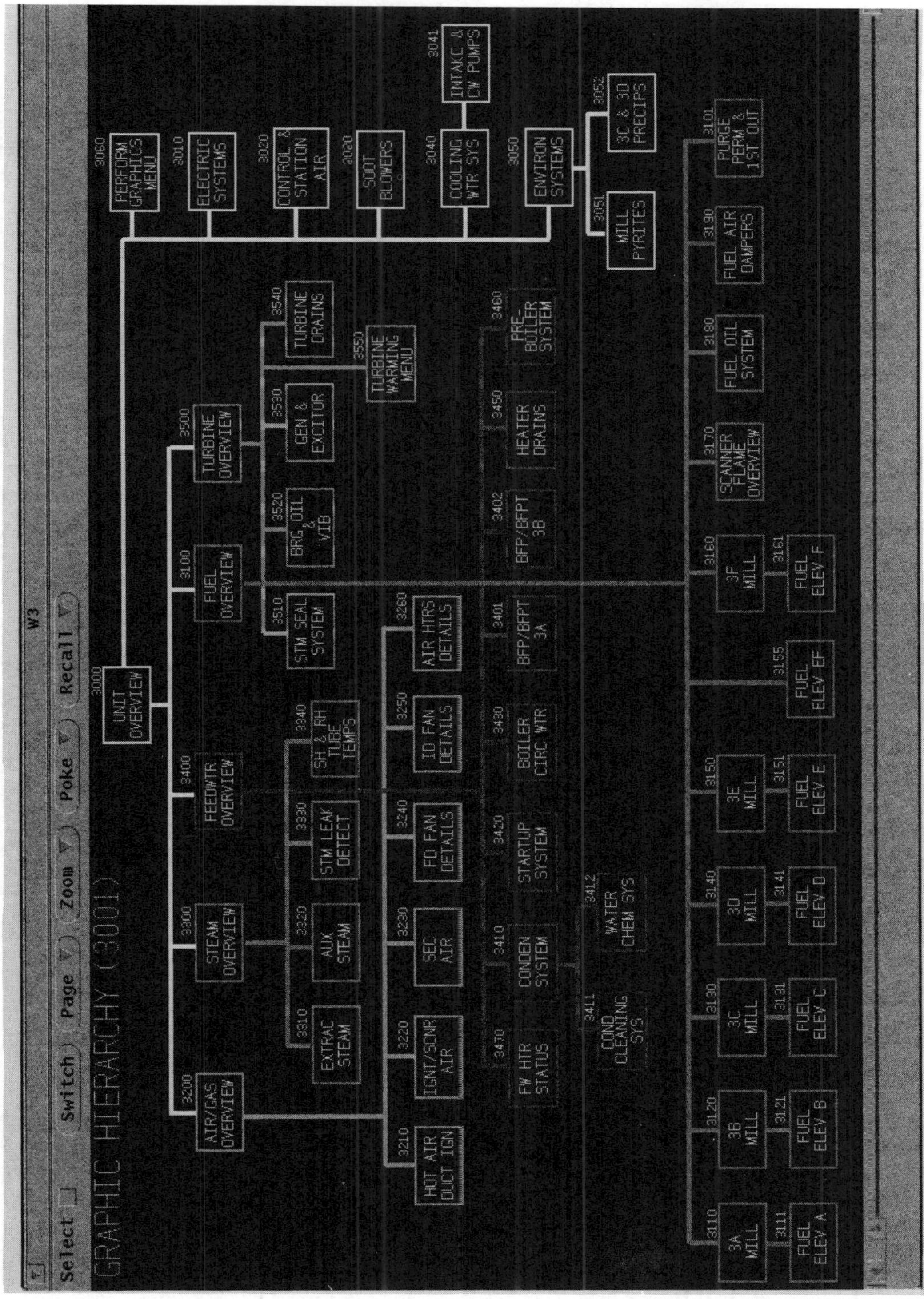

W3
Select
Switch
Page
Zoom
Poke
Recall
GRAPHIC HIERARCHY (3001)
3000 UNIT OVERVIEW
3060 PERFORM GRAPHICS MENU
3010 ELECTRIC SYSTEMS
3020 CONTROL & STATION AIR
3030 SOOT BLOWERS
3040 COOLING WTR SYS
3041 INTAKE & CW PUMPS
3050 ENVIRON SYSTEMS
3051 MILL PYRITES
3052 3C & 3D PRECIPS
3200 AIR/GAS OVERVIEW
3300 STEAM OVERVIEW
3400 FEEDWTR OVERVIEW
3100 FUEL OVERVIEW
3500 TURBINE OVERVIEW
3510 STM SEAL SYSTEM
3520 BRG OIL & VIB
3530 GEN & EXCITOR
3540 TURBINE DRAINS
3550 TURBINE WARMING MENU
3310 EXTRAC STEAM
3320 AUX STEAM
3330 STM LEAK DETECT
3340 SH & RH TUBE TEMPS
3210 HOT AIR DUCT IGN
3220 IGNT/SCNR AIR
3230 SEC AIR
3240 FD FAN DETAILS
3250 ID FAN DETAILS
3260 AIR HTRS DETAILS
3470 FW HTR STATUS
3410 CONDEN SYSTEM
3411 COND CLEANING SYS
3412 WATER CHEM SYS
3420 STARTUP SYSTEM
3430 BOILER CIRC WTR
3401 BFP/BFPT 3A
3402 BFP/BFPT 3B
3450 HEATER DRAINS
3460 PRE-BOILER SYSTEM
3110 3A MILL
3111 FUEL ELEV A
3120 3B MILL
3121 FUEL ELEV B
3130 3C MILL
3131 FUEL ELEV C
3140 3D MILL
3141 FUEL ELEV D
3150 3E MILL
3151 FUEL ELEV E
3155 FUEL ELEV EF
3160 3F MILL
3161 FUEL ELEV F
3170 SCANNER FLAME OVERVIEW
3180 FUEL OIL SYSTEM
3190 FUEL AIR DAMPERS
3101 PURGE PERM & 1ST OUT

Figure 3

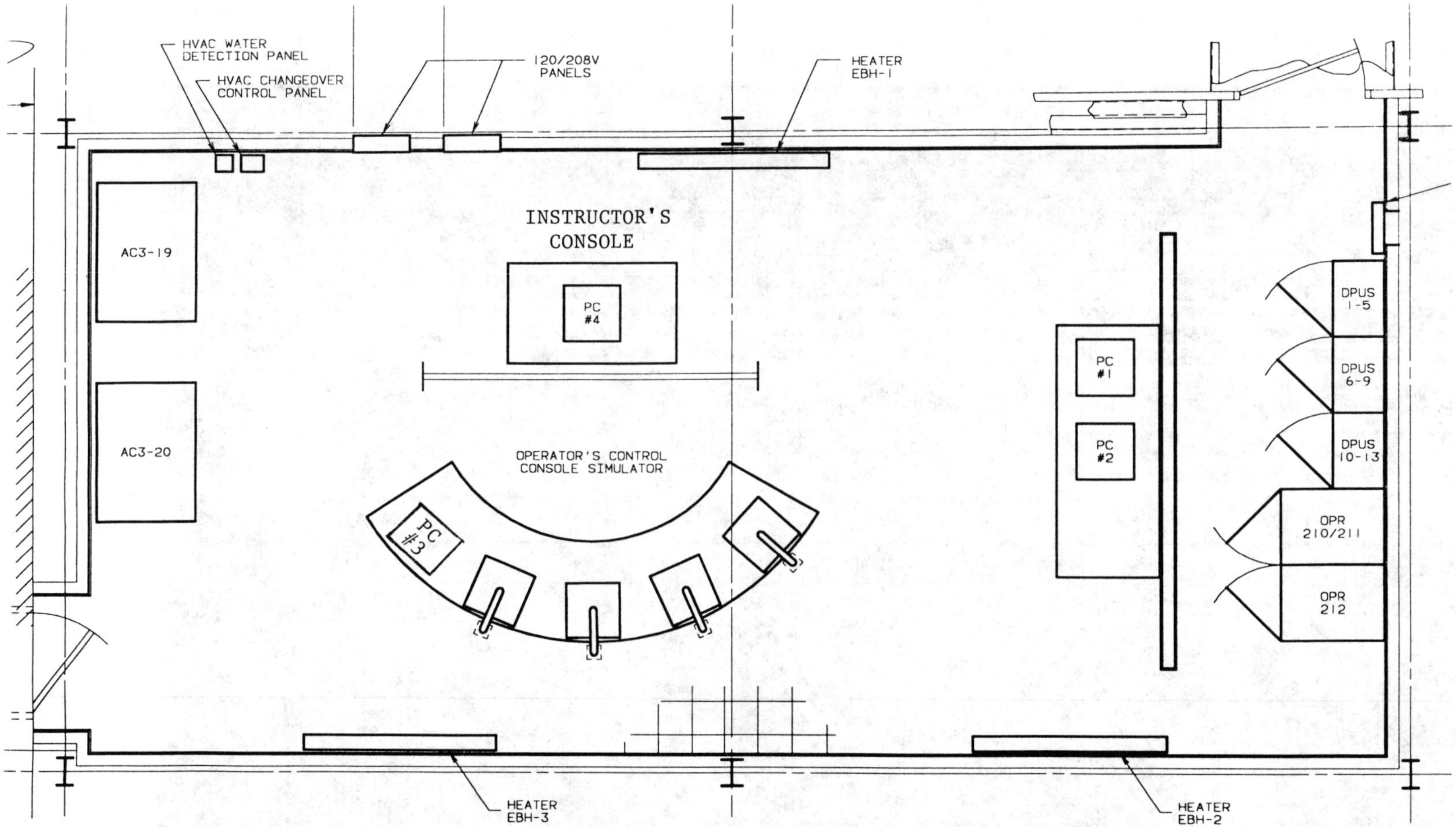

SIMULATOR ROOM
PLAN @ EL. 306'-0"

)074-056X/95/151-159/$0 + .50pp

ADVANTAGES OF DIGITAL GAS TURBINE CONTROLS UPGRADE

A. L. Suszynski
Manager of Electrical Engineering
Energy Services, Inc.
270 Farmington Avenue
Farmington, CT 06032

KEYWORDS

DCS, Turbine, Gas Turbine, Digital, Upgrade, Aero-Derivative, Original Control, Control System

ABSTRACT

Application of a Digital Control System (DCS) to upgrade existing gas turbine peaking plants has resulted in reliability and availability improvements. Turbine protection is improved because of digital software solutions to problems. Analog and digital input redundancy is attainable in a simplex control system. Automatic software sensor failure detection is only one of many available features. The programmable digital controls also permit use of a run-oriented control philosophy instead of the original shutdown-oriented control philosophy. Expanded alarm and event readouts provide definitive maintenance information and data collection.

In summary, if you can think of the control scheme that you need, it can be done digitally.

INTRODUCTION

Many gas turbine peaking plants which were put in operation from the 1960's through the 80's are now unreliable because of recurrent control system problems which are difficult to diagnose and repair. Many of the existing control system components are no longer manufactured or supported by the original manufacturer. Owners of these plants are interested in programs which will extend the life of these plants for another 20 years in order to take advantage of the existing plant infrastructure, but only if the plant reliability can be improved to an acceptable value.

The mechanical and electrical portions of the electrical peaking plant unit are comprised of a gas turbine engine of either the aero-derivative or frame machine type and an electric generator and a main step-up transformer. Multiple gas turbine units are common on the same plant site. The mechanical equipment is modular and repairable and can be restored to normal operating condition as needed through available refurbishment and maintenance programs.

The major consideration for a gas turbine peaking unit is that it be reliable and available when called for by the remote electrical power system load dispatcher. The plants were designed for fast start, within three minutes for aero-derivatives and within fifteen minutes for frame machines, to quickly satisfy the local electrical peaking demands. Typically no one is at the local turbine location. Many times, if the plant fails to start, the peak demand is over by the time someone goes to the local site and locates the problem. In the meantime, it was necessary to purchase more expensive peaking power to satisfy the peak demand. When the peaking units are unreliable it jeopardizes utility power pool availability credit.

The gas turbines are also used to provide emergency power during blackouts or natural disasters. They must also be reliable to fulfill this role.

RELIABILITY AND AVAILABILITY

The original turbine control systems did not have a very good record of reliability and availability even when they were new. This was due in part to analog control hardware problems and to the control design philosophy used. This has been further compounded by the lack of parts and service and the retirement of key personnel with the training and intuitive technical skills which are required to keep the original equipment serviceable.

The positive results of a digital control system (DCS) upgrade can be dramatic and immediate. Original equipment with reliability of 60 percent or less has been demonstrated to be better than 98 percent reliable in a step change. The outage required to complete the upgrade is typically less than a month. Once the control system problems which affect start reliability have been corrected, mechanical problems, which in some cases were masked by the old control system faults, become the center of attention. This should be anticipated and planned for and a reassessment of the mechanical equipment should be conducted using the automatic data acquisition capabilities of the new DCS after recommissioning is complete. Occasionally the mechanical problems will hinder DCS startup and recommissioning activities and require immediate attention.

TURBINE PROTECTION IMPROVEMENTS

DCS software can contain logic which completely replaces the analog fuel control, the relay logic sequencer and the chart recorder data collection functions of the original equipment in one highly integrated package with much simpler hardwire interconnections and with the addition of many new features. The typical fuel control and sequencing portion of the software logic within a DCS contains all of the following necessary turbine control and protection features:

- Variable speed references (for low compressor, high compressor and power turbine)
- Speed control PID loop (for low compressor, high compressor and power turbine)
- Load limiting biased by inlet air temperature schedules (for generator protection)
- Speed limiting biased by inlet air temperature (for turbine protection)
- Load sharing (for dual engine loading)
- Exhaust gas temperature limiting (for turbine protection)
- Exhaust gas temperature spread detection (for turbine protection)
- Exhaust gas temperature over-temperature (for turbine protection)
- Sensor signal overrides (for start-up)
- Starting fuel ramp (gas fuel, liquid fuel for turbine accel to idle)
- Speed switch points (for control of fans, pumps, starters, etc.)
- Isochronous control mode (for isolated operation)
- Droop control mode (for parallel operation)
- Water injection control (if required for emissions control)
- Synchronous condenser mode (for system voltage support)
- Megawatt load limiter (for generator protection)
- Megavar load limiter (for generator protection)
- Automatic VAR shed (for turbine stop sequence)
- Sequential timing logic for start and stop sequences (for turbine protection)
- Automatic on-line fuel transfer (gas and liquid fuel)
- Temperature, pressure, speed and flow data acquisition (for archives)

New software based power limits can be adjusted by the operator from the CRT using a pointing device such as a trackball or mouse to raise or lower the setpoint. The setpoint is digitally displayed on both the local and remote CRT's as well as actual megawatt output power and megavars. Automatic MVAR shed logic is implemented to unload reactive power (VARS) during a turbine stop sequence. This permits a more closely matched voltage on subsequent starts. These features were open loops with local only setpoints in the old system and are now closed loops with local or remote setpoints in the new system.

Dual fuel gas turbine units have more reliability problems than single fuel units because the fuel systems are much more complex. There are more parts to fail and accurate timing sequences are critical. Digital timer accuracy corrected these problems. Originally fuel transfers resulted in large generator power output excursions of fifty percent or more and the probability of tripping during the excursion was great. With digital controls, both fuel control valves can be controlled simultaneously instead of just one so that if a problem occurs during fuel transfer the control system can alarm and revert back to the original fuel or advance quickly to the oncoming fuel instead of tripping. The typical fuel transfer output power excursion is also reduced from 50 percent to about 5 percent. This is achieved partially by separate gain and compensation for each fuel loop.

The new engine start scheme design provides an accurate linear digital starting fuel ramp instead of the original analog time-constant ramp which is subject to drifting. The digital ramp is capable of increasing to a preset maximum value that assures that the engine achieves idle without over-temperature and without overspeed and without exceeding accel rate limits. The start ramp is linear and does not need to be readjusted from season to season. Liquid fuel starting problems were common with the original system during cold weather because of poor combustion of the cold fuel. This has not been a problem with the linear ramp. Gas fuel systems always had better start reliability than liquid fuel except when gas fuel pressure changed. The addition of pressure biasing using a gas supply pressure compensation loop corrected the problem.

Minimum power in the new design is an adjustable digital setpoint that is a repeatable part of a power control loop. The original system relied on pneumatic timers and watt transducer relays and a motor-operated potentiometer in an open loop configuration that was not repeatable. When minimum power is selected by the operator in the new system, the generator is held at a power output which is controlled by the new control loop. This reduces the chance of an anti-motoring trip while in the minimum power mode. With the old system single engine minimum power was not available with a two engine system. With the new system the minimum power set point is automatically selected to the correct value for single engine operation or for dual engine operation.

The new control scheme provides bumpless transfer when switching from isochronous to droop while operating isolated from the power grid. This is accomplished via digital tracking of the droop speed reference when operating in the isochronous mode. When the transfer from isochronous to droop is selected by the operator, the generator output power level has already caused the tracking circuit to preposition the generator power reference setpoint so the transition is bumpless. The original system could cause large step changes in generator frequency when shifting from droop to isochronous or vice versa while carrying an isolated load because there was no tracking at all. The old scheme created the possibility of tripping while running isolated from the power grid during an emergency situation. The new scheme prevents that problem automatically.

The new scheme uses digital synchronizer modules to automatically synchronize and close the generator main breaker and to backup the manual synchronizing procedure. The synchronizer matches generator (speed) frequency and generator voltage to bus frequency and bus voltage and provides an output signal to close the generator main circuit breaker. The "sync check" confirms that speed and voltage are in phase and it blocks the output of the automatic synchronizer if it is not. This new scheme replaces the existing analog auto synchronizer and sync check relays which are no longer manufactured. The digital synchronizing system is more accurate and repeatable because the DCS system can respond to very small, precise setpoint change commands from the new digital auto synchronizing system which are proportional to the deviation between setpoint and actual phase angle.

Load share between two gas turbines driving the same electric generator is accomplished by digital speed matching to a common reference setpoint instead of two separate analog control setpoints. The DCS load share scheme accurately matches both engine rotor speeds or compressor discharge pressures from engine idle through turbine load limit. Separate trimming and adjusting is not required. Because the engine speed references are digital they change setpoint simultaneously and do not drift apart. Compressor discharge pressure load sharing is used as primary control and engine rotor speed load sharing becomes a backup to it. Accurate load sharing is particularly important for dual engine units which exhaust into the same power turbine (expander) to prevent too high a back pressure to either engine.

Flameout detection is important for engines that use water injection because of the increased possibility that the water which is mixed with the fuel would put the flame out. The flameout detector logic detects flameout and trips the engine within one second. The danger of flameout is that the unburned fuel could cause an explosion in the exhaust stack. The flameout detector logic provided in the DCS system is high speed and it is free from false shutdowns which result from normal load excursions such as manual loading and unloading, setpoint step changes and other power excursions such as generator breaker opening and system load rejection. A 10 millisecond DCS input scan rate provides sufficient speed to operate this loop and other high speed loops such as rotor overspeeds. Other inputs which are slow to change can be scanned at slower scan rates typically 20 to 100 milliseconds.

The DCS is DC powered for black start capability. The turbine and its auxiliaries are designed for black start and can be started using station batteries as the only source of power. Upgrading to a DCS has actually reduced the DC power consumed and therefore improved the possibility of a successful black start by making more power available to the DC pump motors.

ANALOG AND DIGITAL INPUT REDUNDANCY

The DCS equipment and system design standardizes the peaker control system and instrumentation to eliminate the many control system design variations which can exist from unit to unit. Even on the same plant site, original equipment can vary drastically in type, input/output (I/O) location and sequential control logic schemes. This standardization is especially helpful when the turbines at a given site are different models.

The engine has many sensors mounted in various locations for speed, temperature, pressure and vibration. These sensors are all subject to vibration and heat and disconnection during engine change-out. Engine-mounted sensor failures are a major cause of unreliability and can result in turbine damage. Engine-mounted thermocouple signal problems are examples. The new control scheme uses each individual probe signal as an input to the DCS in addition to the hardwire average. Therefore, failure of any one signal is not critical to continued operation. The single inputs are digitally averaged and compared to the hardwire average for a deviation alarm. The DCS has logic which identifies most probe or signal failures and it automatically removes the probe with the bad signal from the calculated average and provides an alarm indication of which probe has failed. Multiple failed individual signals can be present without causing a shutdown. The two average temperature signals (hardwire and calculated) are high signal selected to better protect the turbine. Typically, a thermocouple will read low because of bad connections and that problem can also be detected by the spread alarm.

Replacement of some of the existing custom turbine temperature sensors with standard industrial RTD sensors can improve availability of replacement spare parts and improve accuracy of measured parameters. This permits interfacing to standard DCS I/O modules.

Dual Speed Sensors improve reliability when DCS software is used to detect and alarm a sensor failure and switch from a failed sensor to a good one without tripping. The alarm stays on until the operator resets the automatic override. Also, dual sensor signals can be distributed on separate I/O boards to achieve a degree of fault tolerance in a simplex control. Therefore, operation can continue with some boards out of service while an alarm is displayed. Input signal failure is detected for all speed sensors, thermocouples and transducers and is separately alarmed. The original system uses two speed sensor signals which are wired to separate control devices. If one speed sensor signal fails the unit trips. With the new DCS system both signals input the DCS and the unit continues to operate if one signal fails. It is not always practical to add new duplex sensors to existing turbines because the probe housings don't exist on the engines. However, for dual turbines when both turbines must run simultaneously additional redundancy is achieved by using a sensor from each turbine with automatic switching from a failed sensor to a good one even though there is only one sensor per turbine.

The field discrete input contacts can be rewired to a more self-diagnostic state so that a broken wire or loose connections will result in an alarm instead of defeating the alarm or trip protection. This usually involves using a switch which opens-to-trip instead of one that closes-to-trip. Both the field wiring and contact integrity and input circuit board integrity is automatically monitored this way. Redundancy can be achieved by using both the switches and transducers to monitor the same parameters. A further improvement is to eliminate the hardware switches completely and replace them with single or dual transducers and use new soft setpoint switches instead. These soft switches provide several advantages such as the capability to add new pretrip alarm setpoints which did not previously exist and the inherent ability to detect and to alarm transducer signal failure by use of over/under range alarms. On- screen digital readout of the transduced parameters permits trending, logging and precise determination of turbine condition from a remote location. This is very useful when deciding what degree of technical assistance to dispatch to the site.

The CRT's can be used by the operator for soft switch control of all modes of operation currently available on the hardwire control panel except for manual synchronizing. For safety, the soft switch selection always requires two selections by the operator. CRT based soft switches are more reliable because they essentially don't wear out.

The I/O cabinets in the turbine control room use standard 19" equipment racks to house the DCS equipment. These cabinets have interface terminal boards which are used to terminate some of the field I/O cables. The DCS has the capability for future expansion by adding new I/O circuit boards. This is important because various emissions reduction systems, such as water injection and steam injection, can easily be added to the digital turbine controls in the future. The original system used custom circuit boards in a draw-out chassis with no expansion capability.

CONTROL PHILOSOPHY

The original control system consisted of numerous individually mounted devices which were hard-wired into a system which was dependent on acceptable operation of every device in the system. Failure of any one device resulted in a turbine trip. This was considered acceptable by the designers at the time because it was "fail-safe" and the turbine could not run without the protection of any one device. However, this philosophy also made the control system failure prone. Any one device with something wrong kept the turbine from running. A new control philosophy which is "run-oriented" instead of "shutdown-oriented" was implemented in a complete redesign of the turbine controls using the new digital hardware and software.

It is important to exploit the advantages of a digital system in order to fully benefit from the cost of the upgrade. If the original control system logic were simply copied into digital logic with no intent to improve it, then the diagnostic capabilities inherent in the digital system would be wasted, and only a minor reliability improvement would be made since many of the defects and deficiencies of the original design would also be copied.

Digital turbine sequence logic is implemented in reliable proven compact digital software algorithms instead of hard-wired control relay logic. Relays are used mostly for discrete output power switching of solenoids, etc. and these new relays are dust-proof-type. This eliminates the original equipment problem of sequential logic problems caused by "dirty contacts" in the control relays. This is especially important because the original equipment had many control relays which were part of a generic design that did not necessarily apply to the actual installed equipment. For instance, some units are liquid fuel only but gas fuel or dual fuel relay logic contacts are required to complete the liquid fuel logic circuits. These unused circuit contacts pose no problem in the new digital design. The physical room resulting from removal of the several hundred control relays and twenty-five timers can be used for mounting of the new DCS equipment.

Soft setpoints permit the use of a feature called "auto unload" (sometimes called governor runback) which automatically reduces electric generator output power to keep the plant running in an automatic attempt to avoid certain turbine shutdown trip conditions. For example, if a parameter is above the alarm setpoint, but below the trip setpoint, an intermediate soft setpoint will initiate an auto unload. The same type of operation can be programmed for other analog parameters by introducing new soft setpoints. The auto unload feature is designed to keep engines operating without needlessly tripping the unit. This gives the operator an opportunity to diagnose a problem and then reload the turbine if possible.

ALARM READOUTS

Each alarm or trip is separately indicated with individualized text on the DCS operator control CRT in order to specifically identify the fault. The original equipment had many possible causes for the same annunciator alarm window. A typical original annunciator had only forty-eight windows which were used to describe two-hundred-and-fifty (250) alarms and trips. Often it was necessary to start the turbine again and again to try to separate out the many possible causes of the problem. The separation of each individual alarm and trip cause in the DCS design contributes to less down time because of quicker fault analysis. A further benefit of the digital CRT display is that the precise value of the alarm setpoint can be displayed in text. For example: "Fuel pressure less than 5 PSIG" instead of the original "low fuel pressure" alarm.

The operator interface CRT has text and numerical data displayed for all alarms. Digital displayed data and analog trend displays are used by the operator to determine equipment status as well as to control the unit, start/stop, load etc. All turbine "Ready To Start" lockouts are also individually identified in text on the CRT for quicker response by the operator without the need to search through manuals and drawings and without the need to do extensive electrical testing to determine the lockout cause.

DATA COLLECTION

The automatic data collection capabilities of the DCS are impressive. All transducer analog values are available on the operator's CRT and can be saved in an archive at the same instant in time with one second resolution. This replaces and far surpasses the use of three or four separate strip chart recorders which had no common time correlation between them. Entire start sequences can be automatically recorded, stored and archived for future reference to determine the quality of that particular start or as a reference to compare to other start archives. This is a powerful preventative maintenance tool that was previously unavailable.

Automatic totalization of important parameters such as power generation in megawatt hours and fuel consumption in gallons burned also permits automatic heat rate calculations for determination of actual turbine efficiency. Calculated relative efficiency of one turbine unit versus others at the same multi-unit site permits starting the most efficient units first. Even when fuel flow transmitters are not installed automatic performance monitoring calculations based on engine parameters can be used to determine relative efficiency by calculating and comparing deltas from a common performance baseline.

Operators, technicians and engineers can all deal with direct reading of all monitored parameters in engineering units, for example, rotor speed is expressed in RPM instead of gearbox frequency in HZ, temperature is provided in degrees, F. or C. instead of millivolts. Level is expressed in gallons or inches, flow is expressed in GPM or pounds per hour instead of HZ or flowmeter counts. Once the conversion to engineering units has been made in the DCS the same parameter can be displayed simultaneously in several different engineering units as a matter of convenience.

The DCS system documents operator action such as load setpoint changes automatically as an aid to problem diagnosis. Manual start/stop and raise/lower commands as well as trip/reset commands are all automatically logged with date and time.

The control CRT or operator interface is a color CRT which is used for turbine start/stop, raise/lower load and to select operating modes and test modes in addition to the existing hardwire control panel. It is icon-selectable and provides the operator with alarm and event lists and built-in trending capability. A laser event printer is used to print all alarms and events and to provide CRT screen copies.

The simplex DCS system uses a single central processing unit (CPU) to operate the combination of sequencing and fuel governing as well as data acquisition. It is typically a 32 bit microprocessor based design. The main unit also contains a service panel for operation data entry and readout. The DCS, contains a high degree of system integration to incorporate the function of a maximum number of original system devices. Once all the field I/O's are provided to the DCS just about any control scheme can be implemented using any combination of them in any control scheme that you can think of.

MAINTENANCE IMPROVEMENTS

The new DCS can be programmed to reduce maintenance adjustments by eliminating many of the tedious fuel control, speed relay and sequencer calibrations previously required. Some original analog controls had as many as seventy-five (75) adjustable potentiometers in one control system for one turbine. A new design philosophy can be implemented to make the digital control system as free of adjustments and as self diagnostic as practical.

Original sequential timers are pneumatic and are subject to setpoint drifting as well as coil and contact failure. The only way to check the pneumatic timers for accuracy is to energize and de-energize them over and over again and time them with a stop watch. This is a tedious process. The DCS uses an accurate digital clock which provides a time base for software logic timers which have an adjustable digital setpoint. The timing logic is all done in software. There is no maintenance need for manual timing tests of the digital timers. Software program verification tests are conducted at the manufacturer's facility prior to shipment and all timer functions are verified at that time. Debugging a digital system in the field is minimal. However, on screen diagnostic software shows algorithm I/O logic state for ease of troubleshooting if needed.

Turbine output power load trim adjustments are not required because of a new temperature vs. load schedule which prevents overloading the engines. These schedules result from the ability to characterize a curve with the new digital controls and also the ability to plot almost any one parameter versus another once the parameters are assigned to the input cards.

The turbine loading rate is a continuous digital ramp instead of a series of raise pulses. The rate is set by preprogramming a digital rate of change setpoint value instead of timing and retiming a pulser circuit discrete output to a motor-operated reference potentiometer. This permits custom settings to any customer requirement.

The DCS printed circuit boards require no field calibration. They are generic for a given type and interchangeable. The I/O circuit boards can be replaced without powering down the rest of the DCS. The software program is stored on UV erasable PROM's (Programmable Read Only Memory) for program security. These programs are unique to each unit because each customer has site specific requirements. Backup copies of the program are stored off site.

All DCS setpoints have digital accuracy to eliminate the drift inherent in the design of the existing analog controls. This eliminates tedious tuning of old-style analog controls in an effort to stay within an acceptance band.

The existing operator control panels can be left as-is physically if so desired by the owner. The advantages of this concept are that the appearance of the control panels remain the same visually for operator and maintenance personnel familiarity. However the operators can elect to operate from the CRT also. The disadvantage is that the I/O count is increased so initial cost and installation time increase.

Power loss is individually identified for motor and load critical circuits. Loss of power might jeopardize a turbine "Ready to Start" condition. Older controls did not detect the malfunction until after a start attempt was made.

The turbine water wash sequencer logic and timers are also integrated into the new DCS. Ready-To-Wash start lockouts and count-down timers are available on the CRT.

Safety is improved for maintenance personnel because hundreds of 125VDC circuits are operated at 24VDC after the upgrade.

AUTOMATIC TURBINE TESTS

The flexibility of the DCS digital software permits the implementation of built-in automatic turbine test programs which are safer and more complete than the old method of jumpering around internal sequential logic. DCS software can include preprogrammed turbine operational tests for fuel shutoff valves, ignitors, overspeed, engine starter cranking, lite-off flow, etc. These new test programs are automatically programmed to time-out and, in some cases, self-indicate that the test was successful or unsuccessful. Operational demonstration and periodic testing of these turbine functions by maintenance personnel is critical to repeatable reliable turbine starts.

Dynamic overspeed testing is fully automated. When the prerequisites have been completed and overspeed test is selected, the generator speed reference automatically advances toward the trip value. If shutdown does not occur on time, the reference will automatically decrease to a safe value. If the trip does occur, it is automatically archived and alarmed. This minimizes the danger of spending an excessive amount of time in the overspeed condition or the need to repeat the demonstration test. The DCS overspeed test mode also has a reduced speed test built-in in order to functionally prove that the trip circuit works without actually over-speeding the turbine during start.

A manually initiated, but automatically sequenced mode of operation is provided to motor (crank) an engine on the starter. This mode can be used to check speed pickup signal output, to confirm starter air flow settings, to cool the engines off prior to water wash or to purge the exhaust stack after a trip. The operator selects this soft control mode by using a mouse or trackball while observing the CRT.

SELF-DIAGNOSTIC FEATURES

The "first out " cause of any unit trip is automatically and specifically identified in text on the CRT in an event stack listing with date and time indicated. The event stack lists all alarms, trips and operator initiated actions in sequential order as they occurred and it can also be printed out by the operator on demand or at timed intervals. Status and time stamp of critical sequential events is indicated on the CRT and available to the printer.

DCS I/O circuit boards indicate a fault and cause an alarm if they fail. They are provided with fault indication lights.

Turbine sensor failures and transducer failures are detected by the DCS and individually alarmed. Many of these turbine sensor failures cannot be detected unless the turbine is in operation, then they result in auto transfer to a safe backup mode of control instead of trip.

During startup, failure to achieve specific rotor speeds within the allotted time periods are individually identified on the DCS CRT instead of just a single original group alarm "Incomplete Sequence". The time remaining is also displayed on count-down timers so that the quality of the start can be evaluated.

Automatic archive trending of the last sixty seconds prior to any trip provides all analog data with one second resolution. This is extremely valuable and useful to diagnose problems quickly and accurately and an invaluable aid for remote technical assistance.

Because field device switch contacts go direct to the DCS inputs, logic status of any input contact can be observed on the DCS service panel without using any other test equipment.

CONCLUSION

The overall effect of the controls upgrade is to make the turbine control system more straightforward, easier to understand and simpler to operate. The effect on the gas turbine peaking plant is to make it more reliable and more available for operation. These benefits are the direct result of the application of the advantages of the digital control system.

This paper has provided a general description of some of the many advantages of a digital control system upgrade for gas turbines. It is based on the writer's experience as project manager for the design, installation and startup of gas turbine control system upgrades using the Woodward Governor Co. NetCon 5000 digital control system (DCS) and the OpTrend operator control interface when applied to FT4 and GG4 aero-derivative gas turbine engine peaking plants.

0074-056X/95/161-176/$0 + .50pp

ENHANCED CONTROLS FOR TURBINES AND BYPASSES IN A GERMAN COAL-FIRED POWER STATION

Hans Georg Houzer
Bayernwerk AG
Manager Electrical and I&C Equipment
39 Nymphenburger Strasse
Munich, Germany, D-80335

KEYWORDS
Enhanced Turbine Controls, Automatic Start-up, Bypass Controls

ABSTRACT

This article describes the enhancing of turbine and bypass controls of two 150 MW units in 1994 and 1995 which came in service 1962 and 1971. The turbines are equipped with digital controls for fully automatic start-up (and shut-down) i.e. sequence controls for engaging and opening of the stop valves and control valves, disengaging of the turning gear, exitation, starting of the automatic synchronizing device, switching off of electrical lub and governor oil pumps etc. and a controller for start-up of the turbine with a controlled acceleration according to the conditions of the turbine and live as well as reheat steam.
The hydraulic controllers for HP- and IP-bypasses are replaced by digital controllers (part of DCS). The hydraulic actuators of the bypass valves are now positioned by electro-hydraulic servo valves following the output of the DCS. The paper covers design considerations and first commissioning and operating experience.

1. INTRODUCTION

Bayernwerk AG and it's subsidiaries are a group of electric utilities serving an area in the southeast of Germany in the federal states of Bavaria and Thuringia (see map Fig. 1). Bayernwerk group ist the No. 3 power producer in Germany. It is owned by VIAG, a privatized industrial holding company.

The main power source of Bayernwerk is nuclear energy (see map Fig. 1 and demand and supply diagrams Fig. 2). The nuclear plants are operated on base load for economical reasons. Hydro-electric generation, source No. 2, provided by a subsidiary "Bayernwerk Wasserkraft AG" depends on the flow of water in the rivers. Hydro-electric storage capacity is available from this subsidiary and also on a contractual basis from neighboring Austria. The third power source are fossil-fired plants based on different kinds of fuel: hard coal from German mines and the international market, lignite from the Czech Republic, heavy oil (refinery residues), light oil and natural gas.

Aschaffenburg Power Station is a coal-fired power plant (see main data Table 1 and reference I). The two remaining 150 MW units have been equipped with polution control equipment (a common FGD plant, individual high-dust DeNOxes and enhanced electrostatic precipitators) and are supposed to stay in service until 2010.

Unit 21 has been retrofitted in 1994 with a new boiler, a new generator, a new LP-turbine rotor, a new turbine for the main boiler feed pump, a new 400 V switchgear and a new closed-loop control system with a CRT-based operating and monitoring system.

Unit 31 will be retrofitted with the same state-of-the-art DCS system during a major turbine overhaul this year (June 24 through Sept. 26).

Both units have once-through boilers with circulating pumps and separators for start-up and low load operation up to 40%. The furnaces have two half load wet-bottom combustion chambers.

2. DUTIES OF FOSSIL-FIRED POWER PLANTS

Since the nuclear plants run normally on full load (except stretch-out and fuel changing) and hydro generation depends on the flow in the rivers, that means on precipitation fossil-fired plants have to fill the gap (see Fig. 2). Coal- and lignite-fired plants run also on base load or on cycling duty during the high load period in winter, maybe with standstills during weekends. In summer they have start-ups and shut-downs every weekday (see load diagrams Fig. 3).

In order to achieve quick start-ups and load changes with minimum energy losses, the units are normally operated in the sliding pressure/sliding temperature mode and are equipped with fully automatic control systems. The control systems may also be switched to constant-pressure operation.

In order to intercept a possible breakdown of the power system, a special control mode may be switched on (automatically if voltage or frequency are abnormally low or on manual remote command from the load dispatch center). In this mode the load of the turbine is controlled at the previous value and the boiler load increased (surplus steam generation via turbine bypasses) to enable the unit to increase electric output quickly if this is necessary to stabilize the power system.

3. CONTROL CONCEPT OF ASCHAFFENBURG POWER STATION, UNIT 21

Since the design and the duties of both units of the Aschaffenburg Power Station are more or less the same, the overall control concept is essentially the same (see Fig. 4 & 5). Some parts of the control system had to be retrofitted earlier (see reference I and Table 2). These already retrofitted systems meet all essential requirements. Therefore these controls were integrated in the new overall control concept. Unfortunately an interface for exchange of signals between the system buses of the different DCS systems is not available. Therefore all signals are exchanged via hard-wired input/output units.

The overall control concept comprises a hierarchically structured open-loop control system (see Fig. 6) interacting with an also hierarchically structured closed-loop control system. Normal operation and monitoring is done by a redundant PC/CRT based operating and monitoring system.

The control desk is also equipped with a mosaic panel for manual operation of drives and certain control loops.

The sequential unit master control (see Figures 6 & 7) starts the subsequent systems by so-called function group controls. These comprise control logics which will operate the necessary number of components depending on the load, e.g. boiler feed pumps, pulverizers etc. The automatic start of such a component with it's auxiliaries (lub oil pumps, dampers, valves etc.) is carried out by sequential subgroup controls. There are independent subsequent controllers for each final control element and each drive. If master or group controls fail, an operation with a lower level of automation is still possible by means of the group/subgroup controls or drive/single-loop controls.

4 TURBINE MASTER CONTROL

The duties of this automatic start-up and shut-down control are:

- prepare turbine for start-up
- controlled accelleration of turbine according to a predetermined start-up diagram
- automatic exitation and synchronizing of the generator by a subsequent group control
- increase load according to predetermined gradients until turbine bypasses are closed (then closed-loop unit load control takes over)

This master control consists of an open-loop sequential control and integrated closed-loop controls (see Figures 6 & 8).

4.1 Automatic start-up

In steps 1-3 prerequisite conditions are checked, controllers are set in the right mode for start-up, stop valves are engaged and the trip oil valves are opened. In step 4 the digital speed controller will increase the setpoint of the hydraulic speed controller (speed changer motor) until the governor oil pressure is > 3 bar (43.5 psig) or the turbine speed has reached the value of 500 rpm. Then the accelleration is controlled by an integrator. The gradient of the integrator varies depending on the actual value of the speed (see Fig. 9) in steps 4-6. If the actual speed should not follow the increasing setpoint (deviation is beyond a certain limit) the increase of the speed setpoint will be stopped and released again when the deviation is again within the limit.

At a speed of 2700 rpm the mechanical speed controller engages and accellerates the turbine to the nominal speed of 3000 rpm (= 50 Hz). The integrator will be set to this nominal speed. When the speed has reached 2950 rpm the subsequent group control for automatic exitation and synchronizing is started.
This sequential control causes and monitors the following actions (see refernce I):

- switch ON exitation, put voltage controller on AUTO
- start automatic synchronizing device which
  - adjusts turbine speed by commands to speed changer motor
  - adjusts voltage by commands to the tap-changer of the step-up transformer
  - closes breaker sychronously by two channel commands (to avoid damage to the generator in case of a single fault)

- stabilizes operation by commands for minimum load (speed changer motor) reactive power (tap-changer)

Upon synchronizing, the load is increased in step 8 with variable gradients depending on the HP casing temperature. An integrator which is set on the actual value in step 7 provides the gradient. If the actual value should not follow the increasing setpoint the latter will be reduced to avoid big jumps in case of large deviations.

When bypass valves are closed the closed-loop unit load control will increase the load by increasing fuel, combustion air and feed water in the necessary ratio including rate-action of the firing until the preset target load setpoint is reached.

4.2 Automatic shut-down

This section of the sequential control may also be started by a corresponding section of the unit master control or manually.

In this section of the turbine shut-down program the initial pressure is controlled. The setpoint will be the minimum pressure for the boiler (the boiler remains under pressure during overnight and weekend stand-stills). Within the shut-down procedure the closed-loop control will decrease load by reducing firing (fuel and air). The steam generation will decline and the governor valves will close to maintain the above mentioned minimum pressure. When steam flow has stopped the reverse power relay will open the breaker and the turbine will be tripped.

Subsequent controls will start the electrically driven lub and governor oil pumps when the speed drops to a value where the shaft main oil pump does not provide enough oil. When the turbine has reached the turning speed, jacking oil pumps will be startet and the turning gear will be engaged. This is the end of the shut-down program because the turbine must be turned during outages up to approximately one week.

5. INTERFACE FOR TURBINE BYPASS CONTROL VALVES

HP and IP bypasses were equipped with independent hydraulic controllers prior to the DCS retrofit 1994/1995.

An integration in the DCS was inevitable for an optimal control of these bypasses. An example of the complexity of these control tasks is shown in Fig. 10 for the IP-bypass. Details of the interface between DCS system and the control valve may be seen in Fig. 11. The i/p transmitter in Fig. 10 is a positioner, the setpoint of which is controlled by the analog output of the DCS system.

6. COMMISSIONING AND OPERATING EXPERIENCE

The comprehensive retrofit measures have been carried out in a record-braking short period. This may be seen in the following schedule of events:

Shut-down for retrofit	Jan. 14, 1994
On-site integration test of DCS	April 25 - June 20, 1994
Boiler pressure test	Sept. 9, 1994
Trial operation period	Dec. 19 - Jan. 27, 1995

During commissioning the operating and monitoring system with only 4 CRT control stations proved to be a bottleneck, notwithstanding the conventional controls and indicators. The operating staff had to switch to other images very often to carry out all necessary control actions and to monitor the behavior of the plant. The installation of additional makeshift CRT control stations was considered for commissioning. For unit 21 only one additional CRT station was possible but unit 31 will be equipped with more additional CRT's during this critical period.

Such a temporary extension may be easily achieved by connecting PC's to the existing terminal bus of this operating and monitoring system.

The CRT images were developed in close cooperation with the operating staff and provide easy to survey information (see Figures 12 & 13).

The automatic start-up and shut-down control of the turbine was already installed in 1989/90 and commissioned 1990/91. The sequential unit master control was also installed in 1990. It was not put in service before the retrofit of the closed-loop control system in 1994 because it was merely impossible to add all the necessary signal exchange to the old closed-loop controls. Therefore fully automatic start-up and shut-down of the whole unit was only possible after completion of the DCS retrofit.

The first few weeks of operation of unit 21 showed that the new boiler does not fullfill the requirements of the specification. Most urgent improvements were carried out from Feb. 24 through March 6, 1995. Major changes which are necessary to meet the specifications fully may only be carried out this summer.

The basic control circuits were successfully commissioned during the first few weeks of operation. The commissioning of the upper level controls and the final optimizing of all controls was not possible for two reasons:

- The need for power in the high load winter period did not allow the start-ups, load changes and shut-downs which are necessary for commissioning and optimizing of these controls.
- A fully automatic unit operation requires a boiler which meets all specified requirements.

The turbine and bypass controls will be completely commissioned and optimized from March through May, 1995.

An updated version of this paper with the latest results will be presented at the conference.

Reference I: Hans Georg Houzer
RENEWAL OF THE I&C EQUIPMENT
OF A GERMAN 27-YEAR-OLD
150-MW POWER PLANT UNIT AS A
PART OF THE MEASURES TO EXTEND
THE LIFETIME:
ISA90 International Conference and
Exhibit Proceedings. New Orleans, LA
October 14-18, 1990

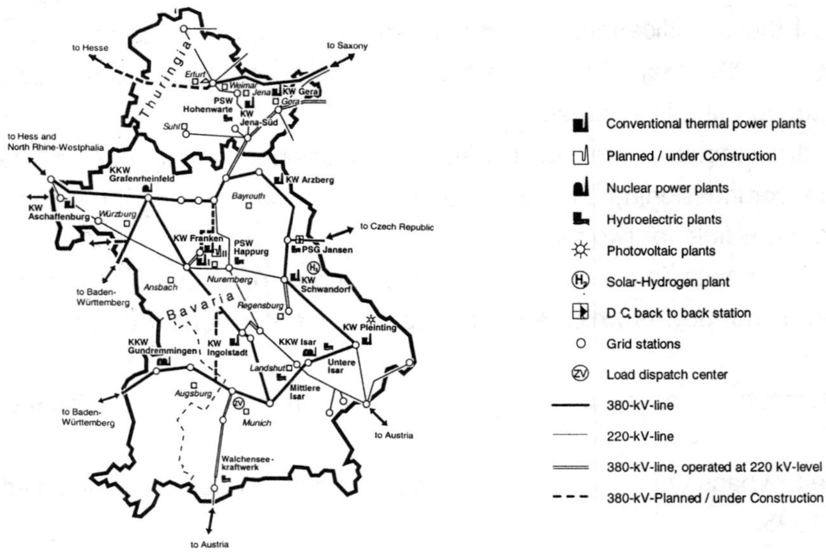

Fig. 1 Bayernwerk Group, High Voltage Grid und Power Plants

Houzer -Kraftwerksleittechnik
950206

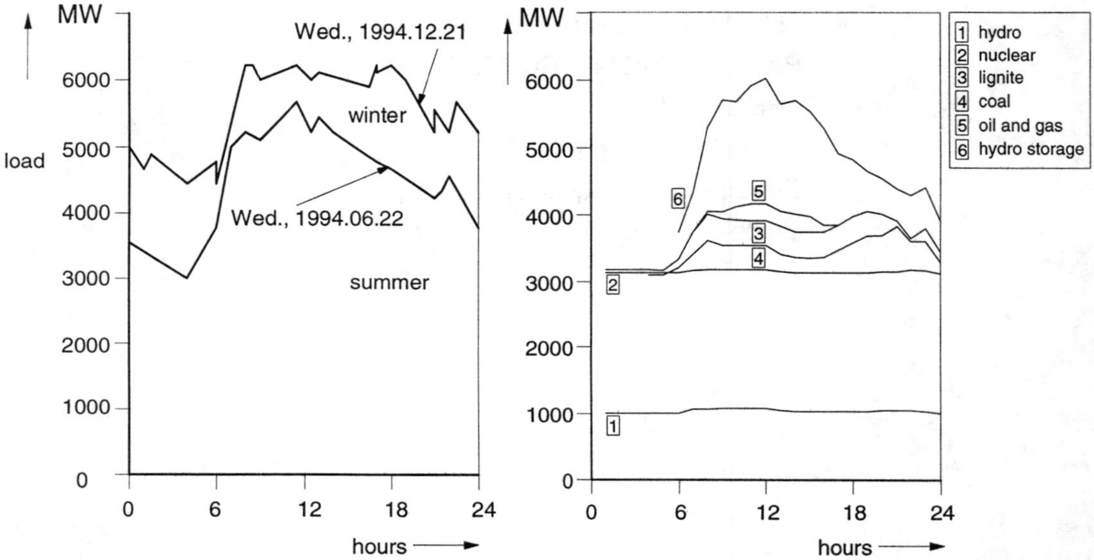

Fig. 2: Bayernwerk Demand and Supply Diagrams

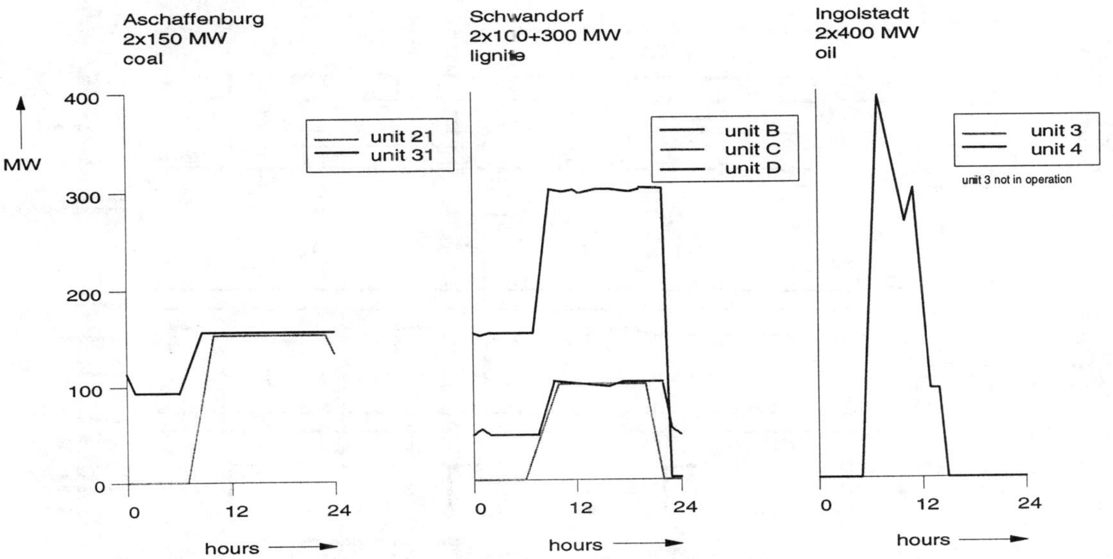

Fig. 3 Bayernwerk Typical Load Diagrams of Fossil-fired Power Plants in Summer

Heuzer -Kraftwerksleittechnik
950206

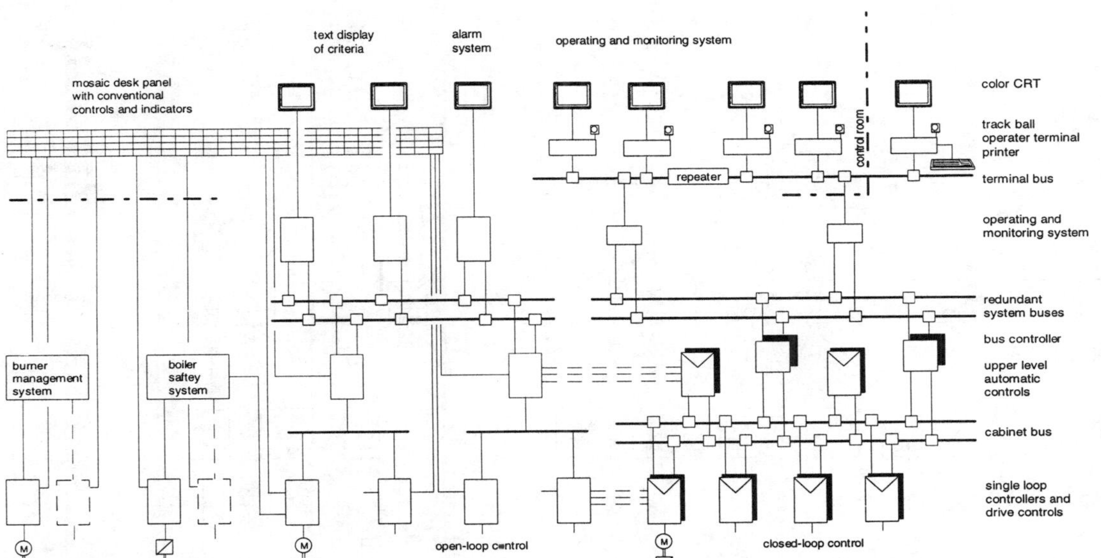

Fig. 4 Aschaffenburg Power Station, Unit 21, Structure of Distributed Control System (DCS)

Heuzer -Kraftwerksleittechnik
950206

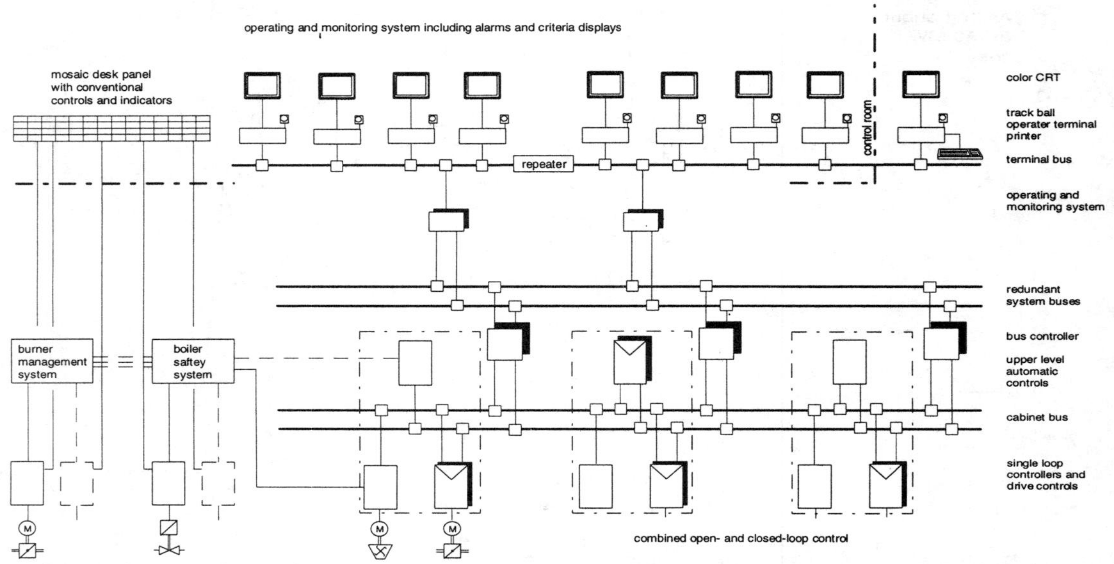

Fig. 5 Aschaffenburg Power Station, Unit 31, Structure of Distributed control System (DCS) Houzer -Kraftwerksleittechnik 950206

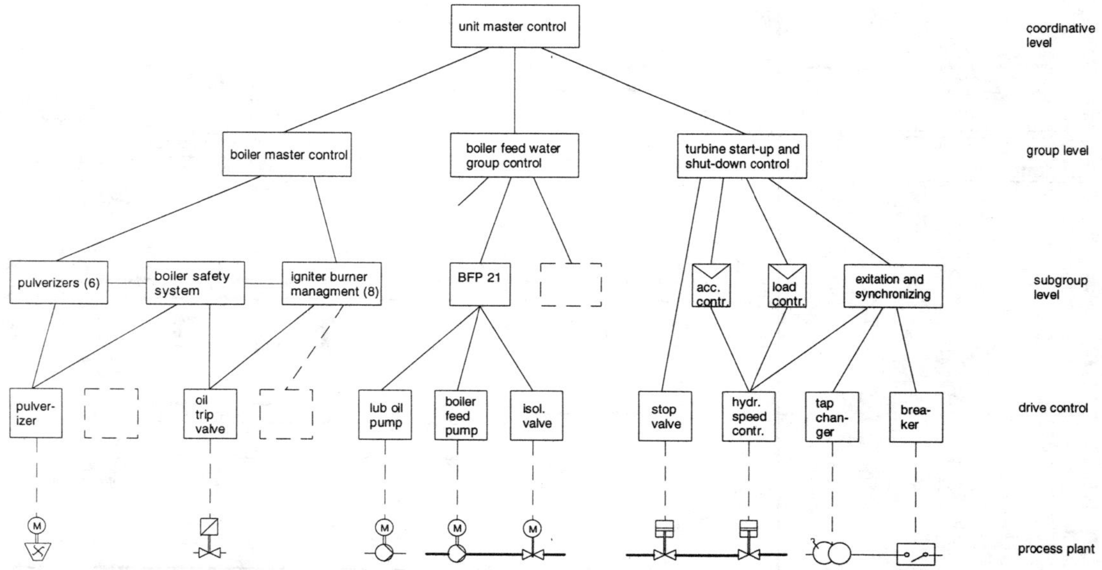

Fig. 6 Aschaffenburg Power Station, Unit 21, Hierarchical Structure of Open-Loop Control Houzer -Kraftwerksleittechnik 950206

Preconditions:
Turbine in turning gear, auxiliary systems and cooling water in operation.

Step	Start Operation	Switches	Wait for
1	NH_3-evaporation for Denox		
2	Main condensate system	Boiler drain system to AUTO	
3	Auxiliary condensate system	Turbine drain system to AUTO	
4	Gland sealing system		
	Slag removal systems		
5	Condenser evacuating system		
6	Rotary air heaters		
7	Boiler feed water system		
8	HP feed water heater system		
	LP feed water heater system		
9	Fly ash removal (to bin) system		
10	ID fan		
11	FD fan 21		
12	FD fan 22		

Fig. 7a Sequential unit master control, start-up program

Houzer -Kraftwerksleittechnik
950206

Step	Start Operation	Switches	Wait for
13	Purge of furnace and flue gas system	Electrostatic precipitator on reduced voltage	
14			Purge complete
15			Igniting conditions (automatic actions of subsequent controls)
16	Fuel oil pump for igniter burners		
17	Firing master control	Spray desuperheaters to AUTO	
18			Steam conditions (fire in furnace, steam temp.>turbine temp., HP-and reheat steam superheated)
19	Automatic turbine start-up control		
20		Electrostatic precipitator on full voltage	Coal fire
21	END of start-up program		

Fig. 7b Sequential unit master control, start-up program

Houzer -Kraftwerksleittechnik
950206

Step	Command	Adjust.	Comment
1	Put setpoint of hydraulic controller to	MIN	
	Switch hydraulic speed controller to	MAN	Initial pressure controlled.
	Switch HP-by-pass control to	AUTO	
2	Engage stop valves		
3	Open trip oil valves		
4	Switch speed controller to	AUTO	Reset speed integrator.
	Switch positioner to	AUTO	Put speed setpoint to 80 rpm.
5	Release speed integrator		
	Release speed limiter		
6	Prepare exitation and synchronizing		
7	Start automatic exitation and synchronizing		Switches on exitation and automatic generator voltage control Starts automatic synchronizing device which - adjusts voltage by tap-changer and turbine speed - closes breaker synchroneously Increases load (turbine speed controller) and reactive power (tap-changer).
8	Switch load controller to	AUTO	Reset load integrator.
9	Wait for automatic closing of HP- and IP-by-passes		
10	Switch HP flange heating	ON	
11	Switch IP flange heating	ON	
12	END of start-up program		

Fig. 8a Automatic turbine start-up control

Houzer -Kraftwerksleittechnik
950206

Step	Command	Adjust.	Comment
51	Switch speed controller to	AUTO	
	Switch pressure controller to	AUTO	
	Switch Aut. exitation and synchronizing	OFF	
	Switch HP- and IP-flange heating	OFF	Reset load integrator.
	Switch positioner to	AUTO	
	Switch speed changer motor to	AUTO	
	Switch voltage controller to	MAN	
52	Trip turbine		Breaker will open by reverse power relay.
	Open 10kV isolator		After opening of the breaker
53	Put speed setpoint to	MIN	
54	Wait for decline of speed and Automatic engaging of turning gear		
55	End of shut-down program		

Fig. 8b Automatic turbine shut-down control

Houzer -Kraftwerksleittechnik
950206

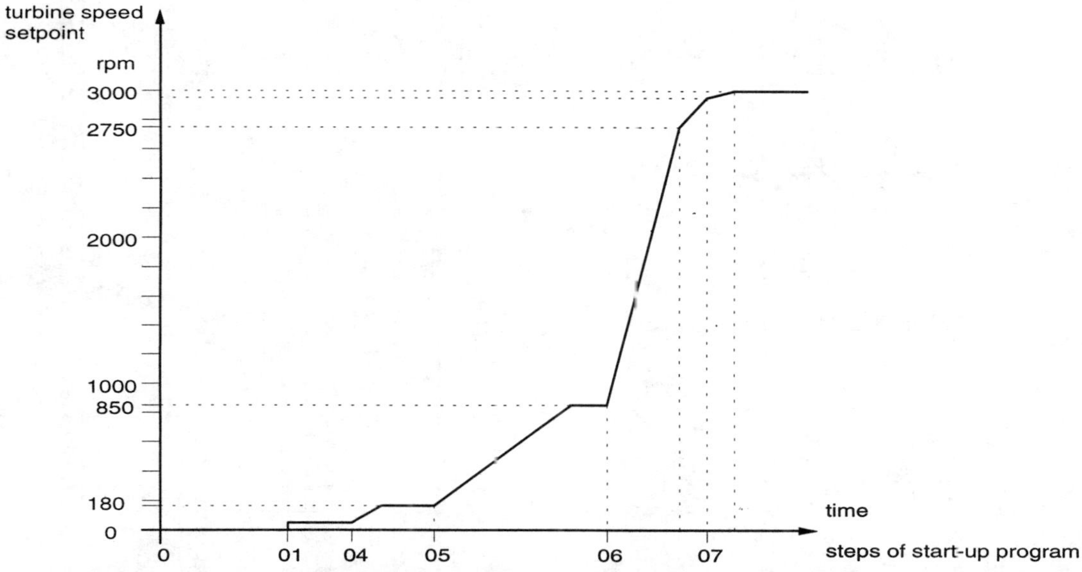

Fig. 9 Aschaffenburg Power Station, Turbine 21, Start-up Diagram

Houzer -Kraftwerksleittechnik 950206

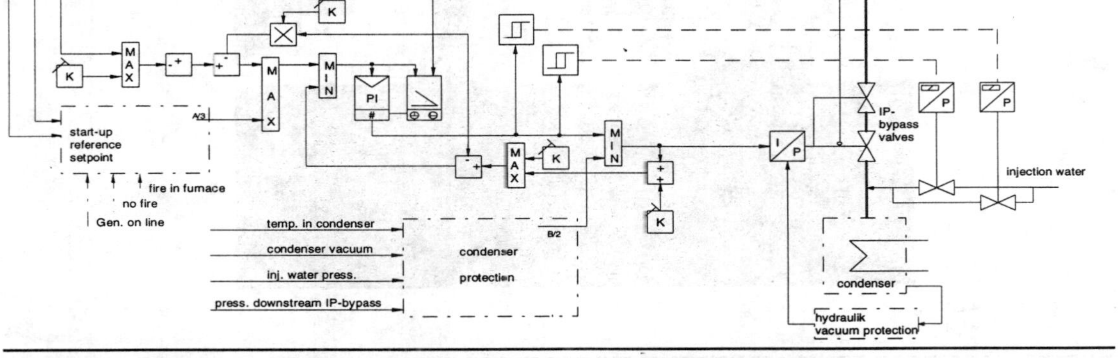

Fig. 10: Aschaffenburg Power Station, Unit 21, Survey Diagram IP-Bypass Control

Houzer -Kraftwerksleittechnik 950206

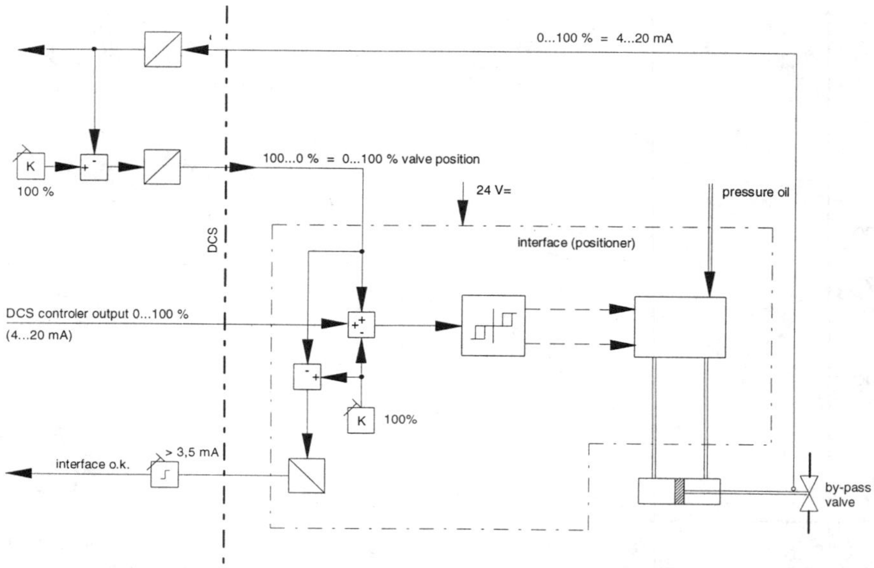

Fig. 11 Aschaffenburg Power Station, Unit 21, Interface for Turbine Bypass Control

Houzer -Kraftwerksleittechnik
950206

Fig. 12 Aschaffenburg Power Station, Unit 21, Control Desk and Panel

Houzer -Kraftwerksleittechnik
950206

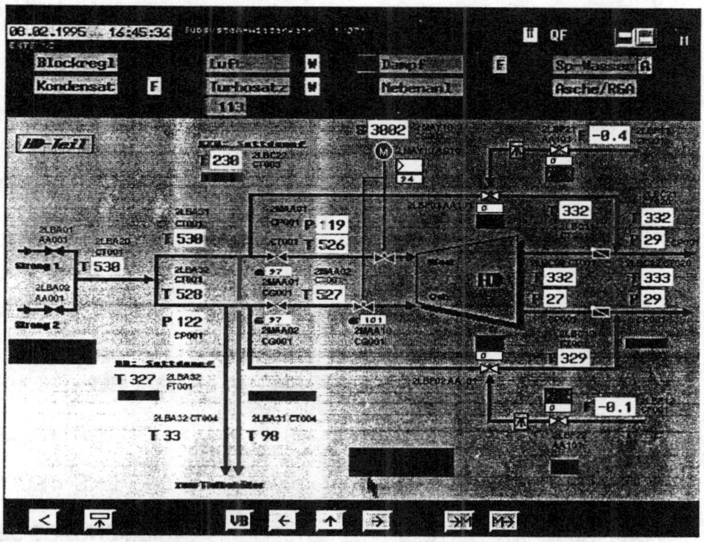

Fig. 13 Aschaffenburg Power Station, Unit 21, CRT-image HP Turbine Control

Houzer -Kraftwerksleittechnik
950206

Unit	**21**			**31**		
Boiler capacity	512 t/h	=	313.2 lbs/s	490 t/h	=	300 lbs/s
Live steam pressure	190 bar	=	2755 psig	186 bar	=	2697 psig
Live steam temperature	535 °C	=	995 °F	535 °C	=	995 °F
Reheat steam pressure	40 bar	=	580 psig	43 bar	=	624 psig
Reheat steam temperature	535 °C	=	995 °F	535 °C	=	995 °F
Generator capacity	160 MW 214 Mvar			150 MW 200 MVar		
Number of pulverizers	6			4		
Number of igniter burners	8			4		
Commissioning of unit	1963			1971		
Commissioning of common FGD plant	1986			1986		
Commissioning of DeNOx plants	1990			1990		

Table 1: Main Data of Aschaffenburg Power Station

Houzer -Kraftwerksleittechnik
950206

	Unit 21			Unit 31		
	Vendor	System	Commiss.	Vendor	System	Commiss.
Original equipment:						
Open-loop-control	SIEMENS	SIMATIC G	1963	BBC	DECONTIC B	1971
Closed-loop-control	SIEMENS	TELEPERM S	1963	SIEMENS	TELEPERM B	1971
Retrofitted and extended equipment:						
Boiler safety and igniter burner management system	MAUELL	ME4002/4002s	1985	MAUELL	ME4002/4002s	1985
Open-loop-control	MAUELL	ME4012	1989	SIEMENS	TELEPERM MI	1995
Closed-loop-control	SIEMENS	TELEPERM MI	1994	SIEMENS	TELEPERM MI	1995
Operating and monitoring system	SIEMENS	OS 525	1994	SIEMENS	OS 525	1995

Table 2: Applied Control Systems

Houzer -Kraftwerksleittechnik
950206

0074-056X/95/177-181/$0 + .50pp

ASSESSING THE ADEQUACY OF RADIATION MONITORS IN STEAM GENERATOR TUBE LEAKAGE DETECTION SYSTEM

V. Willems
Nuclear Instrumentation Consultant
Gilbert/Commonwealth Inc.
Reading PA.

ABSTRACT

Steam generators used in Pressurized Water Reactors are experiencing tube failures resulting in primary to secondary systems leakages. Typically, radiation monitors are used to monitor the radiation levels in the condenser offgas, main steam lines and steam generator blowdown lines. The adequacy of the measuring ranges of these monitors is addressed for all operating conditions. Measurement of N^{16} in the steam lines is also addressed. This paper presents, via a typical example, a methodology allowing to evaluate these monitors as an integrated system. Extremes of operating conditions such as low and high reactor coolant radioactivity, low and high leakage rates and leakage from active or plugged tubes are evaluated.

APPROACH

In PWRs, on-line radiation monitors alert the operator of primary to secondary leaks. For operating conditions such as low primary coolant radioactivity or plug leakage on inactive steam generator tubes, some radiation monitors may not have sufficient sensitivity or range to perform adequately their functions. This situation has the potential to mislead the operator by suggesting a leakage rate lower than actually exist. Assessing the range of detection of the different radiation monitors, as it relates to leakage, is not a straight forward process. It is difficult because of the many operating and equipment factors to consider simultaneously. These factors include the following:

Source term:

- The potential range of fission products in the primary system is bounded, on the low side, by a realistic primary system activity estimated as equivalent to approximately .1% fail fuel. The high side is limited by the design basis failed fuel, usually 1%.

- Depending on the location of the leak along the length of the tube, different time delays are introduced thus affecting the amount of N^{16} decay. This affects the amount of radiation available for measurement by the monitors.
- Leaking plugs in inactive steam generator tubes, while potentially contributing significantly to the total leakage, produce only small amounts of N^{16} in the secondary system because of the significant time delays involved. The highest levels of N^{16} occur when active tubes leak.

- The radiation leaking from the primary system includes gases and solids that are separated at the steam/water interface in the steam generator. The result is that monitors are not measuring the same type of radiation.

Leakage rate:

- The range of leakage rate requiring monitoring is wide and includes, on the low side, realistic leaks of a few gallons per day to leaks that can be classified as Loss of Coolant Accidents.

Process conditions:

- The process conditions such as the boiling in the steam generators and the different flows change the concentration of the primary coolant radioactivity and impacts the amount of activity seen by the monitors.

Radiation monitors characteristics:

- The type of radiation detected, range and energy response of the detectors must be considered in evaluating the range of leakage measurement.

Traditionally, radioactivity levels in plant systems and monitor ranges are expressed in units of concentration. Because process flows dilute differently the radioactivity leaking from the primary system, direct correlation of the performance of different radiation monitors in terms of leakage is not possible.

Assessing the performance of the radiation monitors requires that the envelope of plant operating conditions and monitor characteristics be defined using a common parameter to allow their correlation. This is accomplished by expressing the portion of the source term detectable by the monitors as a release rate and graphing it as a function of the leakage rate. The monitors sensitivities and ranges are also expressed as release rates by taking specific process flow rates into consideration.

RESULTS

In a typical 3 loops Westinghouse PWR, radiation monitors measure the radiation levels of noble gases in the condenser offgas, non volatile in the steam generator blowdown and noble gases or N^{16} in the steam line. These monitors are shown schematically on figure 1. The performances of this typical primary to secondary leakage monitoring system have been assessed using representative design base plant operating conditions. These conditions are typical and were derived from a FSAR submittal. They are summarized in table 1.

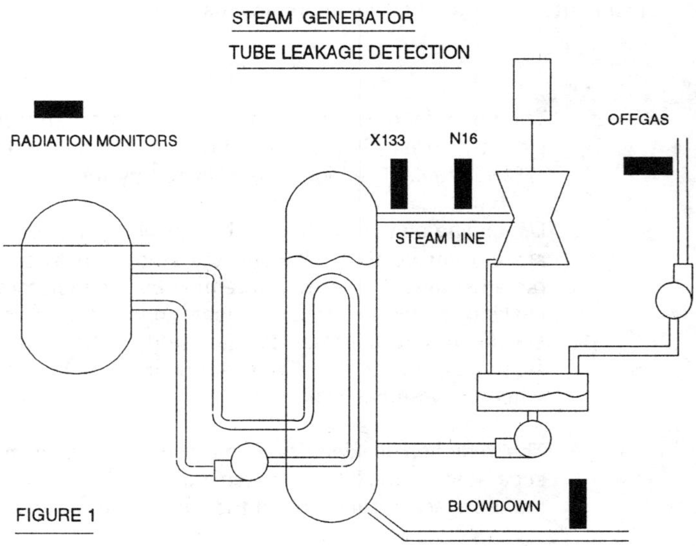

FIGURE 1

The source term data and process conditions of table 1 are used to define an operating envelope for X^{133} as a function of leakage rate. Figure 2 illustrates the results for the steam line and condenser offgas monitors based on Xe133. The area between the realistic and 1% fail fuel lines represents the potential variation of the release rate from the primary coolant to the secondary system using data from table 1. The same operating envelope is applicable to both the steam line and condenser offgas monitors since the same level of radioactivity exist in both paths.

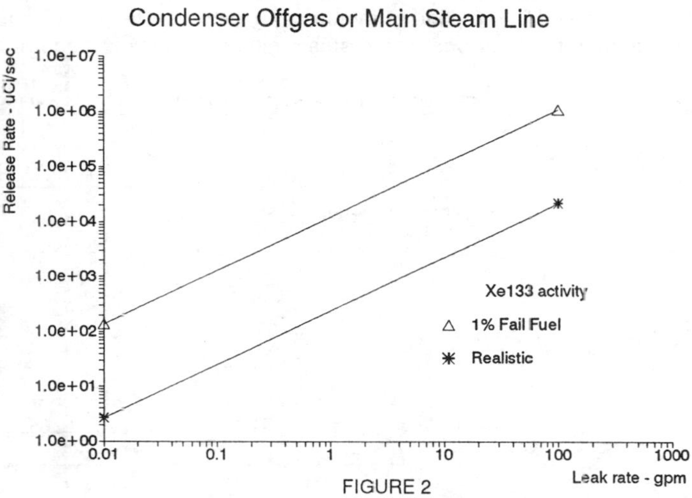

FIGURE 2

The X^{133} measuring range of the condenser offgas and main steam line monitors can be expressed in terms of release rates using corresponding process flow rate conditions and drawn on the same graph as the operating envelope of the source term.

Figure 3 is a composite of the operating envelope and the monitors ranges. The ranges are drawn along the right abscissa. Figure 3 shows that the range of the condenser offgas monitor covers the whole range of the postulated envelope. However, the capability of the main steam line is limited to detecting approximately 1 gpm with 1% fail fuel and approximately 60 gpm under realistic conditions, thus, falling short of detecting the 1 gpm maximum imposed by plant Technical Specifications.

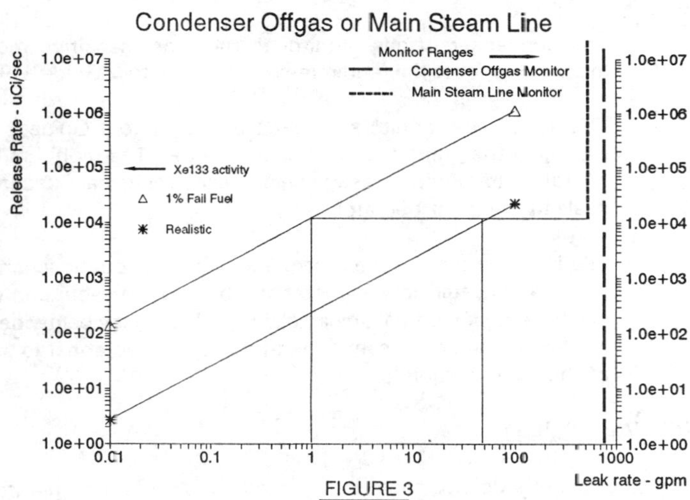

FIGURE 3

Similar graphs have been drawn for the blowdown monitor and main steam line N^{16} monitors. These graphs, when taken in toto, provide a good overview of the advantages and limitations of the different type of monitors for different plant operating conditions.

Figure 4 is a composite of the results obtained by this assessment. It illustrates the relative measuring ranges of the different kind of monitors for extremes levels of both fission and activation source terms over a wide range of postulated steam generator tube leakage rates.

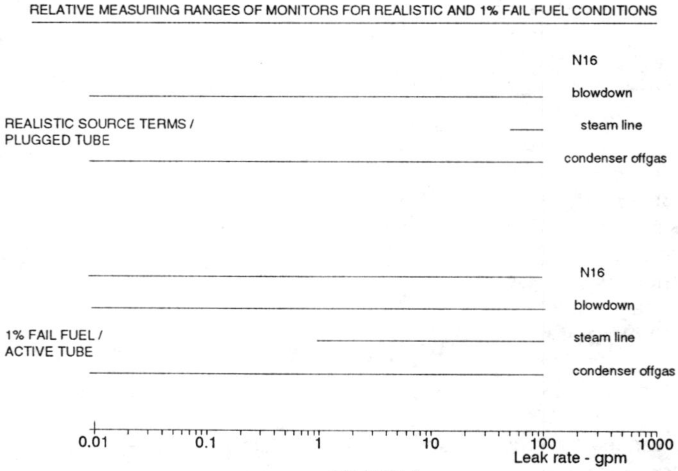

FIGURE 4

This graph confirms what operating experience has shown and, in particular, it suggests the following:

- The condenser offgas monitor is the most sensitive monitor for leak detection and when modified to meet the requirements of Regulatory Guide 1.97 provide adequate extended range.

- The steam line monitors required by Regulatory Guide 1.97 are useful to detect large leaks providing the primary activity is also large. The monitors have little value during normal plant operation to monitor leakage below plant Technical Specefications value and, as a result, could mislead the plant operator.

- The N^{16} monitors on the steam lines have adequate sensitivity to detect small leaks on active tubes during full power operations but are ineffective in detecting leaks from plugged tubes because the added time delay allows N^{16} to decay below detectable levels. Operating with plug leakage in multiple steam generators has the potential to significantly reduce the effectiveness of this type of monitor.

ON-LINE DISPLAY

The above methodology could be adapted to provide the operator with on-line CRT display of each monitor capabilities. A CRT display of each monitor graph could be provided on a demand basis. This approach would complement advantageously existing operating procedures.

CONCLUSION

The above methodology effectively helps analyze the capabilities of monitors installed on different processes lines and present results on a system basis. It can be concluded that the Regulatory Guide 1.97 mandated X^{133} steam line monitors have little value during normal plant conditions. These monitors could benefit from advances in detector and digital electronics technology. Also, it allows to evaluate the benefit to be gained by adding N^{16} monitors.

TABLE 1

TYPICAL MONITOR RANGES	SOURCE TERMS (100#pd)	PROCESS CONDITIONS	LEAKAGE RATES
CONDENSER OFFGAS:	REALISTIC (± .1%FF) RC	CONDENSER OFFGAS FLOW	REALISTIC - 100 #pd
$10E^{-6}$ TO $10E^{-1}$ μCi/cc	Xe^{133} 5.1 μCi/gm N^{16} 40 μCi/gm	30 SCFM	SMALL LEAK - 500 gpd
	Rb^{88} .23 μCi/gm Cs^{137} .019 μCi/gm	BLOWDOWN FLOW	T.S. LIMIT - 1 gpm
STEAM LINE (RG 1.97):		$6.1E^{4}$ #/hr	LARGE BREAK - 100 gpm
	REALISTIC SG STEAM PHASE	STEAM FLOW (ONE SG)	
$10E^{-3}$ TO $10E^{+3}$ μCi/cc	Xe^{133} $1.7E^{-6}$ μCi/gm N^{16} $1.6E^{-7}$ μCi/gm	$1.2E^{7}$ #/hr	
	Rb^{88} $1.4E^{-9}$ μCi/gm Cs^{137} $1.2E^{-9}$ μCi/gm		
BLOWDOWN:			
$10E^{-6}$ TO $10E^{-1}$ μCi/cc	REALISTIC SG WATER PHASE		
	N^{16} $1.6E^{-6}$ μCi/gm Rb^{88} $1.4E^{-9}$ μCi/gm Cs^{137} $1.2E^{-9}$ μCi/gm		
N^{16}:	1% FAIL FUEL RC		
$10E^{0}$ TO $10E^{+3}$ gpd	Xe^{133} $2.6E^{+2}$ μCi/gm Rb^{88} 8.9 μCi/gm Cs^{137} 1.8 μCi/gm		
	N^{16} STEAM:		
	ACT TB $4.2E^{-5}$ μCi/gm max. PLUG TB $7E^{-14}$ μCi/gm min.		

0074-056X/95/183-192/$0 + .50pp

The Reactor Protection and Control Process Instrumentation Digital Upgrade Project Testing Program at the Donald C. Cook Nuclear Power Plant Units 1 and 2

William G. Sotos
Senior Engineer
American Electric Power
1 Riverside Plaza
Columbus, Ohio 43215

Robert C. Carruth
Manager I&C/Nuclear Design Engineering
American Electric Power
1 Riverside Plaza
Columbus, Ohio 43215

KEYWORDS

Digital Control Systems

ABSTRACT

The Donald C. Cook Nuclear Power Plant, located in Bridgman Michigan, recently completed the installation of new digital based process instrumentation in the Reactor Protection System as well as all the major control and monitoring systems for both Units 1 and 2. A major contributor to the success of this project was an extensive testing program developed and executed by the American electric Power (AEP). The major elements of this testing program included qualification, EMI/RFI susceptibility, factory acceptance, site acceptance, pre-installation integration, pre-operational, and calibration testing. In addition, numerous other special tests were performed to address specific issues and/or concerns. This paper will explain the overall program, define the concepts and reasons behind each element, discuss implementation details of each major element, and provide insight into the types of problems anticipated and experienced.

INTRODUCTION

The Reactor Protection and Control Process Instrumentation Digital Upgrade Project at the Donald C. Cook Nuclear Power Plant (DCCNP) Units 1 and 2 located in Bridgman, Michigan has had a rather long and interesting history. The path to completion, while torturous at times, ultimately led to successful installation and operation. This project was arguably the largest I&C project of its kind ever performed on an in-service nuclear power plant. The fact that the new equipment was tested, installed, and started-up off the critical path during routine refueling outages, and operated with no significant technical or operational difficulties, is a significant positive achievement for AEP and the nuclear industry in general.

The Donald C. Cook Nuclear Plant consists of two (2) Westinghouse four (4) loop PWR's, each rated at 1100 MW (nominal). They are somewhat unique because of their ice condenser containments. Unit 1 came on line in 1975 and Unit 2 in 1978. AEP acts as its own Architect/Engineer. The Reactor Protection and Control Process Instrumentation Digital Upgrade Project was conceived in 1988 when it was determined that the existing Foxboro H-Line equipment had to be replaced because of reliability concerns (due to aging), obsolescence, and limited vendor support capability. The formal project was begun in the fall of 1989 and was completed in 1994 with the installation in both units. AEP performed much of the design work in-house which had a significant positive impact on cost containment.

The scope of work was made up of two (2) major parts:

1. Reactor Protection (safety related)
2. Control Systems (non-safety related)

For each unit the Reactor Protection portion consisted of 13 racks of equipment divided into 4 channels or sets. These are referred to as Protection Sets.

The Control System portion consisted of 16 racks of equipment in each unit divided into 4 groups and 2 additional miscellaneous groupings. These are referred to as Control Groups. Approximately 60 front control panel instruments were also replaced in each unit.

Refer to Figures 1 and 2 for a simplified pictorial representation of the project scope.

There were four (4) major areas of this project that required significant attention and resources. These were:

1. Project Management
2. Licensing
3. Engineering/ Design
4. Testing/ Installation

Because of the myriad of tasks and the large number of interfacing organizations (both internal and external), tight project management was very important. The necessity of keeping on a very tight schedule and within budget made project management critical to the success of the project.

The licensing area was one that evolved over the course of the project. The initial intention was to implement this project under 10 CFR 50.59 (no NRC review required). Minimal licensing involvement was anticipated because there was no need to modify any Technical Specifications, there was no impact on the FSAR, and because the upgrade was a direct functional replacement. In response to NRC concerns with digital upgrades in general, we redirected our efforts to support having a full inspection and a subsequent Safety Evaluation Report (SER) issued prior to implementation. As a result, licensing activities increased dramatically. The path ultimately taken was very similar to that now being followed by other utilities undertaking digital upgrades. The NRC performed an inspection in 1993 and a SER was issued in early 1994.

Engineering and Design was the area that consumed the most resources. Required activities included writing equipment specifications, equipment qualification (temperature/ humidity, radiation, seismic, EMI/ RFI, ESD, electrical isolation), equipment/ system design, vendor drawing review, power quality, grounding, setpoints/ dynamic constants, response time, failure modes and effects, functional and equipment diversity, software V&V, configuration management (hardware and software), equipment acceptance testing (factory and on-site) support, procedure writing support, and installation support. In addition, a significant amount of time was spent managing equipment vendors and consultants.

Of these four (4) areas, testing and installation were the most critical activities. Through an extensive and comprehensive acceptance testing program that lasted approximately 9 months, we were able to identify and correct numerous design and equipment flaws, several of which would have had serious consequences for the project had they not been discovered prior to installation and operation. The major elements of this testing program included:

- EMI/ RFI Susceptibility Testing (ST)
- Qualification Testing (QT)
- Factory Acceptance Testing (FAT)

- Site Acceptance Testing (SAT)
- Pre-Installation Integration Testing (IT)
- Pre-Operational Testing (Pre-Op)
- Calibration Testing (CT)

It is important to note that the testing program elements were technically driven. Satisfying regulatory requirements was also a goal, but this was not the primary driving force in the test program development. As a result, the testing program successfully uncovered system design and equipment problems in sufficient time to permit corrective actions. In addition, the program made the plant technical staff intimately familiar with the new hardware, thus making installation and start-up go very smoothly.

Because of the enormous amount of technical detail required for this project, no attempt is made here to present more information than necessary to provide a general description of each major project test plan element. This includes basic concepts, techniques and standards used, as well as examples of problems found and their effects and corrective actions.

TESTING PROGRAM DESCRIPTION

EMI/RFI SUSCEPTIBILITY TESTING

EMI/RFI was an area that received a lot of attention from the NRC during the licensing process. There were, as well, valid technical reasons for us to pay considerable attention to this issue - but not particularly related to the digital aspects of the technology. The NRC was concerned that EMI/RFI might cause some new failure mode or effect with digital equipment. They were also concerned with radiated EMI. The testing we performed showed that:

1. Radiated emissions in our plant were typically several magnitudes lower than those imposed during our testing program (for example 10 v/m tested vs 0.005 v/m actual site level at 1 MHz) and were thus a non-factor on our designs.

2. Digital equipment performs much the same (sometimes superior) as analog equipment when exposed to radiated EMI/RFI. No new failure modes were found.

3. Conducted emissions in our plant were close to and sometimes slightly exceeded the generic test levels specified in the test standard that we applied (MIL STD 461C).

4. Conducted emissions can create real problems as described later in this paper.

5. No special shielding was necessary. However it is prudent and quite often necessary to follow established industry practices and standards when designing a new or retrofit installation.

The testing was formally done using MIL STD 461C as the primary basis. We used CS01, CS02, CS06, RS02, and RS03 as the specific susceptibility tests. For emissions, we used CE01, CE03, RE01, and RE02. The contracted test lab wrote the test procedures which required AEP's approval and, in the case of the Protection Set hardware, the equipment vendor's approval as well. In all cases, AEP specified the equipment performance criteria.

Our testing was performed on a rack (Protection Sets) or racks (Control Groups) built to replicate the actual equipment to be installed. No special materials or gasketing was used to shield the equipment. In our tests, the Control Group racks used for seismic and environmental testing were also used for EMI/RFI susceptibility testing. An actual rack built to be installed was used for the Protection Set testing. During the tests, the equipment was energized with inputs simulated and dummy outputs connected. I/O as well as power supply closed to duplicate operating conditions. External I/O and test instrument lead wires were also shielded to replicate actual installed conditions.

The EMI/RFI tests were actually run by a composite team comprised of technicians and engineers from both AEP and the test lab. The AEP personnel set up and operated the test specimen, including the I/O, while the test lab personnel set up and operated the test equipment.

There were no serious problems that occurred during the testing. The only problem of note was the effects the external I/O and test instrument lead wires initially had on the equipment performance before shielding was added (a test setup inadequacy). Stray EMI was being picked up, somewhat degrading equipment performance. Once the test leads were shielded, the equipment performed within the specified performance criteria.

Formal test reports which contained the test plan, test results, test anomalies, test equipment set-up, test specimen description, configuration, and set-up, and test results, were submitted to AEP for review and approval. All of the equipment performed within its specified performance parameters so there were no significant anomalies that required explanation.

QUALIFICATION TESTING

Equipment qualification was performed according to typical nuclear industry standards with respect to seismic and environmental parameters. IEEE 344-1974 (seismic) and IEEE-323-1975 (environmental) were used as the primary standards.

Actual seismic and environmental tests in replicate racks (Control Group) were performed as well as analytical methods (Protection Sets) to qualify the new instruments in their installed locations. For the Control Group hardware, the contracted test lab wrote the test procedures which required AEP's approval. AEP specified the equipment performance criteria. In the case of the Protection Set hardware, the equipment was previously generically qualified and was subsequently qualified for our specific application by analysis.

As with the EMI/RFI testing, the Control Group seismic and environmental tests were run by a composite team comprised of technicians and engineers from both AEP and the test lab. As with the EMI/RFI susceptibility testing, the AEP personnel set up and operated the test specimen including the I/O while the test lab personnel set up and operated the test equipment.

Environmental tests were performed to show that the equipment could function under extreme conditions. This involved cycling ambient temperatures between 40 degrees F and 122 degrees F over the course of approximately two (2) days. Relative humidity was also cycled between 50 and 90 percent.

Because the new equipment was to be installed in existing racks that are not structurally rigid, we elected to perform shake table testing on the replicate Control Group racks. DCCNP is located in a low seismic activity area that results in a DBE floor response spectra peaking at less than 1.5 G's at 2 Hz. Testing of our equipment was performed at these levels (with margin). While the test racks shook and flexed considerably, there were no equipment failures during or after the testing. We did have some test panel fastener failures which prompted us to replace the fasteners with a more robust type.

Environmental testing that was performed on the replicate Control Group racks had three (3) main purposes:

1. Verify the heat rise characteristics of the racks.

2. Verify that the equipment would perform under extreme temperature conditions.

3. Verify that equipment performance/reliability was not compromised by reasonable temperature cycling.

The racks are constructed such that they have open tops covered with a mesh screen, and have louvers in the rear doors. There is no forced ventilation. Rack heat rise was shown to be minimal. The worst hot-spot occurred directly above the bulk power supply and was approximately 15 degrees F above ambient. Other areas were 5-10 degrees F above ambient. This was consistent over the entire cycled temperature range.

Extreme temperatures proved to not be a problem for any of the equipment. The highest temperatures reached internal to the cabinet (based on 122 degrees F ambient that would be caused by a loss of HVAC under a station blackout scenario) were slightly less than the 140 degrees F design temperature limits specified for the equipment. No adverse effects were found.

Temperature cycling did actually cause a Control Group equipment problem but the root cause was not determined at the time it occurred. Repairs were made and the test resumed without further incident. The problem noted involved a situation where two modules independently went into an unconfigured condition at different times. This was corrected by downloading the correct configuration which resulted in proper operation being restored. Subsequent tests at the vendor were inconclusive as to the root cause. It was not until much later, when the problem reoccurred during the SAT, that the root cause was determined (defective PROM sockets). This situation will be discussed further in the SAT section.

There were no other serious problems that occurred during the testing. Formal test reports, which contained the test plan, test results, test anomalies, test equipment set-up, test specimen description, configuration, and set-up, and test results, were submitted to AEP for review and approval. Other than the previously described problem, all of the equipment performed within its specified performance parameters, so there were no significant anomalies that required explanation.

FACTORY ACCEPTANCE TESTING

Factory Acceptance Testing was pre-installation testing that verified that the replacement equipment, including power supplies, interposing relays, rack wiring, input and output modules, and loop software configurations, performed their intended functions as required by both the equipment vendor and AEP system design specifications prior to acceptance of the equipment for delivery.

Testing at the vendor shop was performed only on the Protection Set racks. Because the Control Group racks were built at a third party fabrication shop using AEP design (no direct equipment vendor participation), formal Control Group testing was performed on-site using procedures similar to the Protection racks'. A less formal checkout was performed at the third party fabrication shop to verify that the basic design requirements were met and that the fabrication and assembly were sound. Control Group testing is discussed in more detail in the SAT section.

The FAT on the Protection racks was performed in the vendor's shop using formal written, meticulously detailed procedures. Test procedure requirements were provided to the vendor by AEP. The actual procedures were reviewed in detail by AEP and eventually approved for use after the review cycle was complete. These test procedures included basic power and functional checks such as inrush current, voltage variations, and power-up/power-down tests. Loop calibration and functional checks were also performed as were dynamic and time response testing (where applicable). The test procedures required documentation of the complete test history including a log of discrepancies, a record of test personnel, test equipment, test set-up and configuration.

In order to perform the FAT, the racks were staged at the vendor shop in as close as possible to the actually installed physical configuration. This included simulating inputs and outputs by use of transmitter simulators, decade boxes (for RTD inputs), analog meters, dummy load resistors, and relays. For any one set of racks, all I/O was connected at once to replicate actual operating conditions. Signal levels were nominally set to those corresponding to normal plant operation at 100% power. In addition, some signals not used during power operation were set at a nominal setting to conservatively add load. The actual software configuration intended to be used in the plant was used for this testing.

The FAT was actually performed and documented twice. The vendor first performed the test without AEP participation. This allowed the vendor to correct any obvious flaws and be reasonably confident of success on the "official" FAT. AEP performed the second test (which took four (4) weeks per unit) with a small team of engineers and technicians and with the vendor's assistance. We chose to do this ourselves for several reasons:

1. Our understanding of the system operation and maintenance was more in depth than the vendor's, thus resulting in a better opportunity to uncover flaws, particularly those of a more subtle nature.

2. The inherent thoroughness that a second, independently performed test can add to the project's success was considered an advantage.

3. Performing the testing was good hands-on training for those individuals expected to later install and maintain the equipment.

4. The extensive testing provided a long equipment burn-in period.

During the <u>second</u> FAT (performed by AEP), several significant problems were found that were corrected at the vendor shop. These would have created major difficulties had they not been uncovered prior to installation.

For example, one problem involved discrete output modules that provide reactor trip, ESF actuation, and alarm functions. These modules each have four (4) DPDT mechanical relays on them. As we were performing the tests, we found an abnormally high rate of problems/failures with these modules. A subsequent investigation by the vendor found that a bad batch of relays provided by a sub-vendor had made its way into the hardware. As a result, all of the relays in all of the modules were replaced and a Part 21 notification was issued.

Other minor items were uncovered but were easily corrected. Overall, no major system design flaws were found. The FAT went smoothly and was successfully completed on schedule. When we were finished, a record copy of all completed test procedures, test results, data, and software configurations was made. Copies were provided to AEP for final approval prior to authorizing equipment delivery.

SITE ACCEPTANCE TESTING

Site Acceptance testing was performed on the Control Group racks in lieu of factory testing. This was done because the racks were fabricated by a third party vendor (to AEP) using design documents prepared by AEP for equipment supplied by AEP. The system design was done by AEP. Once the racks were fabricated and a basic checkout completed, they were shipped to the plant site. A dedicated and secure staging area had previously been established, complete with a regulated power supply of sufficient size, HVAC, proper lighting, etc. Its close proximity to the Control Rooms also helped later, during the installation phase of the project.

Once the equipment was located in the staging area, proper power, grounding, and I/O connections were made, similar to that done with the Protection equipment at the vendor's shop. Inputs and outputs were simulated to replicate actual plant conditions as closely as possible. The actual front panel hardware that was to be installed was also connected. Once powered up, the equipment remained energized until installation, except when corrective actions required de-energization. As we will discuss a bit later, this burn-in time (6-9 months, depending on the Unit) proved crucial to uncovering the most serious equipment flaw found in the project.

The testing concepts were the same as those used in the Protection equipment. Because the Control Group equipment performs many complex control functions, its procedures were actually more complicated than those for the protection equipment. As a result, this testing took considerably more time to complete (approximately 8 weeks per unit). Minor equipment design and procedural flaws were found and corrected as testing progressed. No major system design flaws were discovered.

As mentioned previously, a very serious equipment flaw was uncovered while the on-site testing was in progress. We experienced a number of the microprocessor based modules in the Control Group equipment switching at random into an unconfigured state. In other words, they would unexpectedly lose the stored configuration data. This caused all analog outputs to go to 0 and discrete outputs to be de-energized. Once this occurred, we sometimes could reconfigure them but sometimes could not. This problem was occurring with such an alarming frequency (a dozen or so within a few months, that we began a full investigation in conjunction with the equipment vendor. What we found was that, on some instruments, the vendor had used a lower quality of sockets for the PROM's that hold the instrument configuration. Because of this, environmental conditions, even as benign as those to which the new hardware was being subjected, over the course of time would cause an electrical continuity problem between the sockets and the PROM's. When that happened, the equipment's watchdog timer would time out and the instrument revert to an unconfigured state. In some cases this was only intermittent. In others it was a fixed state. This failure assessment explained the previously unexplained failures that occurred during the environmental testing.

The corrective action for this problem was to have the equipment vendor perform a replacement of the sockets on all socketed IC's. Since this was a lengthy process, we had to proceed with our testing program using the suspect equipment. We did this to prove out our system design as well as the other system components. Once the repaired equipment was returned, additional tests were performed to verify each device worked properly in its intended application. Since the repairs were made, we have not seen a recurrence of this problem.

Documentation requirements were similar to that of the Protection equipment FAT. When all tests were finished, a record copy of the completed test procedures, test results, data, and software configurations was made. The test results were reviewed for final approval prior to authorizing equipment acceptance for installation.

PRE-INSTALLATION INTEGRATION TESTING

Formal Integration Testing was the last effort prior to installation. The concept behind this testing was to replicate the entire installation as closely as possible, including the interconnection of all Protection and Control racks and the connection of all I/O, including all of the new front panel devices similar to that done previously during the FAT and SAT. Once energized, basic functional checks were performed on all loops. This was done to determine if there were any remaining interface problems either between the new Protection Set and Control Group equipment or with the existing devices. Power-down/Power-up tests of various iterations were also performed to verify failure mode/effect assumptions.

Once all of the connections were made, the equipment was energized and the testing commenced. As during the previous tests, we found a significant problem that required correction. The Protection Set equipment uses a digital based module to perform its trip functions among other things. We observed during the integration tests that on certain occasions (seemingly random at the time - later found to be repeatable), certain modules would go into an "error standby" mode, which means that the device would fail to its default state (0 for analog outputs and de-energized for discrete outputs) and would require rebooting to restore its functionality.

This was at first very puzzling because it was not a true equipment failure. We discovered that the problem was a result of coupled electrical noise induced by the inductive kick of the test relays as they actuated during the testing. This was affecting the module's diagnostics, causing the watchdog timer to time out. Once this happened, the module's response, as designed, was to go to the "error standby" mode.

In attempting to reduce this electrical noise problem, we evaluated a number of techniques including careful separation of analog DC wiring from AC wiring, but eventually traced the problem to Metal Oxide Varistors (MOV's) inadvertently omitted from the coils of some of the test set-up output relays. There were no further occurrences during this phase of the testing. Once the equipment was actually installed, the problem reappeared in a few locations while undergoing pre-operational testing.

Prior to actual operation, we isolated the problem to the inductive noise being picked up from various sources by specific wiring tied directly into a specific diagnostic output of certain modules. This particular output was directly connected (internal to the module) with the module's diagnostic circuitry. Since we were not applying this feature (external card failure alarming), the wiring was removed. The problem has not recurred. The vendor later issued an informational notice to all nuclear users regarding this particular concern.

Other than the problem just mentioned, the integration testing went well and was completed, as scheduled, in a period of approximately three (3) weeks per unit.

PRE-OPERATIONAL TESTING

Pre-operational testing was the first of the post-installation checks. Once QC was completed, these tests were formally performed. They consisted mainly of wiring continuity checks, power supply checks, and basic functional testing from the primary sensor to the final control device and/or indication. Special attention was paid to ensure that sufficient overlap was provided in the loop checks so that there were no untested interfaces.

The only problems found were minor things such as an occasional wiring error, configuration changes, etc. Other than the electrical noise coupling problem initially encountered during the integration testing, there were no significant difficulties. This phase of the test program went well. Approximately 2-3 days was spent on any one Protection Set or Control Group.

CALIBRATION TESTING

This was the last step in the testing program. These tests were the initial performance of the routine formal calibration/maintenance procedures that will be performed whenever maintenance is done on the loops. Approximately 180 procedures were generated for each Unit. They consisted mainly of meticulously detailed calibration tests/checks which also verify functionality. Time response testing was also performed on applicable Protection Set loops to satisfy Technical Specification requirements.

Even with the extensive testing that we had performed up to this point, one embarrassing event did occur just prior to operation. The Pressurizer Pressure controller was incorrectly configured to operate the reverse of what was required. When an increase in pressurizer pressure was demanded, the controller incorrectly sent out a signal to decrease pressure and vice versa. This was the only significant design flaw that found its way completely through the engineering, design, installation, and testing process. Very early in the design process we had incorrectly stated how the device was expected to function. This mistake was overlooked and thus carried through in all of the design documents, as well as in all of the test procedures. A minor pressure transient resulted when we first started to heat-up. The problem was easily diagnosed and corrected within hours of discovery. As a result of this experience, we further enhanced the testing program by adding some basic, operational-type functional tests which were performed just prior to turning the equipment over to Operations.

SUMMARY

Thanks in large part to the extensive testing program AEP conceived, designed, and executed, installation and start-up went smoothly on both units with respect to the new equipment. Start-up was uneventful and everything performed as designed. Once a stable power level of 30% was reached, some tuning of the Feedwater Flow/Steam Generator Level loops was performed. After that, operation was routine and the equipment continued to perform well right up to 100% power.

An extensive test program such as described herein, while very resource intensive, is necessary to successfully implement any large I&C retrofit project. As demonstrated at DCCNP, without all of the testing that was performed our installation effort would have been much more difficult, resulting in a significantly large negative impact on outage duration (we were never on outage critical path). In addition, the probability of having two flawless start-ups, such as we had, would have been greatly reduced. The safety and economic benefits of this accomplishment were very significant.

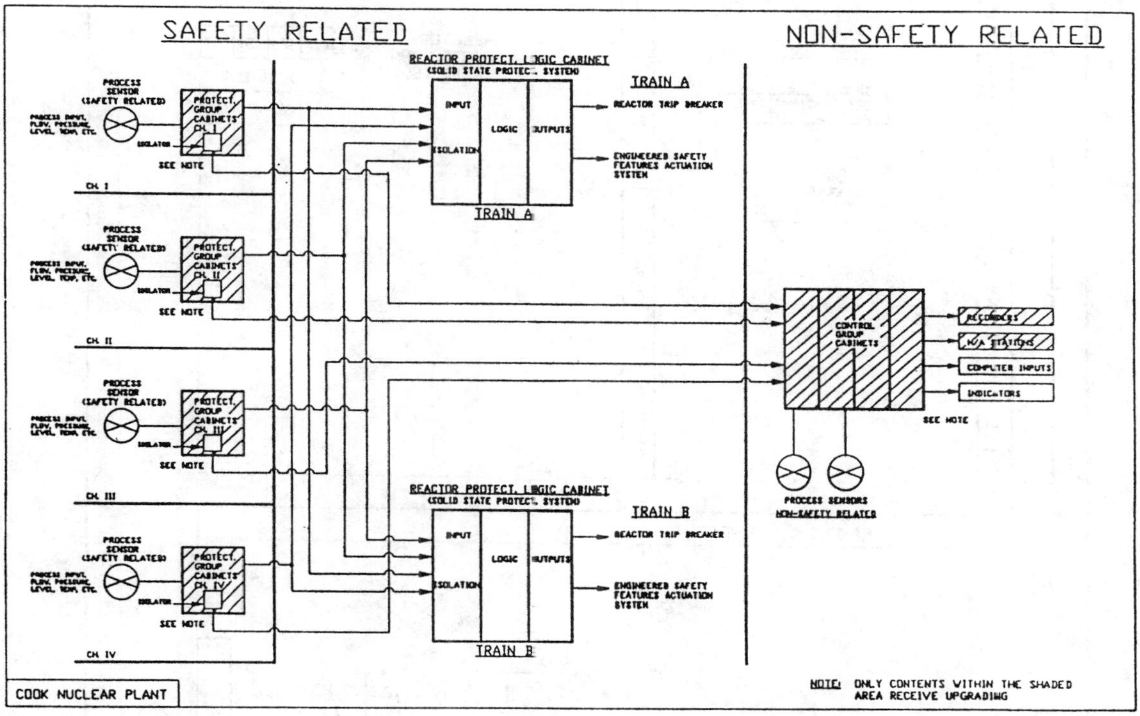

Figure 1

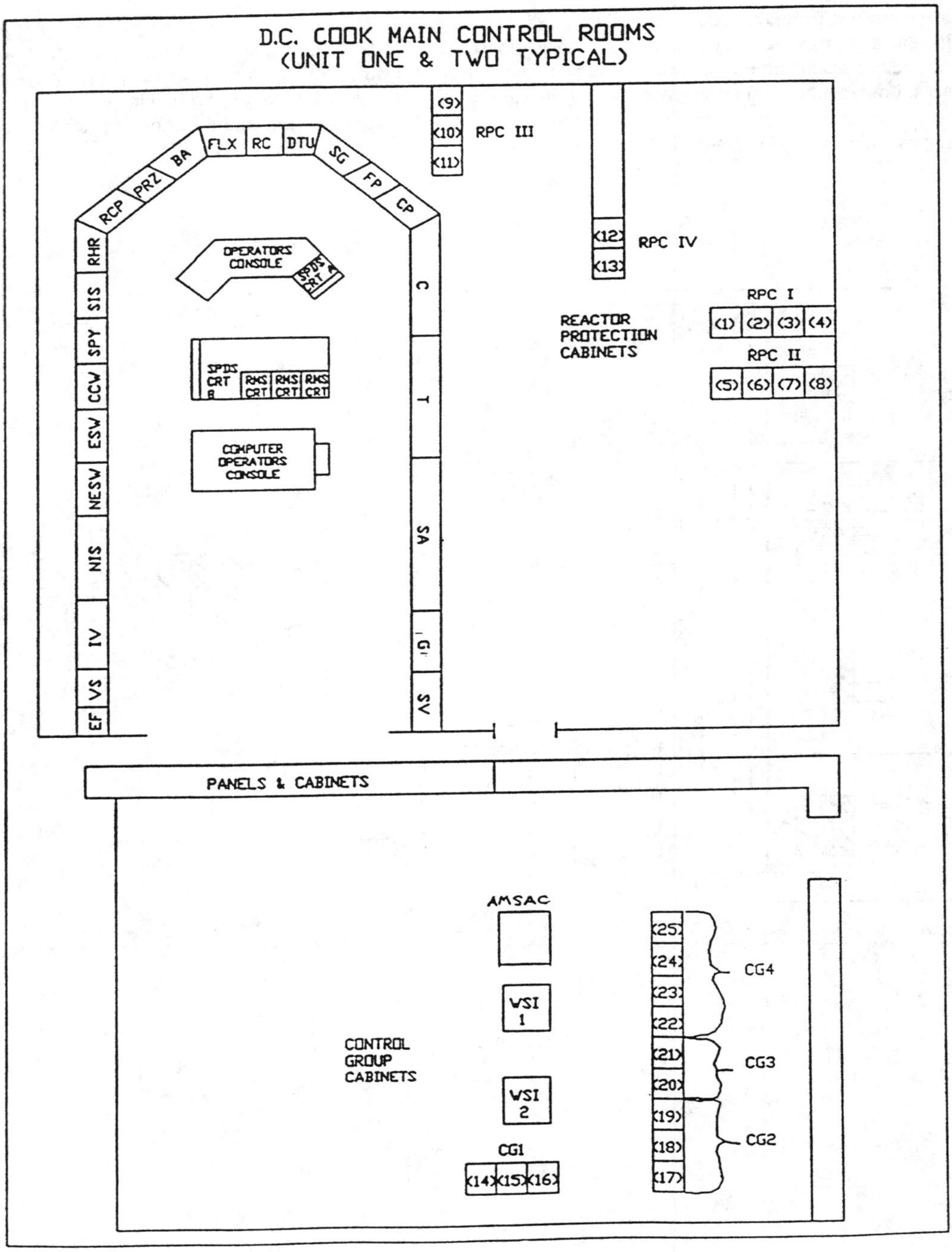
D.C. COOK MAIN CONTROL ROOMS
(UNIT ONE & TWO TYPICAL)
RPC III
RPC IV
RPC I
RPC II
REACTOR PROTECTION CABINETS
OPERATORS CONSOLE
SPDS CRT A
SPDS CRT B
RMS CRT
COMPUTER OPERATORS CONSOLE
RCP
PRZ
BA
FLX
RC
DTU
SG
FP
CP
RHR
SIS
SPY
CCW
ESW
NESW
NIS
IV
VS
EF
C
T
SA
G
SV
PANELS & CABINETS
AMSAC
WSI 1
WSI 2
CONTROL GROUP CABINETS
CG1
CG2
CG3
CG4

Figure 2

074-056X/95/193-200/$0 + .50pp

DEVELOPMENTS IN ELECTROMAGNETIC COMPATIBILITY FOR DIGITAL RADIATION MONITORING SYSTEMS (RMS)

John Meisner
Principal Engineer
Sorrento Electronics
10240 Flanders Court
San Diego, CA 92121

KEYWORDS

Power Plant, Computer Control, Microcomputer, Radiation Instrumentation, Smart Sensors

ABSTRACT

Recently the Nuclear industry has begun to address the Electromagnetic Compatibility requirements for Nuclear Power Plant equipment. European and U.S. commercial and military guidelines exist that address Electrically Fast transients (EFT), Radio Frequency Interference (RFI), Electrostatic Discharge (ESD), and power line surge transients. Radiation Monitoring Systems typically include wide bandwidth/high gain front-end electronics. Additionally, high-speed digital processing opens the way for significant emissions. This paper outlines engineering tests addressing these concerns. Test analysis and subsequent equipment modifications are discussed, including software noise rejection improvements, new filter hardware and installation, grounding improvements, improved cable shield and enclosure terminations, and the addition of smart sensors.

INTRODUCTION

Design philosophy for Nuclear Power Plant Radiation Monitoring Systems has undergone a change. Use of high frequency transmitters has become more commonplace. These transmitters cause spurious radiation alarms. Simultaneously, utility companies are investigating ways to lessen radioactive releases by reducing radiation monitor Minimum Detectable Concentrations (MDCs) and, therefore, lower alarm setpoints. To ensure reduced releases, appropriate time responses must be maintained. Most methods of decreasing MDC and maintaining time response increase monitor vulnerability to conducted and radiated Electromagnetic Interference (EMI). Since EMI and Radio Frequency Interference (RFI) are considered *extraordinary* events, they have not been well defined as a design basis for either software or hardware. The European Union has passed EMC Directives that will affect how U.S. exporters certify and sell electronic instruments, digital computers, and other electrical products. Compliance will be mandatory by December 31, 1995 for suppliers and manufacturers. Because we sell Radiation Monitoring Systems (RMS) to Nuclear Power Plants in the European community, compliance becomes particularly important. In the United States, Electric Power Research Institute (EPRI) and the Nuclear Regulatory Commission (NRC) are specifying Electromagnetic Compatibility (EMC) testing guidelines. To some extent, these guidelines are based on plant field mappings. These guidelines are, therefore, more applicable to the Nuclear Industry. Existing RMS designs should be surveyed to determine if their EMI/RFI susceptibility is acceptable, and retrofitted if necessary to minimize the problem.

OLD VS. NEW DESIGNS

Many utility companies have spent large sums of money to decrease susceptibility in old RMS designs. Many others simply deal with operational anomalies. Additionally, the industry is formulating guidelines for analog-to-digital RMS conversions. The latter has higher noise immunity; however, an RMS digital system failure could be catastrophic (i.e., a monitor may block or lock up and therefore nullify its safety function). To ameliorate the problem of staff having to physically go to an RMS monitor to reset it (an undertaking that could prove formidable), most systems are designed with remote reset capability. This solution addresses only the symptom. To provide a systematic EMC solution, more attention must be paid to high frequency transient control.

Newer RMS designs implement superior Printed Wire Assembly (PWA) design techniques, including multi-layer boards, copious decoupling capacitors, short trace lengths or terminated traces, minimized crosstalk, minimized loops, etc. The advent of higher frequency Switching Mode Power Supplies (SMPS) presents yet further emission and susceptibility issues. The advent of high speed microprocessors, microcontrollers, Digital Signal Processors (DSP), and First-In-First-Out/Analog-to-Digital Converters (FIFO/ADCs) provides a means to implement additional noise rejection in software. This approach presumes that induced transients will not upset data/address buses or reset the processor, and no catastrophic power supply perturbations occur. Use of high speed distributed processors must be designed with RFI emissions in mind. Otherwise, when common mode high frequency currents travel over cables, the possibility for increased RFI emissions exists.

Our old generation monitors were designed in the 70s. The new were designed in the 90s. The skid design (i.e., wireways, conduits, enclosures, pumps, piping, and design philosophy) is intended to be the same. The computer boxes are vastly different in two areas: board design and microprocessor. The old design uses an 8085 (4MHz) microprocessor for central processing, and the new an 80960 (16 MHz). The 80960 communicates with "distributed" processors: Digital Signal Processing (DSP) (TMS32031, 32MHz) and microcontrollers (8051, 16 MHz). A proprietary backplane (mechanically identical to VME) replaces a two-layer motherboard. Barsebäck Nuclear Power Plant in Sweden currently has the only installation of the new monitors. This prototype determines fission product concentrations in the steam (i.e., indication of failed fuel in a Boiling Water Reactor (BWR). Operational evaluation is ongoing.

PRELIMINARY EMC TESTING/INVESTIGATIONS

New European standards and EPRI guidelines, coupled with past plant experience, drove us to investigate EMC for the next generation monitor design. The following questions needed to be answered:

- Are RFI gaskets needed?
- Are emissions excessive?
- What points on the monitor are particularly susceptible?
- To what frequencies is the monitor sensitive?
- Are induced current common mode or differential mode?
- What is the needed Insertion Loss (IL) vs. frequency for added filters?
- Is the intra-skid "grounding" acceptable?
- What process variables show the largest perturbations?

We performed investigative tests on the most complex monitors to help us formulate an improved EMC plan. Tests of this skid-mounted design were performed at a local lab in an "outside area". We chose not to use an anechoic chamber due to the monitor size and complications from Electromagnetic wave reflections. The tests were based on IEC 801-3. Antennae were swept from 27 to 500MHz and located one meter from the monitor. The antennae were calibrated with no monitor in place to yield a field strength of 10 volts/meter. The amplifier was adjusted to obtain a "maximally flat" output over the frequency range.

The significant test results were:

- Skid ground connections using conductors of any significant length (i.e., greater than two feet) did not provide low impedance connections at high frequencies. Instead, they acted as antennae, transmitting energy into the enclosures and wireways!

- Emissions were within limits.
- The flow measurements were most susceptible. Radiation values exhibited no discernible perturbation.
- Using current probes, we determined that significant common mode currents were induced.
- Serial communications to a computer located within a shielded trailer were lost at key frequencies. This problem was eliminated by shield termination modifications.
- Minor flow sensitivity was exhibited at frequencies greater than 400MHz.

Based on these results, we implemented the following modifications:

- Remove external ground wire connections that penetrate monitor enclosures. Instead, use the conduit surrounding external cables to connect the skid to station ground. (More will be said about this later.)
- Limit high frequency energy to the interface junction box, disallowing its penetration into sensitive electronic areas. This requires "bulkhead" mounted filters and suppressors. Based on the results, we specified custom filter connectors and boxes (i.e., AC mains).
- External cable shields must be properly terminated (i.e., to eliminate — or at least reduce — "pigtail" antennae) in the interface junction box.

Formal testing was subsequently performed to a combination of IEC standards and EPRI guidelines. These tests included susceptibility to conducted, RFI, EFT, surge, and Electrostatic Discharge (ESD) transients.

PLANT EMC PROBLEMS AND RESOLUTIONS

During the last year, I have observed two interesting EMC problems on our presently installed radiation monitors. One was related to RFI, and the other to EMI.

Sizewell B Power Plant (England)

The British have formulated their own EMC testing standards. In particular, our RMS monitors were required to meet RFI suppression requirements defined in section 4.20 of SXB-IJ-020804 for class X(mild). The tests were conducted *in situ*. The RMS system would not be turned over to the operations group until it passed these tests. We were informed that successful RFI testing of these monitors was on the critical path for plant startup.

The monitors sample containment air and determine levels of radioactivity. Pumps located on our skids provide sufficient change in detector volume to provide acceptable transient response. Beta scintillator detectors are located within Pb shields to reduce background and, therefore, improve MDC. Light generated by the beta scintillator is then amplified by a

Photomultiplier tube. No further electronic amplification occurs at the detector. Monitor software then calculates radiation levels. Via 4-20 mA current loops, our monitor sends its calculated values to the Plant Protection System (PPS). Additionally, the monitor generates local and remote alarms. To ensure that sample line filters are not clogged, the system implements a "low flow" trip.

Since *in situ* testing was required, the plant RFI group calibrated antennae in an open area to generate the necessary electric field intensity (i.e., 10 V/m). To adequately test RFI susceptibility, the test standard states that the antenna is swept from 20 Mhz to 1 GHz. We went a step farther. Resonant and known communications frequencies were dwelled at, to ensure no alarms occurred.

Radiation front-end circuitry was not susceptible. Once again, however, the flow instrumentation exhibited high susceptibility. The front-end electronics were enclosed in a box connected to the skid, with all external cable connections via conduit. The flow probe is a very sensitive thermal anemometer, with flow control designed for either MANUAL or AUTOMATIC mode.

When we imposed 10 volts/meter fields, calculated flow perturbations occurred. The nature of these perturbations indicated cable/conduit resonances, the largest magnitudes occurring at ~90 Mhz. We expected a previous change in skid grounding to eliminate grounding conductor resonances; therefore, we believed that conduit resonances and RF ingress caused this behavior. Apparent flow would increase as measured by the flowmeter electronics. The control system would, therefore, act to reduce flow. A low flow alarm would occur, requiring operator attention. In MANUAL mode, a worse anomaly occurred — apparent flow increased to the maximum and then folded over to 0.0 liters/minute. With RFI removed, the flow remained fixed at 0.0 volts until powered off/on.
We identified two options: either identify and eliminate the source of RFI ingress; *or* make the flowmeter box more robust with respect to RFI. An analysis of the PWA layout and grounding scheme/shield terminations indicated that the circuit was highly susceptible. Using a "loop antenna", we identified many sensitive conduits and conduit connections. Therefore, we chose to accept the noise and attempted to minimize its effect. An acceptable perturbation was defined as that which caused no "low flow" alarm. If not for the foldover phenomenon, we could achieve this by limiting the extent to which the flow control valve could close; as a result, the control algorithm would not cause the alarm.

We determined that the unusual behavior was caused by the electronics that control the current source driving the bridge circuit. Rectification and filtering by the bipolar input operational amplifier induce large offsets when exposed to RFI. In effect, the high frequency is downconverted to DC. The current loop then turns off. Enough RF attenuation must be added to prevent this condition. In particular, high frequency currents must be highly attenuated before entering the flowmeter box. We used NiZn ferrite beads to accomplish this. Placement around incoming AC, and input from sensor and output wires to the computer box reduced interference to a level at which we could be certain no foldover would occur.

Beznau Power Plant, Units I and II (Switzerland)

Beznau Nuclear Power Plants I and II have experienced blocked monitors (i.e., not communicating with the control room) and spurious radiation alarms directly related to plant EMI. The most severe events are those due to switching of inductive loads and are called Electrical Fast Transients (EFT). These events cause bursts of high voltage pulses. Since RMS cables are routed in the same cable trays as cables to the inductive loads, electric field coupling occurs. These fields couple into RMS cable shields. Since the risetime of a typical induced pulse is ~5 nanoseconds, we needed to consider the transmission line characteristics of the shield and termination. Incoming AC and Data/Communications/Control (DCC) cables all had potential for this type of crosstalk.

We could not resolve the crosstalk problem simply with ferrites, as at the Sizewell Plant. Therefore, we set up a test to simulate the above transients using European standard IEC 801-4. The goal was to meet as high a severity level as possible. The key parameters were as follows:

- pulse risetime = 5 nsec
- pulse width = 50 nsec
- pulse frequency = 2.5 Khz or 5 Khz depending on severity level
- pulse amplitude = 1 KV for signal(2 KV for AC) or 2 KV for signal(4 KV for AC) depending on severity level
- burst duration = 15 msec
- burst period = 300 msec

We injected these test signals onto AC mains and DCC twisted shielded pair (TSP) cables by using a coupling clamp specified in the IEC 801-4 standard. The test pass/fail requirements invoked were:

- no monitor blocks
- no occurrence of inadvertent alarm occurs

We know that a microprocessor upset occurs if a noise generator cable is coiled inside the computer box and 2 KV pulse bursts are coupled. To avoid this situation, high frequency energy must be confined to the Customer Interface Junction Box (CIJB). All non-AC cables enter the monitor skid at the CIJB.

We implemented the new monitor skid design modifications outlined in a previous section to accomplish this. To reiterate:

- Bulkhead mounted filters/suppressor were installed at the CIJB interface to the computer box.
- "Ground" wires connected outside the "Faraday cage" provided by enclosures, with wireways and conduits were removed.
- Shields were terminated immediately upon entering the CIJB to reduce radiating

efficiency.

Unfortunately, results of our testing of these modifications were poor. Further investigation revealed the problem. The low impedance of the shield termination and the coupling clamp at the frequencies of interest result in large noise currents. If these remained on the outside of the "cage", we did not expect them to couple inside. However, the currents flowing back to the ground plane under the monitor, test generator, and clamp *do* couple inside, due to imperfect conduit and wireway fittings. We attempted to direct these currents with ferrites and feedthrough capacitors to overcome parasitic or unintended coupling paths. The improvement was insufficient to satisfy the maximum IEC 801-4 severity level.

Consequently, we investigated a method called "ground window" (used in the computer industry), that prevents high frequency traveling waves generated along the external cable shields from entering the enclosure. We accomplished this by connecting all shields entering the enclosure to a highly conductive ground block. Then, if electromagnetic waves appear they are reflected back. We recommended bonding the conduit between the top of the CIJB and the cable tray. Even if the trays were not well connected to station ground (even though they *must* be for safety), high frequency energy is dissipated as heat in the metal. We terminated the shields as close as possible to the conduit connection point, more to reduce internal antenna currents than for low impedance. Since the shield-cable tray/conduit transmission line's characteristic impedance is between 50 and 100 ohms, at EFT frequencies, even if terminations are 6 inches, most of the energy is reflected. To further reduce susceptibility to any high frequency entering the computer box, we used ferrites to minimize the significant antennae. Final testing indicated no susceptibility at the maximum IEC 801-4 severity level.

FUTURE EMC IMPROVEMENTS

Digital Pulse Processing

As stated above, the most susceptible channels are those that process pulses from the scintillator/Photomultiplier Tube (PMT) detectors. Front end bandwidth and gain are large. Since this is true, induced noise pulses must be less than 1 millivolt with no remote pre-amplification. Our latest design uses digital processing of detector pulses. Since EFT-induced pulses tend to be short, ringing pulses, we could easily discern them from detector pulses. The latter will be essentially unipolar. This assumes that sufficient processing capacity exists to perform shape discrimination as well as the usual pulse processing to support count rates as high as 10^7 cpm. Additionally, it assumes adequate preamplifier bandwidth. Presently, we have implemented a simple rejection algorithm that completely rejects bursts of pulses less than some chosen duration. Other algorithms exist that maintain statistical performance as well as step response times (e.g., Nearest Mean (NEME) filter).

Pulse discrimination also minimizes susceptibility to long duration bursts (e.g., arc welder). Via such techniques, required RFI suppression would be defined by the digital circuit noise immunity level (i.e., greater than 1 volt), instead of by low level analog circuitry susceptibility. This would guarantee that no processor resets, address/data bus upsets, or metastable states occur for all reasonable plant transients.

Smart Sensor

To reduce area monitor EFT susceptibility, we redesigned the analog Geiger Mueller (GM) detector and electronics. This detector utilizes "time-to-count" methodology to achieve high count rates. We implemented a "Smart Detector" using an 8051 microcontroller; it interfaces with front electronics, determines the count rate, and transmits results to the main CPU. The GM tube high voltage supply is located inside the detector housing, and uses a 4-20 ma current loop along with twisted shielded pairs. Consequently, bursts of noise pulses will appear as communications errors in the worst case. Error correction may be required where time response is of concern. Alternatively, fiber optics could be used.

CONCLUSIONS

EMC must be incorporated into RMS designs. Actual experience at Nuclear Power Plants illustrates the resultant problems in systems without EMC. This paper addresses the resolution. Using the results of investigative tests we formulated a systematic approach to modify our designs. We have implemented these design modifications and have passed final tests to European standards and EPRI guidelines. Additionally, this paper outines possible future improvements to RMS monitors to further reduce EMI susceptibility.

0074-056X/95/201-210/$0 + .50pp

A JUSTIFICATION FOR ASSUMING A LAPLACE DISTRIBUTION IN THE COMPUTATION OF DRIFT TOLERANCE INTERVALS

Kim A. Jacklin
VECTRA Technologies Inc.
215 Shuman Blvd, Suite 200
Naperville IL 60563

William E. Jacklin
Illinois Institute of Technology
Electrical & Computer Engr. Dept.
Chicago IL 60616

Abstract
Traditional methods of determining drift tolerance intervals assume that the underlying data has a normal distribution. However, the peaked, heavy tailed nature of the examined bistable drift data distributions are not indicative of a normal, motivating a search for an alternate model. A statistical investigation including chi-square goodness of fit tests, histograms and probability plots was performed to identify the characteristics of the drift data. The results of this investigation, along with a preliminary computation of revised tolerance intervals, indicate that the Laplace distribution better approximates the underlying data distribution. As a result, it is recommended that the Laplace distribution be explored as the preferred underlying distribution for peaked, heavy tailed drift data sets that are not adequately approximated by the normal distribution. Note that the results provided herein are considered preliminary and are presented solely for illustrative purposes to provide motivation for further study. Direct application of the computed tolerance intervals for use in determining reactor protection setpoints has not been investigated and additional research is required before utilizing these values.

1. Introduction

This paper presents a novel method for determining the 95%/95% tolerance interval [1] for bistable drift data[1], with the aim of quantifying instrument loop uncertainty to determine reactor protection setpoints and improving nuclear plant operations via extended surveillance intervals. While the calculation of the required tolerance interval traditionally employs statistical techniques based on the assumption that the underlying data distribution is Normal, we propose instead the assumption that the data has a double-exponential, or Laplacian, distribution. For the data examined, this assumption is supported by the chi-square goodness of fit test and a visual inspection of probability plots and histograms.

The organization of the paper is as follows. First, a brief background concerning the drift analysis methodology and creation of the drift data sets is provided, followed by the motivation for the Laplacian assumption. Next, a preliminary algorithm for computing the required tolerance interval is developed. Finally, 95%/95% drift tolerance intervals calculated using the Laplace assumption are presented and compared to values calculated assuming a Normal distribution. Note that the 95%/95% drift tolerance values calculated herein are considered preliminary. Prior to applying the values in the calculation of reactor protection setpoints, further investigation is required. Therefore, this study represents the first step in the multi-step process of uncertainty and setpoint calculation.

2. Drift Analysis

The definition of instrument drift is "an undesired change in output over a period of time, which change is unrelated to the input, environment, or load." [2] The purpose of estimating drift experienced by

---

[1]The data sets examined in this study were calculated from actual calibration data recorded at one nuclear plant for Foxboro 63U-BC bistables from 1985 through 1992. Due to the preliminary nature and limited scope of this study, the results should not be generalized and are not necessarily indicative of the performance of Foxboro instrumentation in other applications.

instrumentation is to include this effect when determining the uncertainty associated with an instrument loop, thereby allowing for conservative calculation of the trip setpoint for the channel of interest. As-found/as-left data is recorded for an instrument before and after calibration, and is typically recorded on a regular basis. The as-found/as-left calibration data may be used to determine the drift occurring between each calibration interval. For a given instrument the calculated drift values are treated as independent random samples from the same population, and constitute a drift data set. The drift data set can then be statistically evaluated to determine the maximum expected drift of the instrument between calibrations. For determining critical setpoints, industry practice recommends the calculation of a 95%/95% drift tolerance interval which encompasses the drift experienced by the instrument at a 95% probability and a 95% confidence level. When the use of as-found/as-left calibration data is employed to determine drift it represents "the net effects of several uncertainties, including the reference accuracy and drift of the module over the calibration interval." [2]

The first task in performing a drift analysis is to group the observed as-found/as-left values into meaningful data sets. Since the manufacturer and model number of the instruments in this study are identical, it is not necessary to separate the data according to type and classification. Instead, the drift data is grouped together by trip function/process application, resulting in a total of nine groups for the data examined. Outliers were removed using a standard outlier removal procedure [3]. A check was performed to ensure that 95% of the original drift values remained after outlier removal.

Given monthly as-found/as-left calibration data, we are interested in determining the drift that occurs over a one month interval. The monthly observed drift values, D_M, expressed in percent of span are given by

$$D_M = \frac{(AF_M - AL_M)}{bistable\ span} \cdot 100\% \tag{1}$$

where AF_M is the as-found value for the month of interest, and AL_M is the as-left value collected during the previous monthly calibration.

Drift estimation may also be used to justify surveillance interval extension from monthly to quarterly. The quarterly observed drift, D_Q, calculated using monthly calibration data is given by

$$D_Q = \frac{(AF_Q - AL_Q)}{bistable\ span} \cdot 100\% \tag{2}$$

where AF_Q is the as-found value for the month of interest and AL_Q is the as-left value collected during the monthly calibration performed three months prior to the month of interest. Note that Eq. (2) may only be used for data collected during monthly surveillances in which adjustments to the bistable were *not* required. Using Eq. (1) and Eq. (2) on the nine groupings of the data examined in this paper results in a total of eighteen separate data sets, labelled 1M through 9M and 1Q through 9Q, respectively, in the analysis to follow.[2]

3. The Laplacian Assumption

Typically, drift studies for nuclear plant instrumentation employ statistical techniques based on the assumption that the underlying data distribution is normal. However, an examination of the observed

[2]In the analysis and results to follow, the drift values, D_M and D_Q, are recorded to a resolution of 0.025% of span, indicative of measurement resolution for the examined Foxboro 63U-BC bistables.

bistable drift data revealed that the underlying data distribution is more peaked than a normal distribution. Furthermore, the 95%/95% drift tolerance intervals calculated using the normal distribution assumption consistently failed to encompass 95% of the data for each data set, implying that the tails of the observed data distribution are heavier than the tails of a normal distribution. With these considerations, it is proposed that the more peaked, heavier tailed Laplacian distribution be used as a better approximation. For a Laplacian random variable, X, its corresponding *probability density function* (pdf) is

$$f_X(x)=\frac{1}{\sqrt{2\sigma^2}}e^{-\frac{\sqrt{2}}{\sigma}|x-\mu|} \tag{3}$$

where μ is the mean and σ^2 is the variance. The normal pdf, conversely, is given by

$$f_X(x)=\frac{1}{\sqrt{2\pi\sigma^2}}e^{-\frac{(x-\mu)^2}{2\sigma^2}}. \tag{4}$$

By comparing Eq. (3) and Eq. (4) at $x=\mu$ we see that the Laplacian pdf is more peaked than the normal. Furthermore, for x large in magnitude, we see that the Laplacian pdf has heavier tails that decay at a rate proportional to $e^{-|x|}$, whereas the tails of the normal pdf decay at a faster rate proportional to e^{-x^2}.

To support the Laplacian hypothesis, the following statistical tests were performed on the observed drift data.[3] First, two chi-square goodness of fit tests [4] were performed on the data, one which assumed the data came from an underlying Laplace distribution and one which assumed an underlying normal distribution. The chi-square test was performed to test the following hypothesis: Can the observed drift data be regarded as randomly drawn from the assumed distribution? For each assumed distribution the test statistic, χ^2, was calculated and compared to the $\chi^2_{0.050}$ value from [4]. If the observed drift comes from a distribution other than the assumed, i.e., the hypothesis is false, then the computed χ^2 test statistic is larger than the $\chi^2_{0.050}$ value, where the value "0.050" indicates that the probability of falsely rejecting the assumption is less than 5%. The results of the chi-square test for each data set are presented in Table 1.

Table 1 reveals that for 9 of the 18 data sets the chi-square test statistic was less than $\chi^2_{0.050}$, which supports the assumption that the data originated from a Laplacian distribution. However, all eighteen values of the test statistic for an assumed normal distribution exceeded $\chi^2_{0.050}$. In addition, the Laplace χ^2 test statistic for 3Q and 8M are accepted at the 1% and 2.5% levels, respectively. However, none of the normal χ^2 test statistics are accepted even at the lowest tabulated significance level of 0.5%, i.e. there is less than a 0.5% probability of falsely rejecting the assumption (hypothesis) that the data is from a normally distributed population. Therefore, the results of the chi-square test support the assumption that the underlying data distribution is better approximated by a Laplace distribution.

Second, a normalized histogram [4] for each data set was constructed and compared to plots of the Laplace and normal pdfs. Samples of the histograms are shown in Fig. 1 to Fig. 4. A visual, qualitative examination of the plots demonstrates that the histograms follow the peaked, heavier tailed shape of the Laplace distribution. In a few anomalous cases (not shown), the histogram does not exhibit the shape of

---

[3] In the statistical analysis to follow, note that the sample mean and sample standard deviation for a normal population are calculated differently than the same parameters for a Laplacian population. In each case, the corresponding maximum likelihood estimates [4,5] of the mean and standard deviation were used to compute the required parameters.

either the normal curve or the Laplace. For these data sets, it is assumed that process or environmental factors, that are not being measured here, have influenced the bistable performance causing a deviation from the behavior exhibited by the majority of the data sets.

Third, a set of two probability plots [6] was created for each data set, one assuming a normal distribution and a second assuming a Laplacian distribution. Ideally, if the data has the assumed distribution, the probability plot will be a straight line having a slope equal to one and an x-intercept equal to zero. For the majority of the data sets the Laplacian probability plots are approximately linear and have the appropriate slope and x-intercept. However, the probability plots created assuming a normal distribution exhibit marked deviation from a linear pattern at the tails, causing the plots to be "s-shaped". Typical plots for the examined data sets are provided in Fig. 5 to Fig. 8.

4. A Preliminary Method for Computing the Tolerance Interval

A preliminary algorithm for computing a 95%/95% tolerance interval assuming Laplacian data can be derived from the method by Kappenman [5]. In particular, one-sided tolerance intervals of the form (u_1,∞) and $(-\infty,u_2)$ are considered in [5]. However, we are interested in two-sided intervals of the form (v_1,v_2). Thus, as an initial attempt, the two-sided tolerance intervals are formed as the intersection of suitable one-sided intervals, which are chosen to yield the required overall tolerance. The major drawback of this method is that the resulting tolerance interval may not be symmetric about the mean. Further research is necessary to develop an algorithm that produces a symmetric tolerance interval that is more indicative of the underlying Laplace distribution.

To begin, let x_1, ..., x_n be n independent *ordered* samples from a random variable X having the pdf

$$f_X(x)=\frac{1}{2b}e^{-\frac{|x-a|}{b}}, \tag{5}$$

where a and b are parameters of the pdf. The maximum likelihood estimates of a and b are [5]

$$\hat{a}=\begin{cases} x_{(n+1)/2}, & \text{if } n \text{ is odd} \\ \frac{1}{2}(x_{n/2}+x_{n/2+1}), & \text{if } n \text{ is even} \end{cases}, \text{ and } \hat{b}=\frac{1}{n}\sum_{i=1}^{n}|x_i-\hat{a}|. \tag{6}$$

Using the one-sided results in [5], our goal is to compute a two-sided interval of the form $(\hat{a}-q_l\hat{b},\hat{a}-q_u\hat{b})$ where q_l and q_u are chosen so as to yield the required tolerance, e.g., 95%/95%.

Consider the case when n is odd (the case of n even is not examined in this paper). In general, the β/γ tolerance interval (in our specific case, $\beta=\gamma=0.95$) is computed as follows. First, as in [5], define the normalized ordered observations, u_i, $i=1,\ldots,n$, and the function $c(t)$ as

$$u_i=(x_i-\hat{a})/\hat{b}, \text{ and } c(t)=\sum_{i=1}^{n}|u_i-t|, \tag{7}$$

where t is an arbitrary variable. Then, using u_i and $c(t)$, compute the values $B_n(u_1,\ldots,u_n)$ and k, where

$$B_n(u_1,\ldots,u_n)=\left[\sum_{j=1}^{n}\frac{[c(\hat{a})/c(u_j)]^{n-1}}{(2j-n)(n+2-2j)}\right]^{-1/(n-1)} \quad \text{(for } n \text{ odd)}, \tag{8}$$

$$k=\frac{[B_n(u_1,\ldots,u_n)\,c(\hat{a})]^{n-1}}{2\Gamma(n-1)}, \tag{9}$$

and $\Gamma(m)$ is the gamma function [7]. Next, let $d_l=-d_u=\ln(1-\beta)$ and define $Q(h)$ to be the expression

$$Q(h)=\frac{[B_n(u_1,\ldots,u_n)]^{n-1}}{2}\left\{\frac{1}{n}\left[\frac{c(\hat{a})}{c(u_1)}\right]^{n-1}+\sum_{i=1}^{h-1}\frac{1}{n-2i}\left(\left[\frac{c(\hat{a})}{c(u_{i+1})}\right]^{n-1}-\left[\frac{c(\hat{a})}{c(u_i)}\right]^{n-1}\right)\right\}. \tag{10}$$

Using Eq. (10), determine the quantities h_l and h_u as the largest positive integers (≥ 2) such that $Q(h_l)<1-\gamma$ and $Q(h_u)<\gamma$. Finally, letting $p_l=1-\gamma-Q(h_l)$ and $p_u=\gamma-Q(h_u)$, the values for q_l and q_u are given by [5]

$$q_{l,u}=-u_{h_{l,u}}-\frac{c(u_{h_{l,u}})}{n-2h_{l,u}}+\frac{\exp\left(-\dfrac{d_{l,u}(n-2h_{l,u})}{n-1}\right)}{n-2h_{l,u}}\cdot\left[\frac{n-2h_{l,u}}{k\Gamma(n-1)}p_{l,u}+c(u_{h_{l,u}})^{-n+1}\right]^{-1/(n-1)}, \tag{11}$$

yielding the final β/γ tolerance interval: $(\hat{a}-q_l\hat{b},\hat{a}-q_u\hat{b})$.

The actual computation of Eq. (11) can prove quite challenging for large values of n. In particular, $\Gamma(n-1)$ will become very large, and may not be representable in many computer systems. Also, the quantities $[B_n(u_1,\ldots,u_n)\,c(\hat{a})]^{n-1}$ and $[1/c(u_i)]^{n-1}$ may become either extremely large or extremely small depending on whether the argument being raised to the $(n-1)^{\text{th}}$ power is either greater than 1 or less than 1, respectively. As a result, these quantities also may not be representable in many computer systems. Thus, methods of approximating Eq. (11) may be necessary, and are discussed in the Appendix. These methods were used to compute the tolerance intervals presented in the next section.

5. Sample 95%/95% Drift Tolerance Interval Values

Table 2 presents two 95%/95% drift tolerance interval values calculated for each data set. The Laplace 95%/95% values were calculated using the preliminary method discussed in Section 4 for an assumed underlying Laplacian distribution. The normal 95%/95% values were calculated using the sample standard deviation scaled by a 95%/95% multiplier provided in [3]. The Laplace drift tolerance intervals are consistently larger than the normal drift tolerance intervals and, as demonstrated in Table 2, encompass over 95% of the observed drift. Conversely, the normal drift tolerance intervals fail to consistently encompass 95% of the observed drift. The failure of the normal 95%/95% drift tolerance intervals to encompass 95% of the observed drift raises concern about the use of this interval as a valid bound for the maximum expected drift.

6. Conclusion

The results of this paper support the assumption that the underlying distribution of the drift data sets examined in this study is Laplacian. To summarize, the chi-square goodness of fit tests reject the assumption of a normal distribution while allowing acceptance of the Laplace assumption. Figures 1 through 4 provide a vivid illustration of the drift characteristics as the sample normalized histograms closely track the peaked, heavy tailed Laplace probability density function. The linear pattern exhibited in the Laplace probability plots further enforces the selection of the Laplace distribution, while the "s-shaped" normal probability plots demonstrate a deviation from the normal. Application of the preliminary Laplace tolerance interval algorithm presented in Section 4 shows that the use of the Laplace distribution results in larger 95%/95% drift tolerance interval values than those calculated using the normal assumption. Table 2 also illustrates that the Laplace intervals encompass over 98% of the observed drift data, reassuring us that the values are a valid bound for the maximum expected drift.

In conclusion, for the data sets examined in this study, the Laplace distribution is a valid approximation of the underlying data distribution. As a result, it is recommended that the Laplace distribution be investigated as the preferred underlying distribution for peaked, heavy tailed drift data sets that are not adequately approximated by the normal. However, as discussed in the introduction, the preliminary 95%/95% drift tolerance values calculated herein are not intended for direct application in determination of reactor protection setpoints. Rather the results are presented for the illustrative purpose of providing motivation for further study. Prior to direct application of the values an investigation of how the non-normal values should be employed with respect to ISA-67.04, Part I "Setpoints for Nuclear Safety-Related Instrumentation" must be conducted.

Acknowledgements

The authors would like to express their gratitude to Tom Donat, Karen DePodesta, Joe Basak, and Gene Gross for their insight, suggestions and support toward this effort.

Appendix

This Appendix presents methods for approximating the values of q_l and q_u used in determining the β/γ tolerance interval. Let $\hat{q}_l$ and $\hat{q}_u$ be the approximations of q_l and q_u. These approximations should be such that $\hat{q}_l \geq q_l$ and $\hat{q}_u \leq q_u$, thus yielding a larger, and more generous, tolerance interval. To determine an expression for $\hat{q}_l$, we must consider two cases, $(n-2h_l)>0$ and $(n-2h_l)<0$. For the case when $(n-2h_l)>0$, let $z=1/c(u_{h_l})$, $y=[2p_l(n-2h_l)]^{1/(n-1)}/[B_n(u_1,..,u_n)c(\hat{a})]$, and j be the smallest integer (>0) such that the computer-generated quantity $z^{(n-1)/m}+y^{(n-1)/m}$, $m=2^j$, has a finite, non-zero value. Then, using the approximation (from the Cauchy-Schwarz Inequality) that $(z^{n-1}+y^{n-1}) \geq (1/m)(z^{(n-1)/m}+y^{(n-1)/m})^m$, $m=2^j$, the expression for $\hat{q}_l$ becomes

$$\hat{q}_l=-u_{h_l}-\frac{c(u_{h_l})}{n-2h_l}+\frac{\exp\left[-\frac{d_l(n-2h_l)}{n-1}\right]}{n-2h_l}\cdot\left[\frac{1}{m}(z^{(n-1)/m}+y^{(n-1)/m})\right]^{-m/(n-1)} \geq q_l, \quad (n-2h_l)>0,\ m=2^j. \tag{12}$$

Similarly, to determine $\hat{q}_l$ when $(n-2h_l)<0$, let $z=1/c(u_{h_l})$, $y=[2p_l(2h_l-n)]^{1/(n-1)}/[B_n(u_1,..,u_n)c(\hat{a})]$, and j be the smallest integer (>0) such that the computer-generated quantity $z^{(n-1)/m}-y^{(n-1)/m}$, $m=2^j$, has a finite, non-zero value. Then, noting that $(u^n-v^n)=(u^{n/2}-v^{n/2})(u^{n/2}+v^{n/2})$ and that $(u^n+v^n) \leq (u+v)^n$, for u and v positive, the expression for $\hat{q}_l$ becomes

$$\hat{q}_l = -u_{h_l} - \frac{c(u_{h_l})}{n-2h_l} + \frac{\exp\left[-\frac{d_l(n-2h_l)}{n-1}\right]}{n-2h_l} \cdot \left[z^{(n-1)/m} - y^{(n-1)/m}\right]^{-1/(n-1)} \left[z+y\right]^{-\sum_{i=1}^{\log_2(m)} \frac{1}{2^i}} \geq q_l, \tag{13}$$

$$\text{for} \quad (n-2h_l)<0, \quad z>y>0, \quad m=2^j.$$

The two cases $(n-2h_u)>0$ and $(n-2h_u)<0$ must also be considered when deriving an expression for $\hat{q}_u$. For the case when $(n-2h_u)>0$, let $z=1/c(u_{h_u})$ and $y=[2p_u(n-2h_u)]^{1/(n-1)}/[B_n(u_1,..,u_n)c(\hat{a})]$. Noting that $(u^n+v^n)\leq(u+v)^n$ for u and v positive, the expression for $\hat{q}_u$ is

$$\hat{q}_u = -u_{h_u} - \frac{c(u_{h_u})}{n-2h_u} + \frac{\exp\left[-\frac{d_u(n-2h_u)}{n-1}\right]}{(n-2h_u)(z+y)} \leq q_u, \quad (n-2h_u)>0. \tag{14}$$

Similarly, for the case when $(n-2h_u)<0$, let $z=1/c(u_{h_u})$, $y=[2p_u(2h_u-n)]^{1/(n-1)}/[B_n(u_1,..,u_n)c(\hat{a})]$, and j be the smallest integer (>0) such that the computer-generated quantity $z^{(n-1)/m}-y^{(n-1)/m}$, $m=2^j$, has a finite, non-zero value. Then, using the inequality $(z^{n-1}-y^{n-1})\geq(z^{(n-1)/m}-y^{(n-1)/m})^m$ for $z>y>0$, the expression for $\hat{q}_u$ becomes

$$\hat{q}_u = -u_{h_u} - \frac{c(u_{h_u})}{n-2h_u} + \frac{\exp\left[-\frac{d_u(n-2h_u)}{n-1}\right]}{n-2h_u} \cdot \left[z^{(n-1)/m} - y^{(n-1)/m}\right]^{-m/(n-1)} \leq q_u, \tag{15}$$

$$\text{for} \quad (n-2h_u)<0, \quad z>y>0, \quad m=2^j.$$

References

[1] G. J. Hahn and W. Q. Meeker, *Statistical Intervals: A Guide for Practitioners*, New York: John Wiley & Sons, Inc., 1991.

[2] ISA-RP67.04, Part II, Recommended Practice, Approved September 1994, "Method for the Determination of Setpoints for Nuclear Safety-Related Instrumentation."

[3] WAPD-TM-1292 DOE R&D Report, "Statistics for Nuclear Engineers and Scientists", Feb. 1981

[4] G. W. Snedecor and W. G. Cochran, *Statistical Methods*, 7th Ed., The Iowa St. Univ. Press, 1980.

[5] R. F. Kappenman, "Tolerance Intervals for the Double Exponential Distribution," *Journal of the American Statistical Assoc.*, Vol. 72, No. 360, Theory and Methods Sec., pp. 908-909, Dec. 1977.

[6] G. J. Hahn and S. S. Shapiro, *Statistical Models in Engineering*, New York: John Wiley & Sons, Inc., 1967.

[7] R. V. Hogg and A. T. Craig, *Introduction to Mathematical Statistics*, 4th Ed., New York: Macmillan Publishing Co., Inc., 1978.

Data Set	Degrees of Freedom k	Value of $\chi^2_{0.050}$	Laplace χ^2 Statistic	Accept / Reject	Normal χ^2 Statistic	Accept /Reject
1M	8	15.51	13.71	**ACCEPT**	57.14	**REJECT**
1Q	9	16.92	16.70	**ACCEPT**	111.48	**REJECT**
2M	8	15.51	24.44	**REJECT**	128.56	**REJECT**
2Q	17	27.59	19.97	**ACCEPT**	65.87	**REJECT**
3M	10	18.31	31.10	**REJECT**	215.78	**REJECT**
3Q	14	23.68	28.54	**REJECT**	208.57	**REJECT**
4M	11	19.68	75.62	**REJECT**	360.03	**REJECT**
4Q	9	16.92	32.46	**REJECT**	140.18	**REJECT**
5M	7	14.07	37.52	**REJECT**	97.68	**REJECT**
5Q	9	16.92	16.22	**ACCEPT**	78.21	**REJECT**
6M	9	16.92	13.43	**ACCEPT**	165.76	**REJECT**
6Q	12	21.03	17.20	**ACCEPT**	90.64	**REJECT**
7M	8	15.51	10.92	**ACCEPT**	48.50	**REJECT**
7Q	7	14.07	11.96	**ACCEPT**	155.01	**REJECT**
8M	8	15.51	16.92	**REJECT**	48.71	**REJECT**
8Q	11	19.68	12.66	**ACCEPT**	64.69	**REJECT**
9M	10	18.31	25.67	**REJECT**	84.65	**REJECT**
9Q	10	18.31	26.60	**REJECT**	33.90	**REJECT**

Results of the Chi-square goodness of fit test

(Note: The "M" and "Q" used in the data set titles denote monthly and quarterly data respectively.)

Table 1

Data Set	Number of Observed Drift Values	Laplace 95%/95% tolerance interval value (% span)	% of Observed Data Encompassed	Normal 95%/95% tolerance interval value (% span)	% of Observed Data Encompassed
1M	437	-0.175% /+0.225%	99.09%	+/-0.125%	93.38%
1Q	235	-0.200% /+0.300%	97.87%	+/-0.175%	96.17%
2M	433	-0.200% /+0.200%	98.62%	+/-0.125%	93.32%
2Q	255	-0.225% /+0.275%	98.43%	+/-0.175%	94.90%
3M	889	-0.200% /+0.200%	99.55%	+/-0.100%	93.26%
3Q	529	-0.250% /+0.275%	99.62%	+/-0.175%	94.53%
4M	865	-0.175% /+0.175%	99.17%	+/-0.100%	93.19%
4Q	525	-0.250% /+0.200%	99.05%	+/-0.125%	92.00%
5M	889	-0.175% /+0.150%	100%	+/-0.100%	96.06%
5Q	535	-0.200% /+0.175%	100%	+/-0.100%	93.27%
6M	877	-0.175% +/0.175%	100%	+/-0.100%	94.43%
6Q	549	-0.200% /+0.200%	100%	+/-0.125%	91.80%
7M	295	-0.200% /+0.175%	100%	+/-0.125%	96.62%
7Q	183	-0.225% /+0.175%	100%	+/-0.125%	96.74%
8M	897	-0.175% /+0.200%	100%	+/-0.100%	93.65%
8Q	607	-0.200% /+0.250%	99.18%	+/-0.150%	92.43%
9M	893	-0.200% /+0.225%	99.78%	+/-0.125%	94.40%
9Q	583	-0.200% /+0.250%	99.83%	+/-0.150%	95.72%

Comparison of drift tolerance intervals for the Laplace and normal assumptions

Table 2

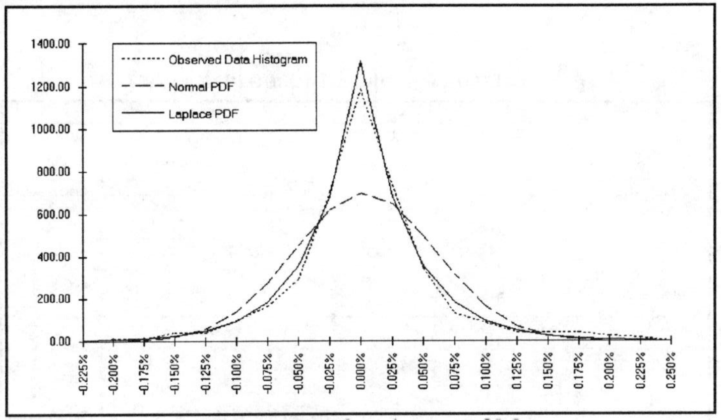

Histograms for data set 3M

Figure 1

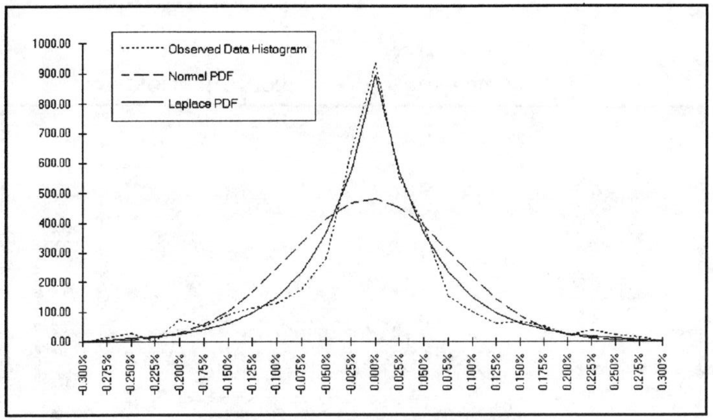

Histograms for data set 3Q

Figure 2

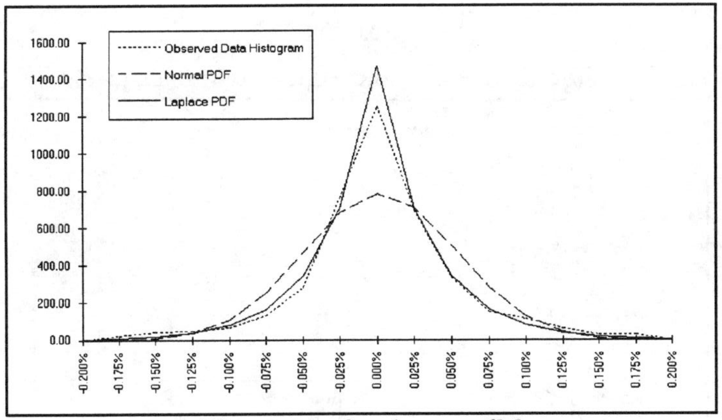

Histograms for data set 6M

Figure 3

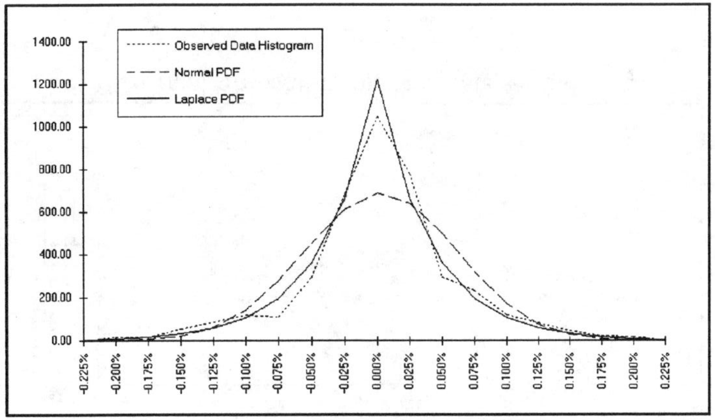

Histograms for data set 6Q

Figure 4

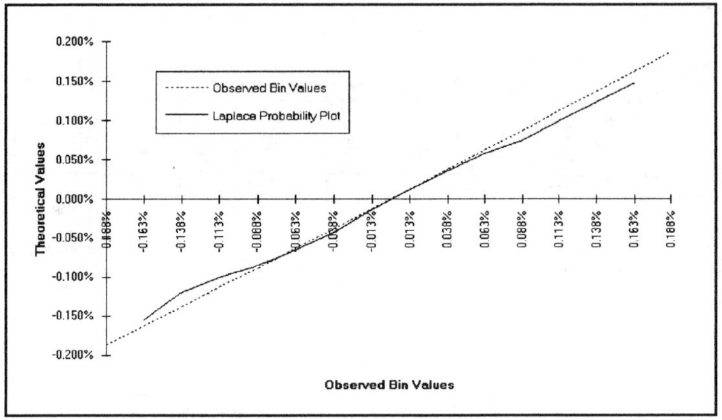

Laplace probability plot for data set 6M

Figure 5

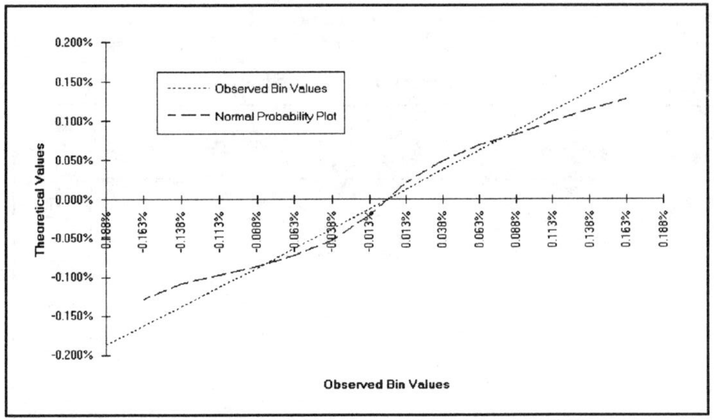

Normal probability plot for data set 6M

Figure 6

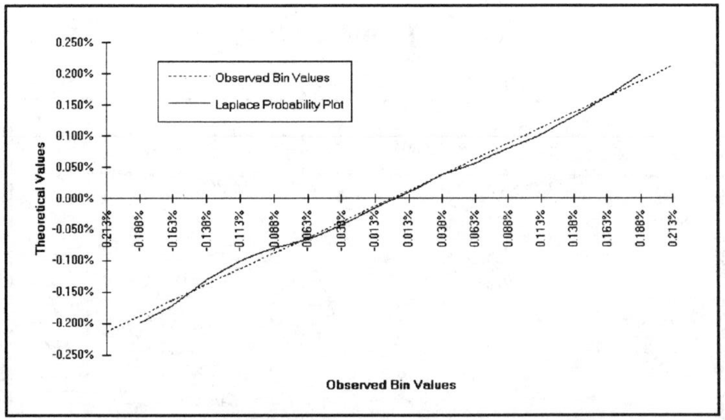

Laplace probability plot for data set 6Q

Figure 7

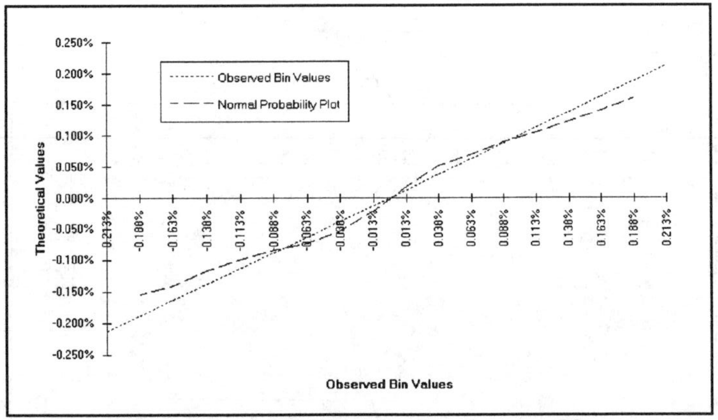

Normal probability plot for data set 6Q

Figure 8

©0074-056X/95/211-221/$0 + .50pp

MAGNETIC FUSION ENERGY DEVELOPMENT AND RELATED INSTRUMENT REQUIREMENTS

C.J. Hamilton
General Atomics
P.O. Box 85608
San Diego, CA 92186-9784

R.T. Snider
General Atomics
P.O. Box 85608
San Diego, CA 92186-9784

KEYWORDS

Fusion, diagnostics, plasma, current, magnetic, confinement

ABSTRACT

Fusion is the process of combining lighter elements to produce heavier elements and energy. The science of magnetic confinement fusion has evolved to a degree that several practical designs exist for fusion devices, which would produce energy on a commercially interesting scale. The current or near-term technology of fusion; however, will not support sustained steady-state operation with the efficiency and reliability necessary for commercially attractive or economically competitive electricity generation. Over the next thirty years several large fusion facilities (domestic and international) are to be built to advance both the science and technology to the point that a quasi-commercial, fusion electricity-generating demonstration plant, can be put into operation. A summary of existing and planned, major experimental programs and devices will be provided.

Fusion research and eventual power production present severe challenges for current and future instruments and diagnostic systems. Much of the progress that has been realized in the quality and performance of the plasmas produced has come about because of the increased understanding of the plasmas, including stability and confinement, made possible by advances in the diagnostics used on fusion devices. The next generation of fusion devices will require new measurement techniques to monitor the nuclear processes such as alpha particle production and transport, and will present severe challenges for existing diagnostic techniques primarily due to the harsh temperature and radiation environment inside the tokamak. A survey of existing fusion instruments will be given along with an overview of the what is expected of the diagnostics for future fusion devices.

INTRODUCTION

Nuclear fusion could be the world's best solution for a long-term energy source. It is an attractive option because its fuel is virtually inexhaustible and the safety concerns and environmental impacts should be much less than other candidate technologies for central generating stations. Harnessing fusion energy will require substantial development in instruments and diagnostics as well as both the science and technology of fusion.

A demonstration fusion power plant will not be built until the next century — a time when new, large sources of clean energy will be needed. From an environmental point of view, fossil fuels may by then be unable to meet this demand in their present form due to the problems of fuel availability and/or atmospheric pollution concerns. A fusion power plant will possess attractive safety characteristics since any significant equipment failure will result in a rapid termination of the process and since the quantity of radioactive material available for release is quite small.

Optimization of the performance and reliability of a fusion power plant will require a sophisticated set of diagnostic instruments, operating in a very harsh environment. These diagnostics will measure a broad range of plasma parameters including the magnetic structure and the plasma temperature and density.

Fusion energy is one of the most challenging and ambitious scientific endeavors ever undertaken. It includes scientific research at the frontiers of human understanding, trying to recreate here on Earth the process that sustains the stars. To attain fusion energy requires the development of new technologies and materials. Several of the technologies already developed for fusion have resulted in significant spin-offs and applications to other areas. Once fusion technology is perfected there should be enough fuel on Earth to meet the world's energy requirements for at least one billion years.

WHAT IS FUSION?

Fusion is the energy process which fuels the stars. When the nuclei of light atoms are sufficiently close together they fuse, generating heavier nuclei and releasing energy in the process. The reacting nuclei must be sufficiently energetic (to overcome the electrical repulsive forces between them) and they must be of high concentration to produce a reaction rate which releases significant quantities of energy.

In a star, the large stellar mass has sufficiently high gravitational force to confine the nuclei at temperatures and densities required for fusion energy production. We cannot duplicate the conditions of temperature and gravitational confinement field in a star, but suitable conditions for fusion energy production can be achieved by heating the hydrogen isotopes to approximately 200 million degrees and by confining a sufficient concentration of the reacting charged nuclei with powerful magnetic fields generated by large electromagnets.

The soup of free electrons and hydrogen nuclei, called a plasma, is confined by the magnetic fields which acts as a barrier, thermally insulating the surfaces of the fusion containment device. The fusion energy is released in the form of kinetic energy of the reaction products, neutrons, and helium ions. The helium ions remain within the magnetically-confined plasma giving up their energy to the plasma to sustain the thermo-nuclear burn. The energetic (principally 14 MeV) neutrons are captured outside the plasma in a shield/blanket where their kinetic energy is transformed to heat. This heat is absorbed by a fluid coolant in the shield/blanket. For electricity production, the resulting high-temperature fluid will be used to spin a turbine that will drive electrical generators.

The great attraction of fusion power is that the basic fuel is deuterium, a heavy form of hydrogen. Deuterium is easily extracted from water and thus the fuel supply is essentially inexhaustible. However, the first generation of fusion reactors will use two forms [isotopes] of hydrogen, namely deuterium and tritium. Tritium, a radioactive form of hydrogen with a short half-life, does not occur naturally in significant quantities. Tritium is currently "manufactured" by exposing the light metal, lithium, to neutrons from fission reactions. The tritium quantities available for fusion fuel will be expanded by utilizing neutrons from future fusion power plants. There is sufficient lithium to supply the world's electricity needs for thousands of years. More advanced fusion reactors may be able to use deuterium alone as fuel.

Although fusion is a nuclear process, it differs from the fission process in that there is no radioactive by-product from the fusion reaction -- only helium gas and a neutron. The structure of the fusion device will, however, become activated by the neutrons from the fusion reaction. By careful selection of structural materials and development of low activation materials, the required isolation of radioactive structural components could be limited to less than a hundred years.

HISTORY OF FUSION RESEARCH

Research to produce a sustained and controlled magnetic fusion reaction for the purposes of energy production began in the early 1950s. In the late 1950s, magnetic fusion energy research was declassified and international cooperation began between the European Community, Soviet Union, and United States. Japan became a major player in the late 1960s.

At present, there are more than 30 countries conducting research in controlled fusion. Of these countries, the most significant programs are in the European Community, Japan, Russia and the U.S.

Over the years each of the various countries have made significant contributions, either in the area of science, new concepts, theory, new technology, or some combination of the above. One of the greatest magnetic confinement innovations was the tokamak concept which was invented by the Russians. The tokamak is now the major magnetic confinement approach being pursued throughout the world, and has demonstrated the greatest success to date.

The plasma produced by a tokamak is torus-shaped with the confining magnetic fields (Fig. 1) wrapping helically around the torus. Large electromagnets produce a magnetic field in the toroidal direction (or long way around the torus), while smaller magnets drive an electric current in the plasma in the toroidal direction. The magnetic field produced by the plasma current combined with the large externally applied toroidal field give the helical magnetic fields needed for stable confinement (Fig. 2).

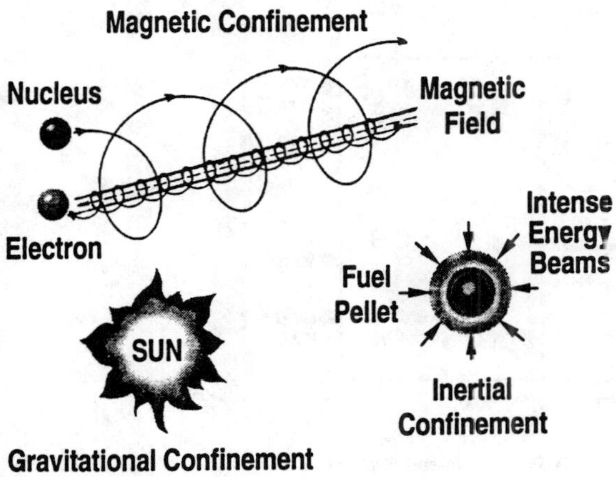

Fig. 1. Fusion plasmas can be confined by magnetic fields as well as gravitational or inertial forces.

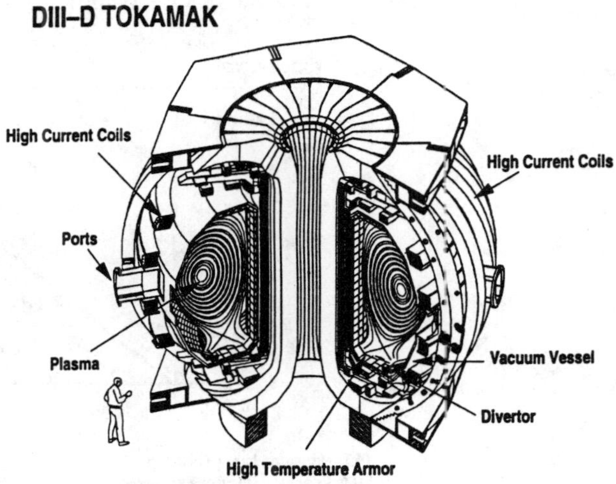

Fig. 2. A tokamak employs toroidal magnetic fields to confine high temperature plasmas.

Over the past 20 years, a 1000-fold increase has been achieved in the overall performance of fusion experiments — a rate of change that surpasses that achieved in increasing the memory density of computer chips. We are now within a factor of five of the fusion performance required for a fusion power plant. The next significant milestone in the development of fusion energy is achieving simultaneously the conditions required for a fusion power plant, i.e., plasma temperature, density and confinement needed to attain a completely self-sustaining reaction.

CURRENT LARGE TOKAMAK EXPERIMENTAL DEVICES

The tokamak concept is the leading candidate design for future magnetic fusion power plants. In fact, two tokamak experiments used deuterium/tritium plasmas to produce the first megawatts of fusion power — a major milestone in demonstrating the scientific feasibility of fusion energy as a source of electric power. In 1991, the Joint European Torus (JET), in Abingdon, England produced two megawatts and, more recently, the Tokamak Fusion Test Reactor (TFTR), in Princeton, NJ produced more than ten megawatts (Fig. 3).

JET. At this time, JET is world's largest and most advanced tokamak. JET is funded and staffed primarily by the European Atomic Energy Community (Euratom) and includes participation in the experiment by 14 countries.

TFTR. The TFTR is the largest US tokamak. The purpose of the current TFTR program is to study burning D/T fuels in short pulsed mode. In 1994, TFTR set a world record with the production of over 10 megawatts of fusion energy.

JT–60U. The most advanced tokamak in Japan is JT–60U, which is operated by the Japan Atomic Energy Research Institute (JAERI). JT–60's experimental program is focused on the development of higher plasma performance. JAERI is planning a substantial upgrade — JT–60 SU.

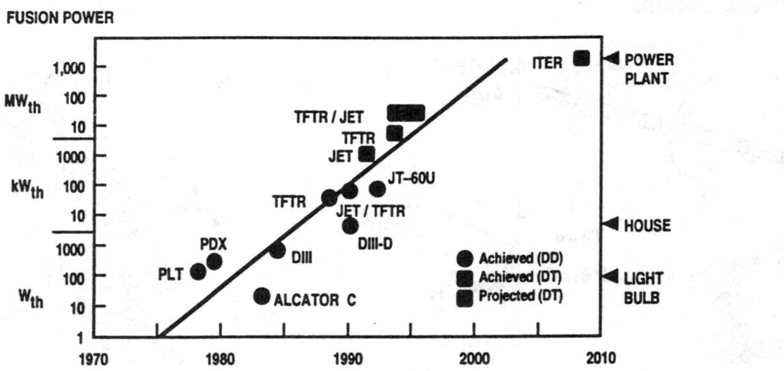

PLT Princeton Large Tokamak
PDX Princeton Divertor Experiment
JET Joint European Torus
DIII & DIII-D General Atomics Tokamak Experiments
TFTR Princeton Plasma Physics Laboratory
ALCATOR Massachusetts Institute of Technology
C International Thermonuclear Experimental Reactor
ITER Japanese Tokamak Experiment
JT–60U

Fig. 3. The power produced in magnetically confined fusion devices is approaching the range of interest for commercial power plants.

DIII–D. General Atomics in San Diego, California operates DIII–D, which is the second largest tokamak in the U.S. DIII–D has demonstrated stable operation with high plasma pressure to magnetic pressure ratios, essential for economical power generation. DIII–D has also demonstrated that the rate of helium heat removal is acceptable for ITER and a fusion power plant.

Tore Supra. Tore Supra is a somewhat smaller tokamak with superconducting toroidal field coils located in Cadarache, France. The main objectives of the project are oriented towards both technology and physics issues in hot, dense, quasi-steady-state plasmas.

NEXT-GENERATION EXPERIMENTS

TPX. The planned centerpiece of the U.S. national fusion program is the Tokamak Physics Experiment (TPX). The TPX design will test advanced steady-state operation, remote handling, and superconducting magnets. This large tokamak will be operated as a national research facility with broad participation by US industry, universities, and national laboratories. TPX will utilize the existing TFTR facility at the Princeton Plasma Physics Laboratory (PPPL).

The TPX conceptual design and research began in 1994. The construction of TPX and its first plasma operation are scheduled for 2000. Seven major industry contracts have been awarded for the design and R&D. Congressional authorization of the construction of TPX is pending. TPX will provide U.S. industry and universities with the experience and capabilities to be competitive in the construction of ITER, and ultimately provide the U.S. with the capability to build fusion utility power plants worldwide.

ITER. Magnetic fusion offers excellent opportunities for cost-effective collaboration with the involvement of international partners who have comparable skills and capabilities as well as an interest in participating on an equal cost-sharing and technology-sharing basis. The world's four major fusion programs are collaborating in the largest international scientific joint venture ever formed — the International Thermonuclear Experimental Reactor (ITER) Project. The world fusion community has agreed that ITER, which will achieve ignition, is the next step on the path to demonstrating the scientific and technological feasibility of fusion energy (Fig. 4).

The ITER Project is in its second phase — the Engineering Design Activities. The objective of the ITER EDA is to produce a detailed, complete, and fully integrated engineering design of ITER, as well as all the technical data necessary for future decisions on the construction of ITER. Such design and technical data shall then be available for each of the Parties to use either as part of an international collaborative program or in its own domestic program.

In addition to the detailed engineering design, the ITER partners will conduct materials research and development, as well as prototype fabrication and testing of components. The inside surfaces of the ITER plasma containment chamber will be designed to withstand temperatures of up to nearly 1,000 degrees centigrade, although normal operating temperatures will be substantially lower. Materials will have to be carefully chosen to withstand these temperatures. In addition, other components of the device will be composed of state-of-the-art metal alloys, ceramics and composites, many of which are now in the early stage of development and testing. ITER is foreseen to be the first fusion experiment capable of producing more than 1,000 megawatts of fusion power, an amount of power that is comparable to today's large utility power plants.

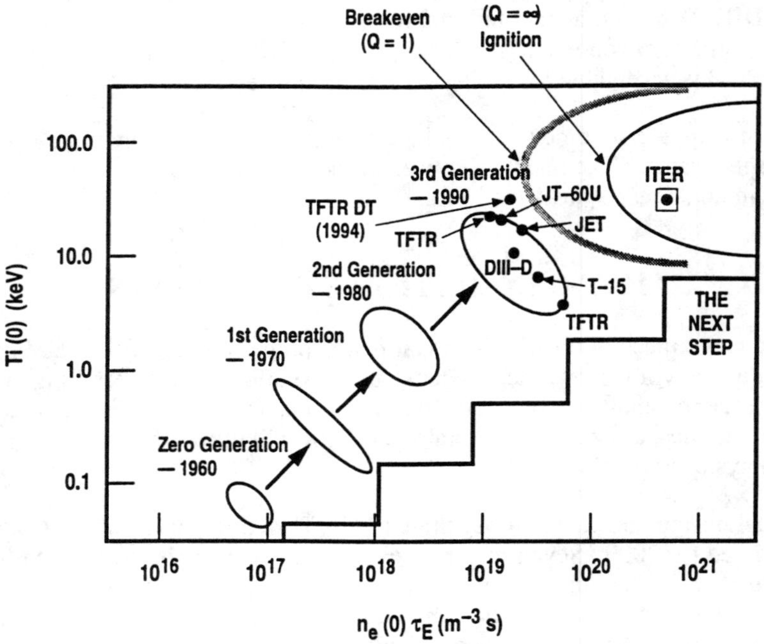

Fig. 4. Tokamak performance has increased by about five orders of magnitude in thirty years.

DEMO. The next specific milestone in the development of magnetic fusion energy as a source of electrical power is the operation of a quasi-commercial Demonstration Power Plant (DEMO) beginning in roughly 2025. DEMO will provide the basis for utility power plants to be operated in 2040 (Fig. 5).

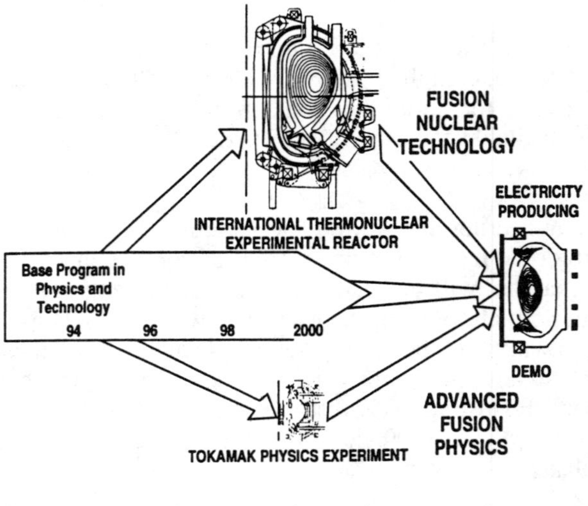

Fig. 5. The fusion program is targeted at a demonstration power plant in 2025.

OVERVIEW OF TOKAMAK DIAGNOSTICS

Advances in the performance of magnetically confined plasmas have been possible because of improvements in the physics understanding of the plasma itself and it's interaction with its surroundings. A large number of diagnostic instruments have been required to sufficiently characterize plasmas produced in modern tokamaks in order to reach the level of physics understanding we now have. Typically more then 50 separate diagnostic systems are installed on each of the existing large tokamaks. While a description of each of these diagnostic systems is beyond the scope of this paper, Table 1 lists classes of diagnostic techniques and the plasma parameters measured by those techniques. Tokamaks offer a number of challenges for diagnostic systems. These include the complicated toroidal geometry, harsh environment and very high temperatures. The very high temperatures (as high as 30 keV, i.e. 100 million °C) in a reactor grade plasma such as in TFTR, limits the use of spectroscopy since most of the naturally occurring elements in a tokamak are fully stripped at those temperatures.

The toroidal geometry of a tokamak requires spatial unfolding of data in order to get radial profile information generally required from the measurements. The unfolding places severe requirements on both the accuracy of the individual measurements and on the number of views of the plasma required. An example is shown in Fig. 6 which shows two arrays of bolometer detectors installed on DIII–D that are used to measure the power radiated from the plasma. Since each detector measures the line integral of the radiated power along its sightline, two or more arrays of detectors are required to invert the data to give the radiated power mapping shown in Fig. 7 [1].

The environment in and around a reactor grade tokamak places severe burdens, i.e. high temperature, high heat flux, radiation dose, etc., on the plasma diagnostics, particularly on the components directly facing the plasma. As the performance of tokamaks has improved, so to has the harshness of the environment in and around the tokamak. This has prompted a trend to locate as many of the diagnostic system components as far away from the tokamak as possible, often most of the diagnostic components are located outside of the biological shield wall that surrounds the tokamak, linked by wave guides or fiber optic cable. Radiated power as high as 50 W/cm^2 impinges on components at the first wall. Active cooling, selection of materials and careful design are some of the ways these large heat loads are handled. Redeposition of material eroded from the first wall (typically stainless steel, carbon or beryllium) coat optical components which have direct lines of sight to the plasma. Careful shielding of mirrors and windows can reduce the coating rate but cannot completely eliminate the problem. The next

Table 1. Diagnostic Class and Plasma Parameters Measured

Diagnostic Class	Parameters/Use
Magnetics	$\dot{B}$, magnetic flux contours
Microwave	
Probing beam	Electron density, electron density fluctuations
Passive detection	Electron temperature spatial profiles
Infrared	
Passive	First wall temperature
Probing beams	Electron density, temperature
Visible Light	
Passive plasma	Impurity concentrations
Emission from particle beam	Ion temperature, plasma rotation, plasma current spatial profile
X-rays	Plasma instabilities

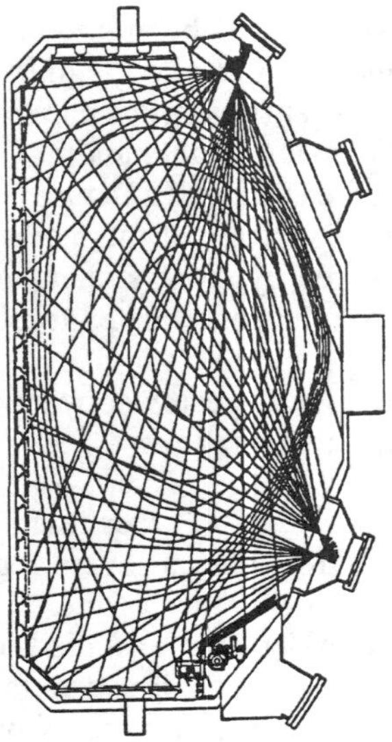

Fig. 6. Two 24-channel bolometer arrays measure radiated power from a tokamak plasma. Crossed arrays allow inversion of line-integrated signals.

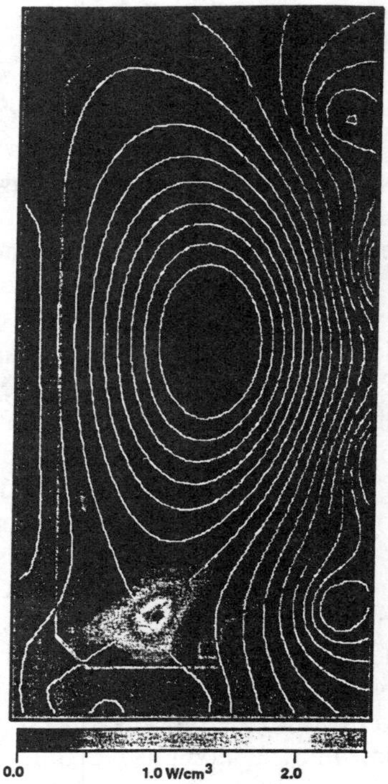

Fig. 7. Inverted radiated power contours in a tokamak plasma calculated from bolometer signals.

generation of tokamaks will have plasma pulse lengths of tens or hundreds of seconds; i.e. two orders of magnitude longer then present day machines. These longer pulses will make coatings on optical components from plasma redeposition a critical issue that has not been adequately addressed for optical diagnostics using wavelengths in the visible or shorter.

Magnetic fields used to contain the plasma can be as high as 6 Tesla in some tokamaks. These high fields place a number of constraints on diagnostic component design, including restricting the materials that can be used near the tokamak. The magnetic fields can be large sources of noise, from direct inductive pick-up in the electronics to mechanical vibration caused by induced eddy currents. Neutrons produced by existing tokamaks, while orders of magnitude less then will be produced in a fission reactor, present challenges to diagnostic design, mostly because of induced noise and permanent damage caused by the neutrons. Luminescent and darkening of fiber optic cable has been observed on TFTR [2]. Solid state detectors, particularly silicon, are susceptible to neutron damage at the radiation levels common on existing tokamaks. This restricts the use of these types of detectors close to the plasma without very large, cumbersome neutron shielding.

Despite the restrictions on diagnostic design that are inherent to the tokamak, very detailed information on tokamak plasmas has been acquired from the diagnostic systems installed on existing tokamaks. In the next section we will discuss briefly a few of the most important systems.

MAGNETIC DIAGNOSTICS

The magnetic configuration of the plasma in a tokamak defines the basic structure of the plasma and to a large extent determines the performance of the plasma. The determination of the magnetic flux contours is the most important information that can be extracted from the tokamak diagnostics, it is the logical coordinate system of any tokamak plasma. The magnetic diagnostics consist primarily of various loops of cable or inductive pick-up coils, located around the plasma (Fig. 8 [1]), that measure changes in the magnetic flux through the loops. These B-dot, or magnetic flux, measurements are integrated over time to give magnetic field measurements at the various locations around the plasma. These measurements are then inverted to give self-consistent reconstructions of the current profiles and their associated magnetic topology using the magnetohydrodynamic (MHD) equilibrium constraint [1]. An example of a reconstructed magnetic equilibria is shown in Fig. 8. The magnetic reconstructions made using the external magnetic coils alone are very good representations of the external shape of the plasma and global plasma parameters (such as total plasma current and internal inductance) but do not give reliable, detailed information in the interior of the plasma. Recently it has become clear from both the theoretical and experimental evidence that the details of the spatial profile of the plasma current play an important role in the stability and the energy confinement of the plasma. Internal measurements of the magnetic field combined with the external magnetic measurements are used to reconstruct both the internal magnetic surfaces and the plasma current profile. While several techniques (Faraday rotation measurements of probing laser beams, for example) for measuring or inferring the internal magnetic field have been developed, the most successful is the application of polarization spectroscopy techniques to emissions from a high-energy neutral particle beam injected into the plasma. High energy (60–120 keV) neutral particle beams are typically used for bulk heating of the plasma (in some cases neutral beam injectors have been built for dedicated diagnostic use). The motional electric field $E = v \times B$, where v is the velocity and B is the tokamak magnetic field, produces a strong Stark effect in spectral lines emitted by the high energy hydrogenic neutral beams [4]. The tilt angle of the magnetic field line, which is directly related to the magnetic field produced by the plasma current, is deduced from a measurement of the direction of polarization of the Stark components. A plan view the DIII–D tokamak showing a layout of the diagnostic using this method is shown in Fig. 9. Very accurate measurements of the internal magnetic field have been made and used in the magnetic equilibrium reconstruction including the current profile.

DIAGNOSTICS FOR FUSION POWER PLANTS

In order to operate a fusion power plant safely, with optimal performance and reliability, it will be necessary to measure a wide range of plasma parameters with accuracy comparable to present day tokamaks. Most of the plasma parameters that are measured on existing tokamaks (Table 1) will probably also have to be measured on a power plant (certainly on ITER), with the addition of several other important measurements such as deuterium/tritium mix in the plasma and neutron emission radial profiles. Many of the measurements that are now used primarily for physics research, will be used for control of the plasma. This places a much more stringent reliability and availability requirement on these diagnostic systems than in present machines. Differences between existing large tokamaks and tokamak power plants place further challenges to implementing diagnostic systems for power plants. To illustrate some of the new issues that power plants will impose on diagnostics we will use the ITER design to compare and contrast with existing large tokamaks [5]. The physical size of ITER is considerably larger, 2.5 times in the linear dimension and 20 times in volume, then existing large

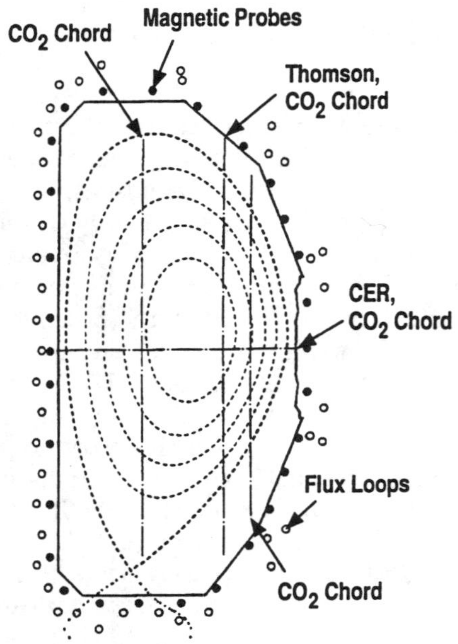

Fig. 8. Layout of some of the diagnostics on the DIII-D tokamak.

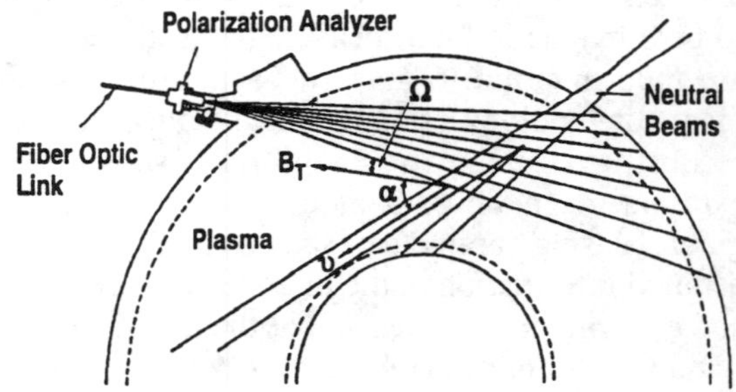

Fig. 9. Plan view of the Motional Stark Effect diagnostic on DIII-D. This diagnostic is used to determine theplasma current spatial profile.

tokamaks [6]. The large physical distances exacerbate alignment difficulties and restrict wavelength choices for plasma probing beams due to refraction effects. The large physical size of ITER will make thermal expansion and machine vibration a much larger difficulty than on present machines. The nuclear environment of ITER places enormous constraints on materials near the first wall. Neutron flux at the first wall for ITER is expected to be around 2×10^{14} n cm^{-2}s^{-1}, roughly 100 times larger then the highest neutron flux in TFTR, and the annual neutron fluence at the first wall of ITER will be around 4×10^{19} n/cm^{-2}. Development of radiation hardened diagnostic components will be a necessity for ITER and is an active area of work. Access to ITER will be severely restricted due to neutron shielding requirements of the tokamak coil system and remote handling needs of components on the interior of the vacuum vessel. Remote handling of all components in and close to the torus will drive major changes to the way diagnostics systems will be designed and operated. Remote handling will be a critical issue that will have to be addressed in the near future. The duration of a plasma pulse in ITER will be at least 1000 seconds. These long pulse lengths have significant impact on many diagnostics but of particular concern are the magnetic field measurements which in existing machines are measured using inductive coils and integrating the measured B dot signal. Very long pulse integrators are being developed and appear promising, but other methods may have to be developed if pulses much longer than 10^4 seconds are required.

Many challenges remain for fusion diagnosticians in the coming years and both the needs of the fusion physics program and the fusion reactor environment will continue to drive this area of research.

SUMMARY

Fusion offers the hope of a virtually inexhaustible supply of energy with minimal environmental impact. Enormous advances have been made in the last two decades that puts the fusion performance today within a factor of five required for a fusion power plant. The next generation of magnetic fusion devices are expected to demonstrate the scientific and technological feasibility of fusion reactors. This next generation of devices pose enormous challenges for diagnostic instruments needed to understand and control these devices.

REFERENCES

[1] A.W. Leonard, et al., "2-D Tomography with Bolometry in DIII–D," to be published in Rev. Sci. Instru.
[2] A.T. Ramsey, H.G. Adler, and K.W. Hill, "Reduced Optical Transmission of SiO_2 Fibers Used in Controlled Fusion Diagnostics," Princeton Plasma Physics Laboratory Report PPPL-2867 (1993).
[3] L.L. Lao, et al., Nucl. Fusion **30**, 1035 (1990).
[4] D. Wróblewski, et al., Rev. Sci. Instr. **63**, 5140 (1992).
[5] K.M. Young, et al., Rev. Sci. Instr. **63**, 4703 (1992).
[6] A.E. Costley, et al., Bull. Am. Phys. Soc. **39**, 1594 (1994).

0074-056X/95/223-233/$0 + .50pp

Nuclear Power Plant Instrumentation and Control Upgrade Activities

Joseph Naser
Electric Power Research Institute
3412 Hillview Avenue
P.O. Box 10412
Palo Alto, California 94303

Albert Machiels
Electric Power Research Institute
3412 Hillview Avenue
P.O. Box 10412
Palo Alto, California 94303

Raymond Torok
Electric Power Research Institute
3412 Hillview Avenue
P.O. Box 10412
Palo Alto, California 94303

Dan Wilkinson
Electric Power Research Institute
3412 Hillview Avenue
P.O. Box 10412
Palo Alto, California 94303

Abstract

Most nuclear power plants are operating with their original analog I&C equipment. This equipment requires increasing maintenance efforts to sustain system performance. Decreasing availability of replacement parts and support organizations for analog technology accentuate obsolescence problems and resultant O&M cost increases. Modern technology, especially digital systems, offers improved functionality, performance, and reliability; solutions to obsolescence of equipment; reduction in O&M costs; and the potential to enhance safety. Digital systems with their inherent advantages will be implemented only if reliable and cost-effective implementation and licensing acceptance is achieved and if the upgraded system supports reduced power production costs. EPRI and its member utilities are working together under the Integrated I&C Upgrade Initiative to address I&C issues. This paper will describe and give the status of EPRI's Integrated I&C Upgrade Initiative.

Introduction

Operating nuclear power plants in the United States were designed 20 to 40 years ago with analog instrumentation and control (I&C) technology. Today, most plants continue to operate with the original I&C equipment. This equipment is approaching or exceeding its life expectancy, resulting in increasing maintenance efforts to sustain system performance. Surveys of licensee event reports (LERs) show that a majority of the LERs are related to I&C issues. Decreasing availability of replacement parts and the accelerating deterioration of the infrastructure of manufacturers that support analog technology exacerbate obsolescence problems and resultant operation and maintenance (O&M) cost increases.

Instrumentation and control systems in nuclear power plants need to be upgraded in a reliable and cost-effective manner to replace obsolete equipment, to reduce operation and maintenance costs, to improve plant performance, and to enhance safety. The major drivers for the replacement of the safety, control, and information systems in nuclear power plants are the obsolescence of the existing hardware and the need for more cost-effective power production. The procurement of replacement modules and spares under current requirements, for hardware that was designed 20 to 40 years ago and is no longer fully supported by the original equipment manufacturer, is costly, time consuming and, in some cases, not even possible. Competition between power producers is dictating more cost-effective power production. The increasing operation and maintenance costs to maintain many of the analog systems is counter to the needs for more cost-effective power production and improved competitiveness.

Technological improvements, particularly the availability of digital systems, offer improved functionality, performance, and reliability; solutions to obsolescence of analog equipment; reduction in operation and maintenance costs; and the potential to enhance safety. Modern digital technology holds a significant potential to improve cost-effectiveness and productivity of nuclear power plants. Digital systems have the potential for solving the utilities' current problems of increasing analog equipment obsolescence; rapidly escalating operation and maintenance costs; lost generation due to system unavailability, spurious operation, and human error; and the inability to increase plant capacity due to equipment limitations. All of these problems contribute to reduced competitiveness with other power production sources and could lead to premature plant closures.

Reliance on proprietary system suppliers coupled with new licensing and design issues have resulted in high implementation costs when digital upgrades have been performed in nuclear power plants. There is a need for a systematic approach leading to the identification, prioritization, and implementation of I&C solutions in nuclear power plants. Viable alternatives range from extending the useful life of existing equipment to the complete replacement of systems in a cost-effective manner when vulnerability to obsolescence or the need for increased productivity so dictates.

Reliable, integrated information is a critical element for protecting the utility's capital investment and increasing availability and reliability. Integrated systems with integrated information can perform more effectively to increase productivity, to enhance safety, and to reduce O&M costs. A plant communications and computing architecture is the infrastructure needed to allow the implementation of I&C systems in an integrated manner. Current technology for distributed digital systems, plant process computers, and plant communications and computing networks support the integration of systems and information.

The test for future digital I&C system upgrades will be whether they are cost beneficial to the plant and if they can offer a payback to the utility in an acceptable time period. Digital systems with their inherent advantages will be implemented in nuclear power plants only if reliable and cost-effective implementation and licensing acceptance is achieved and if the upgraded system supports reduced power production costs.

EPRI Nuclear Power Plant Instrumentation and Control Upgrade Program

Nuclear utilities are confronted by a growing equipment obsolescence problem which is a significant contributing factor to increasing costs for plant operation and maintenance. Plant age combined with the rapid pace of evolution of electronic technology is a significant factor in equipment obsolescence. The flexibility and performance of modern digital technology could be used as the basis for replacing obsolete modules or systems in a cost-effective manner in nuclear power plants. The realization of the benefits of digital technology is currently restrained by the relatively high cost of initial applications of new technology for the domestic nuclear power industry in a highly regulated environment. Work is needed to establish reliable and cost-effective methodologies for the design, qualification, and implementation of digital systems in nuclear power plants. This work should utilize, as much as possible, relevant information and experience from other process industries where digital systems are commonly used. Commercial-grade digital systems have proven reliable in other process industries for applications including safety related systems. Cost-effective approaches are needed to implement and qualify commercial-grade hardware and software for nuclear power plant applications. To address these issues and facilitate the upgrading of I&C systems in nuclear power plants, the Electric Power Research Institute (EPRI) has put together an industry-wide instrumentation and control program.

The Electric Power Research Institute and its member nuclear utilities are working together under the Integrated Instrumentation and Control Upgrade Program on a three pronged approach to address I&C issues. This program is documented in the Integrated Instrumentation and Control Upgrade Plan (1). This plan covers a wide range of I&C activities as demonstrated by the seven technical elements of the program. They are Instrumentation, Control and Protection, Man-Machine Support Systems, Verification and Validation, Communications, Maintenance, and Specifications and Standards.

The three prongs of the program consist of research and development activities, utility demonstration plant activities, and licensing stabilization activities. The research and development activities support the development and implementation of digital systems for cost and performance improvements, as well as providing a technical basis for qualification and licensing responses. It also provides part of the bases for the requirements and methodologies needed to design, develop, qualify, implement, operate, and maintain digital systems. The demonstration plant activities identify utility's needs, provide part of the bases for requirements and methodologies mentioned above, provide a test bed for and feedback on requirements and methodologies for upgrading systems, support the development of specifications for digital systems, and capture experience from implementing new digital systems. The licensing stabilization activities provide technical support, as requested, for the industry licensing positions with the United States Nuclear Regulatory Commission (USNRC) on digital systems which are being developed by utility working groups and facilitated by the Nuclear Energy Institute (NEI).

Part of the research and development activities, in conjunction with the demonstration plant activities, is to define and develop a set of generic methodologies and guidelines that will assist utilities in identifying, prioritizing, and implementing I&C solutions more effectively. These methodologies and guidelines will assist the utilities in performing integrated I&C upgrades in the most timely and cost-effective manner possible.

The EPRI Instrumentation and Control Upgrade Program has developed a life-cycle management program for I&C systems. Life-cycle management involves the optimization of maintenance, monitoring and capital resources to sustain safety and performance throughout the plant life. Life-cycle management for I&C systems and components additionally may require the use of digital technology, when analog equipment cannot be cost-effectively maintained or when an improvement in performance is desired. The main product of the life-cycle management program is a set of methodologies and guidelines that, as part of the utility's overall life-cycle management effort, will enable nuclear power plants to fully consider

I&C cost and performance improvements, including the application of digital technology. Specific examples of system specification and designs will also be developed through the application of the upgrade implementation methodologies to safety-related and non safety-related systems and system prototypes.

Planning Methodologies

Four strategic planning methodologies are being developed under the I&C Initiative. The first two methodologies enable the utility to prepare an I&C life-cycle management program plan (2) and a plant communications and computing architecture plan (3). The last two methodologies enable the utility to perform long-term maintenance planning (4) and detailed upgrade evaluation (5) for I&C systems or components.

The Life-Cycle Management Plan is a long term strategic plan for managing a power plant's I&C systems over the planning period selected by the utility. The Life-Cycle Management Plan Methodology (2) guides a designated team of utility personnel through a comparison of I&C life-cycle management strategies and through existing and planned life-cycle management program activities to identify interfaces and integration options. On the basis of this comparison, the I&C Life-Cycle Management Plan is prepared. This plan includes the identification of systems and components to be included in the program; the development of bases for upgrade or long term maintenance options; the initial cost and performance improvement estimates, prioritization for detailed upgrade evaluation, and deferred-upgrade maintenance planning; and the identification of related programs and organizational interfaces including key personnel and responsibilities. The methodology is accompanied by a workbook which contains various outlines, worksheets, and generic interview questions and topics that aid in the development of a Life-Cycle Management Plan. The document describing the methodology also explains the overall process for planning and implementing the various elements of I&C life-cycle management, and the relationship of the other EPRI planning methodologies and guidelines. A plant-specific version of the life-cycle management plan is given in reference 6.

The Plant Communications and Computing Architecture Plan Methodology (3) provides utilities with a detailed set of instructions for preparing a Plant Communications and Computing Architecture Plan that will allow them to upgrade their I&C systems in a logical, cost-effective, and non-disruptive fashion. The Plant Communications and Computing Architecture Plan Methodology provides all of the information necessary to allow utilities to develop their strategic architecture plans in the most cost-effective manner possible. It guides a designated team of utility personnel through an assessment of the existing plant data network architecture, corporate communications architecture life-cycle management plans, and I&C life-cycle management implementation guidelines with respect to the communications architecture. On the basis of the assessment results, a Plant Communications and Computing Architecture Plan is prepared to address a characterization of the existing network architecture; a characterization of the future network architecture in terms of a network model and communication standards for connectivity and interoperability of network elements; a set of network architecture requirements regarding the physical configuration, network access, network add-on provisions, network performance monitoring, and I&C equipment communications interfacing; and a set of consistent human-machine interface requirements for I&C systems. Some nuclear power utilities have used the need to upgrade their plant process computer (7) as an opportunity to develop a new plant communications and computing architecture. An example of a plant-specific architecture plan is given in reference 8.

The Systems Maintenance Plan Methodology (4) addresses long-term maintenance planning for systems or components where the initial screening in the Life-Cycle Management Plan indicates that detailed upgrade evaluation is not justified, over the planning period, by cost and performance improvement potential. The Systems Maintenance Plan Methodology contains a process for developing a comprehensive System Maintenance Plan for each identified system. The Systems Maintenance Plan will present the most efficient approach for maintaining the operational goals and life expectancy of the system. The Systems Maintenance Plan Methodology will describe how to develop long range maintenance objectives, to baseline and analyze the existing maintenance process, to analyze failure rates, inventory practices, and obsolescence issues, and to implement maintenance related problem solving techniques.

The Upgrade Evaluation Methodology addresses (5) a detailed evaluation of I&C system and components when upgrading is indicated by the cost and performance screening in the Life-Cycle Management Plan. The Upgrade Evaluation Methodology is used to analyze each candidate system upgrade to determine if the upgrade is justified from a cost benefit perspective. The Upgrade Evaluation Methodology is used to produce an Upgrade Evaluation Report for each candidate upgrade. The Upgrade Evaluation Report describes high level system functionality, upgrade alternatives and associated cost benefit evaluations, and the recommended alternative. The upgrade evaluation process includes detailed cost and performance analysis; conceptual design options analysis; cost/benefit analysis; and upgrade recommendations. Analysis of conceptual design options includes the consideration of digital design basis changes, associated technical specification changes, and equipment selection candidates. Where an upgrade is to proceed, the Upgrade Evaluation Report issued as input to the Functional Requirements Specification.

Integrated Plant Systems

While analog equipment is becoming obsolete and more costly to maintain, the requirements on nuclear power plant personnel to improve availability, reliability, and productivity and to reduce safety challenges to the plant have increased. These personnel are working with more complex systems, and responding to increasing operational, regulatory and productivity demands. As tasks become more complex, involving large numbers of subsystem interrelationships, the potential for human errors increases. Therefore, reliable, integrated information is a critical element for protecting the utility's capital investment and increasing availability and reliability. Integrated systems with integrated information access can perform more effectively to increase productivity and enhance safety.

Traditionally systems have been implemented in a stand-alone manner which has resulted in increased operation and maintenance costs. Increased competition in the utility industry makes it essential that these operating and maintenance costs are minimized. Technology can be used to assist plant personnel and reduce the potential for human errors. At the same time, it can support improved productivity and the reduction of overall operating and maintenance costs. The modern technology available for distributed digital systems, plant process computers, and plant communications and computing networks is fully capable of supporting integration of systems and information. In fact, this capability has been proven in other process industries and in nuclear power plants in other countries.

Integration of the plant systems and information are essential to cost-effectively enhance cooperation between systems and to reduce unnecessary duplication of functions and information. The objectives of integrating plant systems and information are to improve plant availability and reliability, to reduce operations and maintenance costs, to reduce safety challenges, and to improve performance with existing and new equipment systems. The plant communications and computing architecture of the plant supplies the infrastructure which allows the integration of systems and information. This infrastructure supports integrated upgrades, provides access to all of the plant's information sources, and facilitates common interfaces between the human and the machine. This architecture will support the interoperability of systems and the interchangeability of equipment. It will also be designed to be easily expandable. This architecture is defined by a plan that includes a migration strategy to get from the current plant architecture to the final, desired architecture.

Productivity Enhancement Systems

Digital technology can support improved power output from nuclear power plants. The improved accuracy of digital systems and the associated reductions in uncertainties can allow the utility to increase its plant's power rating. Digital systems also have the potential to support faster startups for increased power output. They can also support the faster determination of the root causes of an unanticipated trip. At the same time, they can support the faster evaluation of the performance of the equipment and systems during the unanticipated trip. Both of these will allow a faster return to power after an unanticipated trip and; therefore, allow more power to be produced by the plant.

The abilities of digital systems offer many ways to reduce O&M costs. Besides improved reliability and availability, two examples of ways to reduce O&M costs are derived from the continuous monitoring, trending and reporting capabilities, as well as the archival capabilities, of digital systems. The first is the instrumentation calibration reduction program (9, 10) which can reduce the number of instruments to be calibrated and increase the interval between calibrations. Additional reports exist on instrumentation calibration, surveillance, and testing (11-14). The second is a condition-based maintenance program which would allow maintenance to be performed when needed rather than on some predetermined interval. This is the next step in maintenance cost-effectiveness after preventive maintenance (15) and reliability centered maintenance (16).

The technological advances of the last few years have made it possible to develop sophisticated personnel support systems, which can not only process and present information, but can also give advice to the human. With appropriately implemented personnel support systems, humans can be augmented substantially in their capacity to monitor, process, interpret, and apply information; thus reducing errors and increasing reliability and availability. These personnel support systems can increase productivity by eliminating routine human-power-intensive efforts such as recording, collecting, integrating, and evaluating data; and by assisting in monitoring and control activities. These systems can improve the consistency and completeness of decision-making activities by performing the role of diagnostic and decision-support advisors. Personnel support systems can assist in reducing safety challenges to the plant by presenting more complete, integrated, and reliable information to plant staff to better cope with operating and emergency conditions. Reducing safety challenges leads directly to improved reliability and availability and hence productivity. It can also reduce the maintenance activities, which would have been required, for equipment that would have been unnecessarily challenged. Functional requirements for an environment that would support these capabilities is given in reference 17. A Reactor Water Cleanup prototype system using integrated information and soft controls is described in reference 18.

Advances in technological and human engineering offer the promise of helping nuclear power plant staff to reduce errors, improve productivity, and minimize the risk to plant and personnel. A plant-wide infrastructure for coordinated personnel support systems should be created to enhance these systems and to reduce their implementation costs. This infrastructure will include information communication capabilities, database and knowledge base managers, and a unified human-machine interface. This infrastructure, which is the plant communications and computing architecture discussed above, will permit incremental additions of personnel support systems in all domains.

Application of Three Diverse Modern I&C Technologies

The majority of control and protection systems in nuclear power plants are based on analog and electromechanical technologies. However, utilities face a growing unavailability of replacement parts for such systems. Three diverse modern forms of I&C technologies are being studied and prototyped for application to safety systems in nuclear power plants. The first is programmable logic controllers (PLCs). PLCs with appropriate qualification programs appear to be ideally suited for a large number of nuclear power plant applications. PLCs have proven highly reliable in many industrial applications and can be used to replace aging analog and electromechanical equipment to solve the obsolescence problem, improve operation and productivity, and reduce operation and maintenance costs. Areas that must receive careful attention when adopting commercially available PLCs include software verification and validation, hardware qualification, and regulatory acceptance. Standardized designs of PLC-based systems for safety system applications offer the opportunity for increased cost-effectiveness in implementations.

Work is underway to apply modern PLC technology to design, develop, test, and demonstrate a safety-grade control system. It will study various alternative PLC designs for safety systems to develop a cost-effective standardized design. This effort will develop, in conjunction with the commercial-grade dedication working group's activities, a reviewable process for hardware qualification and software verification and validation of commercially available PLCs for safety-grade applications in nuclear power plants. It will demonstrate the applicability of the generically-developed commercial-grade dedication process for digital systems to commercial PLCs. It will develop a generic "off-the-shelf" PC-based simulator for stimulation, and later plant simulation, of PLC-based designs for specific safety-grade nuclear applications.

Due to the stringent and, from past history, costly requirements for licensing digital systems for reactor protection systems, application specific integrated circuit (ASIC) technology is being considered as a solution for reactor protection system upgrades. Cost and regulatory risk are major concerns with licensing a digital reactor protection system. To satisfactorily insure that a microprocessor-based reactor protection system will perform as desired, be highly reliable, and not have unintended functions is very costly. A potentially cost-effective alternative is to develop an ASIC-based reactor protection system. In this case, the ASIC is designed to perform only the needed functionality of the reactor protection system. This reduces the effort required to assure the reactor protection system's performance in protecting the plant and public.

The reactor protection systems at most plants are based on obsolete relay systems. The cost of maintaining these systems to an acceptable safety level is increasing. In addition, the effort to test these systems is substantial. Also, spare parts are hard to find and very costly. The increasing procurement, maintenance, and testing costs are contributing to the decreased competitiveness of nuclear plants. ASIC-based systems could reduce the costs in each of these areas. If designed properly, they would be readily available, highly reliable, require much less test and maintenance time, and allow testing the reactor protection system with the channel for testing in bypass mode rather than out-of-service mode. This last can be very effective in reducing inadvertent trips while testing is underway.

The feasibility of implementing reactor protection system functionality in an ASIC is being evaluated. ASICs are being designed and tested to perform the pressurizer pressure and delta T T average functions in the protection system. Since the latter is the most complex function in the protection system, this demonstration will show the technical and financial feasibility of ASICs. Based on the successful outcome of this feasibility study, an ASIC-based reactor protection system will be designed.

The third technology being studied is the Dynamic Safety System (DSS) technology. Compared to existing static safety systems' technology, DSS offers the advantages of decreased costs and increased reliability. The on-going work includes design, interfacing, and testing an emulator of a BWR reactor protection system using DSS technology at a BWR plant simulator to determine feasibility for use of DSS technology in BWR's. To pursue the objective of eliminating the need for manual surveillance of safety channels, methods for inducing simulated trip conditions in nuclear instrumentation channels that use fission chambers will be investigated. The evaluation will include experiments with detectors used in both PWR and BWR wide range of monitors and in BWR power range monitors.

Areas of Concern about Digital Systems

Design and licensing issues have inhibited access to cost and performance improvements possible with digital technology. Examples of the areas of concern for digital systems in nuclear power plants are licensing, software verification and validation (V&V), hardware qualification including electromagnetic interference compatibility and seismic, reliability, performance, separation, redundancy, fault-tolerance, common-mode failures, diversity, human-machine interfaces, and integration of systems and information through communications networks. Commercial-grade dedication of digital systems is an approach for more cost-effective implementations that is of considerable interest to the nuclear utilities. As part of the EPRI Instrumentation and Control Upgrade Program and other EPRI activities, approaches to address many of these concerns have been developed and the results are given in recent EPRI reports (19-35).

The Guideline on Licensing Digital Upgrades (19) was developed to be consistent with the established 10 CFR 50.59 process. It helps utilities design and implement digital upgrades, perform 10 CFR 50.59 safety evaluations, and develop information to support licensing submittals. It suggests a failure analysis-based approach that encompasses digital-specific issues and other possible failure causes, addressing both according to their potential effects at the system level. Abnormal conditions and Events (ACES) (20), as described in ANSI/IEEE ANS 7-4.3.2-1993 "Application Criteria for Programmable Digital Computer Systems in Safety Systems of Nuclear Power Generating Stations", play an integral role in this approach.

Guidance for electromagnetic interference susceptibility testing of digital equipment (21) and a handbook for electromagnetic compatibility of digital equipment (22) have been developed. These reports integrate the current knowledge and understanding of the electromagnetic issues concerning the installation of digital equipment in power plants. They direct the utility toward practical and economical solutions for dealing with electromagnetic interference. The handbook also helps eliminate some misconceptions that questioned the reliability of digital equipment subjected to the electromagnetic environment of nuclear power plants.

Guidelines and a handbook for software V&V have been developed (23-25). These products describe approaches to categorize the software systems in terms of criticality and consequences of failure. They then identify levels of V&V commiserate with these categorizations. The guidelines for V&V in reference 23 developed a set of 16 V&V guideline packages based on the system category, development phase, and software system component which is being tested. For V&V methods in the guidelines that do not have a good description elsewhere in literature on how to use them, 11 sets of procedures have also been developed. The report identifies 153 V&V methods for software systems which can be used on the 52 identified software defect types. The guidelines developed were based on the attempt to identify the methods which were most successful in finding various types of defects, on the attempt to assure that the different guidelines catered to the different needs of different systems, and on the attempt to emphasize the practicality and cost-effectiveness of the methods recommended.

A process for the commercial-grade dedication of hardware has been developed (26, 27) and proven very successful. The basic concepts of this process are being used as the starting point for proposed commercial-grade dedication processes for digital systems. The use of commercial-grade programmable logic controllers (PLCs) for safety related systems is described in reference 28. Guidelines for evaluating and dedicating commercial-grade PLCs have been developed (29, 30). Considerable concern has been raised about annunciator systems and the magnitude of alarms that an operator must be aware of during a transient. The large number of alarms and the presentation of them make the operator's job more difficult and can potentially contribute to human errors. Work has been done on more intelligent alarm systems and the methods for presenting them (31-33). Additional areas that have been addressed are pressure transmitters (34), radiation monitoring systems (35), and wireless monitoring systems (36).

Demonstration Plant Projects

The utility demonstration plants essentially are the laboratories where I&C cost and improvement options are being researched and developed. There are five utility demonstration plant projects in progress which are providing the primary inputs, as well as testing, validation and refinement activities for the methodology and guideline development under the I&C Initiative.

Activities at each of the five demonstration plants may include the preparation of I&C life-cycle management plans and plant computing and control architecture plans; system screening, deferred-upgrade maintenance planning, and detailed upgrade evaluations; testing, validation, and refinement of various plant-specific methodologies and guidelines; and development of options and plans for integration of I&C cost and performance improvement activities with related life-cycle management efforts.

Demonstration project activities are presently being pursued at the Tennessee Valley Authority's Browns Ferry Unit 2, Baltimore Gas and Electric Company's Calvert Cliffs Units 1 and 2, Northern States Power Company's Prairie Island Units 1 and 2, Entergy Company's Arkansas Nuclear One Units 1 and 2, and Omaha Public Power District's Fort Calhoun. An example of a plant-specific plan developed under the demonstration program is the architecture plan for the plant data network at Browns Ferry (8).

Conclusions

The implementation and integration of digital I&C systems enhances the ability to achieve the goals of improved availability and reliability, enhanced safety, reduced operations and maintenance costs, and improved productivity in nuclear power plants. The plant communications and computing architecture provides the infrastructure which allows the integration of systems and information. The modern technology of distributed digital systems, plant process computers (both monolithic and distributed), and plant communications and computing networks have proven their ability to achieve these goals in other industries and in nuclear power plants in other countries. The use of this modern, proven technology is a key contributor to improved competitiveness in nuclear power plants. EPRI has established an Integrated Instrumentation and Control Upgrade Initiative to support its member nuclear utilities in developing strategic plans and taking advantage of this modern technology to improve nuclear power plant competitiveness.

References

1. "Integrated Instrumentation and Control Upgrade Plan, EPRI NP-7343 Revision 3, December 1992.
2. "Life-Cycle Management Plan Methodology", to be published in 1995.
3. "Plant Communications and Computing Architecture Plan Methodology, Vols. 1&2", EPRI TR-102306, November 1993.
4. "Systems Maintenance Plan Methodology", to be published in 1996.
5. "Upgrade Evaluation Methodology", to be published 1996.
6. "Browns Ferry Instrumentation and Control Upgrade Methodology", EPRI TR-101963, December 1992.
7. "Plant Process Computer Upgrade Guidelines, Vols. 1-3", EPRI TR-101566, December 1992.
8. "Process Data Network Architecture Plan for the Browns Ferry Nuclear Plants", EPRI TR-103445, December 1993.
9. "Instrument Calibration and Monitoring Program, Vols. 1&2", EPRI TR-103436, December 1993
10. "Instrument Calibration Reduction Program, Vols. 1&3", EPRI NP-7207-CCML, March 1991
11. "Guidelines for Instrument Calibration Extension/Reduction Programs", EPRI TR-103335, March 1994

12. "Non-Process Instrumentation Surveillance and Test Reduction", EPRI TR-103457, December 1993
13. "Effects of Resistance Temperature Detector Aging on Cross-Calibration Techniques", EPRI TR-103099, June 1994
14. "Calibration of Radiation Monitors at Nuclear Power Plants", EPRI TR-102644, April 1994
15. "Guide for Determining Preventive Maintenance Task Intervals", EPRI TR-103147, December 1993.
16. "Reliability Centered Maintenance Implementation in the Nuclear Power Industry: Guidelines for Successful RCM Implementation", EPRI TR-103590, April 1994.
17. "Plant-Wide Integrated Environment Distributed on Workstations (Plant-Window) System Functional Requirements, EPRI TR-104756, Draft Report, 1994.
18. "Development of Process Control Capability Through the Browns Ferry Integrated Computer System", EPRI TR-104378, to be published in 1995.
19. "Guideline on Licensing Digital Upgrades", EPRI TR-102348, December 1993.
20. "Abnormal Conditions and Events Analysis for Instrumentation and Control Systems, Vols. 1&2", EPRI TR-104595, to be published in 1995.
21. "Guidelines for Electromagnetic Interference Testing in Power Plants", EPRI TR-102323, September 1994.
22. "Handbook for Electromagnetic Compatibility of Digital Equipment in Power Plants, Vols. 1&2", EPRI TR-102400, June 1994.
23. "Guidelines for the Verification and Validation of Expert System Software and Convention Software, Vols. 1-8", EPRI TR-103331, to be published in 1995.
24. "Verification and Validation Guidelines for High Integrity Systems", EPRI TR-103916, to be published in 1995.
25. "Handbook of Verification and Validation for Digital Systems, Vols. 1-3", EPRI TR-103291, December 1994.
26. "Guideline for the Utilization of Commercial Grade Items in Nuclear Safety Related Applications (NCIG-07)", EPRI NP-5652, June 1988.
27. "Supplemental Guidance for the Application of EPRI NP-5652 on the Utilization of Commercial Grade Items", EPRI TR-102260, March 1994.
28. "Experience with the Use of Programmable Logic Controllers in Nuclear Safety Applications", EPRI TR- 104159, to be published in 1995.
29. "Programmable Logic Controller Requirements and Evaluation Guidelines for BWRs", EPRI TR-103734, November 1994.
30. "Programmable Logic Controller Qualification Guidelines for Nuclear Applications, Vols. 1&2", EPRI TR-103699, October 1994.
31. "Control Room Alarm System Upgrades", EPRI TR-100584, April 1992.
32. "Alarm Processing and Diagnostic System", EPRI TR-100838, June 1992.
33. "Functional Specification Requirements for a Microprocessor-Based Annunciator System", EPRI TR-102872, April 1994.
34. "Review of Technical Issues Related to the Failure of Rosemount Pressure Transmitters Due to Fill Oil Loss", EPRI TR-102908, August 1994
35. "Utility Experience with Major Radiation Monitoring System Upgrades", EPRI TR-104081, November 1994
36. "System Specification for the Wireless Programmable Process Monitoring System", EPRI TR-102287, April 1993

0074-056X/95/235-243/$0 + .50pp

AN EXPERIMENTAL TESTBED FOR ADVANCED DIGITAL NUCLEAR REACTOR CONTROL

Robert M. Edwards
Assistant Professor
Nuclear Engineering Dept.
Penn State University
231 Sackett Building
University Park, PA 16802

Kwang Y. Lee
Professor
Electrical Engineering Dept.
Penn State University
103 EEE Building
University Park, PA 16802

Daniel E. Hughes
Sr. Research Associate
Radiation Science and Engineering Center
Penn State University
University Park, PA 16802

KEYWORDS

Testbed, Nuclear Reactor, Control, Robust, Optimal, Fuzzy Logic, Neural Network, Hybrid Simulation, Boiling Water Reactor, Reconfigurable, Fault-Accommodating

ABSTRACT

A testbed facility for prototyping advanced monitoring and control techniques for commercial nuclear power plants has been established under an NSF/EPRI funded project. The experimental testbed is formulated around a university research reactor. The project has successfully demonstrated optimal, robust, fuzzy logic, neural network, and intelligent reconfigurable control techniques for improving reactor temperature response. A unique hybrid reactor simulation capability has also been implemented that mimics the stability characteristics of commercial BWRs. These control and simulation techniques were initially demonstrated using commercial grade microprocessor-based controllers and PC computers. More recently, UNIX network-compatible controllers and automated C code generation have been adopted.

INTRODUCTION

In response to industry concerns of nuclear plant instrumentation obsolescence, EPRI set a strategic goal to complete modernization of the I&C in at least 10 plants by the year 2000.[1] Although additional functionality is not a priority in current upgrade projects, research to identify the potential of advanced computer-based technologies was also established by an NSF/EPRI intelligent control initiative.[2] The project summarized in this paper is titled *Experimental Development of Power Reactor Intelligent Control*[3] and is comprised of five major tasks:

- I) experimental validation and comparison of advanced direct control algorithms such as optimal, robust, fuzzy logic and neural networks;
- II) intelligent control research emphasizing a reconfigurable supervisory control level to accommodate wide range operation and faulted conditions;
- III) multivariable control;
- IV) hybrid reactor/simulation; and
- V) dissemination of results through such activities as hands-on professional development workshops.

All of these tasks focus on the long-term utilization of the research reactor as a testbed for evaluation and demonstration of advanced controls for nuclear power plants. This paper describes the additions to the research reactor which are common to all the tasks and discusses the experimental program and some of the results obtained thus far.

RESEARCH REACTOR

The Penn State Breazeale Reactor (PSBR) is a General Atomic MARK III TRIGA (**T**raining, **R**esearch, **I**sotopes, **G**eneral **A**tomic). It is part of the Penn State Radiation Science and Engineering Center, RSEC, and is the longest operating university reactor in the United States. The reactor is a light-water cooled and reflected pool type reactor capable of pulse operation to 1000 Megawatts (MW) and constant power operation at 1 MW maximum. The reactor core has a basic shape of a right hexagonal prism containing about 85 fuel elements, Figure 1. There are 4 control rods designated as the safety rod, regulating rod, shim rod, and transient rod.

With periodic upgrades over its 38 year lifetime, the facility has enabled the university to familiarize students with the current instrumentation technology that they would expect to find in industry. In 1991, the original analog-based control and monitoring system was replaced with a modern microprocessor based system. The reasons for this replacement are similar to the problems which face operators of existing U.S. nuclear power plants. Since the control systems for existing U.S. nuclear power plants were designed in the 60s and 70s, spare parts and technical assistance to maintain obsolete technology are becoming more difficult to secure. Modern digital control hardware and software can address the need to replace the obsolete technology and also provide more reliability and increased functionality to improve safety and performance.

Check-out of the new digital system prior to replacement of the analog control system was accomplished with the aid of an experimental procedure created within the existing technical specifications of the TRIGA reactor.[4] The experimental procedure adds a secondary control rod (SCR) to the reactor's standard configuration of 4 rods. The standard 4 rods are always under the control of licensed control, monitoring, and safety systems. The SCR is placed in the central thimble of the reactor, an ideal location for control experiments. The central thimble is a 3.3 cm diameter water-filled tube whose primary function is to provide sample irradiation for materials research and for isotope production. The capability of the SCR to change reactor power (magnitude and rate) is designed within the TRIGA reactor's moveable experiment technical specifications. The inherent ultra-safe characteristics of the TRIGA reactor allow the SCR to be designed with more than enough capability to accomplish the long-term research objectives of a research reactor experimental control testbed. Presently, the SCR is worth $0.35 and can be moved at a maximum rate of 30% of core length per second.

TASK I: Advanced Direct Controllers

Figure 2 summarizes the structure of the advanced direct controllers which have now been experimentally evaluated. All of these advanced controllers are implemented through the manipulation of the power setpoint to an embedded classical proportional control loop. The inclusion of the embedded classical controller has been shown to improve the robustness over an implementation where the advanced algorithm directly manipulates control rod speed demand.[5,6] Furthermore, the embedded classical and optimal control algorithms which were previously formulated with control rod speed as the manipulated variable are now formulated with control rod reactivity rate as the manipulated variable. This current approach provides for more generic control algorithm formulations (gain settings) which are control rod-worth independent. Controller reactivity rate of change demand is converted to a control rod speed

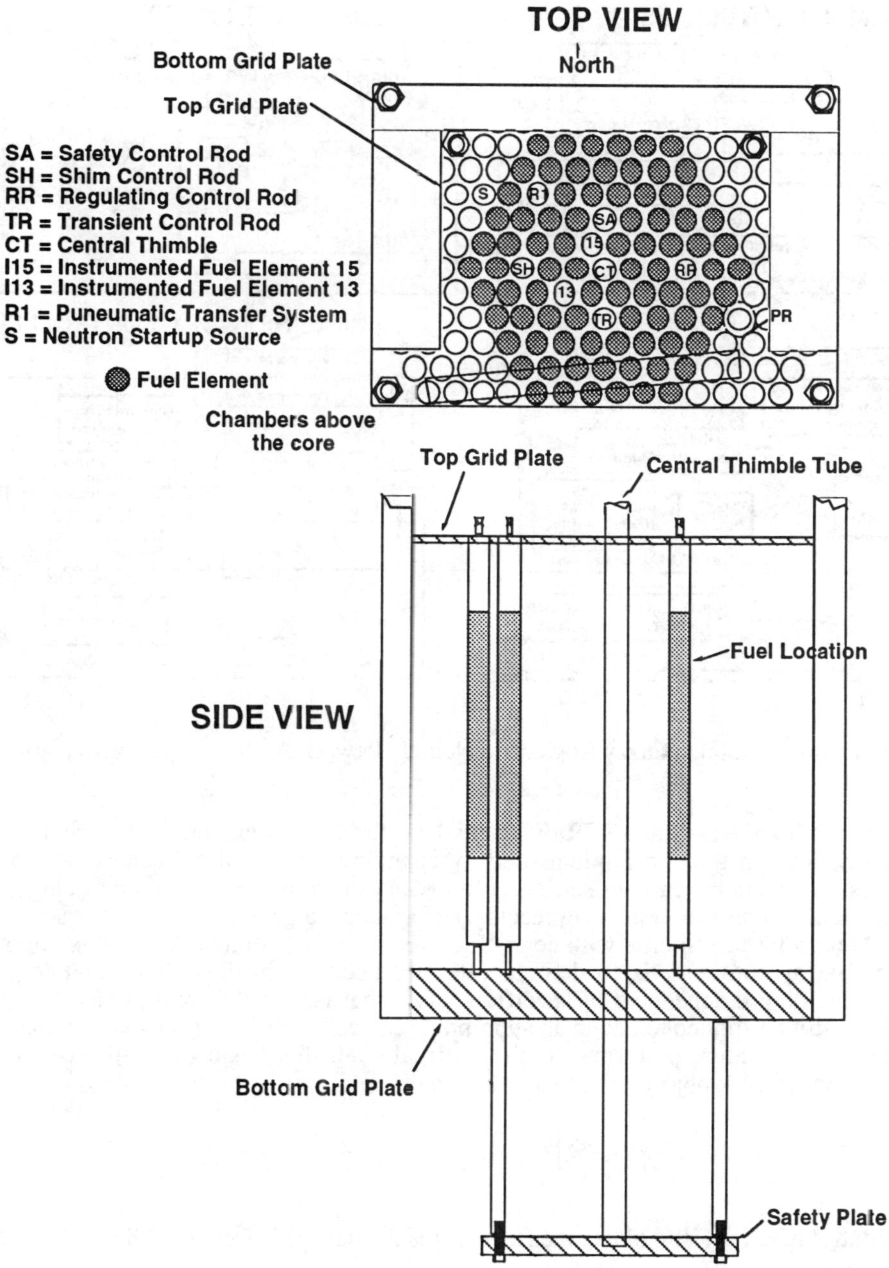

Figure 1: Layout of the Penn State Breazeale Reactor.

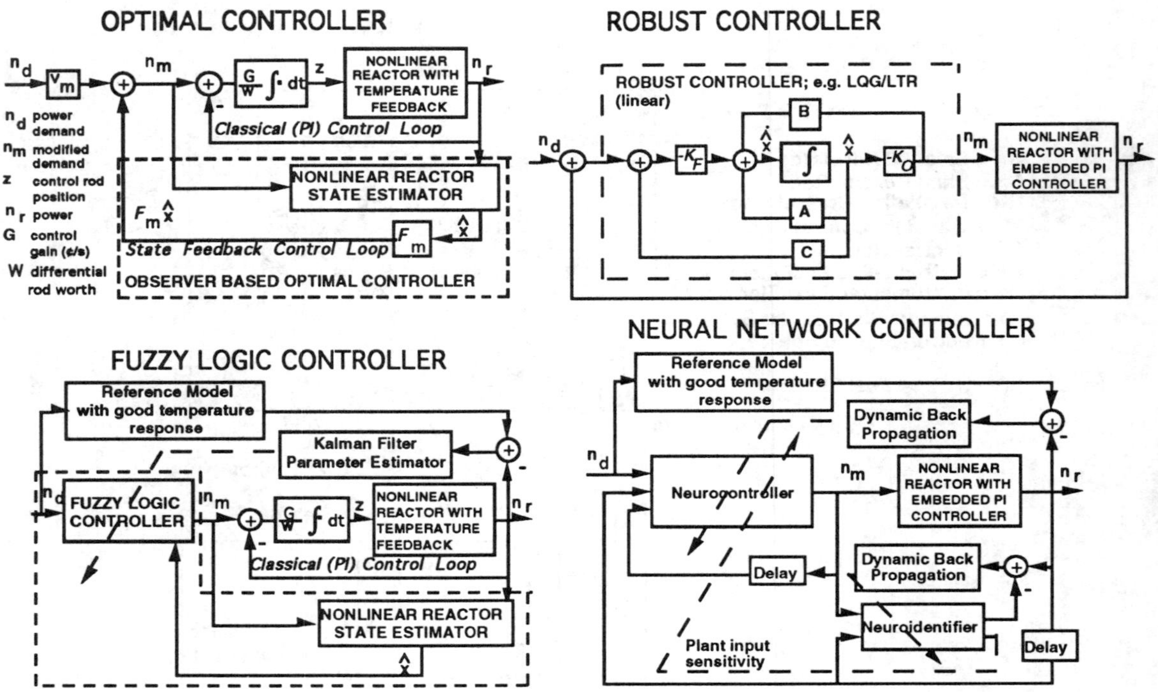

Figure 2: Optimal, Robust, Fuzzy Logic, and Neural Network Architectures for Reactor Control

demand signal utilizing the assigned SCR differential worth. Future utilization of an SCR with different worth characteristics can thus be implemented by changing the speed conversion and not require a controller redesign or change in gain settings. In order to demonstrate the methodology of applying advanced control techniques to improving reactor performance, a goal of improving reactor temperature response to achieve a faster response with negligible overshoots has been pursued. It is important to note that the objective of our work is to demonstrate advanced controller design and implementation methodology which are independent of specific goals. That is, if a different goal were specified, the methodology can still be applied to obtain an appropriate controller design to achieve the goal. To achieve improved reactor temperature performance, the **optimal** controller has been synthesized based upon a linear quadratic performance objective J of the following form:

$$J(\rho_{rr}) = \int_0^\infty \left[0.01\delta T_f^2 + 0.1\delta T_l^2 + 0.1\rho_{rr}^2\right] dt$$

where ρ_{rr} is control reactivity rate of change (¢/s), δT_f is the deviation from equilibrium of estimated core average fuel temperature (°C), and δT_l is the deviation from equilibrium of estimate core average coolant outlet temperature (°C). An important characteristic of this performance objective is that it places no penalties on reactor power deviations from equilibrium and produces a dynamic response where reactor power, through appropriate manipulation of the setpoint to the embedded classical power control loop n_m, is driven appropriately to improve temperature responses. The nonlinear reactor state estimator indicated in Figure 2 is implemented as a numerical integration of a simplified but nonlinear reactor model under the prompt jump approximation with one delayed neutron group and core average fuel and coolant temperatures. Optimal and robust control based on this simplified model have been demonstrated to improve reactor temperature response when applied to very detailed high order simulations of commercial sized pressurized water reactors and liquid metal reactors.[7,8]

Figure 3, presents experimental results for power and temperature from both the embedded classical controller alone and with augmentation by the optimal control. The results are in response to a step reduction in power demand n_d from 100 to 90%. The classical control power response shows a well damped power response to the demand level of 90% with a correspondingly slow temperature response. On the other hand, the power response with the optimal control is driven along a trajectory with significant undershoot of the 90% demand level to as low as 81%. However, the optimal control power trajectory is appropriate for speeding reactor temperature response to the level corresponding to 90% power.

The robust controller architecture shown in Figure 2 exemplifies a linear state-space controller implemented in the forward loop of the controller. A linear quadratic Gaussian with loop transfer recovery (LQG/LTR) [9] has been demonstrated on the research reactor and work continues on developing an H_∞-based [10] robust control algorithm for demonstration. An optimized feedforward with robust feedback control has also been demonstrated.[11] Fuzzy logic [12] and neural network [13] controllers have been implemented for demonstration [14] using the architectures shown in Figure 2. For fuzzy logic and neural network control a reference model, based on the optimal controller designed for improved temperature response, is used for training purposes.

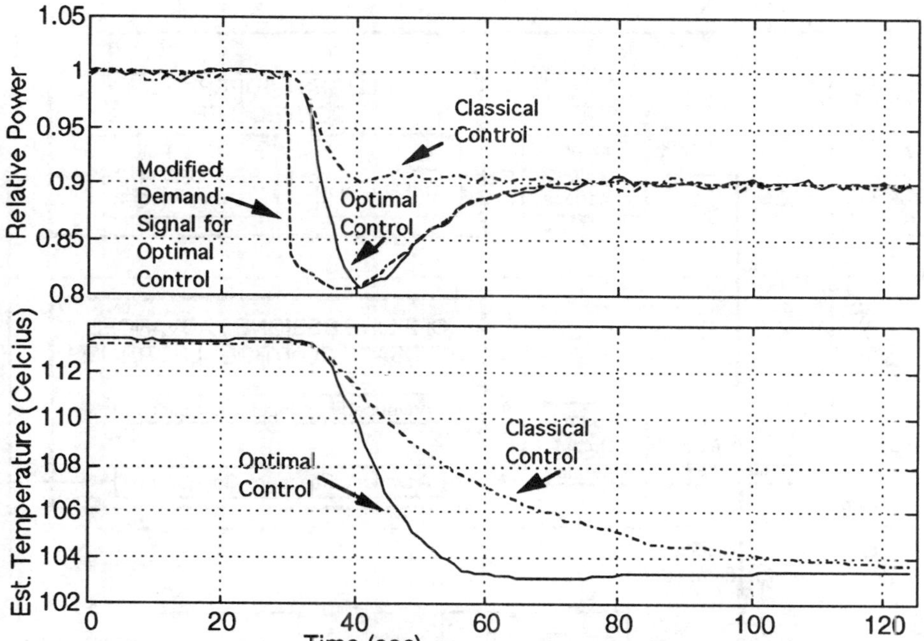

Figure 3: Experimental Results Demonstrating the Potential of Optimal Control For Improving Reactor Temperature Response.

TASK II: Intelligent Reconfigurable Supervisory Control

Although advanced control algorithms based on optimal and robust control can achieve useful system performance characteristics, automated control system reconfiguration may be desirable for wide range operation and for automated response to major component failures and other large disturbances. The overall idea is to provide a bank of preprogrammed controllers which can be individually chosen for enforcement by an automated decision making process. An example of such a controller and its efficacy for power plant applications was demonstrated on the feedwater system of the Experimental Breeder Reactor EBR-II in 1993[15-17]. The components of that controller have since been modified for demonstrating intelligent reconfigurable control on the research reactor testbed. The overall architecture is shown in Figure 4. This architecture consists of three levels. The supervisory level *intelligent supervisory controller* chooses an algorithm from a bank of several *off-line designed advanced feedback control algorithms* in an intermediate level. The enforced advanced algorithm in turn augments the external, or reference, power demand signal n_d to produce a modified demand signal n_m for a *classical feedback controller* to follow, at the lowest embedded level. The *control performance evaluator* indicated in Figure 4 is a critical component of the intelligent supervisory controller and whereas its *control decision* [17-19] component can be generically developed, the performance evaluator must be crafted for a specific dynamical system and its requirements. For the research reactor control objective of improving reactor temperature response an integral quadratic temperature error measure was developed[20]. Similar to the linear quadratic performance objective of the optimal control, a successful quadratic temperature error for demonstrating automated reconfiguration was formulated as a weighted windowed average of the square of the difference between the estimated reactor temperature and reference temperature corresponding to the setpoint power.

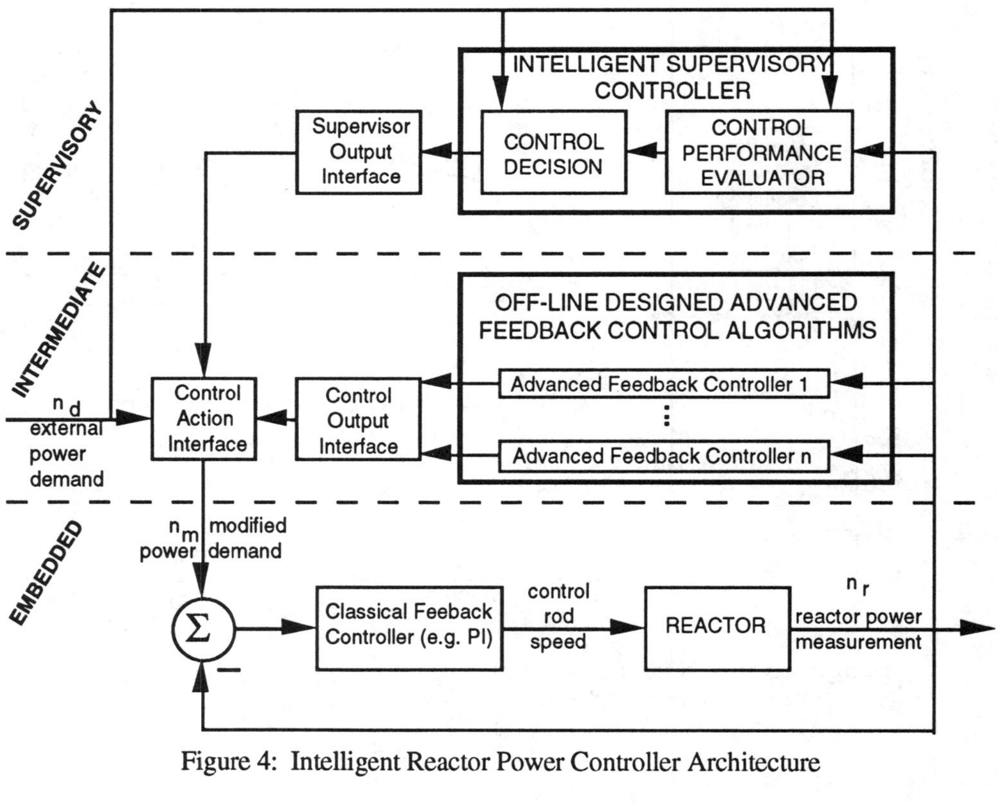

Figure 4: Intelligent Reactor Power Controller Architecture

TASK III: Multivariable Control

In 1995, the reactor control testbed is being expanded to include the capability to manipulate coolant flow rate. Heat is currently removed from the reactor solely by natural circulation of the water coolant. The availability of two manipulated variables, coolant flow rate and control rod positioning, will allow for enhanced demonstration and evaluation of multivariable control algorithms and control system reconfiguration where the failure of one control loop can possibly be mitigated by appropriate compensation or reconfiguration in remaining operational control loops.

TASK IV: Hybrid Reactor Simulation

The hybrid reactor/simulation (HRS) concept was conceived as a means to better enable the research reactor testbed to directly address commercial reactor power plant dynamical monitoring as well as control applications[21]. To implement the HRS, the research reactor secondary control rod is interfaced to a dynamical simulation of a commercial reactor feedback characteristic as shown in Figure 5. With the licensed digital control console in a manual mode of operation, the measured reactor power is used as an input to a digital simulation of a power reactor thermal-hydraulic reactivity feedback mechanism. The simulated reactivity effect is converted to the corresponding speed required by the Secondary Control Rod. The result is that the observed TRIGA reactor power mimics the response of the simulated power reactor. Thus far, the observed stability characteristics of commercial Boiling Water Reactors has been faithfully replicated using the HRS and work is ongoing to develop and validate more efficient stability monitoring methods [22].

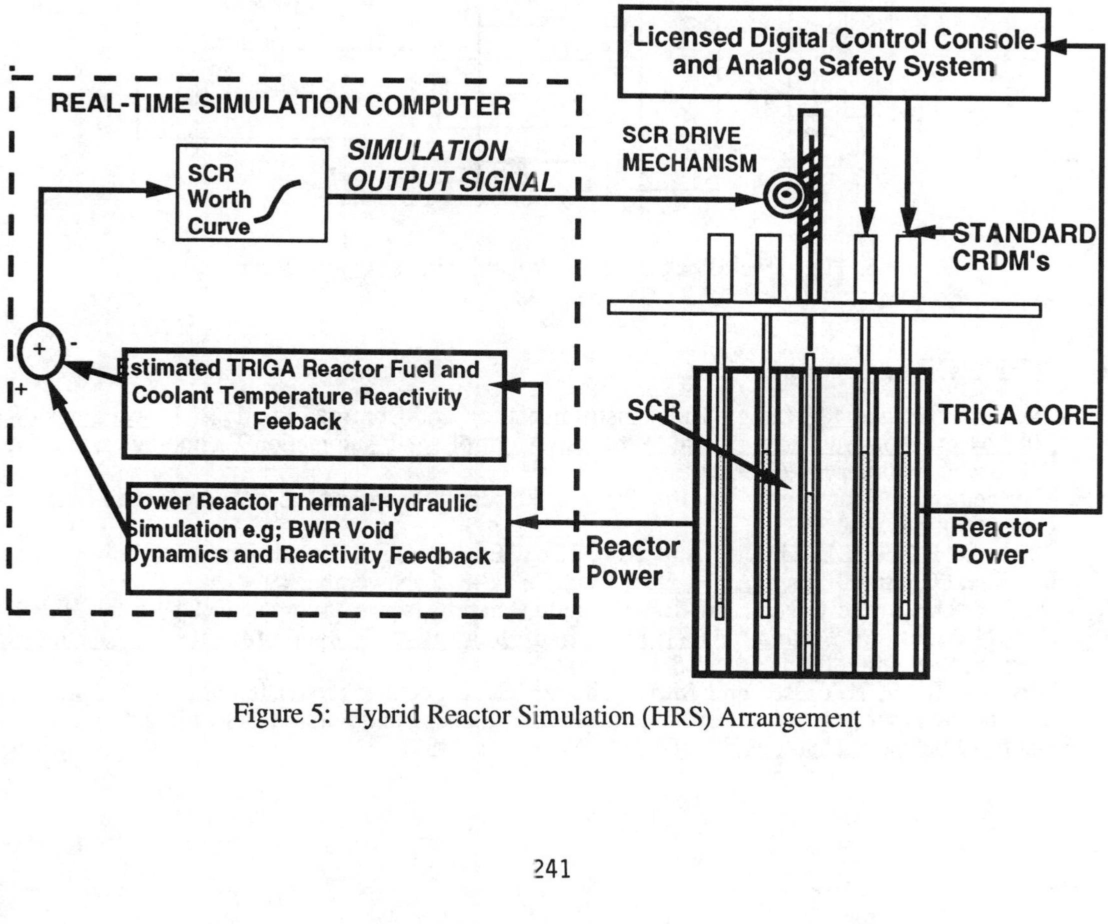

Figure 5: Hybrid Reactor Simulation (HRS) Arrangement

TASK V: Dissemination of Results

The research reactor testbed has established that it can demonstrate the mechanics of designing, implementing, and prototype testing of advanced control algorithms with potential application for improving commercial nuclear power plant operations. Dissemination of results has the objective of documenting the testbed concept and procedures in order to enable subsequent researchers and industry personnel to more easily pursue their advanced and intelligent control applications through an experimental demonstration phase utilizing a physical system. The computer programming of all the prototype controllers was initially conducted using general purpose C language computer programming for a distributed control system actually deployed in current power plants. To improve the transfer of advanced control algorithm technology to industry, an alternative modern controller programming methodology was deployed in the reactor testbed at the end of 1994. The optimal controller discussed earlier can now be implemented directly from a functional block diagram setup, such as shown in Figure 6, through automatic C code generation, controller setup, and execution monitoring in a UNIX network compatible microprocessor-based control system.[23-25] Periodic workshops were also originally envisioned for technology transfer. One was conducted on November 5, 1993 and more are planned in 1995.[26]

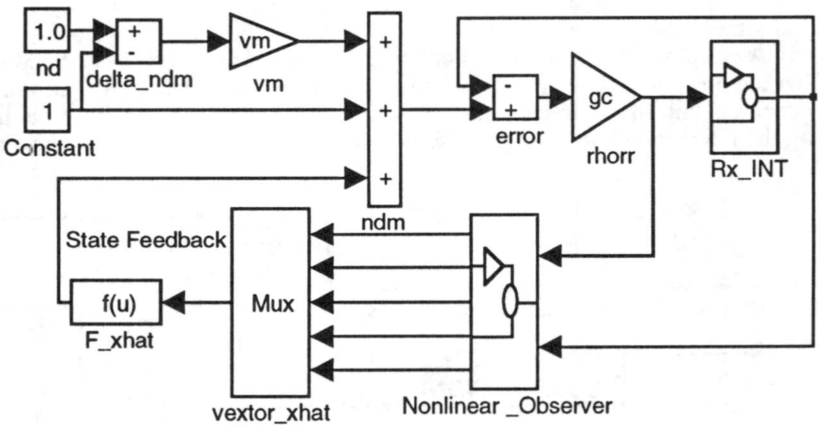

Figure 6: High level Block Diagram Programming of Optimal Control

REFERENCES:

1. Chexal, V.K., and J.F. Lang, "EPRI Instrumentation and Control Initiative," Proceedings of the EPRI Meeting on Advanced Digital Computers, Controls, and Automation Technologies, San Diego, California, (February 5-7, 1992).
2. Maulbetsch, J., "Intelligent Control Systems Research Program," *EPRI Exploratory Research Letter*, No. 14, June 1993.
3. Edwards, R.M., H.E. Garcia, And K.Y. Lee, "Experimental Development of Power Reactor Intelligent Control," Trans. Amer. Nucl. Soc. 66:373-375 (November 1992).
4. Hughes, D.E., Mac Bryan, and Robert Gould, "Safety Issues Involved in Using the PSBR for Control Algorithm Testing," AI 91: Frontiers in Innovative Computing for the Nuclear Industry, Jackson, Wyoming, (September 15-17, 1991).
5. Edwards, R.M., K.Y. Lee and M.A. Schultz, "State Feedback Assisted Classical Control: An Incremental Approach to Control Modernization of Existing and Future Nuclear Reactors and Power Plants," Nuclear Technology, 92:167-185 (November 1990).

6. Edwards, R.M., K.Y. Lee and A. Ray, "Robust Optimal Control of Nuclear Reactors and Power Plants," Nuclear Technology, 98:137-148 (May 1992).
7. Edwards, Robert M., "Robust Optimal Control of Nuclear Reactors and Power Plants," A Ph.D. dissertation in Nuclear Engineering, The Pennsylvania State University (May 1991).
8. Edwards, R.M., and C.K. Weng, "Application of State Feedback Assisted Control Approach to the EBR-II Primary System," FINAL Report on Argonne National Laboratory Contract 31-109-38-9038-W, (December 1991).
9. Ben-Abdennour, A., R.M. Edwards, K.Y. Lee, "LQG/LTR Robust Control of Nuclear Reactors with Improved Temperature Performance," IEEE Transactions on Nuclear Science, 39:2286-2294 (December 1992).
10. Power, M.A., and R.M. Edwards, "Implementation of Robust Control Experiments on a Nuclear Reactor", Proceedings of the Ninth Power Plant Dynamics and Control Conference, Knoxville, TN, (May 1995).
11. Weng, C.K., R.M. Edwards, and A. Ray, "Robust Wide Range Control of Nuclear Reactors By Using the Feedforward-Feedback Concept," Nuclear Science and Engineering, 117:177-185 (July 1994).
12. Ramaswamy, P., R.M. Edwards, and K.Y. Lee, "An Automated Tuning Method of a Fuzzy Logic Controller for Nuclear Reactors," IEEE Transactions on Nuclear Science, 40:1253-1262 (August 1993).
13. Ku, C.C., K.Y. Lee, and R.M. Edwards, "Improved Nuclear Reactor Temperature Control Using Diagonal Recurrent Neural Networks," IEEE Transactions on Nuclear Science, 39:2298-2308 (December 1992).
14. Ramaswamy, P., R.M. Edwards and K.Y. Lee, "Validating Real-time Implementations of Diagonal Recurrent Neural Network and Fuzzy Logic Controllers," Trans. Amer. Nucl. Soc. 71:364-365 (November 1994).
15. Garcia, H.E., A. Ray and R.M. Edwards, "A Reconfigurable Hybrid System and its Application to Power Plant Control," to appear in IEEE Control Systems Technology.
16. Garcia, H.E., A. Ray and R.M. Edwards, "A Reconfigurable Hybrid Supervisory System for Process Control," Proceedings of the IEEE 33rd Conference on Decision and Control, 3131-3136, Lake Buena Vista, FL, (December 14-16, 1994).
17. Edwards, R.M., H.E. Garcia, and N. Messick, "Results from an In-plant Demonstration of Intelligent Control," Trans. of the Amer. Nuc. Soc., 69:153-156 (November 1993).
18. Garcia, H.E., "A Reconfigurable Hybrid Supervisory Control System," A Ph.D. dissertation in Electrical Engineering, The Pennsylvania State University (December 1993).
19. Garcia, H.E. , A. Ray and R.M. Edwards, "Reconfigurable Control of Power Plants Using Learning Automata," IEEE Control Systems Magazine, Vol.11, No.1, Jan. 1991, pp. 85-92.
20. Kenney, S.J., and R.M. Edwards, "An Intelligent Reconfigurable Reactor Power Controller," Proceedings of the Ninth Power Plant Dynamics and Control Conference, Knoxville, TN, (May 1995.)
21. Turso, J.A., R.M. Edwards, and J. March-Leuba, "Hybrid Simulation of Boiling Water Reactor Dynamics Using A University Research Reactor," to appear in Nuclear Technology
22. Turso, J.A., R.M. Edwards, and T.W. Highlands, "Boiling Water Reactor Stability Analysis via Kalman Filter Based Estimation and Maximum A Posteriori (MAP) Detection," Trans. Amer. Nucl. Soc. 71:442-444 (November 1994).
23. SIMULINK, Dynamic System Simulation Language User's Guide, The Mathworks Inc., Natick, MA (1994).
24. VxWorks, Reference Manual and Programmer's Guide, Wind River Systems Inc., Alameda, CA (1993).
25. Stethoscope: Real-time Graphical Monitoring, Performance Analysis, and Data Collection: User's Guide, Real-Time Innovations Inc., Sunnyvale, CA (1992).
26. Second Annual Progress Report on Experimental Development of Power Reactor Intelligent Control, ECS-9216504, Report to National Science Foundation (March 1995).

©0074-056X/95/245-257/$0 + .50pp

Development of an Automated System for Diesel Generator Performance Monitoring

H.M. Hashemian
R.E. Fain
J.L. Riner

Analysis and Measurement Services Corporation
AMS 9111 Cross Park Drive
Knoxville, Tennessee 37923 USA

Phone: (615) 691-1756
Fax: (615) 691-9344
E-Mail: info@ams-corp.com

C.R. Fidler

VECTRA Technologies, Inc.
215 Shuman Blvd., Suite 200
Naperville, IL 60563 USA

Phone: 708-778-4892
Fax: 708-778-4343

KEYWORDS

Diesel Generator, On-Line Monitoring, Station Blackout, Performance Tests, Maintenance

ABSTRACT

Power generation plants typically require the use of diesel generators in such applications as emergency power sources and cogeneration units. Most plants employ periodic maintenance checks and perform periodic maintenance runs as a means of verifying the equipment's ability to perform as required. Failures which may occur between these maintenance checks can go undetected using the current test procedures. Reducing the costs of diesel generator maintenance and operation, as well as providing a remote means of determining the operational status of each diesel generator can be accomplished through an on-line performance monitoring system.

A diesel generator performance monitoring system has been developed that will provide plant personnel with accurate, historical, and real-time data from critical system parameters. It also enables plant personnel to use the data for routine predictive maintenance evaluations. The system presented here was designed based on a modular approach that decreases the complexity of the system's individual components, and increases the flexibility of the performance monitoring system. The system presented here is tailored for emergency diesel generator units, however, the system can also be used on all types of diesel generator units. The system design also allows future expansion to control various operating parameters.

1. INTRODUCTION

An emergency diesel generator unit must have the capability to: 1) start and accelerate a number of large motor loads in rapid succession while maintaining voltage and frequency within acceptable limits, 2) provide power to safety systems if a loss of power and an accident occur at the same time, and 3) supply continuous power to plant safety systems if an extended loss of offsite power occurs. In recent years, reliability and performance have been the issues for which many of the regulations and guidelines have been written relating to emergency diesel generator units.

The ability to maintain a high reliability and availability in the emergency diesel generators will result in a lower probability of a total loss of AC power (station blackout) with a specified coping duration. Most plants employ periodic maintenance checks and perform periodic maintenance runs as a means of verifying the equipment's ability to perform as required. One of the key tasks facing the plant engineering staff in developing a comprehensive diesel generator maintenance program, and reducing operation and maintenance costs, is to determine ways to obtain accurate, historical, and real-time data on critical system parameters. A diesel generator performance monitoring system will provide those needs and enable plant engineering to remotely determine the operational status of each diesel generator, and use the data for predictive maintenance.

For nuclear power stations, the emergency diesel generators are an even more critical component in the overall operation and maintenance activities as compared to other power generation facilities. Ensuring the availability of these diesel generators requires extensive maintenance and operations support.

An enhanced monitoring system which continuously logs the performance parameters of the diesel generators would greatly reduce the amount of work currently performed in obtaining this information during routine runs, and would also allow the data to be acquired in the event of any automatic diesel startup in response to an emergency. This kind of on-line information will be useful in trending operational parameters, scheduling maintenance activities, increasing the reliability and availability of diesel generators, and reducing the amount of maintenance and the scheduled runs of the diesel engines. By adding an expert system to analyze the on-line monitoring data, system degradation can be detected at the onset of any problems that might occur preventing or minimizing the effects of a system failure.

2. BACKGROUND

Diesel generators are widely used throughout the power industry as emergency power sources and cogeneration units. In recent years, the nuclear industry has issued several new requirements and guidelines to ensure that diesel generator units intended for use as onsite emergency power sources be selected with sufficient capacity, be qualified for use, and have the necessary reliability and availability for station blackout and design basis accidents. An understanding of the background of these requirements is important to plant personnel developing a comprehensive diesel generator operation and maintenance program. The pertinent guiding documents include the following:

1. Institute of Electrical and Electronics Engineers (IEEE) Standard 387-1984

   The IEEE standard, "Criteria for Diesel-Generator Units Applied as Standby Power Supplies for Nuclear Power Generating Station", provides principal design criteria and qualification and testing guidelines that, if followed, will help ensure that selected diesel generators units meet performance requirements. The 1977 version of this standard was endorsed by Revision 2 of Regulatory Guide 1.9, "Selection, Design, and Qualification of Diesel Generator Units Used as Standby (Onsite) Electric Power Systems at Nuclear Power Plants", for which Revision 3 was issued in July 1993. The newest revision to Regulatory Guide 1.9 integrates Revision 1 of Regulatory Guide 1.108, Generic Letter 84.15, and appropriate guidelines of IEEE Standard 387-1984. It provides specific guidance in the areas of preoperational testing, periodic testing, reporting and recordkeeping requirments, and valid demands and failures.

2. Title 10 Code of Federal Regulations Part 50, Section 50.63

   This requirement is entitled "Loss of all alternating current power", but is better known as the Station Blackout Rule in reference to Regulatory Guide 1.155 on the subject. The draft regulatory guide was first issued in March 1986 to focus on the reliability of emergency diesel generator units. Procedures and equipment depended upon in a station blackout must be set up to ensure that necessary decay heat removal systems perform their intended functions for the duration of the station blackout coping period. For this reason, high reliability is an issue and must be designed into the emergency diesel generator units and maintained throughout their service lifetime. The expectations placed on all plants to be able to withstand a station blackout of a specified duration, and be able to recover from that, are listed here.

3. Title 10 Code of Federal Regulations Part 50, Section 50.65

   On July 10, 1991, the final maintenance rule, "Requirements for Monitoring the Effectiveness of Maintenance at Nuclear Power Plants" was published by the Nuclear Regulatory Commission (NRC). The thrust of the maintenance rule was to provide regulatory framework as a mechanism to evaluate the overall effectiveness of nuclear power plant maintenance programs. The objective is to ensure that structures, systems and components (SSCs) are maintained so that plant equipment will perform its intended function when required. Full implementation is required by July 10, 1996.

   The NRC provided guidance in Regulatory Guide 1.160 for the industry on how to implement the maintenance rule. Additional guidance was added to the Regulatory Guide specifically addressing the use of monitoring to ensure emergency diesel generator reliability and availability.

   Each guiding document is designed to provide assistance to the plant to ensure the reliability and performance of emergency diesel generator units. This can be achieved by instituting appropriate testing, maintenance, and operating programs and by institution of a reliability program that implements effective root cause analysis of all failures, and is designed to monitor, improve, and maintain the desired reliability and availability.

3. CURRENT TESTING

An emergency diesel generator unit must be able to continuously supply power to the equipment needed to maintain the plant in a safe condition if an extended loss of off-site power occurs. Therefore, the availability of this emergency power is crucial and periodic maintenance testing of each diesel generator is required by plant technical specifications.

The reliability of emergency diesel generators is considered by the NRC as one of the main factors affecting the risk of core damage from a station blackout event. Thus, diesel generator maintenance has been a high priority issue in nuclear power plants. Examples of some of the required surveillance tests are shown in Table 1, in addition to the frequency of those tests. Emergency diesel generators are required to pass each of the listed performance tests.

If there are any failures in these routine tests, the consequence is to increase the frequency of testing. The test frequency is increased to a weekly surveillance if there are two or more failures in the last twenty test runs. This amplified testing must be continued until seven failure free tests have been run, and the number of failures drops to one in the last twenty test runs. Additionally, if an individual emergency diesel generator experiences 4 or more failures in the last 25 demands, the utility must demonstrate the restored emergency diesel generator performance by conducting seven consecutive failure free tests. This form of accelerated testing must be conducted at a frequency of no less than 24 hours and no more than seven days between each demand. A problem emergency diesel generator is defined as an individual emergency diesel generator that has experienced 4 or more failures in the last 25 demands. Most nuclear plants with standard technical specifications have accelerated testing requirements.

Any failures of these tests are logged and compared to the values listed in Table 2 to determine the targeted reliability of the emergency diesel generator. Current nuclear power plant licensees have been given the option in the NRC station blackout rule (10 CFR Part 50, Section 50.63) to declare the reliability of their diesel generators to be either 0.95 or 0.975 dependent upon what their historical monitoring data shows.

If a means of obtaining more accurate and timely data on a more consistent basis were available, the number of possible test failures due to reasons other than actual operational problems could be reduced. Also, future failures could be detected and corrective action taken prior to the actual failure occurring. As a result, accelerated testing could be prevented.

4. PROPOSED TESTING METHODS

In order to reduce the current accelerated testing requirement, new maintenance programs are being implemented to comply with the new maintenance rule (10 CFR Part 50,

Section 50.65). The maintenance rule must be implemented by July 10, 1996. It requires that emergency diesel generator availability and reliability be monitored, and compared against goals set under the maintenance rule, or established as performance criteria in the plant's preventive maintenance program.

Under the maintenance rule the utility will establish performance criteria for balancing emergency diesel generator reliability and availability. Emergency diesel generator performance criteria for reliability would be met by the absence of a maintenance preventable failure, or the occurrence of a single maintenance preventable failure, followed by an appropriate root cause determination and resulting corrective action. The goal is to develop and implement a program that will eliminate the unnecessary accelerated testing currently performed. This accelerated testing (running the diesels once a week) could be causing excessive stress and could even aggravate failures so that it would be difficult to perform its required function well enough to discontinue the weekly testing.

The new maintenance and reliability programs will need to include a more comprehensive performance monitoring program including better data acquisition of critical parameters, root cause analysis of failures, and a good predictive maintenance program to prevent failures. Examples of critical elements of these programs are:

1. Surveillances that identify emergency diesel generator support systems and subcomponents, frequency and scope of testing, and incorporate manufacturer's recommendations

2. Performance monitoring of critical system parameters on an ongoing basis to obtain information on the condition of the emergency diesel generator and key components so that problem conditions can be identified prior to failure

3. Maintenance activities designed for both preventive and corrective actions based upon operating history and past maintenance activities, vendor recommendations, and the results of surveillance testing

4. Failure analysis and root cause investigation to assist in developing effective corrective actions to prevent recurrence of failures

5. Emergency diesel generator reliability data system to ensure the availability and retrievability of important data and information relating to emergency diesel generator reliability

5. ON-LINE DIESEL GENERATOR PERFORMANCE MONITORING SYSTEM

Emergency diesel generator surveillance testing is performed at least once a month or more frequently based on the plants technical specifications requirements. The data obtained from this testing is used to determine emergency diesel generator reliability and to spot potential problems.

The first consideration should be to ensure that the surveillance efforts cover appropriate emergency diesel generator subcomponents, as well as portions of systems which support the emergency diesel generator. In many cases the portion of the support systems used by the emergency diesel generator can be more important to diesel generator reliability than to support system reliability. It should be noted that these support systems are vendor and plant specific. A list of typical emergency diesel generator support systems is provided below:

1. Speed Control
2. Environmental Control
3. Start Control
4. Starting Air
5. Instrument Air
6. Service Water Engine Cooling
7. DC Power Source
8. Lube-oil System
9. Fuel-oil System
10. Engine including combustion air supply/exhaust
11. Generator Electro-mechanical including voltage regulator
12. Electrical and I&C (including field flash)

Performance monitoring is usually applied to equipment that is run on a continuous, or a near-continuous, basis. The purpose is to monitor critical system parameters on an ongoing basis in order to obtain information about the conditions that may impact the operability of the particular piece of equipment, and which could be used for trending purposes. These trends may often signal a degradation in a particular condition. Evaluation of these conditions may enable a utility to predict the onset of failure and allow corrective actions to be taken before failure occurs.

Equipment that is normally in a standby condition, such as an emergency diesel generator, can be monitored on a limited basis. Monitoring of critical operational parameters is usually performed only during monthly operational testing. This testing is also performed for a limited time which restricts the amount of data that can be collected. However, some benefit should be realized by continually recording critical operational parameters and comparing the results to that of previous testing, as well as assist in performing troubleshooting and root cause analysis of system failures.

In order to obtain the information needed to trend the performance of the diesel generators, a data acquisition system was designed to allow high speed data transfers between all diesel generators. Figure 1 shows a block diagram of the complete diesel generator performance monitoring system. Data acquisition hardware is installed at each diesel generator and is connected to all critical signals to record the performance of the generator (Figure 2). Many data points can be acquired and used for trending. Table 3 lists some of the parameters that may be selected to monitor. Once the data is acquired and stored, the engineer can review the performance, either on-line or historical data, of any diesel generator by plotting the data, historical trends, or allowing the expert system to help diagnose problems and/or degraded conditions.

The diesel generator performance monitoring system is a modular design that decreases the complexity of it's individual components, and increases the flexibility of the diesel generator performance monitoring system package. The system consists of the following components:

1. Local/Remote Unit - A seismically qualified rack is installed near each diesel generator that will contain a module to acquire the data and control local system signals. Each unit is capable of acquiring data on more than 100 signals at scan rates of 0.1 seconds or less. Signal conditioning is also provided for current loops, strain gauges, pressure transducers, thermocouples, RTDs, magnetic pick-up units, and relay contacts.

2. Centralized Operator Interface - The module at each diesel generator is then connected to a central unit with an embedded personal computer (PC) that will serve as a data concentrator. The central unit with an embedded PC acts as a bridge between the time critical data acquisition and the interactive presentation and analysis. The central unit will receive the data from the local/remote units and buffer it using RAM buffers. In the event that the database server (#3 below) is temporarily unavailable the embedded PC has hard disk capability that is available for additional buffering. This centralized unit can be linked to as many as seven remote units using either a fault tolerant token-ring network or a high-speed ethernet network.

3. Database Server - The data is then transferred by the central module continuously over a network line to the database server. The database server makes the data available over a local or wide area network, and will be located in an office environment. Archiving of historical data also takes place at the database server.

4. Engineering Workstation - An engineering workstation(s) contains the graphical user interface (GUI) software used to monitor the operational status of each diesel generator. The GUI runs under the Microsoft Windows™ operating system. The software may be installed on existing plant PCs, the database server, or an optional dedicated PC supplied by AMS. This workstation(s) will be located in an office environment.

5. Portable Industrial PC - An optional piece of equipment that may be supplied for use by the plant engineering and operations staff is a portable industrial PC. The portable industrial PC will contain the GUI software, and would be utilized by plant personnel as a local operator interface at the individual diesel generator locations when troubleshooting problems.

6. Software Capabilities - The GUI software includes graphic screens showing diagrams of the diesel generators and their components. In addition, the real-time operational status of each diesel generator may be viewed. The GUI will contain graphical representations of the signals being acquired with digital readouts.

This comprehensive diesel generator monitoring and maintenance package will increase diesel generator availability, as well as reduce operating and maintenance costs in the process.

6. SENSOR SELECTION AND INSTALLATION

The local data acquisition hardware, or the portion of the performance monitoring system that is located at the diesel generator, is capable of accepting inputs from a variety of field sensors such as: strain gauges, pressure transducers, thermocouples, RTDs, MPUs, proximity probes, relay contacts, and software contacts. In addition to meeting the standard requirement of being able to operate satisfactorily for all possible process temperatures and pressures and meet specific accuracy requirements, all of the field mounted sensors have been carefully selected to meet the following additional requirements:

1. Ability to withstand the high vibrations inherent to diesel engine operation (usually much higher in intensity and bandwidth than seismic events)

2. Sized so that they will be unobtrusive, and able to be installed in confined spaces

3. Ability to be easily removed for maintenance, calibration, or replacement

In addition to the requirements for the sensors themselves, the location and way that these sensors are mounted to the engine also requires special consideration. None of the sensors themselves are actually mounted directly onto the engine. Instead, thermowells or special fittings are welded into the piping systems, exhaust manifolds, bearings, etc., such that these fittings, not the sensors, provide the pressure boundary where the penetration into the subsystems are made. This type of mounting has several benefits, including:

1. Physical damage to the sensor will not place the diesel in an inoperable status

2. The sensor does not need to be installed on the engine for the engine to be operable

3. The sensor can be removed and, depending on subsystem pressure, replaced while the engine is running

4. The sensor can be removed so that it will not interfere with engine or generator maintenance activities

For the performance monitoring system data to be meaningful, the new sensors installed with the system must be located as close to existing engine mounted instrumentation as possible. This may mean utilizing the existing penetrations into the system, or making a new penetration within a few inches of the existing instrumentation. In some cases, utilizing the signals from the existing sensors is also possible.

The signals from the field mounted sensors are sent along an armor-braided cable, which either enters an instrument wireway, or is routed directly to the local data acquisition hardware. The armor-braided cable provides excellent protection for the shielded, twisted-pair conductors carrying the signals. The shielded, twisted-pair conductor provides for adequated shielding of the signal. The armor-braided cable also allows for the flexibility necessary to prevent interference during maintenance activities.

Several diesel generator control circuit signals are also sent to the local data acquisition hardware. Examples of control circuit signals that are used by the performance monitoring system are: engine starting and shutdown sequence signals, engine and generator trip signals, output breaker status, and various subsystem pumps and heater status. These inputs to the performance monitoring system are typically provided by spare relay or switch contacts on existing diesel generator control circuit components.

The input signals from the diesel generator control circuit are critical data points for troubleshooting or root cause analysis activities after an engine or generator failure. Some of these signals are also used by the performance monitoring system to determine when or how often to archive the data being received by the local data acquisition hardware.

The data acquired by the data acquisition hardware for each diesel generator is then transmitted back to the Centralized Operator Interface via a high speed network (i.e., LAN or Fiber Optic Highway). Since the Centralized Operator Interface can be located at remote locations (i.e., separate building) or locally at any one of the diesel generator units, the physical routing and hardware required for the high speed network are dependent on the location chosen for the central unit.

7. SUMMARY

A diesel generator performance monitoring system installed to continuously monitor and archive performance data from all diesel generators will provide the advantages listed below, as well as reduce operating and maintenance costs.

1. Provides historical maintenance information to allow for the relief of accelerated testing

2. Allows for proper trending of the performance of a diesel generator

3. Captures performance data at any time the diesel generators are run, for any reason

4. Predicts potential failures, with help from an expert system and the stored historical data to plan maintenance prior to failures.

TABLE 1

Surveillance Testing of Diesel Generators

Item	Maintenance Test	Monthly	6-Month	Refueling Outage
1	Start Test	X		
2	Load-Run Test	X	X	
3	Fast-Start Test		X	
4	Loss-of-Offsite-Power (LOOP) Test			X
5	Safety Injection Actuation Signal (SIAS) Test			X
6	Combined SIAS and LOOP Tests			X
7	Single-Loaded Rejection Test			X
8	Full-Load Rejection Test			X
9	Endurance and Margin Test			X
10	Hot Restart Test			X
11	Synchronizing Test			X
12	Protective-Trip Bypass Test			X
13	Test Mode Change-Over Test			X

TABLE 2

Reliability of Diesel Generators

Selected Emergency Diesel Generator Reliability	Failures in 20 Demands	Failures in 50 Demands	Failures in 100 Demands
0.95	3	5	8
0.975	3	4	5

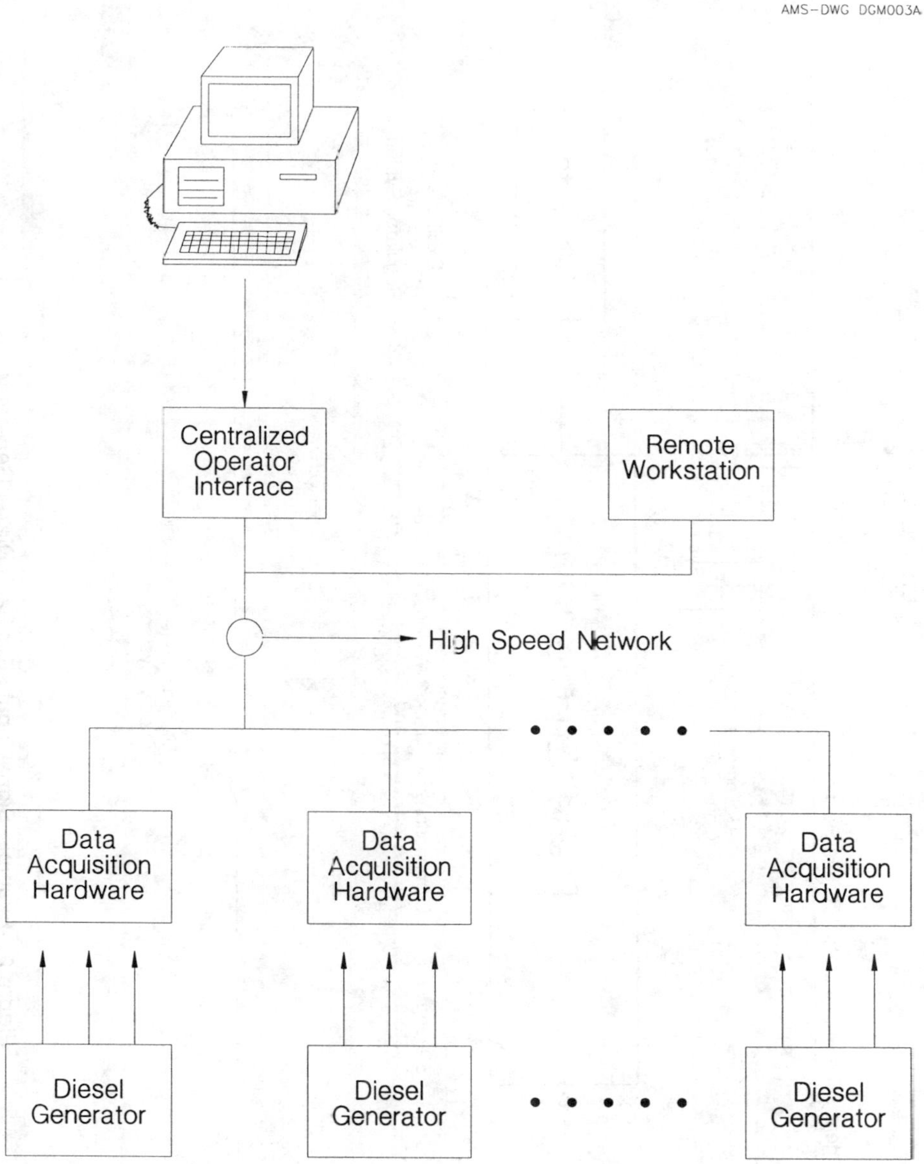

Figure 1. Block Diagram Of The Diesel Generator Monitoring System

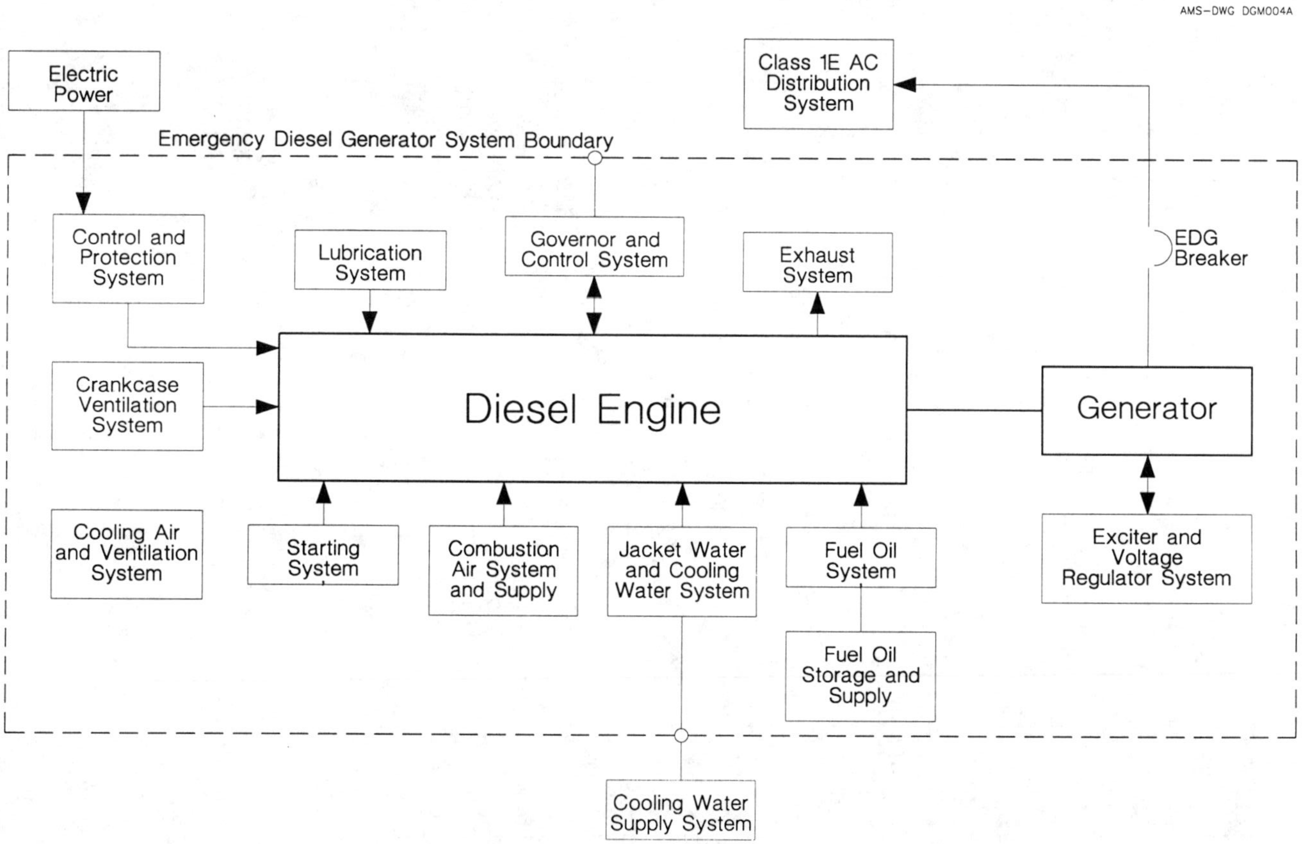

Figure 2. Example Signal Inputs For Each Diesel Generator

TABLE 3

Example of Diesel Generator Performance Parameters

Item	Description
1	Generator Voltage, Current, Frequency, Kilowatts, Kilovars
2	Engine Inlet and Turbocharger Lube Oil Pressure
3	Generator Stator Winding Temperature
4	Crankcase Pressure
5	Air Start System Pressure
6	Jacket Water Pump Discharge Pressure
7	Intercooler Service Water Outlet Temperature
8	Lube Oil Cooler Service Water Outlet Temperature
9	Jacket Water Cooler Service Water Outlet Temperature
10	Turbocharger Combustion Air Discharge Pressure
11	Turbocharger Inlet and Outlet Air Temperature
12	Air Intake Manifold Pressure
13	Emergency diesel generator Room Ambient Temperature
14	Engine Lube Oil Temperature - Cooler Inlet and Outlet
15	Jacket Water Temperature - Engine Inlet and Outlet
16	Turbocharger Blower and Turbine Bearing Outlet Oil Temperature
17	Main Bearing Temperature
18	Exhaust System Temperature
19	Jacket Water Standpipe Level
20	Engine and Turbocharger RPM
21	Engine Vibration
22	Crankcase Sump Level
23	Turbo Thrust Bearing Wear
24	Full Flow Lube Oil Filter Differential Pressure
25	Lube Oil Strainer Differential Pressure
26	Turbocharger Lube Oil Filter Differential Pressure
27	Fuel Oil Filter Differential Pressure
28	Fuel Oil Priming Pump Strainer Differential Pressure
29	Starting and Shutdown Sequencing
30	Emergency Start

0074-056X/95/259-271/$0 + .50pp

Experience With Measurement of Drop Time of Control and Shutdown Rods and Testing of CRDMs in Nuclear Power Plants

H. M. Hashemian
R. E. Fain

Analysis and Measurement Services Corporation
AMS 9111 Cross Park Drive
Knoxville, Tennessee 37923 USA

Phone: (615) 691-1756
Fax: (615) 691-9344
E-Mail: info@ams-corp.com

KEYWORDS

PWR, VVER, Control Rod, Shutdown Rod, Control Rod Drive Mechanism, Automated Testing, Remote Testing, New Equipment

ABSTRACT

Rod drop time testing is usually performed at each refueling outage in Pressurized Water Reactors (PWR). This involves dropping one or more groups of rods while simultaneously collecting data from the Rod Position Indication Systems. These data are then analyzed to calculate the drop time for each rod. Control Rod Drive Mechanism (CRDM) testing is typically performed in conjunction with the rod drop time tests. The purpose of this testing is to verify the proper timing and sequencing of the CRDM system as the rods are withdrawn and inserted.

Rod Drop and CRDM timing tests are performed during plant start up and are usually in the "Critical Path" schedule. Any reduction in the time associated with these tests would greatly minimize the cost of these required tests. Conventional rod drop time testing involves manually dropping one rod at a time and sometimes requires personnel to be stationed in containment during the tests. In order to reduce radiation exposure as well as allow for any number of rods to be dropped and tested simultaneously, computerized test equipment was designed and implemented for automatic testing of rod drop times. A computerized system was also developed to verify proper timing and sequencing of multiple CRDMS in a single test to provide a more detailed and systematic analysis of the data. These developments and experience with their use in nuclear power plants are discussed in this paper.

1. INTRODUCTION

A PWR is equipped with control and shutdown rods which are inserted into and withdrawn from the reactor core to control the power level in the reactor. The CRDM, which is a magnetically coupled positioning system, is provided to raise or lower the control and shutdown rods under manual or automatic control. The CRDM is designed so that, in the event of a reactor trip, all withdrawn control and shutdown rods will fall into the reactor core to shut down the nuclear reaction.

1.1 Rod Drop Time Testing

Rod drop time testing is performed in PWRs to measure the length of time required for control and shutdown rods to travel from their fully withdrawn to their fully inserted positions. The test is generally performed at each refueling outage and is accomplished by interrupting the electrical circuits for one or more groups of rods causing them to drop, while at the same time monitoring and recording their positions in real time.

Rod drop time testing is classically performed manually on one rod at a time, and sometimes requires access inside containment for the duration of the tests, resulting in radiation exposure to the test personnel. In addition, the interruption of the electrical circuits may require manually pulling fuses in order to drop the rods. To resolve these and other problems associated with conventional rod drop testing, computerized test equipment was designed, tested, and validated for automatic testing of the rod drop times in PWRs. This equipment consists of two portable test units with up to sixty-two channel data acquisition capability. The equipment allows all of the rods to be dropped at the same time, if so desired, thereby reducing the time and manpower required for the test. This system is designed to remain in containment during the tests, and eliminates the problems of pulling fuses to initiate each rod drop. The system produces digitized data that can then be used to provide a computational history of the test results to supplement the plant's aging management program.

To perform the rod drop time testing in plants equipped with a Westinghouse Digital Rod Position Indication (DRPI) system, the AC excitation voltage to the rod position indication coils is turned off and the voltage across all the coils in the stack is monitored while the rod is dropped. The motion of the rod drive shaft through the coil stack induces a voltage in the coils. This voltage is directly related to the drive shaft velocity. Figure 1 shows a typical output of the rod position indication coils and the stationary gripper voltage during a control rod drop. This output is commonly referred to as a rod drop trace. This figure also highlights three important points on the rod drop trace. The first point shows the initiation of the rod drop as the decay in the stationary gripper voltage and the coil stack voltage (control rod signal) increasing as the rod's velocity increases. The next point in the trace shows the entry of the rod into the dash pot and a steep decline in voltage as the rod slows down when it enters the dash pot. The last point of the rod drop transient shows the rod reaching the bottom of the dash pot, and then oscillating as it bounces in the dash pot.

1.2 CRDM Timing Tests

The control and shutdown rods in a PWR are moved in or out of the core using a magnetic-jack CRDM. Each Westinghouse CRDM consists of three electromagnetic assemblies and two grippers (called the Stationary Gripper, Movable Gripper, and Lift Coil). When power to the three coils is properly sequenced, a control or shutdown rod can be withdrawn from or inserted into the core one step at a time.

The CRDM testing is performed by monitoring the stationary gripper, movable gripper, and lift coil currents for a bank of control or shutdown rods as they were moved in and out of the reactor. All coil current signals are sampled for each rod and then analyzed to ensure that the CRDMs are operating properly. The signals are analyzed by calculating various time differences in the data as well as voltage amplitudes to verify that the CRDMs exhibit the expected timing and sequencing characteristics. Figure 2 shows a typical CRDM test trace for one withdrawal step of a control rod. The notable features of the trace are numbered and described below:

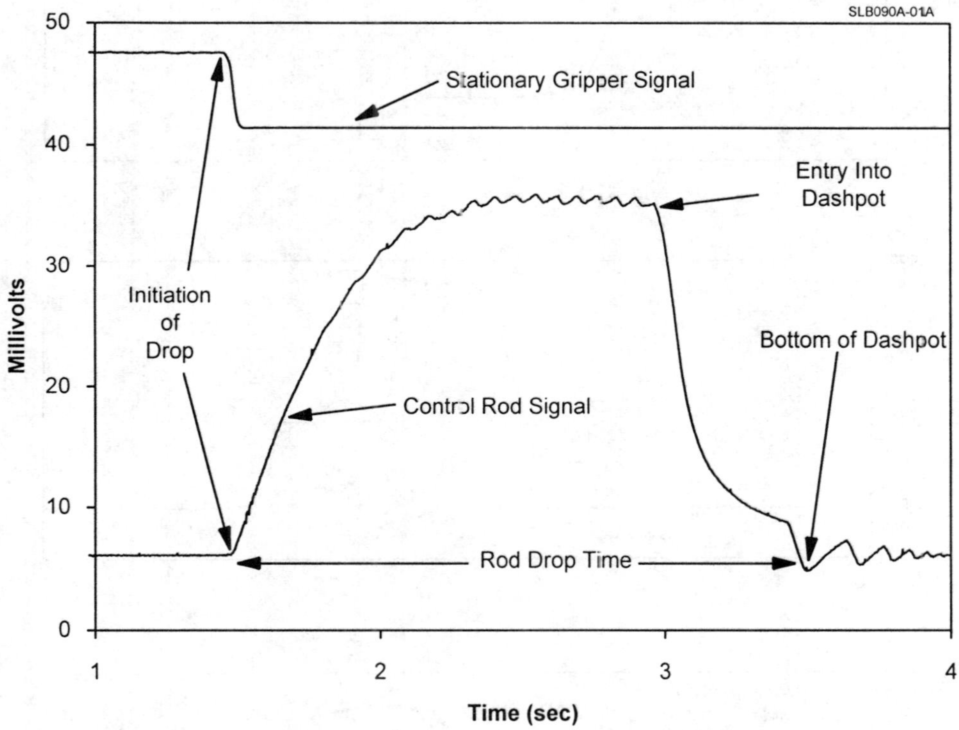

Figure 1. Typical Control Rod Drop Timing Trace

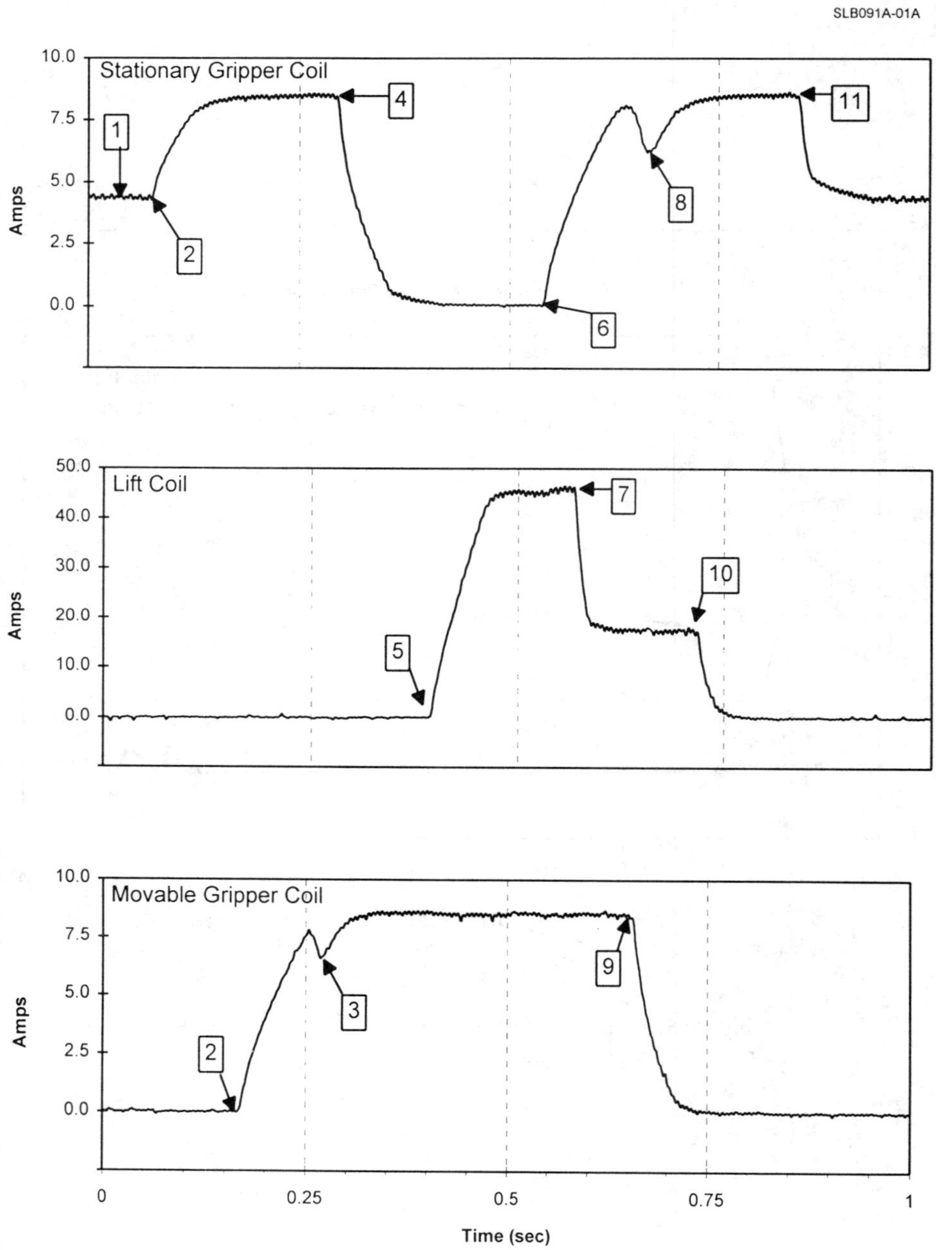

Figure 2. Example Of A CRDM Timing Trace

1. In the normal "hold" mode, the stationary gripper coil is energized with 4.4 Amps (nominal).

2. At the beginning of the withdrawal step, the stationary gripper coil is energized to full current (8 Amps nominal). Shortly thereafter, the movable coil is energized to latch the rod with the movable grippers.

3. The "dip" in the current level to the movable coil indicates proper latching of the rod by the movable grippers. The rod is now supported by both the stationary and the movable gripper latches.

4. The stationary coil current is turned off to release the stationary gripper latches. The rod is only supported by the movable gripper latches at this point.

5. The lift coil is energized which raises the movable gripper (along with the rod) one step (⅝").

6. The stationary coil is energized to engage the stationary grippers and hold the rod in the raised position.

7. The current to the lift coil is reduced to 16 Amps (nominal) to reduce coil heating while still holding the rod and movable gripper in the raised position.

8. The "dip" in the current level to the stationary coil indicates proper latching of the rod by the stationary grippers.

9. The current to the movable coil is turned off to release the movable grippers. The rod is now supported by the stationary gripper alone.

10. The current to the lift coil is turned off which allows the movable gripper to return to its lowered position.

11. The current to the stationary gripper is reduced to 4.4 Amps (nominal). This is the normal "hold" mode for the CRDM.

The analysis of the CRDM data involves identifying the time differences between the various events noted above to determine whether or not the CRDM is functioning properly. The CRDM timing tests are classically performed by recording the three signals for each rod using a chart recorder. They are usually tested one rod at a time due to the fact that a complete bank would require 24 channels of information which is out of the range of most recorders. This causes the operators to have to exercise the bank of rods up to eight times to test each individual rod.

2. EXPERIENCE WITH ROD DROP AND CRDM TIMING TESTS

In order to minimize the cost of both rod drop and CRDM testing, both the duration and manpower for the tests need to be reduced. This is accomplished by using two customized data acquisition systems to automate most of the testing. A more efficient procedure for performing the rod drop time and CRDM timing measurements is to withdraw a complete bank of rods and measure both the CRDM timing and the drop times of all the rods in the bank at the same time using a multi-channel data acquisition system. This can be performed simply by withdrawing the bank of rods and dropping them using the reactor trip breaker. A sample procedure for this type of testing is as follows:

1. CRDM timing information is recorded while the control room operators withdraw one bank of rods.

2. Operations personnel turn power off to the DRPI cabinets.

3. Upon command from the test personnel, data is automatically acquired by the data acquisition system in containment on the bank of rods when the control room operators open the reactor trip breaker.

4. Data is stored for analysis.

5. Operations personnel restore power to the DRPI cabinets.

6. Steps 1 through 5 are repeated until all the rods have been tested.

Using this method, no fuses must be pulled and the test is much more efficient in that a complete bank of rods is tested in the time that is required to test one rod using the conventional methods. In addition, a trigger signal from the reactor trip breaker is acquired with the data to obtain an event initiation for the calculation of rod drop times. This is a slightly more conservative analysis method and better represents an actual reactor trip in the fact that the stationary gripper coil current which is holding the rod in place decays after the reactor trip breaker opens. Figure 3 shows a comparison between using the decay of the stationary gripper coil versus the reactor trip breaker contact closure for the initiation of the drop event. Table 1 provides typical results for both control and shutdown rods that were tested using these automated methods. The use of the reactor trip breaker contact provides a slightly more conservative result by approximately the amounts labeled **Delay** as shown in Table 1.

Another advantage to dropping multiple rods at the same time is the ability to view the traces together to see if the rods all started moving at the same time. Figure 4 plots a group of control rods together in order to see any differences in the plots. In the event of a reactor trip, this is an actual representation of the group of rods falling into the reactor. To further automate the rod drop testing process, the AMS equipment was designed to allow additional interfacing to both the DRPI and rod control systems. The additional interface to the DRPI system allows power to the DRPI system to be turned on and off by command from the control computer. This avoids the need for operations support to cycle the DRPI power breakers. The interface to the rod control power cabinets allows the control computer to initiate the drop of banks of rods without having to cycle the reactor trip breaker. Figure 5 shows a shutdown rod which was tested in this manner. The trigger signal in this case is not the reactor trip breaker but the initiation of the drop by the control computer.

Rod drop testing is performed in all pressurized water reactors prior to startup. Figure 6 shows a rod drop transient obtained with the AMS equipment in a Russian designed VVER (PWR). The output of the rod position indication system is different than in the previous tests, however the principle is the same. The initiation of the drop was recorded (trigger) and the last step of the trace indicated the rod was fully inserted. These tests were performed by dropping one bank at a time and

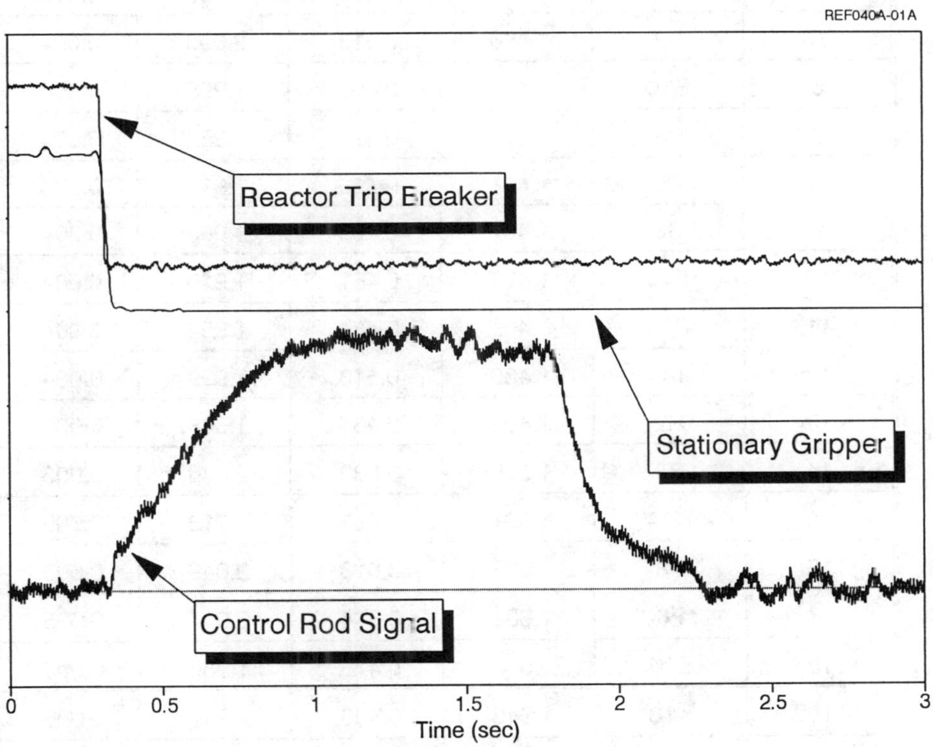

Figure 3. Comparison Between Stationary Gripper Vs. Reactor Trip Breaker Signal

Table 1
Example Drop Times For Control Rods

Item	Rod ID	T_1 (sec)	T_2 (sec)	T_{TOTAL} (sec)	Delay (sec)
1	H6	1.490	0.468	1.958	0.003
2	H10	1.491	0.486	1.977	0.003
3	F8	1.488	0.491	1.979	0.003
4	K8	1.487	0.499	1.986	0.003
5	F2	1.490	0.513	2.003	0.004
6	B10	1.475	0.493	1.968	0.004
7	K14	1.473	0.479	1.952	0.004
8	P6	1.491	0.483	1.974	0.004
9	B6	1.497	0.486	1.983	0.004
10	F14	1.465	0.451	1.916	0.004
11	P10	1.469	0.484	1.953	0.004
12	K2	1.486	0.513	1.999	0.004
13	H2	1.422	0.465	1.887	0.005
14	B8	1.518	0.530	2.048	0.005
15	H14	1.531	0.488	2.019	0.005
16	P8	1.536	0.513	2.049	0.005
17	F6	1.503	0.514	2.017	0.005
18	F10	1.513	0.473	1.986	0.005
19	K10	1.549	0.498	2.047	0.005
20	K6	1.500	0.516	2.016	0.005
21	D4	1.469	0.448	1.917	0.003
22	M12	1.475	0.508	1.983	0.003
23	D12	1.487	0.497	1.984	0.003
24	M4	1.509	0.464	1.973	0.003
25	H8	1.515	0.448	1.963	0.003

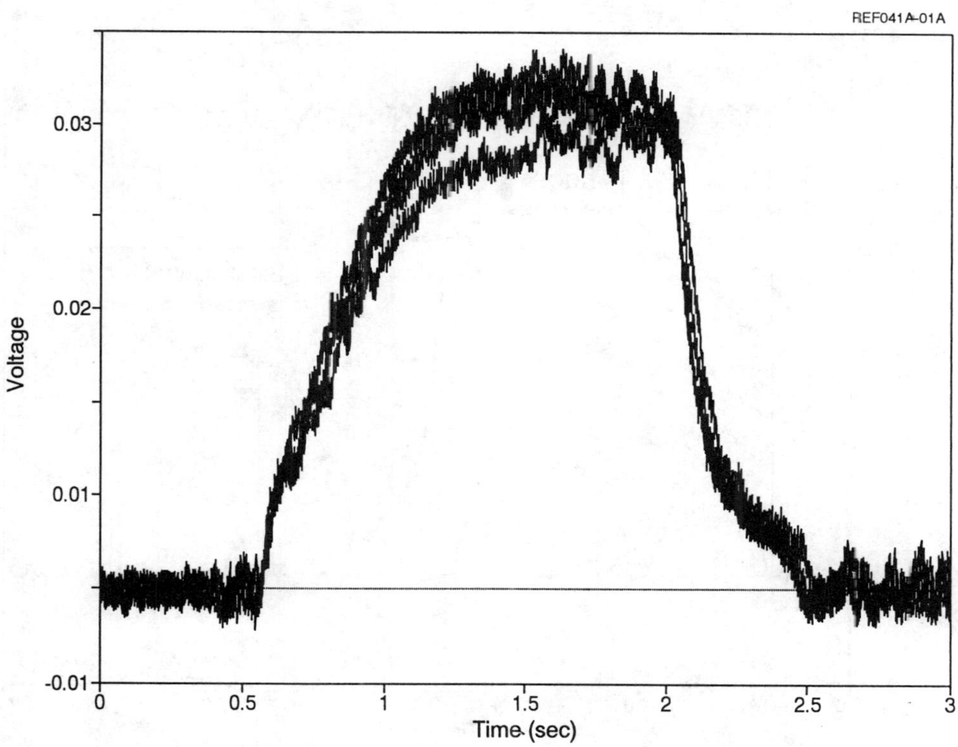

Figure 4. Multiple Control Rod Timing Traces

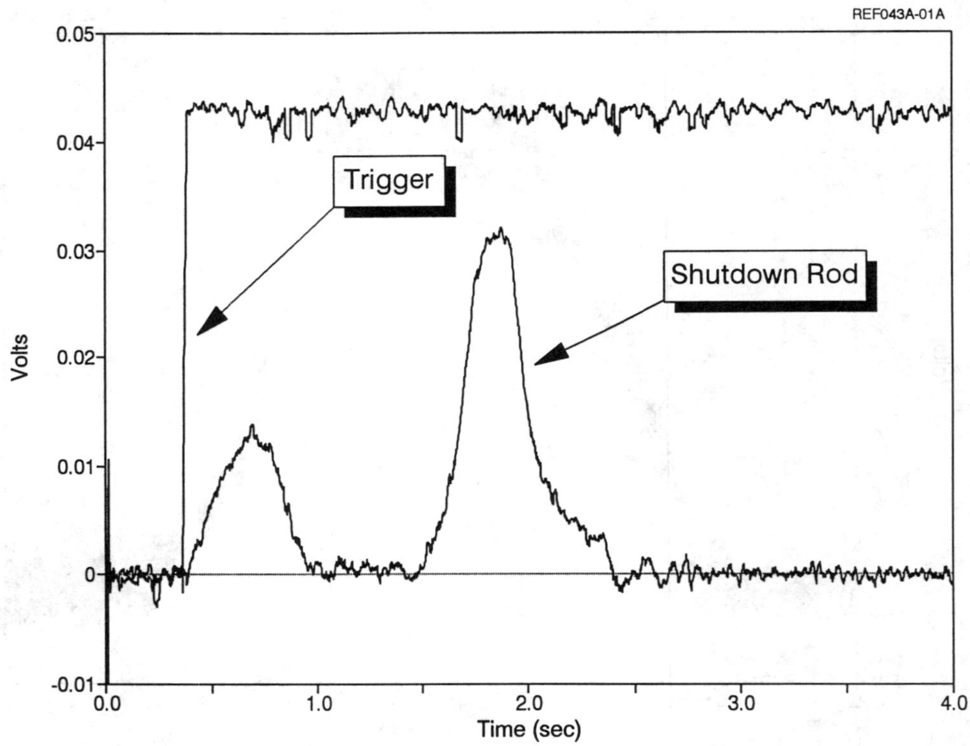

Figure 5. Shutdown Rod Timing Trace Acquired with Complete Automated System

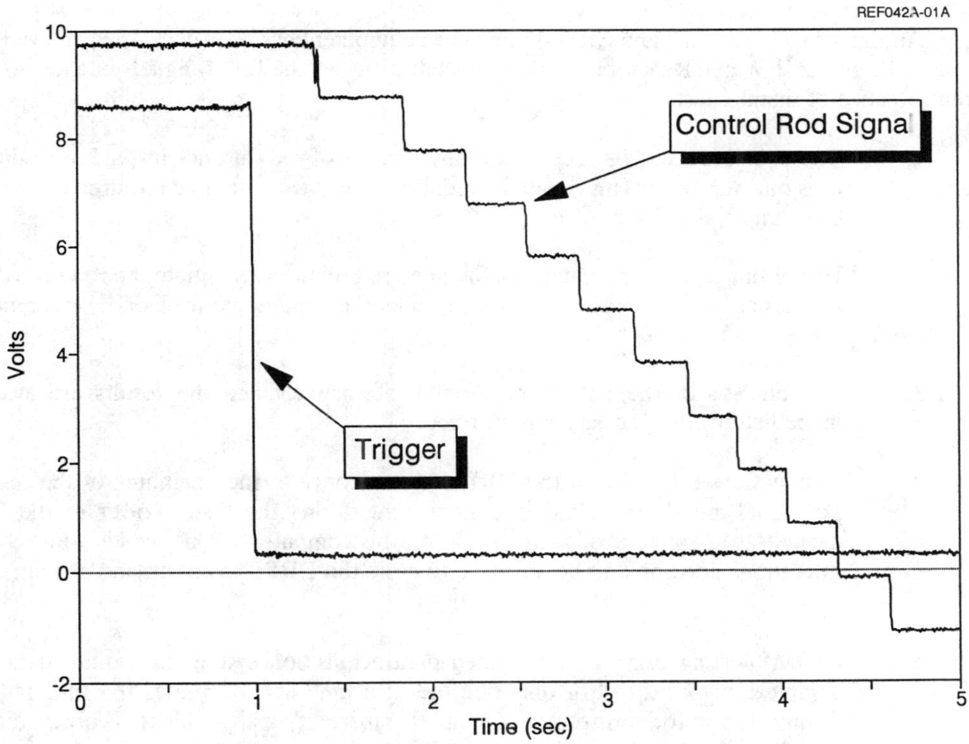

Figure 6. Typical Control Rod Drop Trace
In A Russian VVER (PWR)

is an improvement over the current method of using a timer to obtain rod drop times individually. It also provides multiple testing along with a visual output.

By sampling the CRDM timing traces with a multichannel data acquisition system several different analysis calculations can be performed. The timing information can be determined to the resolution of the sampling rates, and can be zoomed in to allow very accurate fitting of the data. Figure 7 shows an example of multiple steps for the lift, stationary, and movable coils. Also, multiple rods can be viewed together to allow intercomparisons and calculation of the time differences. These functions could not be performed with conventional procedures using a chart recorder for data acquisition.

3. CONCLUSIONS

Automatic rod drop time and CRDM timing test equipment has been developed and validated for use in Pressurized Water Reactors. This equipment provides the following advantages over the current practice of manual tests:

1. The equipment can be used to test any number of rods at once instead of testing the rods one at a time. This capability reduces the critical path time required for the tests by as much as a factor of ten.

2. Use of this equipment eliminates the need to pull fuses to initiate the rod drops. The rod drops are initiated automatically under computer control or by opening the reactor trip breaker.

3. The analysis of the data is performed interactively and the results are available immediately after the test is performed.

4. Connections are made to the DRPI cabinets once at the beginning of the test and personnel are not required in containment during the test. Power to the DRPI cabinets can be controlled remotely by the computer. This avoids the need for operations personnel to be available to cycle the DRPI power on and off during the test.

5. CRDM testing using an automated data acquisition system to acquire data on a complete bank instead of only one rod at a time also decreases the time required during startup for control rod testing. Compared to using a chart recorder, digitally acquiring the data allows for easy analysis and proper storage of the data.

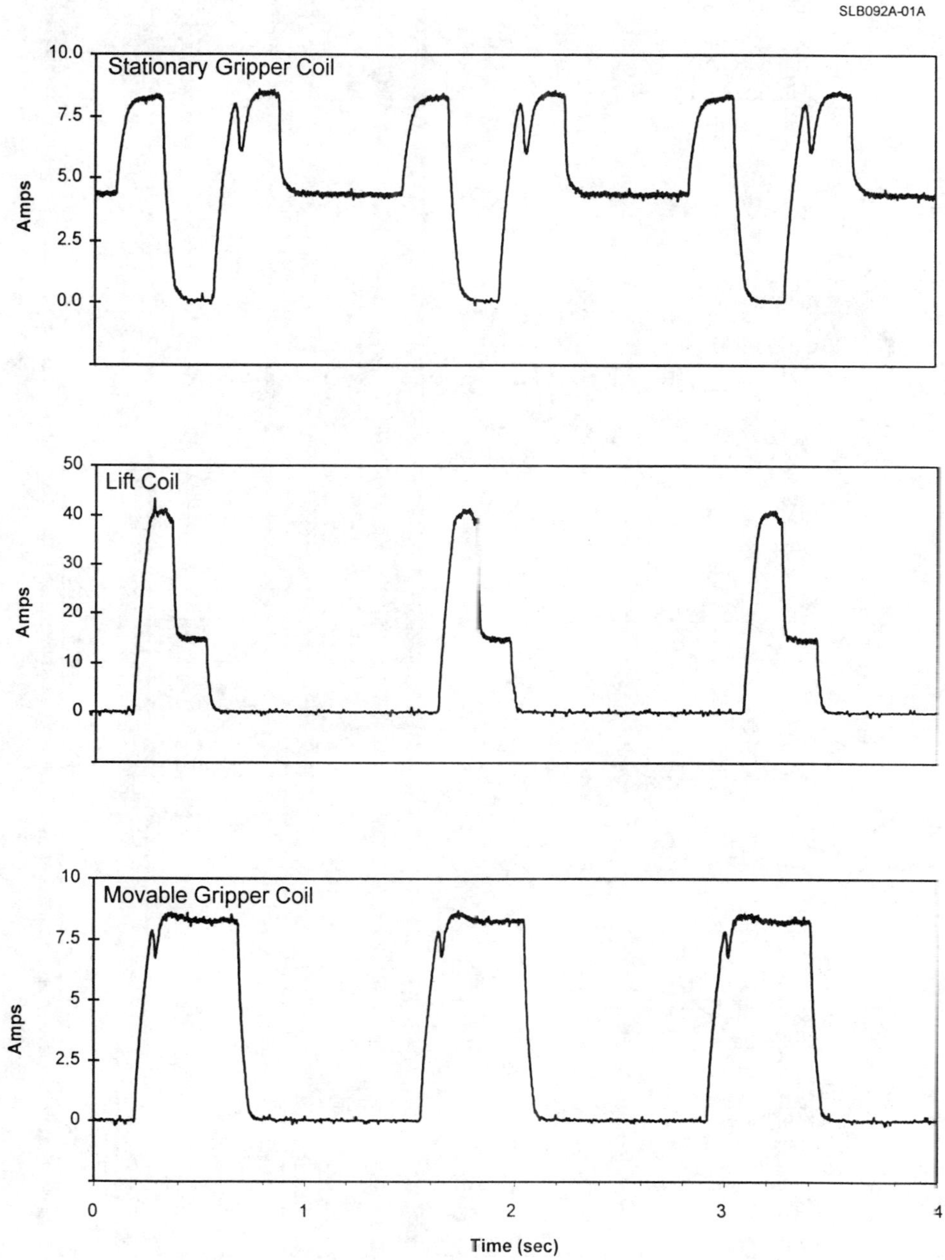

Figure 7. Multiple CRDM Timing Steps

©0074-056X/95/273-282/$0 + .50pp

HIGH TEMPERATURE FIBER-OPTIC STRAIN AND TEMPERATURE SENSOR FOR STRUCTURAL HEALTH MONITORING.

Nadarajah Narendran
Mechanical Technology Incorporated (MTI),
968 Albany-Shaker Road, Latham, NY 12110.
Joe Weiss
Electric Power Research Institute
3412 Hillview Avenue, Palo Alto, CA 94303.

ABSTRACT

An experimental project is being conducted to evaluate the feasibility of using fiber-optic sensors to measure strain and temperature of high temperature structures. These sensors can be used to monitor the health of high temperature structures such as boilers, tube headers, steam pipes etc. during their operational lifetime. The target application is steam pipe strain and temperature measurements. In this study the temperature is limited to 1000°F. This study uses a low cost fiber-optic sensor based on a white-light interferometric technique and utilizes a halogen lamp as the light source and multimode optical fibers for sensing and light transmission. The wide-band optical signal propagates down the fiber to a localized sensing cavity that modulates the optical spectrum. A simple optical correlator demodulates the return optical signal and relates it to the absolute strain sensed by the cavity. The glass fiber-optic sensors are packaged and ruggedized so that they can be spot welded to the structure of interest. The sensors were installed on a steam pipe at Tennessee Valley Authority's Kingston Power Plant for testing with promising preliminary results.

INTRODUCTION

Many utility companies are concerned about possible structural or component damages that can lead to catastrophic failure of systems. Thus, the need for real time automatic health monitoring in the power plant environment has been growing very rapidly. The ability to sense incipient failures reliably and plan corrective measures in advance can be very cost effective. However, currently available sensors are not suited for many of the applications because of the electromagnetic or thermal environments present. In recent years, fiber-optic sensors have opened up applications for the sensing of many physical parameters because they possess a variety of advantages over their conventional counterparts. These advantages include immunity electromagnetic interference; high temperature capability, immunity to corrosive or other hostile environments; geometric versatility; and compatibility with fiber-optic data links [1,2]. Strain and temperature are two important parameters that need to be monitored for many power plant components, especially high temperature structures such as boilers, tube headers, and steam pipes. Lack of appropriate sensors is a major concern in monitoring high temperature structures. Conventional strain gages are generally accurate up to 400°C, but have difficulty maintaining reliability or accuracy at temperatures above this value. Therefore, power plants rely on thermocouples and numerical methods to indirectly estimate the conditions of these structures. This paper describes the feasibility of using fiber-optic sensors to measure strain and temperature on high temperature structural components.

FIBER-OPTIC SENSOR

Fiber-optic sensor development first started in ernest the mid seventies and has been rapidly growing. In recent years the technology has matured to the state where these sensors are becoming commercially available. Several sensing techniques based on intensity, phase or frequency (or wavelength) modulation [1,2] have been reported for strain measurement. Each of these techniques have their own strengths and weaknesses. Most intensity-modulated sensors are simple to construct, and low cost. However, the loss in light intensity due to couplers, microbending of the fiber and other attenuation factors along the length of the optical fiber directly impact the measured signal accuracy and reliability. Therefore, these sensors not suitable for long term monitoring applications requiring accurate measurements. On the other hand, phase-modulated sensors (or interferometric sensors) have shown high sensitivity, and larger bandwidth and are not affected by light intensity fluctuations. Intrinsic Fabry-Perot [3] and Extrinsic Fabry-Perot [4] fiber-optic interferometric sensors are two phase-modulated sensors that have shown promising results for making localized strain measurements (gage length 1-10 mm). However these sensors are useful for measuring relative strain and not absolute values. This requires that the monitoring system be connected all the time or that recalibration be done prior to making measurements. Furthermore such sensors are also not useful for long term monitoring applications. Recently, two sensor techniques, Bragg gratings and white-light interferometers, have shown promise for making absolute strain measurements while being insensitive to light intensity loss factors [5,6]. The Bragg grating sensors do not work well at temperatures above 700°F because the holographically etched gratings on the fiber degrade at high temperatures[5] and loses efficiency. The most attractive candidate sensor for this application is the white-light interferometric sensor described below.

The schematic diagram of a white-light interferometer is shown in Fig. 1 [6]. Light from a halogen lamp is coupled into a multimode fiber coupler/splitter (2 x 1). The beam travels down the fiber and enters into an extrinsic fiber-optic Fabry-Perot sensor shown in Fig. 2. Depending on the length of the air gap, a particular wavelength component of the optical beam is strongly reflected back. A suitable spectrum analyzer placed at the output end can detect the wavelength of the reflected component which represents a unique air gap length. Straining the Fabry-Perot sensor changes the length of the air gap which results in a wavelength change that is detected by a spectrum analyzer. The schematic of an extrinsic Fabry-Perot sensor is shown in Fig. 2 [6]. As shown in Fig. 2, two optical fibers are inserted into a hollow glass tube from both ends and are attached to the glass tube. A small air-gap between the end faces of the two fibers forms the Fabry-Perot cavity. A strain applied to the glass tube changes this gap length which is detected by the interferometer.

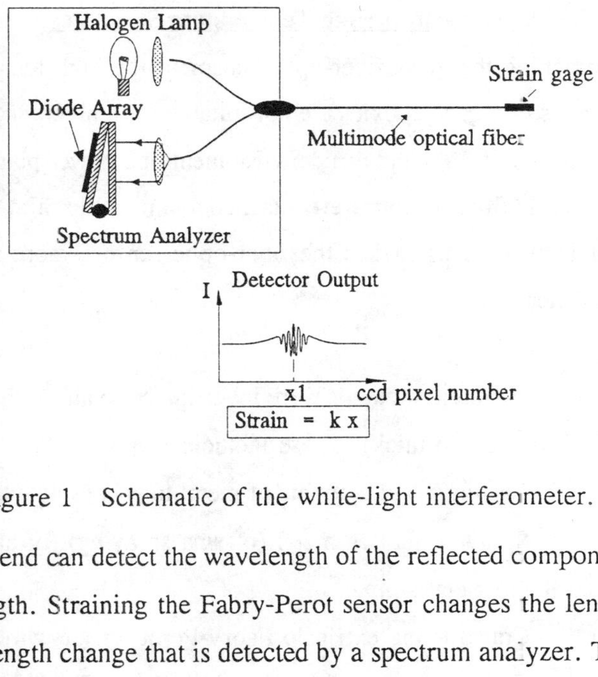

Figure 1 Schematic of the white-light interferometer.

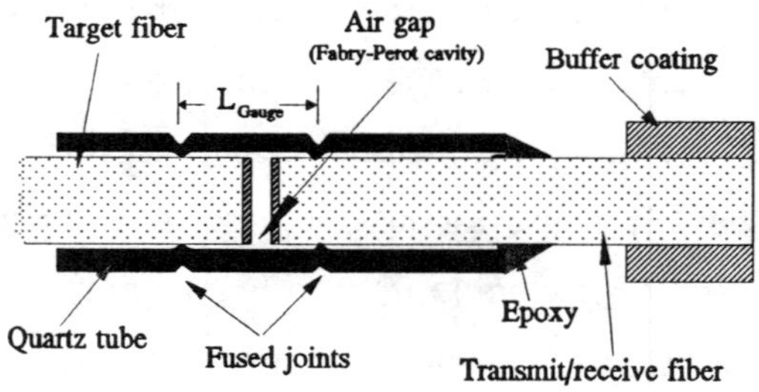

Figure 2 Extrinsic Fabry-Perot Sensor.

Gage Attachment to High-Temperature Structures

Attaching the glass fiber-optic sensor to the high temperature structure is a challenge because most strain gage epoxies are not suitable for temperatures higher than 600°F. These sensors must withstand the hot and dirty environment of a power plant and also be easily attachable. Therefore the glass fiber sensors were attached to a metallic shim which can be spot-welded to the structure of interest. This makes the fiber-optic sensors more rugged and the attachment method very practical.

Several different methods were investigated to attach the fiber-optic sensors to a steel metal shim (125 to 375 μm thick). These include:

1) Casting small amounts of metal to encapsulate the glass fiber and the metal shim.
2) Spraying molten metal (plasma spraying) for attaching the fiber-optic sensors to the metal shim.
3) Constructing extrinsic Fabry-Perot sensors directly on a metal shim.
4) Using ceramic adhesives to attach the fiber-optic gage to the metal shim.
5) Metalizing (gold plating) the fiber-optic sensor and bracing it to the metal shim.

Most of these methods showed very little promise except the ceramic adhesives. In order to identify a suitable adhesive to attach the fibers to the metal shim, several different high temperature adhesives were tested. Most ceramic adhesives delaminated off the attached surface and some of them cracked during the temperature cycle. This is mainly due to the thermal expansion mismatch between the ceramic adhesive and the metal surface. A high temperature metal coat that had metal oxide mixed in a ceramic slurry showed no adhesive cracking or debonding. Based on this test the high temperature metal coat was chosen to bond the fiber-optic sensor to a the metal shim.

Sensor Calibration

To study the strain transfer characteristics of the selected high temperature epoxy, a fiber-optic strain gage was attached to a test specimen The material chosen for this calibration was Invar. This is mainly to keep the thermal strain within the fiber-optic strain sensor measurement range, $\pm$ 3000 $\mu\epsilon$. The specimen was setup on a MTS machine inside a heater coil. A 0.5" gage length high temperature extensometer was placed next to the fiber-optic sensor as shown in Fig. 3. A J-type thermocouple was place on the test specimen for temperature measurement. The fiber-optic sensor and the extensometer were zeroed prior to data acquisition. The test specimen was heated and the data from the thermocouple, extensometer and the fiber-optic sensor were recorded simultaneously. Figure 4(a) illustrates the thermocouple reading and Fig. 4(b) illustrates the corresponding strain data from the fiber-optic sensor and the extensometer. Figure 4(c) shows the calibration plot. The fiber-optic sensor response is fairly linear over the measurement range however some amount of hysteresis was observed during this test.

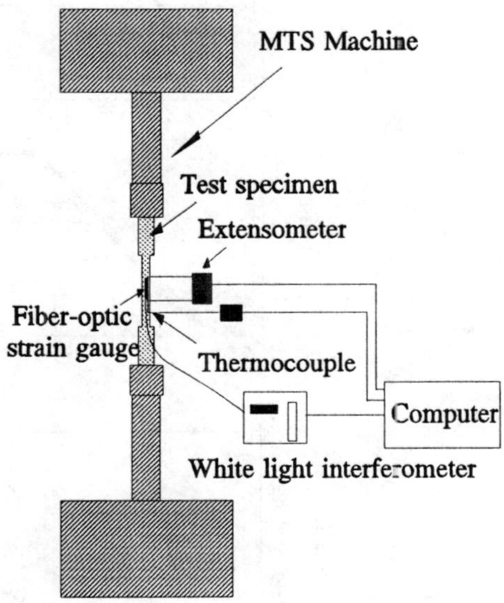

Figure 3 Calibration Setup.

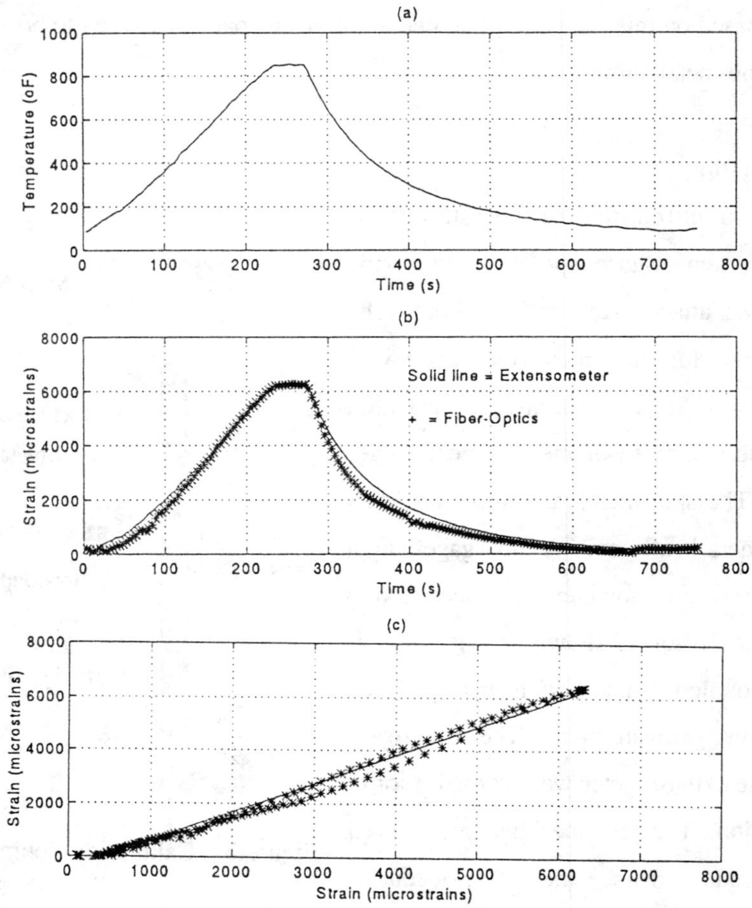

Figure 4. a) J-type thermocouple reading; (b) corresponding strain data from the fiber-optic and the extensometer; c) the calibration plot.

High-Temperature Strain Gage Packaging

The fiber-optic strain gage was packaged for high temperature strain measurement as shown in Fig. 5. The glass fiber-optic sensor was attached to a 1cm by 5cm by 250μm thick stainless steel shim using the high temperature coat. The cure procedure included a room temperature cure followed by 150°F and 650°F temperature cures. The lead-in fiber is housed in a stainless steel tube and the gage section is protected by a cover plate as shown in Fig. 5 for ruggedness. The sensing part can be spot welded to the structure of interest.

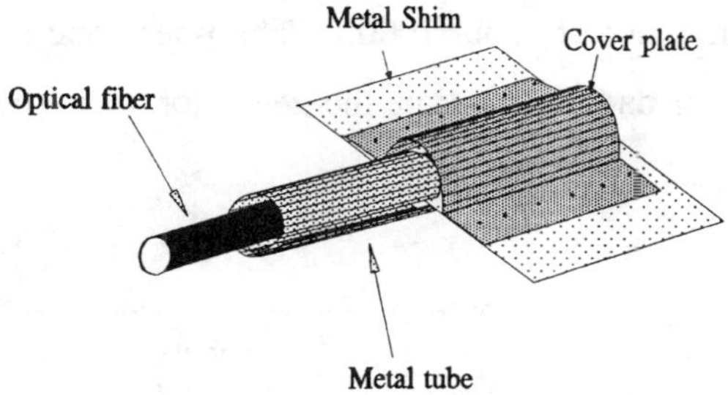

Figure 5 High temperature strain sensor.

Host Utility Testing

Several gages were constructed and calibrated in the laboratory. The laboratory calibration data for one of the strain gages installed at the host utility site is shown in Fig. 6. The average thermal expansion coefficient of this sensor is 6.65 x 10^{-6} oF^{-1}. The average thermal expansion initially is lower than this value. Experiments are currently being conducted to explain this effect. Several of these fiber-optic sensors were installed at the TVA's Kingston Power Plant to study the feasibility of gage installation and long term durability of these fiber-optic gages on an actual power plant component. Two fiber-optic strain gages, similar to the one shown in Fig. 5, were spot-welded to a main steam line and two more gages were placed inside a thermowell to measure the internal temperature of a cold reheat steam line. Installation was performed during a routine plant shut down period. Strain data was gathered as the boiler was fired and the super heated steam was allowed to pass through the steam lines. The common practice is to warm up all the components over a 10 to 12 hour period and bring them to an equilibrium temperature. Then the turbines are rolled and gradually brought up to maximum load. The strain data gathered by the fiber-optic sensor attached to the steam line and the corresponding thermal strain estimated from the plant temperature sensors are illustrated in Fig.7(a) and Fig. 7(b) illustrates the generator load data during this period. The temperature information was gathered at the plant control room from a J-type thermocouple, and at the TVA diagnostics center from a high temperature RTD temperature sensor. The estimated thermal strain is obtained by multiplying the temperature readings by average thermal expansion coefficient, 6.65 x 10^{-6} $\mu\epsilon/^{\circ}F$, determined in the laboratory. It is evident from Fig. 7(a) that the fiber-optic sensor data lags the estimated strain data during the initial warm up period and slowly catches up to the final steady state value at about 1000°F. The fiber-optic sensor data matches well with the other two plant sensor data by the time the generator was producing maximum load.

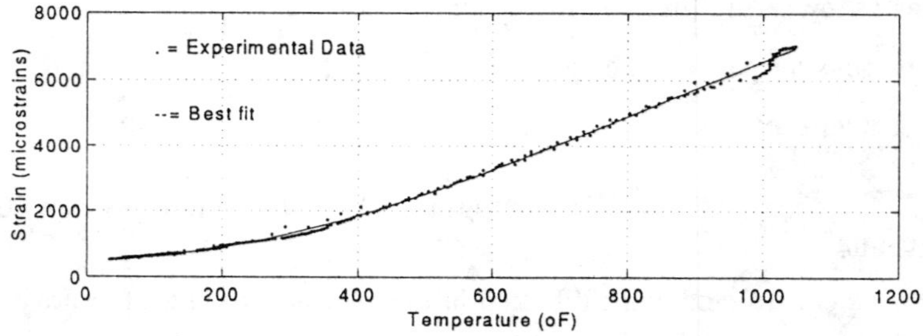

Figure 6 Calibration data

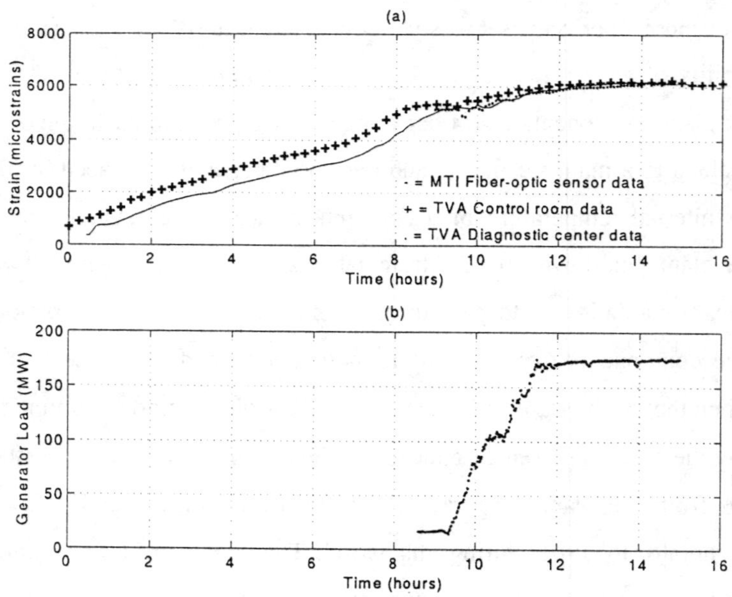

Figure 7 a) Strain data from the fiber-optic sensor and the corresponding values from the temperature sensor data, b) The generator load data during the same period.

A J-type thermocouple was placed next to the fiber-optic sensors inside the thermowell. The temperature values obtained from the fiber-optic sensors and the thermocouple match very well. The sensors will be inspected at regular intervals to study the durability of these fiber-optic gages.

SUMMARY

This study indicates that fiber-optic sensors can be used for health monitoring of high temperature structures. Actual plant testing of these sensors showed excellent correlation with predicted values. The maximum temperature of the target application was 1000°F. However these fiber-optic gages can be used at higher temperatures with some modification to the current sensor demodulator. Some of the sensors tested well up to 1200 °F in the laboratory. The key feature of this white light interferometric technique is the absolute measurement capability. This versatile sensor can be used for sensing many parameters including strain, temperature, pressure, vibration, and index of refraction of liquids. Power plants can significantly benefit from this sensor, especially for applications that cannot find suitable sensors due to the extreme nature of the environment.

ACKNOWLEDGEMENTS

The support of EPRI and TVA is gratefully acknowledged. MTI support for additional instrumentation purchase, and the support of Claude Belleville (FTI), Micheal Tomaszweski (MTI), Ralph McKosky and Steve Halcomb (TVA Diagnostic center) are also gratefully acknowledged.

REFERENCES

1) Brain Culshaw and John Dakin, "*Optical Fiber Sensors*",Vol.I&II, Artec House Inc., Norwood, MA, (1990).

2) E. Udd, "*Fiber-optic sensors: An Introduction for Engineers and Scientists*", John Wiley and Sons, New York, (1990).

3) C.E. Lee, J.J. Alcoz, Y. Yeh, W.N. Gibler, R.A. Atkins and H.F. Taylor,

"Optical fiber Fabry-Perot sensors for smart structures", Smart Mater. Struct. (1992), pp 123-127

4) Murphy et al., "Quadrature phase-shifted extrinsic Fabry-Perot optical fiber sensors", Opt. Lett, Feb. 1991, Vol. 16, No.4, pp 273-275.

5) Morey et al., "Multiplexing fiber Bragg grating sensors", Fiber and Integrated Optics, Vol. 10, 1992, pp 351-360.

6) Belleville et al., "White-light interferometric multimode fiber-optic sensor", Opt. Lett., Vol. 18, No.1., January 1993, pp78-80.

© 0074-056X/95/283-291/$0 + .50pp

RUGGEDIZED HIGH TEMPERATURE 100 Ω PLATINUM RTD FOR POWER PLANTS

Yasushi "Mike" Nozawa, Sc.D.
Engineering Manager
RdF Corporation
Hudson, NH 03015-0490

Dana Brennick
Product Engineer
RdF Corporation
Hudson, NH 03015-0490

KEYWORDS

Temperature sensor, RTD, High temperature, Platinum

ABSTRACT

A new ruggedized high temperature (HT) Resistance Temperature Detector (RTD) probe was specifically developed for power generating plant applications. The RTD probe uses a reference grade, wire-wound 100Ω platinum sensing element which is encased in a 1/4" metallic sheath. A hot zone sheath length which can be exposed safely to the maximum operating temperature is 6 to 24 inches. Actual sheath length can be up to 50 feet with the use of an extension sheath. The probe will withstand continuous vibrations of up to 50 G. It can be operated continuously at HT (up to 1200°F) for a prolonged period with an accuracy of ± 2.5°F. The RTD probe can maintain an insulation resistance of 1 MΩ minimum during HT operations. This paper will focus on design philosophy, some experimental results to support design philosophy, major differences between HT probe and a standard industrial platinum RTD, and some performance data.

INTRODUCTION

The utility industry requires better temperature sensors for more efficient power plant operations. Both RTDs and thermocouples are used in these applications, but the RTD is generally considered to be more accurate and reliable. The present practical upper limit of reliable RTD operation is around 800°F [1]. Many groups are working to overcome this barrier, but have not succeeded yet. The purpose of this paper is to describe the authors' experience with development of a 100Ω ruggedized Platinum (Pt) RTD, which was specifically designed for HT operations.

Many people are confused about the present state of the industrial HT RTD, because a Standard Platinum Resistance Thermometer (SPRT) has been used to measure temperatures up to 660°C (1220°F) since 1927 (IPTS-27) and its upper range was recently extended to the silver point (961.78°C, 1763.204°F) in ITS-90 [2,3]. Also, there is a move to extend the SPRT's capability to the gold point (1064.18°C, 1947.524°F) [4,5].

Many manufacturers have been supplying reliable industrial Pt RTDs (RTD hereafter) for more than 30 years. In our opinion, there is no reliable industrial HT RTD available now, despite the fact that many good HT SPRTs and laboratory type HT RTDs are available. The challenge is to build a ruggedized HT RTD which can withstand an industrial environment for a significant period of time.

It is not an academic research problem, but a practical engineering design problem whose difficulty will be explained below.

SUCCESSFUL CONVERSION OF SPRT TO INDUSTRIAL RTD

Currently, an industrial RTD is reliable and can operate in a severe industrial environment in a temperature range up to 500°C (932°F). Lets review solutions for major problems encountered during the conversion of SPRT technology to successful industrial RTD design for the past 30 or more years.

The basic idea of SPRT is that the purest Pt wire in a clean enclosure under stress-free conditions produces the best SPRT. Therefore, minimum (quantity as well as number of types) material should be present near the Pt wire, since any other material is a potential source of contamination of the Pt wire. The strain free construction is very fragile and cannot withstand an industrial environment. Many different designs for industrial RTDs have been developed [6]. Every successful design has introduced more material compared to an SPRT for mechanical and electrical protection. Material may be cement or glass to attach the Pt wire, or ceramic tube to house the Pt wire.

Even in an SPRT, there is a sheath assembly to protect the Pt wire from the outside environment. A typical SPRT uses a closed end silica or metal tube in which platinum lead wires are used to connect a sensing Pt wire to an outside instrument [7,8,9]. Again, the basic idea is minimization of different materials. Any potential source of contamination must be eliminated from inside the SPRT. Use of the SPRT should be restricted to a specifically acceptable external environment.

In an industrial RTD, a sheath must be metal (usually stainless steel) for installation, and for protection from handling, pressure and damaging atmospheric conditions. Lead wires are made of copper, nickel or other low-cost material. Since the sheath is metal, lead wires must be positively insulated. Also, the wire end of the sheath is often sealed to prevent introduction of outside air and moisture. This means more materials are used in an industrial RTD, compared to an SPRT.

The question is whether the same design philosophy can be used to produce a successful industrial HT RTD.

DESIGN PHILOSOPHY OF INDUSTRIAL HT RTD PROBES

The first criteria for HT RTD design is operating temperature range. It must be above the current standard industrial RTD. Based on studies and testing, current industrial RTDs are best suited for operating temperatures up to 800°F. Above 800°F their accuracy drops, they become unreliable, and are often permanently damaged. An upper limit of 1200°F was chosen. This is low enough to be easily achieved and high enough to satisfy an immediate demand from power stations. At around 1200°F platinum goes through a re-crystallization process. If the operating range is set above 1200°F, problems with platinum wire re-crystallization may occur [10].

In HT RTD design for a temperature range of 800 to 1200°F, two major problems are known: a) decreasing Insulation Resistance (IR) and b) contamination (poisoning) [11]. Both problems are aggravated by the current philosophy of industrial RTD design which uses more materials for protection.

The next decision in HT RTD design was selection of a base resistance value or R_{32} (or R_0). In HT SPRT, a low R_{32} is often selected in order to minimize the effect of decreasing IR at HT . Fig. 1 shows the effect of base resistance values when decreasing IR is equivalent of a 1 MΩ shunt resistor. When R_{32} is 100Ω, the error is 0.5°F at 1100°F. If R_{32} is 25Ω, the error is reduced to 0.13°F, and 2.5Ω reduces it to less than 0.01°F. Because of this effect, an HT SPRT intended for use above the silver point often uses less than 1Ω for base resistance. In an industrial HT RTD, the preferred value is 100Ω. The reason for the relatively high 100Ω selection is resolution or accuracy in the resistance measurement. This effect is shown in Fig. 2. When the resistance measurement accuracy is 0.01Ω, the resulting temperature error is less than 0.1°F in a 100Ω thermometer, but 1.8°F in a 10Ω thermometer. The error will increase to 3.6°F in a 1Ω thermometer. In an industrial environment, it is unreasonable to expect a resistance measurement with better than 0.01Ω accuracy.

It was decided to conform to a published accuracy of ±2.5°F at 1200°F. At temperatures up to 932°F, this specification is nearly identical to DIN Class A, and between 932° and 1200° is the linear connection of both terminal points.

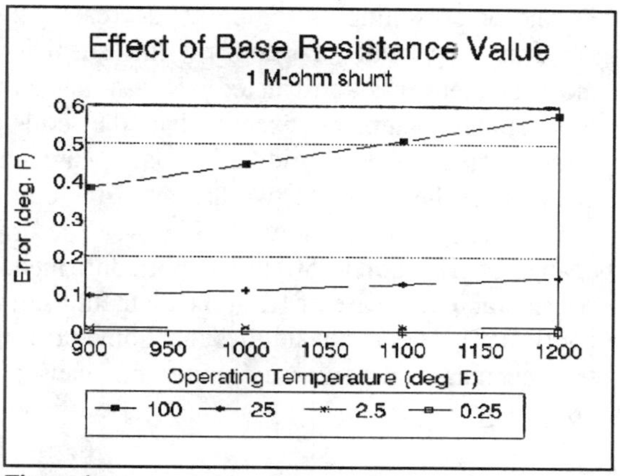

Figure 1

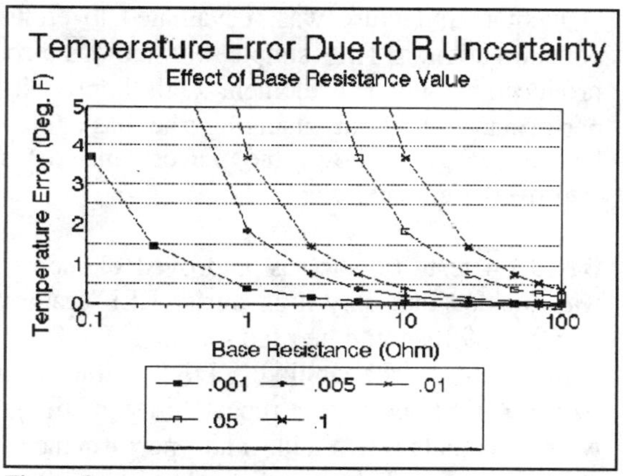

Figure 2

DEVELOPMENT OF HIGH TEMPERATURE PT SENSING ELEMENT

Platinum wire temperature sensing elements for standard temperature range (below 950°F) are manufactured by several companies and most perform well [12]. One type of element design is generally considered to be the best in reliability and accuracy: internal coil with a powder packed element [13]. Furthermore, a test showed that this element design can withstand vibrations up to 25 G. When the same element was packaged into a powder-packed vibration buffer, it could withstand vibrations up to 50 G without affecting its operation. This construction was adopted as the basic structural design for HT RTD elements.

Initially, several elements from different manufacturers were evaluated for their HT characteristics. Fig. 3 shows an example of results from such an evaluation. Around 950°F, the accuracy of Pt wire elements drops drastically in both types of Pt wires (REF grade and DIN grade wire). This occurred in all elements tested. Deterioration of accuracy in Pt elements is considered to be due to decreasing

IR and/or poisoning. Originally, decreasing IR at HT was suspected. To confirm this suspicion, a mock element was constructed. It was the same as the standard element except that the coil was disconnected in the middle. Using the mock element, IR between the two halves of the coil was tested. The result is shown in Fig. 4. The drastic change of IR around 950°F was not confirmed, but considerable decrease of IR in the element itself was confirmed. This result implies that some portion of the decrease in accuracy may be caused by poisoning.

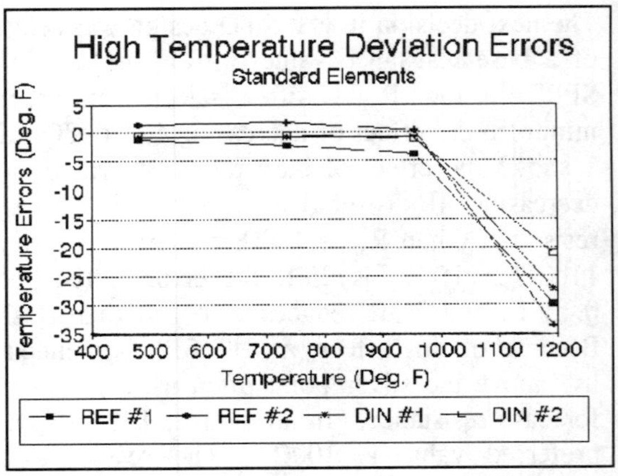

Figure 3

Every material in the element was reexamined for its HT IR characteristics, and every process of element manufacture was reevaluated to eliminate contamination. This simple but tedious process produced a new HT element with improved HT performance without changing the basic element design. Fig. 5 shows the performance of such improved elements.

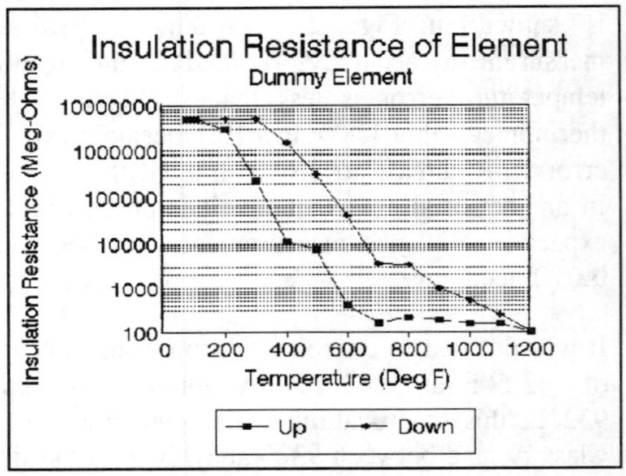

Figure 4

Based on tests of various improved elements, it was decided that only high purity REF grade wire (α=0.00392) should be used for the HT RTD. The reason for abandoning DIN grade wire (α=0.00385) lies in poor repeatability in DIN grade wire elements. Yield and performance are unpredictable. This is believed to be due to self-poisoning.

The industrial Pt element includes more material than the SPRT or laboratory type Pt element, so there is increased potential for contamination and decreased IR. There are several other manufacturers who also successfully market HT elements. At present, none of the elements from other manufacturers are considered adequate to be used in the HT RTD.

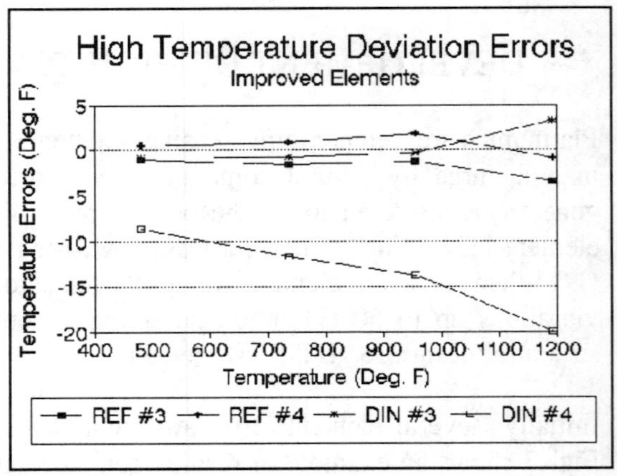

Figure 5

DEVELOPMENT OF HT PROBE ASSEMBLY

The development of a good ruggedized HT probe assembly is more difficult than that of a HT element. The effect of decreasing IR is more prominent in the sheath assembly than in the element as shown in Fig. 6. Since every solid material decreases its IR as temperature rises [14], there is no way to eliminate this effect. The only thing than can be done is to control the amount of IR decrease and to minimize its effect.

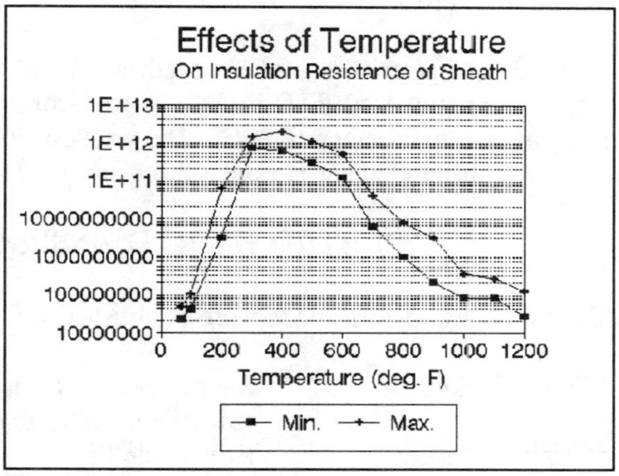

Figure 6

In the authors' design, the major impact of IR decrease is the shunt effect between sheath and lead wires due to a decrease of resistivity in insulation powders: usually aluminum oxide or magnesium oxide.

As seen in Fig. 1, a 1 MΩ shunt in a 100Ω probe will cause a temperature error of 0.5°F at 1100°F. The actual value of shunt resistance depends on insertion length of sheath into the hot zone as shown by Fig. 7.

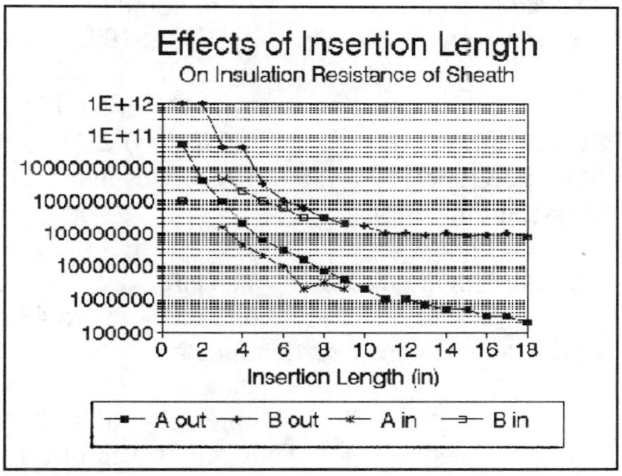

Figure 7

The insertion length in the hot zone is a unique problem in the HT RTD. As the sheath is inserted deeper into the hot zone, the corresponding value of shunt resistance becomes smaller. As the insertion depth approaches infinity, the shunt resistance approaches zero. Any HT RTD may become inaccurate if it is inserted beyond a pre-determined hot zone. Therefore, it is necessary to specify the maximum allowable insertion length for every HT RTD. The specification for insertion length necessary to prevent degradation of accuracy in our probe is 6" to 24", with exposure up to 1200°F.

Another necessary specification for the HT RTD is the maximum temperature for the terminal end of the sheath and lead wires which are not expected to withstand HT. The maximum temperature for terminal end and lead wires in our probe is 350°F. In the case of a probe with a long sheath, an additional specification is needed for the maximum temperature gradient for the transient area.

The IR in the sheath assembly decreases permanently at high temperatures due to poisoning. Cooling of the sheath assembly will not result in recovery of the IR. Poisoning is substantially more severe in an industrial RTD, since it requires the use of a metal sheath and much more material is used. In HT applications, Inconel (80% Ni, 13% Cr, and 7% Fe) is used for standard sheath material. The sheath is one of the biggest sources of potential contamination, and it requires special treatments such as pre-oxidation in order to seal off its surface.

Vibration characteristics of the ruggedized HT RTD probe are not significantly different from those of a standard ruggedized RTD probe. A well-constructed ruggedized RTD can maintain its mechanical integrity at vibrations up to 50G. The weakest link in a ruggedized RTD is the element, therefore use of a ruggedized element is essential for any ruggedized RTD.

COMPLETED HT RTD ASSEMBLY AND ITS PERFORMANCE

The following specifications were established for a ruggedized HT RTD for power station use:

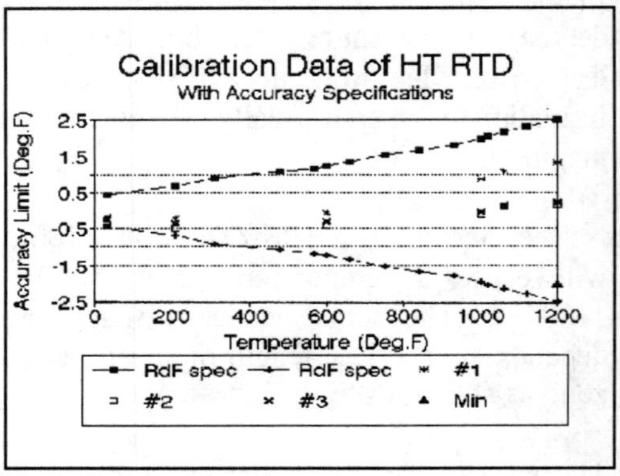

Figure 8

*Configuration: Ruggedized, 4-wire RTD, sealed in metallic assembly

*Sensing Element: REF grade Pt (α=0.00392)

*Base resistance (R_{32}): 100 ± 0.1Ω

*Operating temp range: 32 to 1200°F

*Accuracy : ± 2°F at 1000°F
± 2.5°F at 1200°F
± 0.5°F at 32°F

*Repeatability: 0.05Ω or 0.25°F at 32°F

*Response time: 10 seconds max

*Physical dimensions:
- Sheath O.D. 1/4"
- Sheath length tip to transition: standard - 22"
  optional - 6" to 48 ft (See note, below)
- High temperature insertion length:
  standard - 6"
  optional - 6" to 24"
- Sensing area: within 3 inch of tip
- Transition: 5/8" O.D., 3" long

*Maximum allowable temperature for transition, terminal end, and lead wires: 350°F

*Vibrational Environment:
- At Room Temperature: 3 Hz to 200 Hz: amplitude = 0.01" and 200 Hz to 4000 Hz 50 G
- At 1000°F: 3 Hz to 500 Hz: 3 G, 300 to 5000 Hz: 1 G

Note: Temperature gradient zone: When sheath is longer than 48", the maximum length of temperature transition region (1200 to 350°F) is 24" from the end of HT zone.

The HT calibration data are shown in Fig. 8. The outline drawing is shown in Fig. 9 (following page).

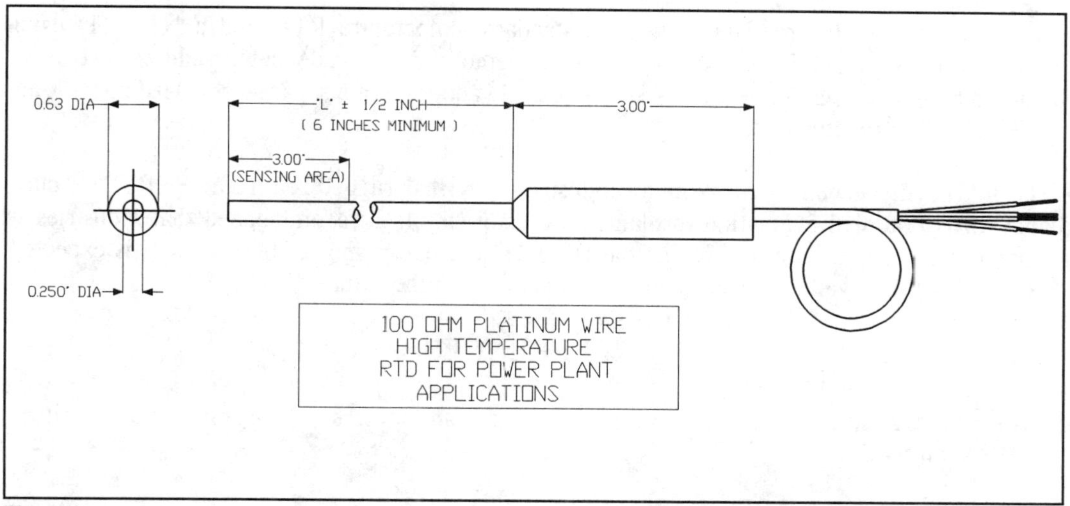

Figure 9

DISCUSSION

Over 3 years of development, technical problems in building a ruggedized industrial HT RTD have been identified and successfully resolved. These problems were decrease of IR and poisoning. Both problems affect the element and the assembly independently, and both cause the same apparent effect, namely decrease of IR and Pt sensor resistance at HT (R_t). Both problems are successfully solved through strict material and process control.

Successful implementation of these solutions can be verified by HT measurement of IR and R_t. HT IR measurement reveals the effectiveness of insulation materials and the probe assembly method. HT sensor resistance measurement verifies the suitability of the element as long as the IR is satisfactory.

Simpler test methods such as repeatability check of R_{32} and IR after HT exposure(s) will not verify the design for HT RTD. Such a test only verifies successful contamination control in element and assembly. A solution to problems which are caused by intrinsic decrease of IR at HT can be verified only at HT.

It is believed that a reliable ruggedized industrial HT RTD will contribute significantly to the improvement of operating efficiency in many industrial facilities.

CONCLUSION AND RECOMMENDATION

A ruggedized 100Ω HT platinum RTD which can operate accurately and reliably at 1200°F for long periods has been successfully developed. This has been accomplished by solving problems of decrease in IR and of material contamination at high temperatures. These problems have been solved through strict material and process controls which are much more extensive than those for the standard temperature RTD.

It is very difficult to differentiate between a standard temperature RTD and a HT RTD using the presently accepted standard test methods. Such methods do not adequately address the problems encountered at high temperatures. The soonest possible adoption of a separate standardized test method for the HT RTD is recommended.

The HT RTD will be beneficial to many industries. A real ruggedized 100Ω Pt RTD is currently available for operational evaluation in plants and factories. It is recommended that industries which could potentially benefit from the use of the HT RTD consider using a HT RTD. It is expected that the HT RTD will be available from many manufacturers in the future.

REFERENCES

[1] Michael Valenti, "A new generation of power plant sensors", Mechanical Engineering, Sept 1994. pp 54-58.

[2] H.G. Nubbemeyer, "High Temperature Platinum Resistance Thermometers and Fixed-point Cells for the realization of the ITS-90 in the range 0°C to 961.78°C", Temperature, Vol 6, pp.199-202, 1992.

[3] Li Xumo et al, "Realization of the international temperate scale of 1990 between 0°C and 961.78°C at NIM", Temperature, Vol 6, pp.193-197, 1992.

[4] G.F.Strouse et al, "Investigation of high-temperature platinum resistance thermometers at temperatures up to 962°C, and, in some cases, 1064°C", Temperature, Vol 6, pp.389-394, 1992.

[5] M. Arai and H. Sakurai, "Development of industrial platinum resistance sensors for use up to the gold point", Temperature, Vol 6, pp.439-442, 1992.

[6] B.W.Mangum, "Platinum Resistance Thermometer Calibrations", NBS Special publication 250-22, U.S.Dept.of Commerce, 1987. pp.44-81.

[7] J.P. Evans, "Experience with high temperature platinum resistance thermometers", Temperature, Vol 5, pp.7171-781, 1982.

[8] H. J. Jung and H. Nubbemeyer, "The Stability of commercially available high temperature platinum resistance thermometer of 5Ω silica cross type up to 961.93°C", Temperature, Vol 5, pp.763-770, 1982.

[9] Long Guang & Tao Hongtu, "Stability of precision high temperature platinum resistance thermometers", Temperature, Vol 5, pp.783-787, 1982.

[10] Robert J. Berry, "Oxidation, Stability, and insulation characteristics of Rosemount standard Platinum resistance thermometers", Temperature, Vol 5, pp.753-762, 1982.

[11] J.R.Saffell, "Designing accurate platinum RTD measuring systems for industry", Temperature, vol. 5, pp.733-737.

[12] J. V. McAllan, "Practical high temperature resistance thermometry", Temperature, Vol 5, pp.789-793, 1982.

[13] D. J. Curtis,"Thermal hysteresis and stress effects in platinum resistance thermometers", Temperature, Vol 5, pp. 803-812, 1982.

[14] Thomas Kerlin and Robert Shepard, "Incustrial Temperature Measurement", ISA, 1982.

0074-056X/95/293-312/$0 + .50pp

Real-Time Measurement of Unburned Combustibles and Combustion Characteristics in Utility Boilers

Mark L. Umberger
Dairyland Power Cooperative
Genoa 3 Generating Station
Rt. 1, Box 276
Genoa, WI 54632-9740

H. Randy Carter, P.E.
Applied Synergistics, Inc.
3831 Old Forest Road, Suite 6
Lynchburg, VA 24501

Jeffrey S. Kite, P.E.
Applied Synergistics, Inc.
3831 Old Forest Road, Suite 6
Lynchburg, VA 24501

Abstract

Federal, state and local emission limits imposed on utility stationary boilers are leading to changes in operating practices, hardware, and measurement systems. Many coal-fired plants have modified existing burners or installed new burners to achieve the required reductions in NO_x emissions. Initial adjustment of the burners and maintaining the required conditions for low NO_x and minimum unburned carbon operation can be difficult because of conflicting goals, interaction of boiler parameters, and lack of real-time process measurements.

This paper describes innovative applications of optical techniques which provide real-time operator information for monitoring and optimizing the combustion process. Field test results are presented for two devices used to measure unburned combustibles and combustion characteristics. The studies indicate that these measurement devices provide critical process information on combustion efficiency (unburned combustibles) and emissions (NO_x) as an aid in adjustment and long term maintenance of the combustion conditions.

Introduction

Most utilities are planning to meet the NO_x reductions mandated by the Clean Air Act Amendment of 1990 by front-end combustion modification. A number of combustion modification techniques are being evaluated and used, including: low NO_x burners, over-fired air, and low excess air.

Satisfying both the requirement to reduce NO_x emissions and the need to maintain low levels of unburned carbon in the flyash is difficult to achieve. Existing measurement systems do not provide the unit operator on-line real-time information to achieve the desired goal. As a result, plants encounter swings in unburned carbon and NO_x that require corrective operator action to avoid exceeding emission compliance limits and equipment capacity limits.

This paper discusses new measurement systems that provide real-time information on unburned carbon levels in the flyash, combustion characteristics, flame shape, size, location and spatially resolved temperatures. Testing conducted at two utility plants to evaluate these measurement systems is presented.

Description of Measurement Systems

Achieving the desired goals for unburned carbon levels and NO_x emission requires finding and maintaining the correct balance between high temperature oxygen rich combustion and low temperature oxygen lean combustion. The measurement devices described provide direct indications of unburned combustibles in the gas stream (Unburned Carbon Module) and spatially resolved combustion temperatures and fireball images (Combustion Module). These measurements provide key on-line information for adjusting the combustion process for achieving the desired goals.

Unburned Carbon Module

The carbon and other combustible particles in pulverized coal ignite in the burner region of the furnace. Depending on the furnace temperature, the residence time of the particles in the furnace, and the availability of oxygen to support combustion, the particles, as they exit the furnace, are completely consumed, smolder or extinguish. The measurement system detects the combustible particles that are smoldering and just extinguish as they pass from the furnace to the boiler region where the temperature is below the particle ignition temperature (~1000 °F for carbon in coal). In this transition region, the combustible particles are hotter than the inert particles, surrounding gas or boiler structure.

The measurement system uses a near infrared (NIR) imaging camera which is optically filtered to enhance resolution between hot and cool particles and surfaces. Figure 1 shows the Unburned Carbon Module camera installed in a side wall opening in the convection region of a utility boiler. A hot unburned combustible particle will be observed as a white spot traversing the camera's field of view. The camera images are processed to produce a quantitative reading of the number (CPM-counts per minute) of white spots passing through the field of view, providing a signal which is proportional to the level of unburned combustibles in the flyash. With correlation, the system provides continuous real-time indications of unburned carbon or LOI (loss on ignition). The measured level of unburned combustibles is from a volume in the boiler convection region defined by the camera lens tube field of view and optical depth of field.

Combustion Module

There have been a number of papers presented on the formation of NO_x in coal fired boilers [1,2,3]. It is beyond the scope of this paper to present an in-depth review of these studies.

---

[1] Rawdon, A. H., and Johnson, S. A., "Control Of NOx Emissions From Power Boilers," Presented at the Annual Meeting of the Institute of Fuel (Australian Membership), Adelaide, Australia. November 7, 1974.

[2] Weiss, M. D., Bauer, F. W., DuBrul, R. S., "Approach To Utility Boiler Acid Rain Compliance Modifications."

However, each notes that for a given coal, the key variables for NO_x control are combustion temperature and oxygen concentration in the devolitization region. Other studies [456] have shown that NO_x control is highly dependent on both burner design/adjustment and boiler operating parameters, such as furnace cleanliness, burner tilt, etc. The studies indicate that on-line measurement of combustion temperatures, as well as overall combustion characteristics (fireball size, shape, and position) could provide an operator important information on the trend and magnitude of NO_x variations resulting from burner adjustment and changes in boiler operating conditions.

The Combustion Module sensor consists of a visible imaging camera with integral scanning multi-wavelength pyrometers. Two pyrometers at different wavelengths are used to determine combustion temperature by ratioing radiant intensity at the different wavelengths. Because a ratio is used, the measurement is not affected by the smoke and ash that can obscure portions of the flame. Relatively short measurement wavelengths allow the pyrometer to respond only to the hottest object in the field of view[7]. The pyrometers are scanned across the field of view to provide both peak and volume average combustion temperatures.

Special colorization schemes are used to digitally enhance the combustion image. This aids in rapid identification of combustion conditions, such as fireball shape, location, as well as possible mechanical problems with introduction of fuel or air. A camera installed in a tangentially fired boiler is illustrated in Figure 2.

Utility Plant Testing

Description of First Test Site

Genoa Unit 3 is a tangentially-fired Combustion Engineering supercritical boiler with combined circulation and double reheat. The unit has a single furnace fitted with LNCF Level 3 low NO_x burners. Full load is approximately 370 MW with main and reheat steam temperatures of 1010 °F and a steam pressure of 3800 psia

---

[3] Carter, H. R., Larson, Wayne K., "NOx Emission Reduction By Furnace Cleanliness and Combustion Management," Diamond Power Specialty Company. 1993.

[4] Levy, E., Eskenazi, D., D'Agostini, M., Cramer, D., Williams, S., Petrill, E., Leyse, R., "Application of EPRI's Plant Monitoring Workstation To NOx Control And Performance Optimization," Presented at the EPRI Heat Rate Improvement Conference. May 3-5, 1994.

[5] Balakrishnan, A., Edwards, D. K., "Radiative Flame Cooling For Reduction Of Nitric Oxide Emissions," Presented at the ASME-AIChE Heat Transfer Conference, Atlanta, Georgia. August 5-8, 1973.

[6] Hill, D. L., "Results From Installation Of Waterlance Soot Blowers At Dairyland Power Cooperative's JPM Station," Presented at the EPRI Heat Rate Conference. May 3-5, 1994.

[7] James, R. K., "Imaging Pyrometry For The Process Industries," Sensors The Journal Of Machine Perception, Volume 9, Number 1. January 1992.

An Unburned Carbon Module was installed on the south side wall between the finishing superheater and low pressure reheater. The installation location is illustrated on Figure 3. A four inch opening in the boiler side wall provides access for the image camera. The camera was located so that the lens tube provided a view of the gas stream as entrained particles exited the finishing superheater. The desired view was obtained with the lens tube outside of the plane of the boiler wall. Services to the unburned carbon sensor included 110V power and plant service air for cooling and purging of the lens tube. A single video cable connected the imaging camera to the image controller for determination of CPM.

The system was installed to evaluate the capability of the system to provide continuous on-line information for the unit operators and to provide insight on unburned combustibles levels during tuning of the low NO_x burners. Flyash samples from selected precipitator hoppers were taken for laboratory determination of LOI and comparison to the Unburned Carbon Module measurements. Limited isokinetic samples were also taken for comparison to the Unburned Carbon Module.

Genoa Test Results

The Unburned Carbon Module was used for eight days at the Genoa Station. Each day the measurements from the Unburned Carbon Module were compared and trended with unit load, O_2, NO_x, and laboratory determined LOI. Figure 4 shows the results from one of the test days. During the 24 hour period presented, the unit was operated in two configurations. In the early morning and late night hours, the boiler was operated with the SOFA (separated overfired air) and CCOFA (close coupled overfired air) dampers closed. From roughly 0900 to 1600 hours the unit was operated with the overfired dampers open and burner tuning in progress.

During the periods that the SOFA and CCOFA dampers were closed (outside noted period of burner tuning), both NO_x and CPM varied directly with load and inversely with O_2. Burner tuning was performed with the load constant at approximately 370 MW and O_2 at 3.5%. Initial adjustments to the SOFA and CCOFA dampers resulted in a reduction in NO_x and an increase in the unburned combustibles CPM. As adjustment of the dampers continued from roughly 1000 to 1600 hours, no further reductions in NO_x were obtained. However, the unburned combustibles CPM increased an additional 20-25 %.

Results from laboratory analysis of the precipitator and isokinetic flyash samples are correlated to the Unburned Carbon Module CPM on Figure 5. The time delays associated with the collection and removal of flyash samples from the precipitator hoppers and from the isokinetic samples were estimated. With this information, the real-time CPM values could be compared and correlated with the laboratory determined LOI values. The relationship between CPM and LOI is linear and repeatable.

Description of Second Test Site

The unit was a tangentially-fired Combustion Engineering supercritical boiler with single reheat. The unit has twin furnaces fitted with LNCF Level 3 low NO_x burners. Full load is approximately 585 MW with main and reheat steam temperatures of 1000 °F and a steam pressure of 3500 psia.

The Unburned Carbon and Combustion Modules were tested during consecutive but separate tests periods. The Unburned Carbon Module was installed on the side wall of the north furnace. The field of view was between the vertical superheater and reheater surfaces. For a ten day period, the measurements from the Unburned Carbon Module were compared and trended with plant parameters and flyash isokinetic samples taken hourly for laboratory LOI determination.

The Combustion Module was installed in the side wall of the south furnace. The imaging camera was installed similar to the arrangement shown in Figure 2. The system was used to observe combustion conditions for a 24 hour period. During this period, the unit returned to operation from a cold shutdown. Load varied from start-up conditions to ~80%. The measured peak and spatially resolved average combustion temperatures were recorded and trended against selected plant parameters.

Second Site Test Results

Unburned Carbon Test Results

The Unburned Carbon Module was used for ten days. Each day the measurements from the Unburned Carbon Module were compared and trended with unit load, O_2, opacity, CO_2, tilts, and NO_x. In addition, the measured CPM were compared and correlated to the LOI data from isokinetically obtained flyash samples. The samples were taken each day and time stamped to allow direct comparison and correlation with the Unburned Carbon Module CPM. Data was acquired during periods of standard boiler operation and low NO_x burner tuning.

The Unburned Carbon Module was responsive to both large and small variations in boiler operations. Figure 6 shows the Unburned Carbon Module measured CPM compared to unit load and total coal flow. Unit load increased from under 300 MW to ~ 475 MW in one hour. Increased unburned combustible levels were measured through the power ascension. The module's sensitivity to small variations in boiler operations was also observed. Figure 7 shows that small changes in load and total coal flow are mimicked by the measured variations in unburned combustibles.

Variations in unburned combustibles, as determined by laboratory analysis of the isokinetic flyash samples and by the Unburned Carbon Module, were similar. Figure 8 compares the measurements obtained from the two methods.

Over the ten day test, dozens of isokinetic flyash samples were taken and analyzed for comparison to the on-line measurements from the Unburned Carbon Module. Figure 9 shows the results from each measurement method plotted against O_2. Both show the same inverse trend with O_2. For each data set, the best fit linear curve is included. The slope for both methods is nearly identical (within 0.4%) indicating that both methods measured the same trend and magnitude of variations in unburned combustibles. The measurements were made over a wide range of unit operations (> 330 MW) and with large changes in unburned combustibles (~20%). Also during the ten day test period, coal stocks were varied. The extremes of the coal properties were (based on as-received conditions): volatility 19.2 to 27.6%, fixed carbon/ash 4.0 to 6.3, and BTU content 12519 to 13308.

Combustion Module Test Results

The Combustion Module was used to record combustion images and spatially resolved temperatures for ~ 24 hours. During this period, unit conditions varied widely from just over 250 MW to ~550 MW. The operating conditions also varied rapidly. Unit load varied, during one period, from 250 MW to 550 MW to 275 MW in 6 hours.

The Combustion Module measured temperatures at six spatially resolved locations in the field of view. The measured temperatures were averaged for comparison to other plants' variables, including load, opacity, O_2, CO_2 and NO_x. Figure 10 shows the measured combustion temperature and selected plant parameters. In general, combustion temperature varied directly with load, opacity, CO_2 and inversely with O_2. However, from 2100 to 0400 hours, furnace combustion temperature increased by ~70 °F while the other parameters shown on Figure 10 were nearly constant.

Figure 11 shows a similar increasing trend in the measured NO_x. The upward trend in combustion temperature and NO_x during a period when the principle combustion conditions are constant is indicative of fouling of the furnace heat surface with corresponding higher furnace combustion temperatures and higher NO_x. The impact of the slagging caused heat transfer reduction was observed only on the combustion temperature and NO_x measurements.

Load ranged between ~270 MW and 410 MW from 1400 to 0400 hours. Combustion temperature and NO_x varied directly during this time period. As illustrated by Figure 12, not only did NO_x and combustion temperature vary directly, but the changes in the combustion temperature resulted in corresponding and repeatable changes in NO_x.

The measured NO_x and combustion temperature varied directly and repeatably in all time periods, except for 0700 to 1300 hours. During this period, the unit operating conditions varied from ~250 MW to 550 MW to 270 MW. Combustion temperature increased and decreased with load, and NO_x appears to lag the changes in combustion temperature.

Conclusions

Information obtained from the Unburned Carbon and Combustion Modules provides unit operators and engineers real-time combustion information. This information allows on-line combustion adjustments not possible without the real-time information. In addition, the information allows concise decisions to trade off emissions and efficiency to achieve desired performance results.

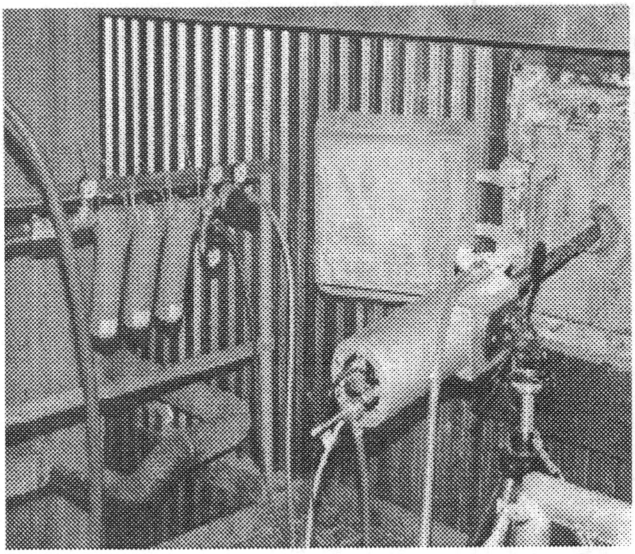

Figure 1 Unburned Carbon Module Camera Installation

Figure 2 Combustion Module Camera Installation

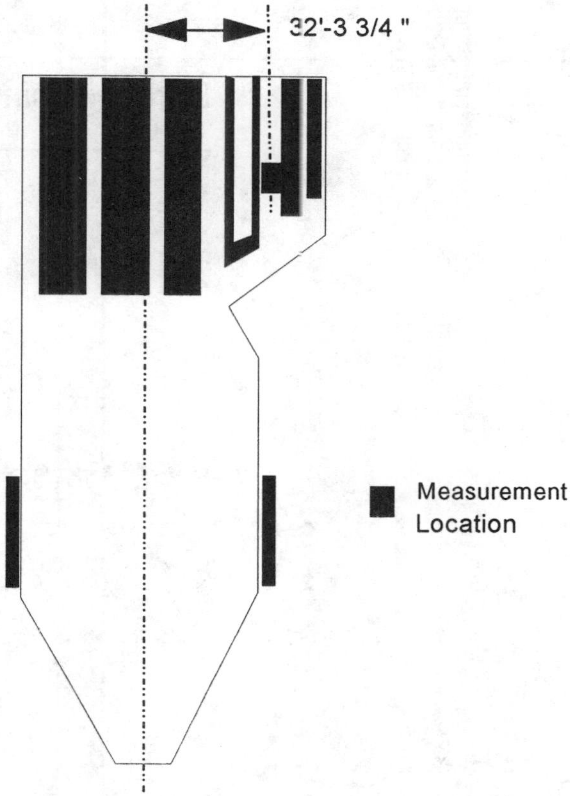

Figure 3 Location Of Unburned Carbon Camera At Genoa

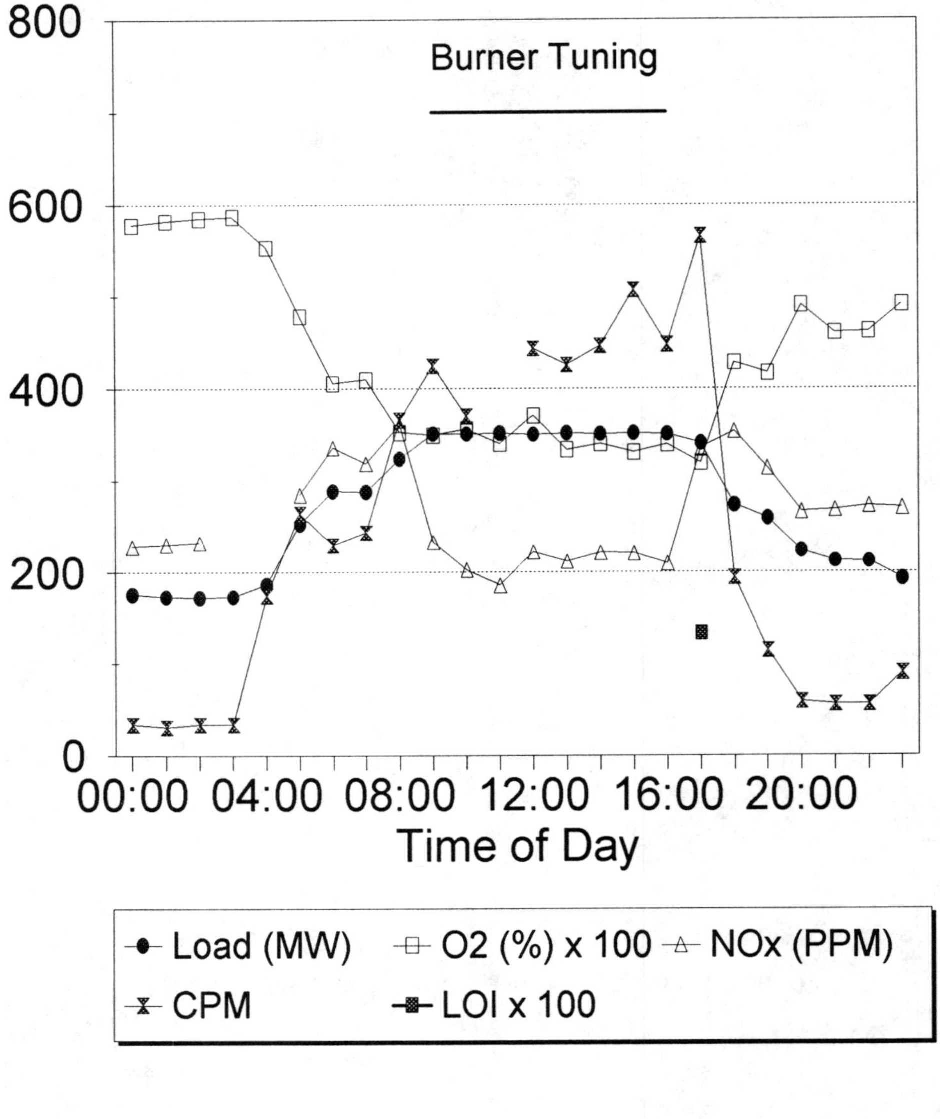

Figure 4 Genoa Unburned Carbon Test Results

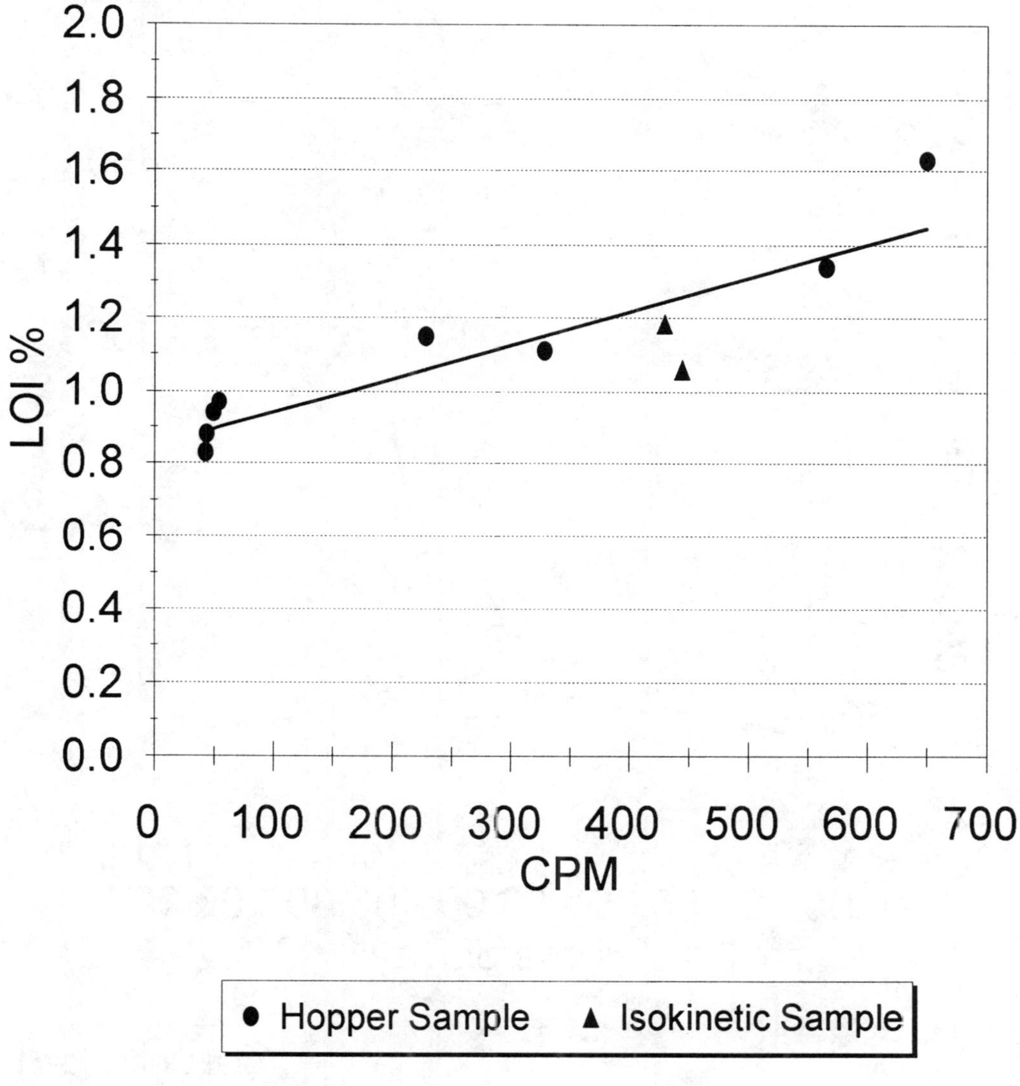

Figure 5 Comparison of Lab LOI and Unburned Carbon CPM

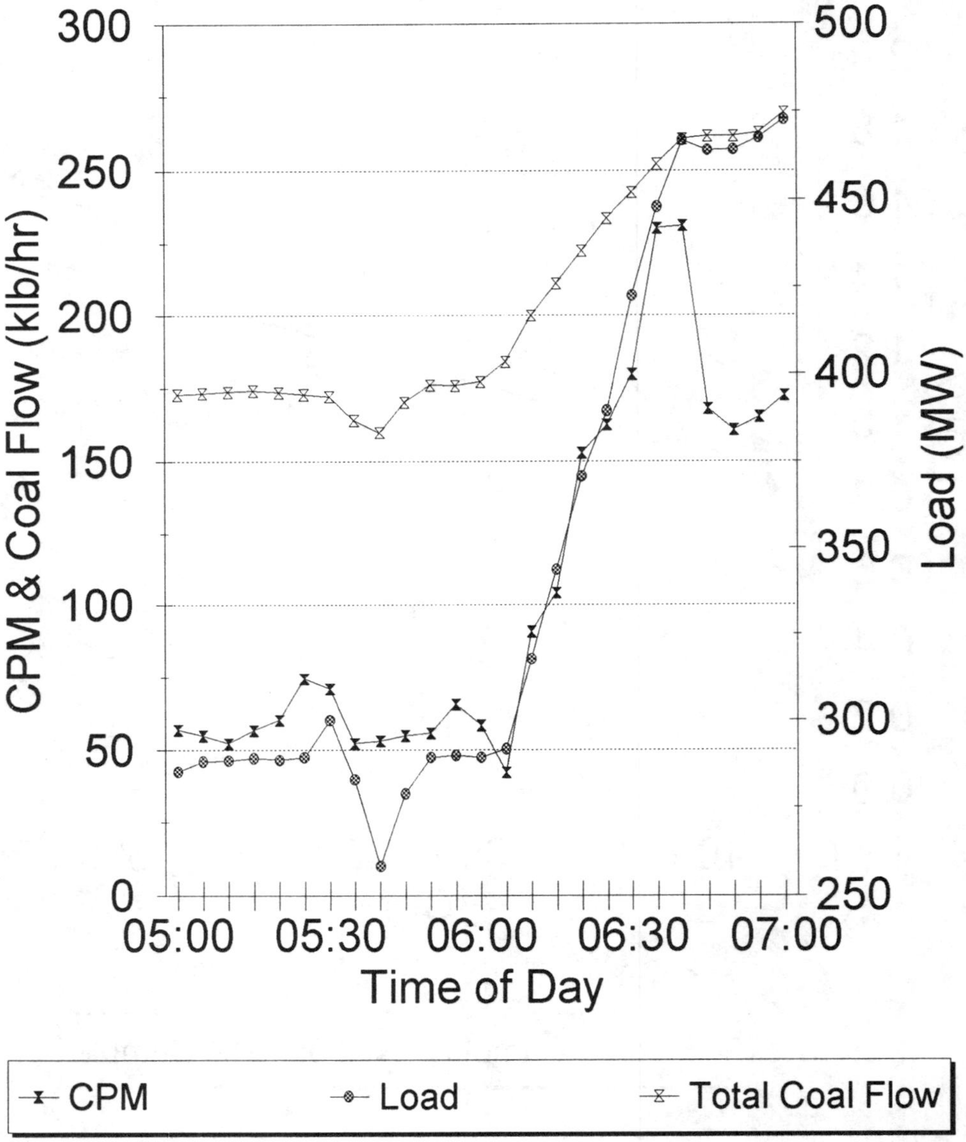

Figure 6 Comparison of Unit Load and Measured Unburned Combustibles

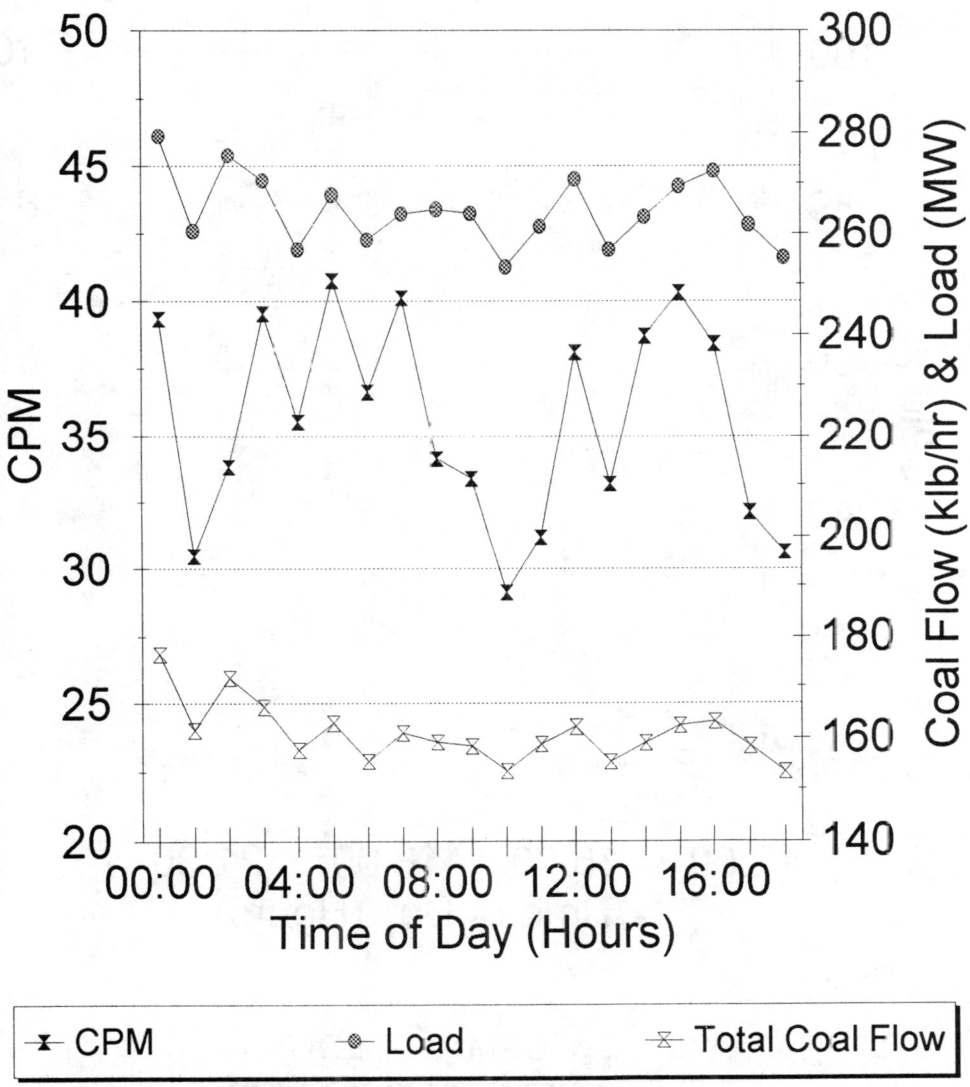

Figure 7 Load and Unburned Combustible Variations

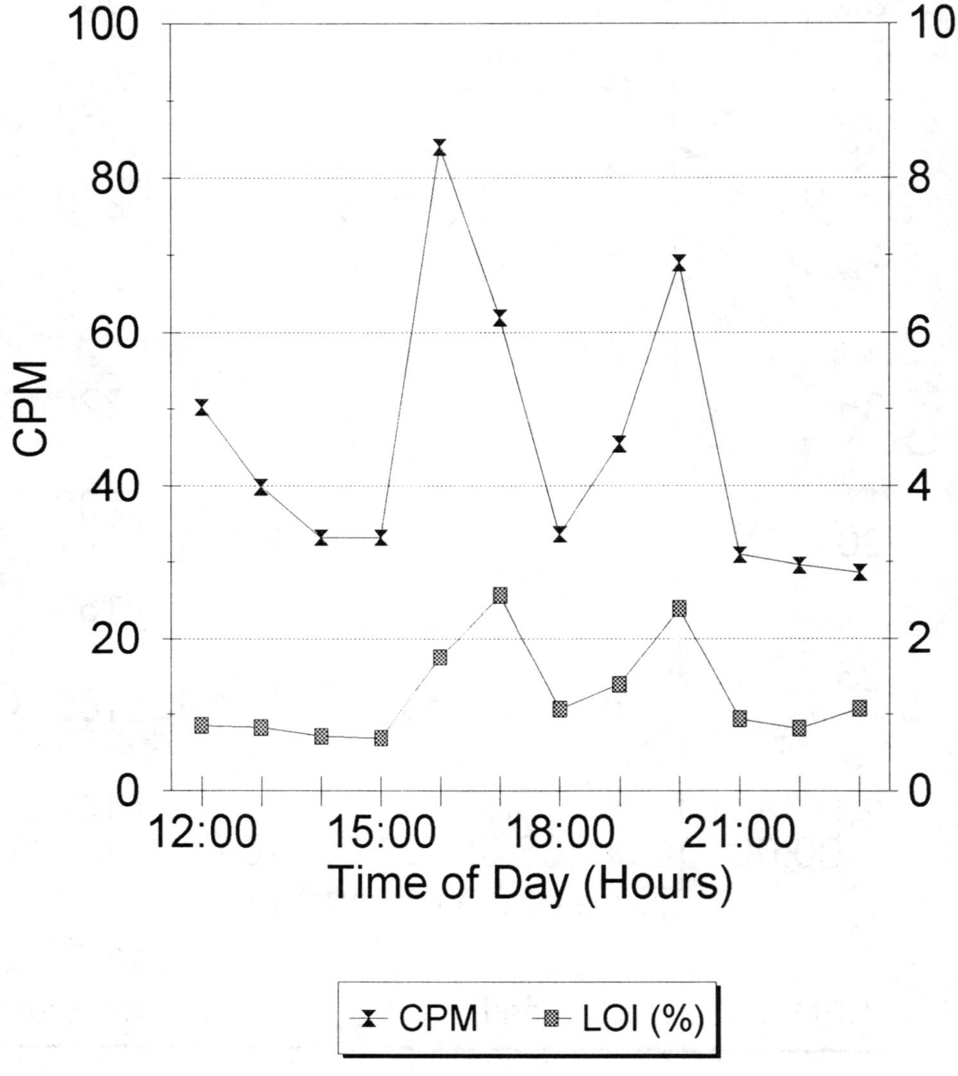

Figure 8 Unburned Carbon Module and Lab Measurement Trends

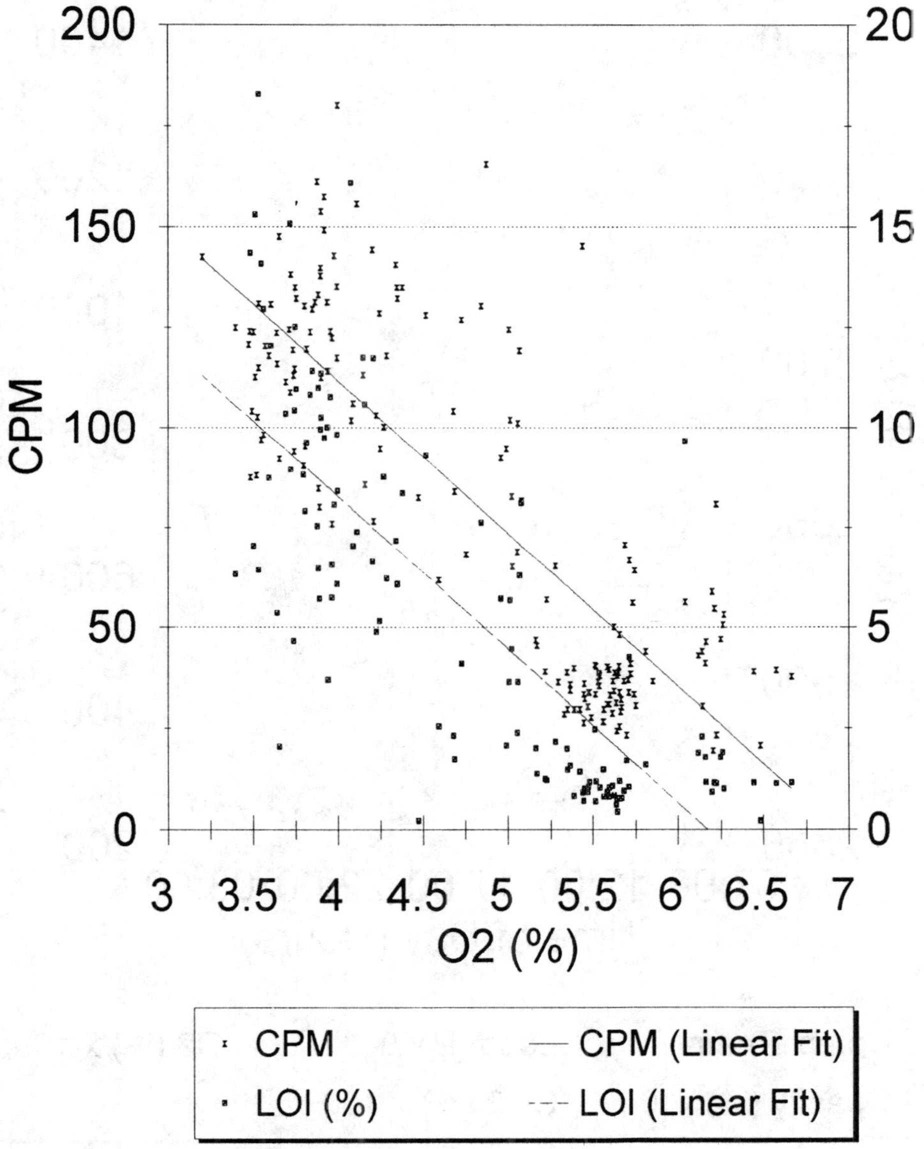

Figure 9 Lab and Unburned Carbon Module Determined Unburned Combustibles

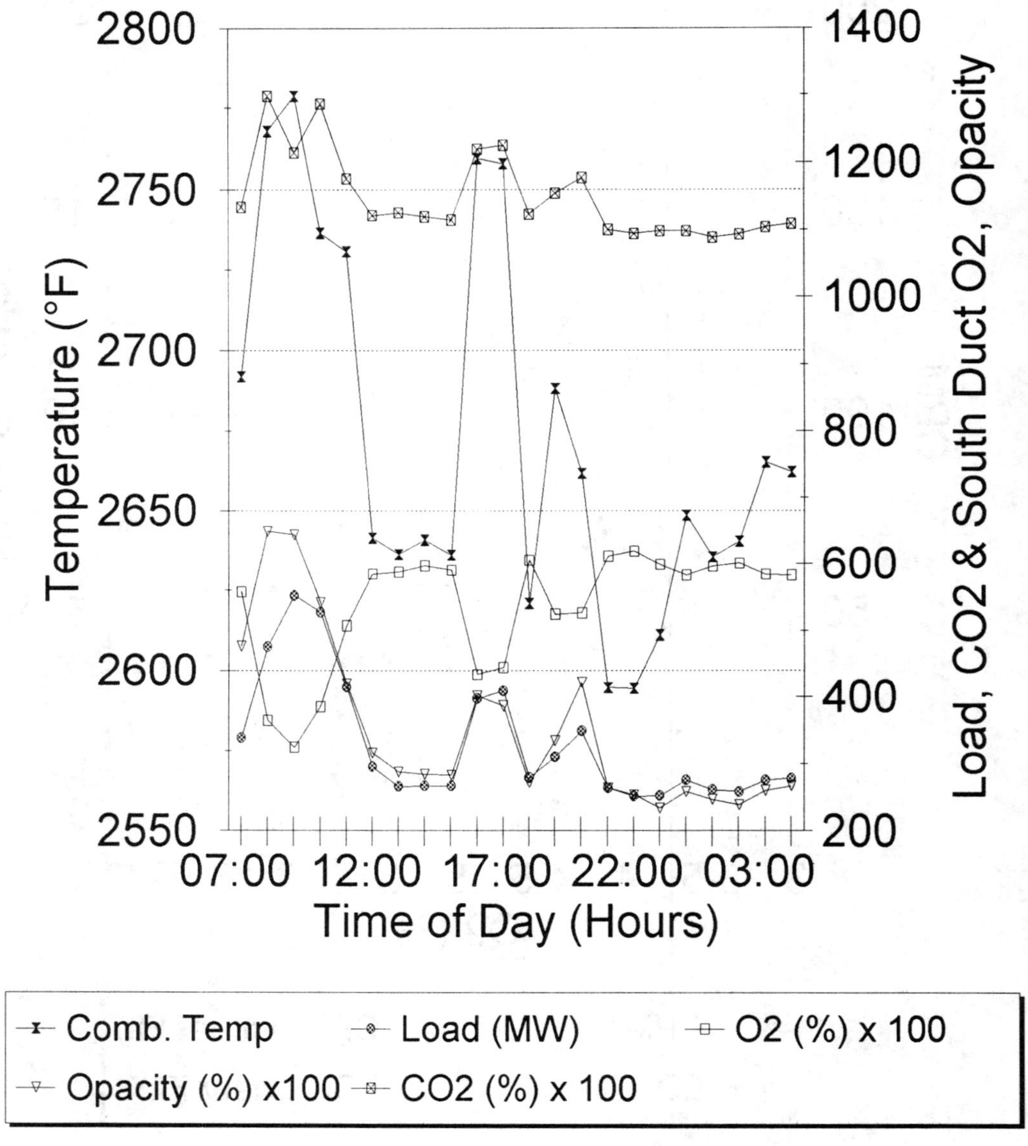

Figure 10 Plant Parameters and Measured Combustion Temperature

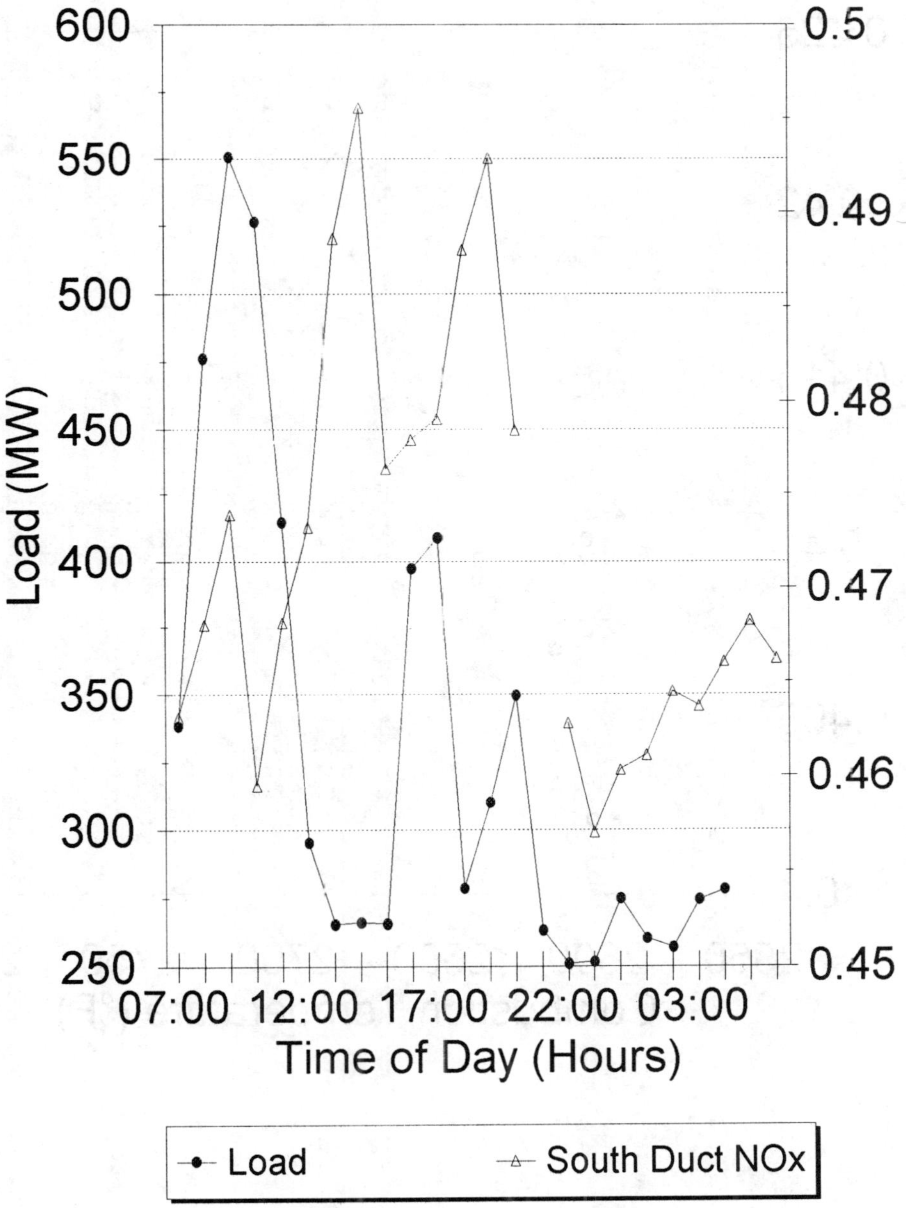

Figure 11 Comparison of Load and NOx

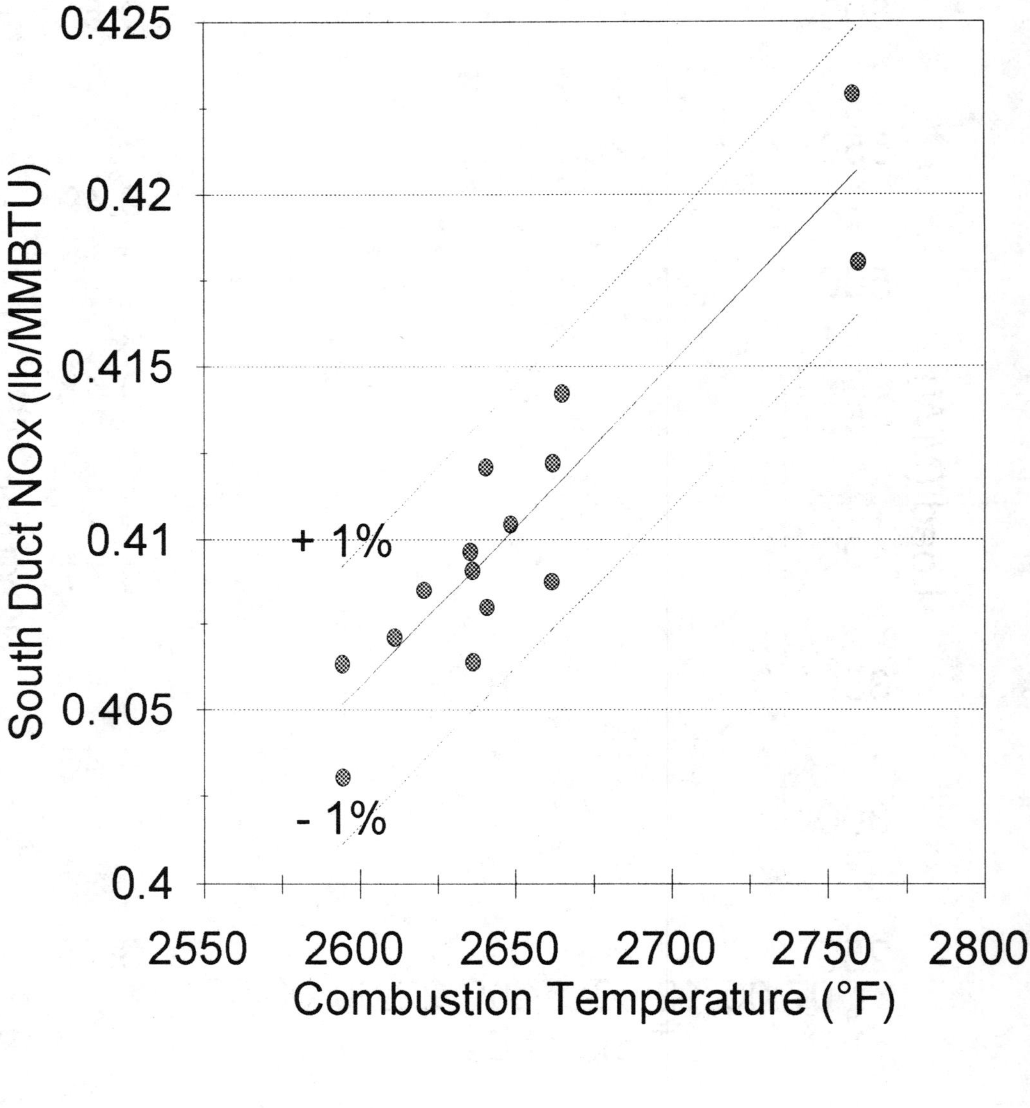

Figure 12 Correlation Between NOx and Combustion Temperature

©0074-056X/95/313-324/$0 + .50pp

Obtaining an Accurate Steam Analysis

Eugene Rommelfaenger
Sentry Equipment Corp.
856 East Armour Road
Oconomowoc, Wisconsin 53066

KEYWORDS

Steam, Sampling, Boilers

ABSTRACT

An accurate chemical analysis of steam requires more than a reliable analyzer. This article discusses system design factors that affect the accuracy of steam analysis. Criteria for obtaining and transporting the sample are presented along with a discussion of the impact on the validity of the sample. Recommendations on sample line sizing and selection of sample system components are included. Sampling of both saturated and superheated steam are discussed.

INTRODUCTION

Sampling is an important aspect of water chemistry control within power plants. Most power stations route samples to a central sampling station which has resulted in many sampling lines exceeding 500 feet in length. Some stations use 1/4" or 3/8" diameter tubing while others use lines of 1/2" or 3/4" pipe for comparable samples. Frequently steam samples are fully condensed prior to reaching the sampling panel. Due to high pressure drop and velocities encountered with two phase flow many low pressure steam sample lines deliver only a fraction of the desired flow rate. Over sized lines promote rapid condensation with slow liquid velocities. Slow moving condensed samples deposit contaminants on the tube walls.

Frequently plant chemists maintain flow and other parameters to their analyzers without significant attention to system design flows and other factors which can affect the sample. Electric Power Research Institute (EPRI) Report "Guideline Manual on Instrumentation and Control for Fossil Plant Cycle Chemistry" (1) states. "In many instances considerable attention is devoted to obtaining the most accurate analytical results possible while the sampling system is ignored, assuming that representative samples will be provided under all conditions."

SAMPLE BEHAVIOR

Design of a sample line requires an understanding of the sample behavior. A saturated steam sample originates at the sampling nozzle as vapor with entrained liquid droplets. As flow proceeds down the tube heat loss causes a liquid film to form on the inside surface of the tube. The liquid film moves significantly slower than the steam vapor. The surface of the liquid has moving waves which vary with the liquid and vapor velocities. If the steam velocity is sufficiently high then droplets of liquid are entrained into the moving steam from the wave crests. Simultaneously droplets carried by the steam flow impinge on the liquid film and become entrapped. The film thickness gradually increases with additional condensation. When the film reaches sufficient thickness the flow develops to slug or churn flow where large bubbles of steam flow faster than the accompanying liquid and bypass the liquid between the bubbles. Gradually the size of the bubbles decreases until all steam condenses and single phase liquid flow results. If the sample line is short then all phases may not be encountered. A superheated sample originates with a portion of dry wall tube then progresses through the same sequence discussed above. Figure 1 shows a conceptual picture of steam sample flow with heat loss from the tube wall.

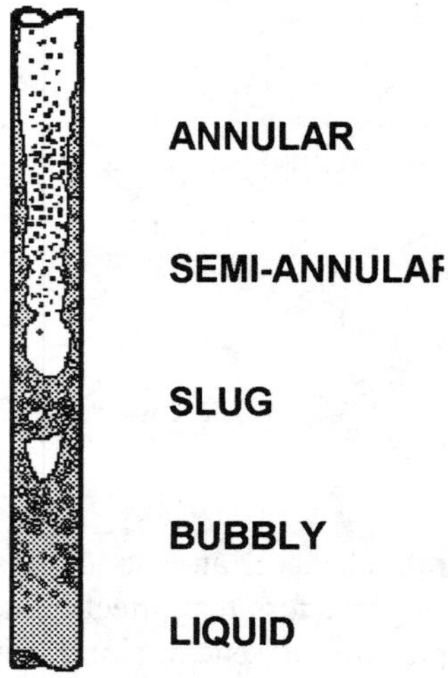

Figure 1. Conceptual Flow Model

ISOKINETIC SAMPLING

The steam sampling rate should be compatible with the steam sample nozzle(s). The Performance Test Code of the American Society of Mechanical Engineers (ASME) (2), the EPRI report (1) and the American Standard for Testing and Materials (ASTM) Standard Practice for Sampling Steam (3) all require isokinetic sampling for saturated steam. The ASME Test Code (2) states "Isokinetic sampling requires that the velocity of the fluid entering the sampler port(s) shall be exactly the same as the velocity of the stream being sampled at the location of the sampler. . . Because saturated steam is normally a two-phase fluid, made up of steam and small droplets of water, isokinetic sampling must be employed." Since the chemical constituents found in the water differ from those in the steam any legitimate sample requires that the proportions of steam and water in the sample be representative of those in the bulk flow.

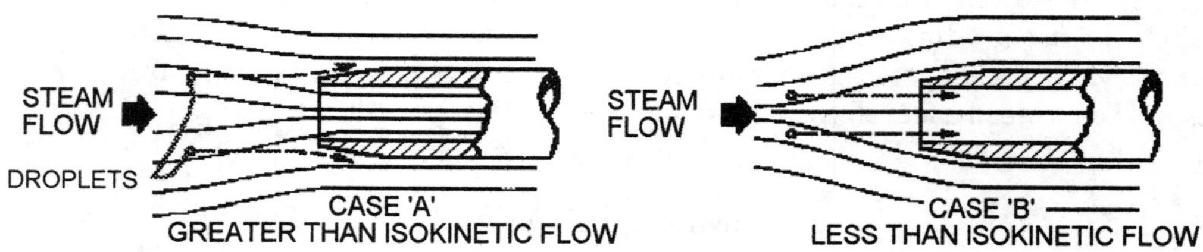

Figure 2. Effects of Non-isokinetic Sampling
Source Reference 4

Wide differences in density between the liquid droplets and the steam cause sampling errors when isokinetic sampling is not practiced. Momentum causes the heavier droplets to maintain their flow path in the vicinity of the sampling nozzle while the lower density steam deflects to accommodate the sampling requirements. When the velocity inside the sampling nozzle throat is greater than the velocity of the bulk flow the steam will divert towards the nozzle entrance while the entrained droplets pass the outside of the nozzle. Reduced sampling velocities cause the steam to diverge around the nozzle while the droplets maintain their path.

Superheated steam is considered to be a mixture of solid oxides (mostly magnetite), solid precipitated salts and dry superheated vapor. (5) Studies have shown that large amounts of impurities can be retained on the surface of the oxides. The densities, and inertia, of the oxides is significantly different from the bulk steam. Isokinetic sampling is recommended (1)(5) in order to obtain a representative sample of steam and the heavier oxides.

Experimenters have determined that samples obtained from slow moving saturated steam lines indicate low levels of impurities. They believe that many impurities are contained in a liquid film which adheres to the surface of the pipe and bypasses the sampling nozzles. With higher steam velocities waves are produced on the film and droplets from the film are entrained into the steam. Present standards (2)(3) for sampling saturated steam have established minimum recommended steam line velocities for representative sampling.

TRANSPORT

In addition to obtaining a representative sample at the nozzle it is important to maintain the composition of the sample within the sample tube. The loss of contaminants by deposition on the tube wall can cause erroneous results from chemical analysis. This condition has been recognized in the ASTM Standard (3) which states "Beyond the inlet port, the velocity should be kept high to minimize the loss of impurities." There are a multitude of factors which contribute to deposition on the tube wall. The most prominent mechanisms include crystallization resulting from solubility changes, settling resulting from gravity and hydrodynamic forces, and electrostatic attraction of charged particles to the tube wall. Reference (6) discusses in detail a number of theories related to deposition on fluid boundaries.

Deposition of impurities from superheated steam samples has been studied and documented since the early 1960s. This phenomenon is attributed to the decrease in solubility of the compounds in superheated steam resulting from the reduction in temperature. Several experimenters have found that a major portion of the steam borne impurities did not reach the analyzers. These same researchers found that a significant portion of the Na based contaminants would be readsorbed when the lines were cooled and filled with condensate. With saturated steam samples crystallization resulting from supersaturation should not take place since condensing occurs with minor temperature changes and most contaminants are more soluble in the condensate than the steam. Therefore deposition from saturated steam samples will be more dependent upon settling and electrostatic attraction.

Settling results when particles are transported to the vicinity of the wall by fluid eddies and across the boundary layer by their inertia, diffusion or gravity. Some of the particles which reach the wall stick. A theoretical 'sticking probability' is used to determine whether the particle will deposit on the wall. Some particles which stick are later eroded from the wall by the fluid drag force. These eroded particles become reentrained in the fluid stream. A common theory is that both the rate of deposition and the rate of erosion increase with increasing velocity. Several researchers have found that with liquid velocities near 6 ft/sec the steady state deposit weight is at a minimum. Minimum deposits are desirable because they reach equilibrium in a shorter time and are considered less susceptible to 'crud bursts' where an area of previously deposited material is swept into the stream. EPRI (1) currently recommends 6 ft/sec as the transport velocity for liquid samples. Condensate resulting from steam samples have not been specifically addressed; however several power plant sampling experts believe that velocities should be maintained near 6 ft/sec. A primary justification for this conclusion is that they have found considerable crud buildup in condensate lines when they were disassembled. Additionally, operators frequently experience significant amounts of particulate during blow down of their low pressure steam lines.

Since the velocity of a steam sample line is changing along the tube the equilibrium conditions of deposit weight and fluid velocity will vary with location. Most often the sample velocities are slower as the sample condenses. Slower velocities and lower incremental pressure drop favor lower erosion rates with corresponding higher equilibrium deposit weights.

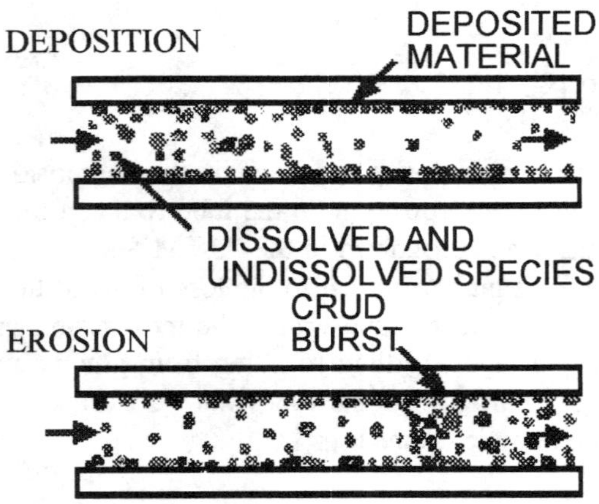

Figure 3 Deposition vs Erosion Within The Sample Line

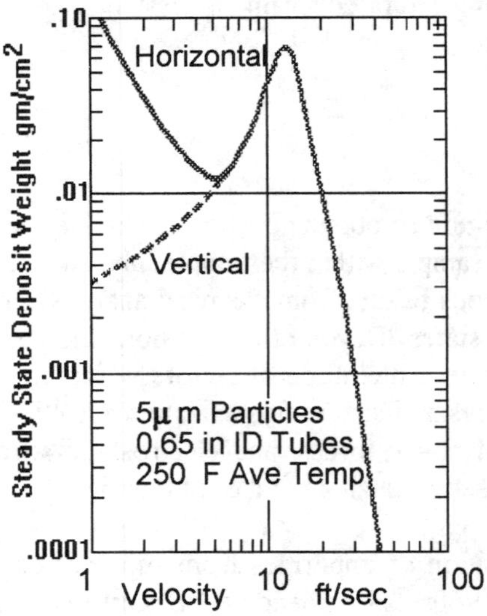

Figure 4 Influence of Fluid Velocity on Deposit Weight
Source: Reference 6

There have been recent discussions that steady state deposit weight is more dependent upon fluid Reynolds number than fluid velocity. This theory is presently advocated by several parties including one EPRI consultant. Since Reynolds number is proportional to the product of velocity times hydraulic diameter this theory would indicate that slower velocities could be tolerated in larger diameter lines. The author has been unable to find any published results showing a Reynolds number dependency.

FLOW RATE CHANGES

Changes of liquid sample flow rate result in velocity changes proportional to the flow rate. A change in flow rate of a steam sample will result in a change in the condensing length. Various regions of two phase flow will then shift along the sample line. Areas of tubing which have liquid flow at one flow rate will experience two phase flow at a higher flow rate. Calculations indicate that flow rate changes of 10 percent can cause velocity changes of a factor of two or three in regions near the fully condensed length.

Most regions of two phase flow have stable velocities. The liquid and steam velocities in annular and bubbly flow regions are both reasonably constant over time. Regions of slug and churn flow experience frequent changes in velocity as large bubbles pass. Studies of these flow regions show liquid movement constantly accelerating and decelerating. Two phase flow theory indicates that slug flow will originate when the steam volume occupies approximately 50% of the tube cross section. Void fraction data from Thom's (7) experiments show that this occurs at low steam quality and in the last fraction of the condensing length. Slight increases in flow will result in relocating the area of slug flow to that previously handling pure liquid. Reductions in flow rate will result in liquid flow occurring in the area of previous slug flow. Deposition theory maintains that there is a balance of deposit weight and velocity for equilibrium of the line. Changes of flow regime would upset this equilibrium and result in either increased erosion or provide for additional deposition.

STEAM AND WATER VELOCITIES

Conservation of mass requires that for a constant diameter sample line the velocity of the steam entering the line will be the ratio of the water density to the steam density (ρ_f / ρ_g) multiplied by the velocity of the condensate at the sampling panel. For some representative steam pressures this ratio is shown in Table 1. From the table it could be calculated that a steam velocity near 1500 ft/sec is required to produce a 6 ft/sec condensate velocity from a 100 psia steam sample. It is apparent that the velocities needed for lower pressure steam will not be obtained such that slower liquid velocities will result.

Table 1

Density Ratio of Water to Steam

pressure (psia)	ρ_f/ρ_g
100	250
200	124
300	82
400	60
500	32
1000	21
1500	12
2000	7.3
2500	4.6
2800	3.3

TWO PHASE FLOW MODEL

Since the composition of the sample is continually changing any analysis of the sample flow requires evaluation at multiple locations. The author developed a computer program which uses two phase flow theory to model the flow. The sample line is divided into increments of several inches in length. Fluid properties, pressure drop and heat loss computations are made at each increment.

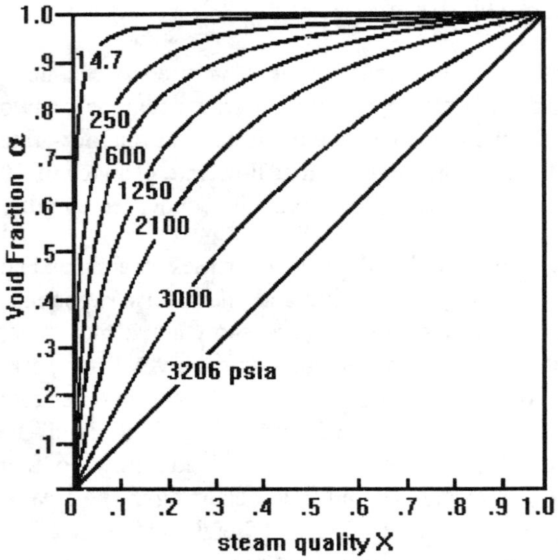

Figure 5 Void Fraction Versus Quality
Source: Reference 7

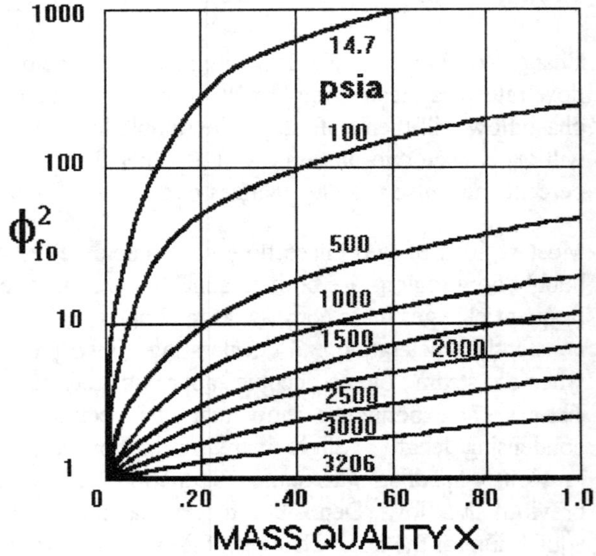

Figure 6 Two Phase Friction Factor Calculated Using Butterworth's Formula

Thom (7) used experimental data to determine friction factors for steam and water mixtures flowing in pipes. He established that steam flows significantly faster than the water and consequently occupies a smaller portion of the tube cross section than would be predicted using steam quality. Thom conceived a slip ratio which is the ratio of the steam velocity to the water velocity as a function of steam pressure or density. Later Butterworth (8) developed an expression for the void volume which fit Thom's data. Butterworth also produced an equation which fit Thom's pressure drop correlations. A two phase friction factor is used to determine two phase pressure drop due to friction as a multiplier of the pressure drop resulting from liquid flowing alone. Butterworth's equation is shown plotted in figure 6.

LINE SIZING

Table 2 summarizes a series of calculations performed to size steam sample tubing. Three relatively standard tube sizes were evaluated at nine separate pressures for three typical flow rates of saturated steam and three flow rates of superheated steam at each pressure. All sample flow rates refer to cooled condensate at a temperature of 77°F (25°C).

It is recognized that steam borne impurities deposit on the dry surface of the sampling tube within the superheated region. Therefore it is highly recommended that superheat be removed from steam samples within or as close to the sampling nozzle as possible. However, a large number of superheated steam samples are still obtained without condensate injection or other desuperheating mechanisms. Therefore tube sizes for superheated steam are included in the table. The sizes, lengths and velocities listed are based solely on mechanical considerations and should not be interpreted as endorsing the practice of handling superheated steam without a mechanism to flush contaminants from the tube wall.

Table 2 includes the following:

1. The maximum recommended length (max L) of tube for that pressure and flow rate. An N.R. in this column indicates that the size of sample tube is not recommended for any length.
2. The length where all steam has been condensed (cond L). An * in this column indicates that the steam was not fully condensed in the maximum recommended length.
3. The maximum velocity of the steam (steam V_{max}) in the recommended length. An • after the value in this column indicates that the steam velocity is increasing due to expansion. Values shown with the • then are taken at the maximum recommended length. The required initial steam velocity is shown for all values where the tube size is not recommended for any length.
4. The velocity of the condensate at a 500 foot length (cond V_{min}). Since the density of water varies with temperature any sampling tube shorter than 500 ft would have terminal velocities slightly higher than this value, longer tubes would have slower velocities. As above an * in this column indicates that the steam is not fully condensed in the recommended length or has not condensed within 500 ft.

SAMPLING GUIDELINES

The steam sampling rate should be established to maintain isokinetic sampling within the steam sample nozzle. The sample line must be adequately sized to deliver this flow rate but not significantly oversized. Oversized lines result in increased area for heat loss with rapid condensation of the sample. If the line is long the sample may fully condense with extremely slow liquid velocities resulting. Slow velocities promote deposition on the tube walls and cause a long time interval for the sample to reach the sample analysis station. Undersized tubing can result in choking of the flow with the maximum obtainable flow rate below the desired value. Choking results when the expansion of the steam due to pressure drop exceeds the volume loss due to condensation. The net increase in volume requires increasing velocities as the sample proceeds along the sample line. Increasing velocities are accompanied by increased pressure drop with increased expansion.

When the steaming rate increases the sample rate should increase to maintain isokinetic sampling at the sample nozzle. However, if the sample rate increases then the region of slug flow will move along the sample tube to a region which previously had single phase liquid flow. This change in flow regime will upset the equilibrium established within the sample tube and cause increased erosion of previously deposited material. Much of the deposits are porous which tend to sorb appreciable amounts of dissolved species. These species are released with particulate and colloidal material. The release of the dissolved species can cause fluctuations in the output of analysis instruments. Similarly decreases in steaming rate should be accompanied by a reduction in sample flow rate. This would result in a decrease of the condensing length and provide for additional deposition of material on the tube wall. Systems which have

Table 2
Recommended Sample Tube Sizes

			1/4" OD x .049" wall				3/8" OD x .065" wall				1/2" OD x .083" wall			
pressure psia	condition	sample flow cc/min	max L ft	cond L ft	steam V_{max} ft/sec	cond V_{min} ft/sec	max L ft	cond L ft	steam V_{max} ft/sec	cond V_{min} ft/sec	max L ft	cond L ft	steam V_{max} ft/sec	cond V_{min} ft/sec
2800	saturated	1300	>700	190	39	7.1	>700	120	15	2.5	>700	90	8	1.3
	"	1000	"	140	30	5.2	"	90	12	1.9	"	70	6	1.0
	"	700	"	90	21	3.4	"	60	8	1.3	"	50	4	0.7
(19 MPa)	1000°F	1300	>700	600	100	*	>700	250	39	2.7	>700	190	21	1.4
	"	1000	"	310	78	5.6	"	190	30	1.9	"	150	16	1.0
	"	700	"	200	55	3.5	"	130	21	1.3	"	100	11	0.7
2400	saturated	1300	>700	310	53	7.4	>700	170	20	2.6	>700	130	11	1.3
	"	1000	"	210	41	5.2	"	130	16	1.9	"	100	8	1.0
	"	700	"	140	29	3.5	"	90	11	1.3	"	70	6	0.7
(16 MPa)	1000°F	1300	150	*	120•	*	>700	300	47	2.7	>700	220	25	1.4
	"	1000	>700	400	93	5.9	"	220	36	2.0	"	170	19	1.0
	"	700	"	240	65	3.6	"	160	25	1.3	"	120	14	0.7
2000	saturated	1300	250	*	87•	*	>700	230	27	2.6	>700	170	15	1.3
	"	1000	>700	290	55	5.4	"	170	21	1.9	"	130	11	1.0
	"	700	"	180	38	3.5	"	120	15	1.3	"	90	8	0.7
(14 Mpa)	1000°F	1300	N.R.	-	150•	-	>700	350	57	2.8	>700	260	31	1.4
	"	1000	300	*	100•	*	"	260	44	2.0	"	200	24	1.0
	"	700	>700	290	80	3.6	"	180	31	1.3	"	140	17	0.7
1750	saturated	1300	150	*	140•	*	>700	270	33	2.6	>700	200	18	1.4
	"	1000	>700	420	66	5.6	"	200	25	1.9	"	160	14	1.0
	"	700	"	220	46	3.5	"	140	18	1.3	"	110	10	0.7
(12 MPa)	1000°F	1300	N.R.	-	170•	-	>700	400	67	2.9	>700	290	36	1.4
	"	1000	150	*	130•	*	"	300	51	2.0	"	230	28	1.0
	"	700	>700	340	93	3.7	"	200	36	1.3	"	160	19	0.7
1500	saturated	1300	N.R.	-	110•	-	>700	320	40	2.7	>700	240	22	1.4
	"	1000	180	*	110•	*	"	240	31	2.0	"	180	17	1.0
	"	700	>700	270	56	3.6	"	170	22	1.3	"	130	12	0.7
(10 MPa)	1000°F	1300	N.R.	-	200•	-	>700	450	79	3.0	>700	330	42	1.4
	"	1000	N.R.	-	160•	-	"	330	61	2.0	"	250	33	1.0
	"	700	500	470	110	3.9	"	230	42	1.3	"	180	23	0.7
1250	saturated	1300	N.R.	-	130•	-	>700	390	50	2.8	>700	280	27	1.4
	"	1000	80	*	130•	*	"	280	39	2.0	"	220	21	1.0
	"	700	>700	390	70	3.7	"	190	27	1.3	"	150	15	0.7
(8.5 MPa)	1000°F	1300	N.R.	-	250•	-	>700	550	96	*	>700	380	51	1.5
	"	1000	N.R.	-	190•	-	"	390	74	2.1	"	290	40	1.0
	"	700	150	*	130•	*	"	260	51	1.3	"	200	28	0.7
1000	saturated	1300	N.R.	-	170•	-	>700	500	65	3.0	>700	340	35	1.4
	"	1000	N.R.	-	130•	-	"	350	50	2.0	"	260	27	1.0
	"	700	N.R.	-	91•	-	"	230	35	1.3	"	180	19	0.7
(7 MPa)	1000°F	1300	N.R.	-	320•	-	400	*	120	*	>700	440	65	1.5
	"	1000	N.R.	-	240•	-	>700	470	93	2.2	"	330	50	1.1
	"	700	N.R.	-	170•	-	"	300	65	1.3	"	230	35	0.7
750	saturated	1300	N.R.	-	230•	-	150	*	98•	*	>700	420	48	1.5
	"	1000	N.R.	-	180•	-	>700	480	68	2.2	"	320	37	1.0
	"	700	N.R.	-	130•	-	"	290	48	1.3	"	220	26	0.7
(5 MPa)	1000°F	1300	N.R.	-	420•	-	N.R.	-	160•	-	>700	560	88	*
	"	1000	N.R.	-	330•	-	300	*	130	*	"	400	68	1.1
	"	700	N.R.	-	230•	-	>700	380	88	1.4	"	280	47	0.7
500	saturated	1300	N.R.	-	360•	-	N.R.	-	140•	-	450	*	73	*
	"	1000	N.R.	-	270•	-	N.R.	-	100•	-	>700	420	56	1.1
	"	700	N.R.	-	190•	-	500	440	73	1.4	"	280	39	0.7
(3.3 MPa)	1000°F	1300	N.R.	-	640•	-	N.R.	-	250•	-	300	*	130	*
	"	1000	N.R.	-	500•	-	N.R.	-	190•	-	>700	550	100	*
	"	700	N.R.	-	350•	-	200	*	130	*	"	350	72	0.7

changing sampling rates should maintain the condensate velocities at the 6 ft/sec value or above to reduce the amount of deposit subject to scrubbing from higher velocities.

The optimum sampling line is short such that full condensing will not occur. The higher velocities during two phase flow effectively scrub the tube walls preventing excessive deposition and plate out However, many sampling lines have lengths of 500 feet and longer. Frequently these long lines result in fully condensed flow. To reduce deposition in the sampling tube it is desirable to maintain the condensate velocity near 6 ft/sec. Since the tube size must be selected based on the steam flow the condensate velocity often is below the desired values.

Another preferred method is to condense the sample as near to the source as possible. Then size the condensate portion of the line to maintain the desired 6 ft/sec velocity. Due to the lack of sufficient cooling water near the sample source few plants have employed source cooling.

Saturated steam samples at pressures over 2000 psia can be transported with 1/4" OD sampling tube. Since high pressure steam samples have a lower condensing length these samples are frequently fully condensed within the lengths of normal sampling lines. However if sample flow rates are maintained near 1200 cc/min then the condensate velocities will be near the desired 6 ft/sec with 1/4" OD tubing.

Saturated steam samples at pressures of 750 psia to 2000 psia can be transported in 3/8" OD sampling tube at reasonable flow rates of 1000 - 1300 cc/min. To maintain a condensate velocity of 6 ft/sec a flow rate of 3300 cc/min would be required. This high flow rate would exceed the steam handling capacity of 3/8" tube and result in additional expense due to increased energy costs, increased cooling water requirements, larger sample coolers etc. The table indicates that the condensate velocity at normal sample flow rates is significantly below the recommended 6 ft/sec. Optimum sample handling would require either the sample tube be less than the condensing length, or a reduction in sample tube size to 1/4" OD tube be made beyond the condensing length.

Most saturated steam samples of 500 psia or less will require tube or pipe of at least 1/2" OD. These samples will have extremely slow condensate velocities if the sample is allowed to fully condense. As above either the length of the line should be less than the condensing length or a reduction in tubing size made to maintain reasonable condensate velocities.

When the sample is fully condensed before it reaches the sampling panel then selection of components at the panel will have very little effect on flow capacity. If the flow entering the panel is two phase then components upstream of the primary sample cooler should be selected based on handling steam. Components located downstream of the primary sample cooler will handle only condensate. When possible tubing and components should be sized to maintain the condensate velocity at six feet per second.

At some location upstream of the chemical analyzers any steam sample must fully condense. Subsequently the liquid velocity will either be slow, or the tube must be sized to result in an acceptable velocity. Calculations of sample coolers using the same program indicate that sample coolers of reduced tube size compared to the sample line can frequently handle the sample flow. This occurs due to a combination of effects. Reduced velocities result in the cooler since some condensing has already occurred in the sample line. If the pressure drop within the sample line is nominal then there will be a significant volume reduction prior to the cooler entrance. Secondly, high heat transfer coefficients on the outside tube surface cause rapid condensation. Most steam samples fully condense within the first six to eleven feet within the sample cooler. The most representative samples should result when samples are not fully condensed until within the sample cooler and the cooler and downstream tubing have been sized to maintain a condensate velocity of near 6 ft/sec.

Testing has substantiated computer calculations that indicate a 1/4" OD tube diameter sample cooler can condense slightly more than 500 cc/min of 115 psia steam. The limiting factor is that all available pressure drop is expended within the sample cooler. A thin wall, 3/8" OD tube cooler has approximately four times the tube cross section area so it should handle over 1200 cc/min at 115 psia. Larger tube sample coolers can provide additional flow capacity for low pressure steam; however the resulting liquid velocities will almost always be less than would be obtained using a 1/4" OD tube sample cooler. The selection of the optimum sample cooler depends on required flow rate, line size and length.

The following are methods which will increase the attainable flow rate of low pressure steam samples:

1. Size the sample line to deliver the sample with nominal pressure drop. Maintain the maximum steam velocity at 100 ft/sec or less.
2. Use large radius tube bends instead of 90° fittings and elbows.
3. Use a full port (ball or plug) isolation valve at the sample panel.
4. Maintain the sample line size right up to the primary sample cooler.
5. Minimize the number of tube bends upstream of the primary cooler.
6. Use a thin wall 3/8" tube sample cooler for primary cooling. (See other discussion of sample coolers.)

Insulation of sample lines may be required for personnel protection. The addition of insulation will reduce heat loss from the tube surface and increase the tube length required to completely condense. Insulation will usually result in a higher pressure drop within the sample line but may be a method to prevent the slow velocities resulting from a fully condensed sample. Evaluation of possible benefits of insulation is continuing.

It has been difficult to obtain operating data to verify theoretical results. The current table is considered an initial effort at sizing recommendations. Pressure drop and heat transfer models currently correlate reasonably with available field data. It is believed that satisfactory performance will result with lines sized per the table recommendations.

ASSUMPTIONS

The data reported in this paper is based upon computer analysis. Many assumptions have been implemented in this work. The most significant of these include the following.

Tubing is considered to be at the average inside diameter resulting from nominal sized outside diameter and average wall. (1/4" OD x .049" average wall has .152" average inside diameter) Commercial tubing is specified by outside diameter and average, or minimum, wall thickness. The standard tolerance on wall thickness is ±10% for average wall; +20%, −0 for minimum wall. For sizes typically encountered with sampling the tolerance on outside diameter is ±.004". Therefore the range of inside diameters for 1/4" OD x .049" average wall tubing is .1382" to .1658". This represents a variation in inside area of +19% to −17% from nominal without accounting for ovality. Conservation of mass requires that velocities for a given flow rate vary by the same percentage.

Local properties of steam and water (density, enthalpy, viscosity, thermal conductivity, Prandtl Number and surface tension) are obtained from values published in "NBS/NRC Steam Tables" (9). Interpolation of data at 80 pressure values was used to result in all calculated values within 1% for temperatures more than 15°F removed from the critical temperature. Additional data has been added in the critical temperature region to provide improved properties. However, due to rapidly changing values of Prandtl Number, conductivity and viscosity near the critical region some errors over one percent exist. All interpolations are based on properties at the five nearest values of pressure.

All properties are considered to remain at the inlet value for the entire increment length. Increment length used for these computations has been 4 inches. Comparison to calculations which used increment lengths of 2" and 1" did not show a significant variation. For a typical 4 inch length pressure drops are normally less than 0.2 psi. Within the condensing range incremental temperature changes are significantly less than 0.1°F with changes in quality less than .01.

Isothermal properties are assumed across the entire tube cross section. Literature demonstrates that for even low quality steam the majority of the area is occupied by steam. Since the water fraction is low the film thickness on the inside of the tube is thin. Steam velocities normally encountered in the sample tube are significantly above the entrainment velocity such that a large portion of the liquid will be entrained as droplets in the steam flow. The liquid droplets provide a large heat transfer area for temperature equilibrium. A high rate of droplet entrainment and reabsorption in the liquid film will result in the liquid and vapor temperatures remaining close.

Condensing heat transfer coefficients typically have large values. Therefore for sample lines normally encountered in power generation stations heat loss is governed by the outside surface. Heat transfer from the outside surface is by combined free convection and radiation. Free convection coefficients have been calculated for a horizontal tube with no adjoining obstructions. Radiation losses are considered for an oxidized stainless steel surface with an emissivity of .85. Ambient conditions are considered at 90°F for all conditions. Internal heat transfer coefficients were calculated based on annular flow with thin liquid films. Condensing heat transfer coefficients are determined according to Kosky and Staub (10).

Pressure drop calculations are based on experimentation done by Thom during the 1950s and 1960s. This testing was for pressure drop of steam and water mixtures during boiling. However, several noted experts in the field of two phase flow believe that this is the most accurate method currently available for steam/water mixtures during both condensing and boiling. Formulations used are based on curve fits by Butterworth. Pressure drop calculations are based on straight lengths and do not include localized losses for root isolation valves, elbows or tubing bends. The effects of insulation or heat loss from hangers have not been considered. Systems which will have high heat losses or pressure drops resulting from nonstandard designs should consider possible impact on the over all sample line design.

REFERENCES

1. EPRI Report No CS-5164, "Guideline Manual on Instrumentation and Control for Fossil Plant Cycle Chemistry", Final Report, April 1987
2. ASME Performance Test Code PTC19.11, "Steam and Water Sampling, Conditioning, Analysis and Leak Detection in the Power Cycle", Tentative Revision, Draft XI, dated October 1992
3. ASTM D1066-82, "Standard Practice for Sampling Steam", 1982
4. Coulter, E.E. "Sampling in Thermal Power Plants" Presented at the Electric Utility Workshop: University of Illinois at Urbana-Champaign, 1982
5. Jonas, O., Mathur, R.K., Rice, J.K. and Coulter, E.E "Development of Steam Sampling" EPRI Research Project RP 2712-8
6. EPRI Report No NP-522, "Survey of Corrosion Product Generation, Transport, and Deposition in Light Water Nuclear Reactors", Final Report, March 1979
7. Thom, J.R.S. "Prediction of Pressure Drop During Forced Circulation Boiling of Water", 1964 International Journal of Heat Mass Transfer, 7, pp 709-724
8. Butterworth, D.A. "A Comparison of Some Void-Fraction Relationships for Co-Current Gas-Liquid Flow" 1975 International Journal of Multiphase Flow, pp 845-850
9. Haar, L., Gallagher, J.S. and Kell, G.S. "NBS/NRC Steam Tables", Hemisphere Publishing Corporation, 1984
10. Kosky, P.G. and Staub, F.W. "Local Condensing Heat Transfer Coefficients in the Annular Flow Regime", 1971 AIChE Journal,17, pp 1037-1048

1995 POWID/EPRI AUTHOR INDEX

0074-056X/95/327-342/$0 + .50pp

Beta Testing of a Neural Network Based Process Optimization System

Brad Radl and Willie Roland

Pegasus Technologies Corporation
397 Manhattan Parkway
Painesville, OH 44077

Abstract

The process optimization system is application software that complements current performance monitoring technology. It is a state-of-the-art application of neural network technology using a combination of AI (artificial intelligence) software tools in a real-time environment. The beta test projects took place at Penn Power's Bruce Mansfield Station and Niagara Mohawks's Huntley Station. Bruce Mansfield Unit #3 is a 800 MW coal fired plant commissioned in the early 1970's. The furnace is a Foster Wheeler front and rear fired supercritical unit with 8 ball mills, 32 burners and 8 overfire air ports. Huntley Unit #65 is a 100 MW coal fired plant commissioned in the 1950's. The boiler is a B&W roof fired, drum unit with 4 ball mills and 8 burners.

The optimization system simultaneously reduced NOx emission levels and improved plant heat rate without extensive testing or large complex physical models. It does not replace performance monitors or engineering judgement, but is a tool that, when used properly, should generate target values that optimize power plant performance. This report covers results of the beta testing activities through March of 1994.

Project Description

The process optimization system is a unique combination of advanced AI techniques imbedded in a real-time software system. The core of the system is a neural network; however, it combines advanced statistical data processing and an expert system to condition data, recognize patterns and train the neural network. The optimization system represents a specific set of neural network concepts that are adaptable to modeling physical processes. These concepts have been extended by modifying the neural network architectures, making them more applicable to power plants. The system can be used to analyze many complex engineering problems and has a batch mode interface to accomplish this task. During the beta test demonstration at Penn Power's Bruce Mansfield Station, the dynamic targets were evaluated for controllable parameters to minimize heat rate while keeping NOx emissions within established guidelines. These goals were tested separately and simultaneously.

Utility data sets were analyzed off-line in the batch mode to verify the ability of the identified neural network architectures to model the plant heat balance and provide realistic recommendations. The data has generated models that accurately predict outputs, and can make recommendations that improve heat rate from 0.5% to 5.0%. Actual testing at the Bruce Mansfield beta test project indicated that heat rate improvements at full load are between 0.2% and 2.0%, with minimal penalty in NOx levels. The NOx emission level on Unit #3, which does not have low NOx burners, was easily maintained at 0.55 lbs/MMbtu and under. Under some circumstances, NOx emission have been reduced to 0.35 lbs/MMbtu.

On-line testing for Niagara Mohawk's Huntley station did not start until March of 1994. Preliminary results indicate that sliding pressure operation and better O_2 control will lower heat rate and improve CO levels. The degree of improvement has yet to be determined.

Heat Rate Improvement

The optimization system for Heat Rate Improvement is designed to be a complementary package to traditional heat rate monitoring packages. Traditional packages incorporate static models based on design conditions and correction curves. They can be complex and labor intensive. The process optimization system in contrast is based on neural network technology and is continually learning from the process it monitors. Many models are built, representing different 'patterns' of inputs which reflect different operating conditions. Conditions such as changing load levels, equipment out of service, poorly performing equipment, and environmental conditions are analyzed to determine the appropriate model to be used. The model developed is then run through many 'what if' scenarios to determine the best combination of controllable parameters to minimize heat rate. These recommendations reflect the tradeoffs that will occur from adjusting parameters. The recommendations are presented to operators in user friendly graphic displays providing feedback on how to reach the best achievable heat rate. This target heat rate is a dynamic target determined by the current operating status of the plant.

NOx Optimization

The Clean Air Act has augmented interest by utilities in how best to minimize NOx levels. In many cases, expensive retrofits are being made to meet the established guidelines. Once in place, many of these retrofits (e.g. low NOx burners) do not guarantee continued compliance. Sophisticated hard coded models can be built to accommodate lowering NOx levels. Factors affecting NOx include: O_2 levels, burner tilt positions, burner balancing, fuel/air damper settings, mill bias and fuel composition. These factors also affect: unburned carbon, steam temperature, opacity and slagging. Trying to model all these factors with traditional methods takes a tremendous amount of engineering time and money. The optimization system, because it empirically models data, is a tool that can quickly provide insights to the optimum combination of parameters. In addition, the system is continually learning, in contrast to static, design-based models. Consequently, as equipment degrades, slagging occurs, etc. the system is able to identify shifts in these combinations of parameters and makes recommendations to lower NOx emissions.

System Description

The process optimization system was initially installed on a dedicated SUN IPX Workstation at Penn Power's Bruce Mansfield Unit #3 during September 1993 and on a Foxboro I/A AP processor at Niagara Mohawk's Huntley Unit #65 in December of 1993. Bruce Mansfield Unit #3 is a 800 MW coal fired plant commissioned in the early 1970's. The furnace is a Foster Wheeler front and rear fired supercritical unit with 8 ball mills, 32 burners and 8 overfire air ports. Huntley Unit #65 is a 100 MW coal fired plant commissioned in the 1950's. The boiler is a B&W roof fired, drum unit with 4 ball mills and 16 burners

Data Link

A RS-232 data link connecting the optimization system to the Honeywell TDC 3000 gateway module has been installed. The link has access to all the data on the DCS loop for Units 1,2 and 3. The system was originally installed on Unit #3. Utilizing the RS-232 ports on the SUN workstation and the Honeywell system, data transfer was accomplished in two weeks. Improvements to the link software currently allow the read of over 200 points, analog and digital, per minute. The communication link utilizes very little CPU time on the SUN IPX and the number of points processed by the system could be increased to 2000 points per minute or more.

In response to the variation in process data collected at one minute intervals, averaging has been implemented. This allows the user to define a time period to average collections over. Currently, collections are made once a minute and averaged over 5 minutes.

Enhancements added since initial installation include: passing data back to the DCS and splitting incoming data into four separate data bases. There is one data base for each units real-time model and an expanded engineering data base for parallel, off-line analysis.

At Niagara Mohawk's beta test site, the process optimization system was installed directly onto a Foxboro I/A AP 50 applications processor. Utilizing the library of I/O routines available with the AP 50, communication occurs between the optimization system and a resident data base on the processor. The processor to loop communication occurs in the Foxboro system; therefore, requiring no additional hardware.

Real-Time Optimization

The real-time optimization has been running since early October. On Unit #3 it drives one top-level operator display for positioning the overfire air ports and adjusting O_2. Continued testing with different I/O lists and parameters to be optimized have not affected the ability to generate on-line recommendations.

CPU loading (utilization) generated by running the on-line optimization has been a problem. In particular, the number of inputs allowed to vary during an optimization and the number of outputs to optimize have significant impact. Increasing the number of inputs has an exponential effect on the optimization's execution time. The impact of the increased number of outputs is linear.

Originally, the plan was to allow only three to six input parameters to vary. Because more parameters are controllable and because some need to vary in order to reflect the changes in other parameters, this list has been as large as fifty inputs. When running with ten varying parameters the optimization executes in less than one minute. When running with fifty floating parameters the optimization takes over twenty minutes to complete.

These numbers are approximate and vary depending on such factors as: 1) other tasks running and 2) how close to optimum the system is currently running.

The outputs to be optimized were originally net unit heat rate (by input/output loss method) and the average NOx signal. To assist testing and validate some test results, additional parameters were added: unit heat rate (boiler/turbine coupling method), boiler efficiency, turbine efficiency, firing rate error and the individual NOx ppm signals. Effectively these increased loading linearly, so that an optimization for two outputs might take one minute, but for eight outputs the run time would be four to five minutes. Of course, if the two outputs took twenty minutes, then eight outputs effectively took forever, (i.e. two hours plus not being real-time). Fortunately, testing revealed that the additional outputs investigated are not necessary. However, new outputs such as CO and perhaps a term to quantify slagging are to be added to the list when they become available.

Real-Time Training

Testing revealed the importance of up to date trained models, consequently, the on-line periodic training was implemented. The time to train a model in the on-line mode is taking 5 to 8 minutes for 88 inputs, 2 outputs and over 100 passes of data. Since, it takes 10 to 20 data sets to recognize new relationships, a training cycle time of less than 2 hours is expected. However, the dynamic nature of NOx control has raised questions concerning this assumption.

Unaddressed issues related to training the neural network are: 1) how to weight new data versus old data and 2) distinguishing between changing setpoints due to system drift and changing setpoints due to changing conditions. In addition, there is a conflict between the time period of data required for NOx optimization and heat rate improvement. The optimal setpoints for NOx control can vary from day to day, such that leaving a given setting based on old data can create a 10% penalty in NOx levels within 24 hours. Consequently, data older than 24 hours should carry less weight. In contrast, heat rate is affected more by the equipment in service and ambient conditions which do not vary dramatically day to day. Therefore, when analyzing heat rate, data retained over a long period and reflecting various conditions provides the information necessary for pattern recognition to choose the appropriate model and minimize heat rate accordingly.

An algorithm under consideration would allow degradation of only those portions of the neural network which are based on recent data. The portions of the neural network based on old data would remain essentially unchanged.

User Display System

A top level display has been completely integrated with the real-time optimization and training. To date, only one display for the overfire air damper positions and O_2 has been implemented. Additional operator displays will be implemented on the DCS system.

The menu system facilitates I/O list changes made by plant personnel, as well as, timer adjustments for data collection and averaging, optimization cycle and training cycle can also be accomplished from the menu system.

Neural Network Configuration

Many combinations of node arrangement and functional link enhancement (FLN) terms have been tested. For modelling heat rate at Bruce Mansfield, it is important to maintain 25 to 35 hidden nodes. For Niagara Mohawk's models the number of hidden nodes may be as high as 115. The ideal number varies with the particular list used, but ranges from a 1/1 ratio up to a 3/1 ratio based on the number of inputs. For NOx, it is critical to have at least one layer of FE (sin/cos type nodes) terms. Without at least one FE layer, the NOx is modelled only to +/- 20% and misses most peaks and valleys. A second layer is sometimes desirable depending on the input list. In some cases, this second FE layer generates an overfitting of data. For both NOx and heat rate, the effect of the JA (polynomial type node) term is marginal. It does seem to help in some areas, but adds significant CPU loading to the system.

When time allows, more predictive testing of the neural network should be done to verify the long term stability of the network to accept continual training and expansion of the model set. Testing to date, indicates that the underlying neural network structure remains stable even while the weights for the individual nodes continue to evolve.

Test Results - Penn Power, Bruce Mansfield Station Unit #3

Testing was conducted under two different scenarios. The first scenario is when the unit is held at steady load and parameters are purposely varied to generate a range of data for developing the model. Then recommendations from this 'calibrated' model are tested to see if the heat rate and NOx respond per the model. The second test scenario involves using only on-line data collected during normal operation of Unit #3 and making recommendations even while the unit is varying load by +/-40MW. Obviously, under the later conditions, results of testing are much less certain.

Testing so far has only been extensively done at high loads, extended controlled testing at lower loads need to be done to determine the gains involved.

O_2 Testing

The process optimization system has resulted in several findings related to the O_2 setpoint.

First, initial testing recommended a setpoint of 4.0% for O_2, which was significantly different than the controllable parameter value of 3.2% Increasing, the O_2 bias to generate 4.0% in the control system resulted in lowering heat rate 50 to 60 btu/kWh. After retesting confirmed this result, the engineer checked the O_2 probes calibration and found one sensor to be off by approximately 2%; therefore, biasing the average signal by at least 0.5%. After calibrating the probes, it took the optimization system two hours to collect sufficient data to determine a new setpoint between 3.2% and 3.5% (depending on conditions).

Testing two weeks after the initial finding, generated poor results for the O_2 setpoint. While initial testing generated an O_2 model with a high linear correlation coefficient (>0.5) and consistent results, testing in this phase had a low correlation coefficient (approximately 0.2) and recommendations were often in error. This was traced to two problems. One problem was data for this run came from normal operation, which is a +/- 40MW load swing, because the unit is on automatic dispatch.

Load variance affects the O_2 value in a couple of ways. One, when load drops the control system takes time to catch up, therefore the O_2 value jumps up; at the same time the heat rate will jump down from the latent heat in he boiler and lag time in the combustion control system. When the unit reaches steady state (even for a few minutes) the O_2 will drop back to its value and the heat rate will climb to a plateau higher than before, (reflecting the lower efficiency of lower loads). When load increases, the exact opposite effects happen. Consequently, when the unit is cycling within a band of unit load, there is a cycle time for O_2, and a cycle time for heat rate. The O_2 and heat rate cycles are not synchronized. This causes the low correlation for O_2 versus heat rate, which in turn results in an increased possibility of error in recommendations. In particular a high recommendation of approximately 6% for O_2 occurred, probably due to extrapolation.

Methods to correct this problem include time averaging of data and the introduction of lead/lag terms. Testing with averaged data has improved the O_2 modelling but some inconsistencies still exist.

Another problem with the O_2 setpoint and control signal appears to be stratification of air flow past the four O_2 probes that make up the average O_2 reading. To some degree, optimization system appears to pick this up by changing the setpoint recommendation +/- 0.2% depending on the values in the I/O list. It appears to be related to the overfire air ports and/or the mills that are running. Further testing to sort out the effects can be made or it may be desirable to await the installation of eight new O_2/CO probes.

Testing of O_2 versus NOx confirms the straight forward relationship, that when O_2 goes up so does the NOx reading and vice versa. Consequently, it is a matter trading off heat rate versus NOx in setting up what to optimize priorities.

Overfire Air Ports

Testing has revealed one not so obvious relationship and one that was intuitive, but not tested.

The surprise finding is that when the unit is running with a high heat rate above 10,000 btu/kWh, opening the overfire ports can actually lower heat rate and NOx simultaneously. Testing has shown the following:

Affect of opening overfire air ports ~5% from ~0%

Heat Rate (Input/Output)	Heat Rate	NOx Level
Above 10,000 btu/kWh	Lower (~40 btu/kWh)	Lower ~15%
9,500 to 10,000 btu/kWh	Indeterminate	Lower ~15%
Under 9,500 btu/kWh	Higher (~40 btu/kWh)	Lower ~15%

The above numbers require further testing and are definitely affected by day to day operating conditions. Several physical explanations are under discussion. The addition of the CO signal may help the understanding of the above results.

In every case, opening the overfire air ports beyond an average of 10% resulted in heat rate degradation. Increasing the opening resulted in lowering NOx, at least, 50%.

A finding more in line with the operating knowledge is that NOx levels are affected by which overfire ports are open and by how much. Under controlled testing it appeared that by adjusting ports individually a decrease of NOx by an additional 10% may be achievable without penalizing heat rate. Further testing is still required for confirmation. One difficulty with modelling the overfire air ports, is the fixed position of the air port dampers over long periods of time. Without purposeful variation in the ports, the models have low correlation results and sometimes conflicting results. The results can be particularly misleading if the ports have been in one position for 12 hours and then altered at around the same time a change in the fuel's higher heating value (HHV) occurs, since the neural network has no means to separate the step change in HHV and overfire air ports with the change in the plateau for the heat rate.

One option being considered is to have the optimization system directly control the overfire port positions via the DCS from the closed position to 35% open. This would allow the NOx and heat rate to be minimized without continual operator adjustments to the overfire air ports. At the same time, the responsiveness would generate more data for better modelling.

Also, the change in one mill flow rate because of the temperature split did not explain the large 200 btu/kWh change due to a stuck feeder. This would suggest a differential temperature alarm, or no coal flow alarm on the feeders may be desirable, providing an earlier alarm signal and maintaining a lower heat rate. In all cases, a setpoint of 175 degF should be maintained more consistently. Once, this is done a refinement of more precise setpoints per mill may become appropriate.

Because of the problems associated with the input/output method, the above tests will be repeated using the boiler/turbine coupling heat rate as an additional optimization target.

Primary Air (PA) Pressure

Testing with the optimum value for the primary air pressure generated mixed results. Tests were run three times, adjusting the pressure to the value recommended by the optimization system. In one case, the heat rate was lowered ~50 btu/kWh, with the heat rate inflection point directly related to the time PA pressure was changed and the full gain achieved in 30 minutes. This was repeated 4 times during the test run. During the second test, the heat rate improved only 20 btu/kWh and the inflection points were not as well correlated. The results of the last test complicated the issue more, the heat rate either had no response or went in the opposite direction by 20 btu/kWh. As with the other parameters, the lower the correlation coefficient the less accurate the recommendation would be. In the case of the PA pressure a correlation below 0.6 results in questionable recommendations. With the PA pressure no purposeful variations have been made to generate a more valid model. This should be done in future testing.

With regard to NOx, the optimum PA pressure setpoint for NOx is 1 to 3 psia lower than for the optimum setpoint for heat rate. Fortunately, they overlap in an area where a significant penalty need not be incurred by either parameter.

Other Parameters

Other parameters which the optimization system has identified as being in off optimum setpoints, but which are not easily affected include:

Ball Mill Outlet Temperatures

Testing with the ball mill outlet temperatures has generated dramatic results which initiated investigating the validity of the heat rate calculated via the input/output method.

Early on it was found that the ball mill outlet temperatures tended to drift away from their setpoint of 175 degF and several would fall into a range from 140 to 165 degF. The optimization system models consistently showed that raising these values to the 175 to 185 degF range would result in a 200 to 400 btu/kWh improvement in heat rate.

During the first test, raising the ball mill outlet temperatures decreased the heat rate 550 btu/kWh. Because of the large drop, it was assumed a fuel change (to a higher HHV content coal) was in progress and testing was suspended.

Testing two weeks later showed several ball mill outlet temperatures had drifted to the 155 to 170 degF range. Adjusting the setpoints to raise the outlet temperatures up to the 175 to 185 degF range resulted in the heat rate dropping 400 btu/kWh, which was in line with the predictions of the models. During the later part of the test the heat rate started to climb by 200 btu/kWh. An investigation showed ball mill #3 had a feeder problem and this was reflected by the temperature on side A increasing to above 190 degF and side B decreasing to below 140 degF. This was brought to operations attention who just received the high temperature alarm on side A. Once the feeder problem was corrected the temperatures returned towards normal, and the heat rate again dropped towards the 400 btu/kWh improvement. The feeder stuck again and the cycle repeated itself.

Because of the huge changes in heat rate, the performance engineer reviewed the input/output method calculation and the inputs, particularly those directly related to the ball mill. His investigation found that the flow rate was uncorrected for temperature and water content. Since the moisture content was unknown he could only back calculate the affect of temperature. For the temperature changes made, the error in the flow signal could have accumulated 3% error in the resultant calculation, therefore accounting for the majority of movement in the heat rate value. Consequently, while optimization system modelled the data properly, the input data was misleading.

With the assumption that the moisture content was reasonably consistent and not at an excessive value over the course of testing it does suggest that the heat rate was improved 30 to 150 btu/kWh, while NOx increased no more than 5%.

Superheat Attemperation Flow
Feedwater Inlet Temperature
Furnace Exit Temperature
Furnace Pressure
Secondary Air Flow Balance

Of these the secondary air flow balance would be the most significant as it would appear to allow for a 5% to 20% reduction in NOx with little or no heat rate penalty. Since the variation is strictly from process variance, this has to cautiously interpreted. On the plus side, this is consistent with off-line data from plants that are able to adjust flows.

The Superheat Attemperation Flow and the Furnace Exit temperatures appear to both be high with penalties in the range of 20 to 50 btu/kWh. This is appears to be driven by slagging as the penalty has been increasing over time. This suggest that these might be used in a soot blowing type model or perhaps as slagging indicators.

Main Steam and Reheat Steam temperatures generally reflected no correlation (when the data is for steady loads) or when correlated match up to the design setpoints. It will be interesting to see how they match up at lower loads.

Results of Testing

1) The optimization system models well for data provided under the following conditions:.

   - Recommended setpoints vary with time and unit load.
   - Models require 2 to 3 hrs to 'calibrate' after changes to instrumentation.
   - Models require variation in inputs to model effectively.
   - Time averaged data improves correlation between inputs and outputs.
   - Dynamic setpoints for O_2, ball mill outlet temperature, primary air pressure and overfire air dampers have been tested at full load conditions with moderate success.
   - NOx and heat rate improvements are not contradictory goals under all circumstances.

2) Data provided the optimization system requires extensive review.

- Ball mill outlet temperatures affected the input/output heat rate and error term in calculation of heat rate, making conclusions on degree of improved performance difficult. Improvements in heat rate have been made but we cannot verify how much because the classifier differential pressure signal drifts with temperature and moisture.

- Large variations in minute to minute heat rate calculations (largely unexplained) force the system to accumulate larger quantities of data to filter out process noise.

- The optimization system has identified parameters affecting NOx and heat rate which are not easily changeable under automatic control or have no control at all (e.g. feedwater outlet temperature, furnace outlet temperature, air flow balancing).

- Without CO measurement unable to quantify affect on unburned carbon.

- Have no direct feedback as to how these operating changes are affecting the slagging rate, other than visual inspection. Experimenting with ball mills has shown that leaving upper levels on while forcing air through lower burners leads to cleaner combustion. NOx increases dramatically, but can be controlled through overfire damper settings to required levels.

- With the unit on automatic dispatch, continual load variations introduce lead/lag effects into data sets which constantly affects the heat rate value, diminishing the ability of testing to quantify results.

3) The optimization system requires more horsepower and completion of non-unix interfaces.

- With large data sets and multiple outputs, manual runs of data sets take too long. (up to two hours).

- For one unit model running in real-time, optimization can be done every 10 minutes and training at least once an hour. However, the current CPU would be overloaded for three units running simultaneously.

4) The optimization system has provided the ability to lower heat rate and control NOx.

- Given the limitations of the current heat rate calculations, the optimization system does model data provided and points out directions to improve heat rate. Without resolution of data problems, can not quantify improvement. Range of improvement could be from 10 to 200 btu/kWh, with expectation of 20 to 50 btu/kWh most likely for 2 to 5 parameters being controlled under normal circumstances.

- NOx can be held down to 0.50 lbs/MMbtu at full load. However, insufficient data is available to identify tradeoffs versus slagging and to a lesser degree heat rate.

5) The optimization system is useful as general tool for engineering and performance engineers.

- Useful to identify instrumentation problems. Through use of the system, problems with O_2 probes were identified and corrected. Imbalances in secondary air flow have been highlighted.

- Useful to look at correlation of large amounts of data. In particular, have loaded coal analysis for previous month to identify key parameters necessary for on-line analyzer to provide input for heat rate calculations.

- Useful for model validation. The heat rate model is effectively being validated by verifying the accuracy of the systems recommendations and checking for basis of deviations. (i.e. actual heat rate change, instrumentation change or assumptions from calculation).

- Will want to load data sets applicable to particular pieces of equipment, for purpose of correlating data to plant performance, equipment maintenance and signal evaluation.

6.) Enhancements to the optimization system have been required to support the real-time system.

- Constraints on inputs, both hard limits and equations, have been added to keep an optimization run from extrapolating. This also is used to ensure recommendations which do not violate operating practices.

- A variable desirability curve has been implemented which lets the desired targets for NOx, CO and heat rate (et. al.) vary with load or other specified parameters.

- Training needs to take into account differing periods of learning required for different outputs.

Conclusion

The optimization system is currently making useful setpoint recommendations, and identifying data and instrument problems, which can lead to NOx and Heat Rate improvements. The system is not ready to be put on-line for use by operations. Significant engineering remains on how to handle the limitations of the heat rate calculations. NOx improvement can be made, but without measurement of the tradeoffs, these changes may not be desirable. Inclusion, of a CO signal is in the spring of 1994, will facilitate this testing. Also, NOx improvements are limited by the inability to adjust air flows.

References

1) Yoh-Han Pao, Fellow, IEEE,"Process Monitoring and

Optimization for Power Systems Applications." Invited paper for the IEEE World Congress on Computational Intelligence, Orlando, FL for June, 1994.

2) CAD-CHEM, a neural network based chemical product formulator (product literature and descriptive material), AI WARE, Cleveland, OH.

3) F.W. Bauer, M.E. Weiss, "Approach to Utility Boiler Acid
Rain Compliance Modifications." , Proceedings of the 1992 EPRI Heat Rate Improvement Conference, Birmingham, Alabama, November, 1992.

4) "Functional-Link Computing: Theory, System Architecture,
and Functionalities." Y.H. Pao and T. Takefuji, Computer Magazine, Vol. 3, pp 76-79, 1991.

5) "Learning and Generalization Characteristics of the Random
Vector Functional-Link Net." Y.H. Pao, G.H. Park, and D.J. Sobajic, Neurocomputing, Vol. 6, No. 1, Elsevier Press, 1994.

6) "Mathematical Concepts Underlying the Functional-Link Approach." Y.H. Pao and B. Igelnik, Proceedings of the World Congress of International Neural Network Society, San Diego, CA, June, 1994.

0074-056X/95/343-350/$0 + .50pp

CHANGES NECESSARY TO THE IMPLEMENTATION OF MODERN CONTROLS IN NUCLEAR SAFETY SYSTEMS

D. L. Damon
Manager of Engineering Technologies
Research & Development
Bechtel Corporation
50 Beale Street
San Francisco, CA 94105-1895

KEYWORDS

Digital Controls, Nuclear Safety, Nuclear Licensing, Power Plant Life Extension

ABSTRACT

There is widespread agreement that upgrading from analog to microprocessor-based digital controls can enhance the safety of operating nuclear power stations. However, modern control technology has evolved faster than the nuclear power industry's ability to assimilate it for safety applications. This has manifested itself in terms of licensing uncertainties and excessive costs. The licensing uncertainty is created by a perceived lack of definitive regulatory requirements, primarily in the areas of software reliability, common mode software failure and equipment qualification. The excessive costs are created both by the cumbersome regulatory approval process and by the use of highly specialized equipment developed specifically for nuclear safety applications. The net result is an increasing hesitancy on the part of utilities to implement digital upgrades. This creates an anomalous situation whereby the very standards and processes initially intended to enhance safety in nuclear power facilities now seem to inhibit the use of modern digital technologies, which offer the potential for improvements in performance and reliability. This situation is as untenable for the regulators as it is for the utilities and is in need of correction. A more efficient method is needed to secure timely regulatory understanding and acceptance of technologically complex and multifaceted issues. It is unlikely that this is possible under the existing fractionalized docket-by-docket reviews of the subject matter, or by the relatively slow industry-based committee approach to guideline development. A new forum is needed where all interested parties can proactively express their concerns, offer suggestions and efficiently develop mutually acceptable generic methods. This paper reviews the history leading up to the current situation, examines the root causes, and outlines a utility-endorsed program plan for resolution of Nuclear Regulatory Commission (NRC) concerns. It also addresses prerequisites to application of modern controls technology to operating units and its application to advanced light water reactor design certification.

BACKGROUND

While there is widespread agreement between regulators and plant owners that upgrading from analog to microprocessor-based digital controls will enhance the overall safety of operating nuclear power facilities, it is also evident that digital systems introduce new equipment (microprocessors) and new dependencies (software). This new equipment and its dependencies have created concerns within the Nuclear Regulatory Commission (NRC) that have resulted in unreviewed safety questions (USQs). When USQs are at issue, NRC review and approval is required prior to implementation. This initiates an expensive and largely indeterminate docket-based resolution process. The absence of clear regulatory criteria specific to digital systems, together with a lack of reasonably well-defined methodologies by which these criteria can be met, has created an uncertain licensing situation. Utilities are understandably hesitant to pursue digital upgrades under these conditions, despite the potential for safety enhancement. Also to be considered are the excessive costs of highly specialized, specifically dedicated equipment which, by its very isolation from the commercial sector, can lead to the same kind of obsolescence characteristic of the existing systems.

These developments may be contrasted with other industries, including fossil power, where the transition to superior digital technologies has occurred without incident. Even in the nuclear power industry, microprocessor-based digital controls have been used successfully in non-safety applications. There have also been some isolated applications in systems related to safety, such as Core Protection Calculators, Safety Parameter Display Systems (SPDS), and various vital power source load sequencers. To support these implementations, many guidelines were developed. However, since none of these applications exercised control over the Emergency Core Cooling Systems (ECCS), design criteria unique to such systems never became a question.

The implementation of a Foxboro Spec-200 system at Connecticut Yankee's Haddam Neck station initiated a series of regulatory challenges to digital upgrades attempted under 10 CFR 50.59, the NRC regulation governing changes to licensed nuclear facilities. In response to that situation, the NRC drafted a Generic Letter to alert other utilities to the NRC position, that any digital upgrade of a safety-related system involved USQs, and therefore required prior approval. Understanding of the NRC's concerns first requires an understanding of the safety related design criteria that gave rise to them.

DESIGN ISSUES

There are multiple systems in nuclear power generation providing what is termed *defense-in-depth* against accidents. These multiple systems employ conservative design criteria that include requirements to protect against natural phenomena, such as earthquakes and electromagnetic/radio frequency interference (EMI/RFI), to ensure operational safety. In addition, extended qualification testing and rigorous assurance procedures, including software verification and validation (V&V), is required to reduce the probability of equipment failure to the lowest level possible. Completely redundant systems are employed to further reduce the statistical probability of failure. To achieve this, the redundant systems cannot be mutually vulnerable to an external hazard. Such vulnerability is termed *common mode failure* and is a key element of nuclear design.

Common mode failure can occur with both hardware and software. Redundant hardware can be physically separated and isolated from the common hazard. The redundant hardware can also be of different type or manufacture to achieve what is termed *diversity*. Redundant safety systems using analog controls have been a successful and accepted practice since the beginning of the industry. Diversity was not required, however, since it would have introduced an unrewarding level of complexity. As a result, separation and redundancy from a common hazard was considered sufficient.

Microprocessor-based digital controls systems, however, have control codes that are common to the redundant systems, . Since this code cannot be so easily separated, it creates the problem of *common mode* software *failure*. While rigorous V&V techniques can enhance the reliability of the code, such approaches are quality assurance techniques and do not address separation requirements. To demonstrate statistical reliability equivalent to the redundant and separated analog systems, one would have to prove that V&V achieved a million-to-one probability against failure. For the types of process systems we are discussing here, this has proven to be a somewhat foreboding task. These factors have combined to create the USQs presently postulated by the NRC.

EVOLUTION OF CONCERNS

Following the effort at Haddam Neck, an implementation at TVA's Sequoyah station consisted of a more or less programmatic review of software reliability and V&V. Common mode software failure, equipment qualification, etc., seemed incidental issues by comparison. This was not the case, however, with the subsequent implementation at Commonwealth Edison's (CECo) Zion station. On the surface, it would appear that Zion, which used essentially the same equipment as Sequoyah, could have been addressed by the same level of review. Instead, the NRC conducted a relatively rigorous "thread audit" of the software. Also, unlike Sequoyah, CECo found it necessary to prepare substantially detailed responses on common mode software failure, equipment qualification, susceptibility to EMI/RFI, separation criteria, grounding, power quality, testability, etc. At Zion, it became apparent that the NRC was getting into issues which transcended existing guidelines. This caused Zion to be more costly and protracted than originally anticipated, creating a significant cost/benefit question for the utility. This question was apparently answered by a return to old technology in a later refurbishment. In that case, rather than implement digital Programmable Logic Controllers (PLCs) in the Zion emergency load sequencing system, CECo elected to simply replace old equipment with similar analog relays.

Pacific Gas & Electric's (PG&E's) Diablo Canyon, which used essentially the same equipment as Sequoyah and Zion, was the next docket. At Diablo, the NRC took notice of the Zion and Sequoyah V&V reviews and shifted their emphasis to common mode software failure. This required a demonstration of adequate diversity to establish defense-in-depth. The NRC's Safety Evaluation Report (SER) defined adequate defense in depth as:

> Defense-in-depth is considered by the staff to be a combination of design features incorporating overlapping and redundant capabilities such as diversity, redundancy, reliability, and performance in order to compensate for the postulated safety system weaknesses. For a defense-in-depth approach to be effective, the echelons of defense must be available to provide a backup function when faced with postulated failures. In determining when an adequate level of defense-in-depth has been achieved, an evaluation is performed of the degree of interdependence that is acceptable and what means are available (for example surveillance, maintenance, quality, and reliability) to maintain an adequate level of safety.

Definitions, such as that given above, are actually statements of general objectives and present more questions than they answer. The following are examples of questions likely to arise for a utility considering a digital safety system upgrade:

n What regulatory guide outlines the parameters of acceptable defense-in-depth?

n What constitutes adequate diversity?

n At what point does dissimilar equipment and software add such complexity that it produces a diminishing return?

n If dissimilar systems provide conflicting information, which one takes precedent?

n Where is this clearly defined for redundant safety systems?

n What are the acceptance criteria?

n How can a utility prepare in advance to demonstrate sufficient defense-in-depth and diversity?

Use of this definition at Diablo lead to a detailed review at the electronic component level to establish that the equipment evaluated under the Diablo Canyon SER was sufficiently "diverse." In doing this, components were assessed that had long been qualified for analog systems. Even with the results of the Diablo Canyon SER in place, it remains unclear what constitutes an appropriate "combination of design features" for a digital safety system.

In the more recent evaluation at D.C. Cook, which used the same equipment as at Haddam Neck, the NRC accepted reliability of the software on the basis of the previous review. This acceptance of precedence is warming, at least, to all except those who paid the initial price. But, it also has the effect of excluding other manufacturers and the "vendor capture" problems of high cost, narrowed support base, and specialized equipment. Consistent with previous reviews, however, the Cook SER introduced some new issues, such as an assessment of the "interfaces of the new digital equipment with the old equipment or circuits" and configuration control. Diversity and defense-in-depth were not mentioned. What is most important, however, is not the answers to these new issues but that fact that they seem to arise on a docket-by-docket basis. As long as this is the case, and in the absence of definitive generic answers to the kinds of questions posed above, nuclear power utilities will simply avoid the use of modern controls technology. This is now becoming a major concern to the entire industry, not only for operating units, but also for Advanced Light Water Reactors (ALWRs) which will have to address the same matters in their first-of-a-kind design certifications.

SITUATION ASSESSMENT

The issues on each docket seem to be raised in an exploratory and somewhat unpredictable manner, indicating that difficulty exists in formulating the necessary positions, requirements, and acceptance standards by which reasonable guidance is possible. At least two major issues are evident.

First, the assessments seem to be based on an analog rule structure that, while well established, is no longer suitable. Clearly, digital technology cannot be objectively assessed by analog-based rules. New rules are necessary and they are not in place. Uncertainty is the result and utilities are naturally, and properly, hesitant. Since technology is changing rapidly, a new method of revising the rules must be created to keep pace. Moreover, the method must be proactive in nature. The current approach of developing the necessary rules on a trial and error or docket by docket basis simply will no longer work.

Second, it seems that the digital systems are required to meet higher standards than the analog systems they replace. The high reliability required of analog systems was necessary primarily because they were neither self-diagnostic nor fault tolerant. This is not the case with digital systems. Such capabilities should be factored into a new rule base that will allow the overall improvements in safety to be the determinant. These considerations did not seem to be in evidence in any of the SERs outlined above. What is worse, many of the issues presented in these SERs were not unique to digital technology. If the utilities sense that each attempt at improvement is considered an opportunity to review the operating license, such improvements will be avoided.

The NRC itself, has strong incentives to see these issues generically resolved as soon as possible. After all, its basic charter is to regulate the operation and development of nuclear power in a manner that protects the health and safety of the public. In pursuing this charter, it would seem that the NRC should embrace every technological evolution that would improve operational safety. However, if the use of modern technologies is frustrated by expensive docket-based reviews that use a rule base inappropriate to the need, and this leads to new requirements in the absence of stated guidelines, then compliance with the basic charter becomes difficult. It creates a situation where the approach initially intended to ensure operational safety now appears to subvert its enhancement.

ROOT CAUSE ANALYSIS

In addressing this matter one issue at time, the industry seems to be dealing merely with effects rather than root causes. Several essential questions arise. Why did the present concerns come into being so long after the initial digital systems were implemented? Why wasn't a rule base and guidance appropriate to modern technology in place before the actual needs were there? Why is it that such concerns are not shared by other industries that have been using digital controls for many years? Why can't the experience and requirements of other industries be translated to nuclear power? Why must the newer technology demonstrate a level of safety higher than that of the technology being replaced? How can the regulators effectively regulate if they haven't published regulations sufficient to the needs? Such questions make it obvious that microprocessor-based controls technology developed faster than the nuclear industry's ability to assimilate it. It would seem, therefore, that our primary objective must be to first analyze what caused this somewhat generic situation to occur and then to take appropriate actions to correct it.

One of the root causes may be the practice of exploring complex and multifaceted subjects on an inherently fractionalized docket-by-docket basis. The purpose of docketed proceedings is to gain approval of a defined implementation at a specific facility. Being facility-specific, only weak precedents can be set for guidance to subsequent applicants/licensees. Moreover, individual dockets do not offer the exploratory freedom to deliberate all the possibilities necessary to reach the best set of solutions. For these and many other reasons, the docketed review process is inherently not a proper forum to develop generic guidelines.

Unfortunately, the only other forum is the various industry committees. At this time, there are several such committees working on different aspects of the basic problems outlined herein. However, due to the large cross-section of issues, several committees seem necessary. This creates duplication, consistency and coordination problems. Committee-based outputs are inherently fractionalized and sufficiently conservative to preclude significant change. They are also conducted on a volunteer basis. As such, they do not meet with sufficient frequency, lack centralized focus, and are subject to a cumbersome internal review and consensus process. So, while they are very helpful, they are also somewhat slow. If this were otherwise, the required analog to digital upgrades rule base and guidelines would be available.

For these reasons, a forum to allow undocketed and timely development of the necessary guidance does not exist. The only industry wide forum where a new rule base and the associated guidelines could be developed is the ALWR program. The NRC's draft Standard Review Plan (SRP) for these units, however, closely parallels those of the operating units. Unfortunately, these SRPs do not provide definitive guidance on such essentials as how one determines adequate diversity or sufficient defense-in-depth. This was essentially admitted when the SRPs had to defer such determinations to a "case-by-case basis." Since docketed reviews are the only mechanism for case-by-case assessment, the ALWR SRPs essentially conclude that the USQs are irreconcilable generically and docket-by-docket reviews are always going to be necessary. Is such an approach consistent with the ALWR design certification concept? By taking such a position, will we be forever locked into the two step licensing process? It would seem that some changes are necessary.

NECESSARY CHANGES

For the reasons outlined above, this will this require a forum where the issues can be generically defined, reviewed, and resolved without the constraints of unique facility and specific purpose evaluations, and without the pressures associated with docketed reviews.

Predictable generic guidelines must be developed on a rule base appropriate to modern technology to allow digital systems to be evaluated on the basis of their overall contribution to safety. Guidelines are necessary to properly evaluate and qualify commercial grade equipment, as well as the support infrastructure it affords, to bring implementation costs within acceptable limits, to avoid early obsolescence of specialized approaches, and to eliminate dependency on single sources of supply. Since the subject is complex and multifaceted, its solutions need to be comprehensive, well integrated and complete. Experience shows that such things are best managed from a centralized point to sustain focus on the salient issues and draw relevance between the relationships. Failing this, the cohesiveness of the final result is often subverted.

This will all necessitate some basic changes. Change invites uncomfortable risks and reassessments which provide strong incentives to stay with the status quo. It is axiomatic, however, that when the status quo is clearly unacceptable, change must be attempted. Such attempts are best guided by acceptance criteria that make meaningful results possible. Each recommendation, therefore, should be judged in terms of improvement to the existing situation. In the present case, that improvement would be a better way of communicating the evolving technological needs of the industry and securing more timely acceptance. Both regulators and licensees have the same incentive to do this – a cost-effective net improvement in overall safety.

INITIAL EXPLORATIONS

In response to growing utility interest in digital upgrades, the Nuclear Management Resource Council (NUMARC) and the Electric Power Research Institute (EPRI) have recently issued a document entitled, Guidelines on Licensing Digital Upgrades. This guideline represents an important step forward in the evolution from analog to microprocessor based digital safety systems. Still, while the guideline redefines what needs to be done, it does not address the more complicated how to do it cost effectively. Nor does it pretend to. Rather, it defers such matters to a different, and perhaps, more appropriate forum. In spite of what would appear a missed opportunity, however, it is inescapable that the very existence of this document plows new ground in the guideline development process. Essentially, it established a new method by which the utilities, operating together rather than on fractionalized dockets, can assist the NRC in proactively developing new guidelines. It would seem that this approach could ultimately improve or even replace the present, and quite cumbersome, formulation of guidelines through industry committees. To this end, EPRI and NUMARC have provided an excellent new paradigm. What is now needed, is a program that capitalizes on this opportunity and offers some workable, practical solutions.

RESTATEMENT OF PRINCIPLES

The preferred approach should be to first agree in principle and then apply such principles to the solutions. Some suggested principles might be:

- Safety depends upon the existence of timely, predictable, uniform and definitive guidelines
- Safety is a relative matter that is most appropriately applied on graded basis
- Improvements should be evaluated on the basis of their relative overall contribution to safety
- As technology changes, so must the rule structure assessing its value and application
- New technologies should not have to meet a standard higher than the approaches they make obsolete

REQUIRED ACTIONS

Guidelines must be prepared that include qualification of commercial grade equipment to bring implementation costs within acceptable limits, to avoid early obsolescence of specialized approaches, and to eliminate dependency on single sources of supply. The following things have to occur: 1) a dedicated program must be developed to integrate the fractionalized efforts of various industry committees, action groups, and researchers currently working on the many elements of the overall problem; and 2) these integrated results will then have to be augmented with additional research to fill in any gaps that exist in these fractionalized efforts. Specifically, this research will have to:

- Develop the acceptance criteria defining adequacy of defense-in-depth in terms that allow compliance without guesswork
- Establish limits whereby diversity produces diminishing returns and then define the parameters by which such diversity can be shown on a structure, system, component, AND software level
- Develop methods by which commercial-grade, microprocessor-based digital systems can be accepted and utilized for safety grade applications on a scale suited to the exposure

In addition, a method has to be developed to allow the necessary guidelines to evolve with reasonable assurance of ultimate acceptance. In particular, a new forum must be created to permit the industry as a whole to have direct interaction with the NRC to review evolving technical matters before they become USQs.

PROGRAM PLAN

IMPLEMENT A DEDICATED PROJECT

Since the subject is complex and multifaceted, its solutions need to be comprehensive, well integrated, and complete. Experience shows that such things are best managed from a centralized point to sustain focus on the salient issues and draw relevance between the relationships. Failing this, the cohesiveness of the final result is often subverted. To bring focus and integration, a centrally managed, specifically dedicated work program will be created that, under one umbrella, satisfies NRC concerns to the point that prior approval should not be necessary and use of commercial grade equipment is possible. Therefore, a single-purpose team should formed and managed on a scheduled task, defined deliverable, dedicated project basis.

INTEGRATE EXISTING EFFORTS

There are substantial but uncoordinated activities under way to address various aspects of the overall problem. This is best evidenced by the variety of references in the NRC's SERs and SRPs. In addition, there is also a substantial bibliography of independent papers, some even commissioned by the NRC as NUREGs. The first task of the project team should be a comprehensive review that reduces this information base to uniformly address the salient concerns of the NRC. This should be attempted by formulating the NRC's concerns in terms of basic attributes and accurately defining them in the technical positions discussed below. From this, overlaps should be eliminated and gaps identified for additional research. In the process a meaningful correlation from the substantial experience base of successful applications in other industries should be correlated to nuclear safety to expose the myth that nuclear power requires highly specialized hardware.

ESTABLISH AN ACCEPTANCE METHOD

A phased development approach is suggested wherein agreement in principle would be established with the NRC prior to development of specific guidance. These agreements in principle would be formulated as a series of cogently argued technical positions. An example would be the concept of classifying levels of defense-in-depth adequacy on the basis of a weighted scale involving the probability of failure, the seriousness of its consequences and the availability of mitigating actions. The question being, is it realistic to apply the same requirements to equipment that must act automatically in the first few seconds of an event to that which is manually initiated and operated hours after? Upon securing NRC agreement in principle, the attributes of each technical position would then be translated into a series of specific "how to" guidelines. In the example given, the specific criteria would be developed to assign the weighting factors for systems, structures, components, and software. By this method, guidelines can be developed faster, with less guesswork and more certainty of ultimate acceptance. Even drafts of the such agreements in principle would be informative to a confused and hesitant industry.

ESTABLISH AN INDUSTRY FORUM

A new forum should be created that builds on the successful work of the EPRI/NUMARC collaboration outlined above. In fact, EPRI presently has a program positioned to extend the "what should be done" issues of the initial EPRI/NUMARC/NRC collaboration into a series of guidelines on "how to do it." At an industry meeting in November 1993, this program outlined above was honed to a point that it received near unanimous endorsement. (Of the 27 utilities in attendance, only 1 deferred endorsement.) Unfortunately, due to budget reductions and the need to husband funds for committed programs, EPRI had to arrest activity on this project for 1994. Based on the interest displayed, level of participation and endorsement of the utilities at the November 1993 meeting, EPRI decided to reassess priorities and associated funding. It is possible that the significance of the proposed program has been overlooked as regards its importance to the nuclear power industry. A top priority of the nuclear power industry is the ALWR design certification effort. To be sure, if the issues discussed herein are not resolved for the operating units, they will have to be resolved for ALWRs. Since these issues were not specifically a part of the ALWR design criteria design, they were not funded as part of the effort. As such, the proposed program should be considered not only important to operating reactors, but also an element of the ALWRs. At this writing, it remains to be seen whether this logic is persuasive and the proposed program will be allowed to go forward.

0074-056X/95/351-362/$0 + .50pp

DESIGN AND APPLICATION OF A PC/PLC UTILITY BOILER CONTROL SYSTEM

David A. Conrad
Instrumentation Engineer
Nova Scotia Power Inc.
P.O. Box 910, Scotia Square
Halifax, Nova Scotia
B3J - 2W5

Daniel P. Muldoon
Plant Engineer
Nova Scotia Power Inc.
P.O. Box 69
Glace Bay, Nova Scotia
B1A - 5B1

KEYWORDS

PLC Boiler Controls , Utility Boiler, Modulating Controls, DCS

ABSTRACT

The Utility Industry has been replacing aging control systems in their plants with Digital Control Systems. Use of Programmable Logic Controllers (PLC) has generally been restricted to stand alone Sequential or Digital Systems and has not been widely used for Boiler Modulating and/or Plant Wide control. With ongoing development of PLC I/O, firmware, operator interface options and configuration flexibility, PLC's are now capable of performing plant wide control including boiler modulating control. In addition to the PLC developments, Personal Computers (PC) based operator interface platforms have evolved to the extent that they rival the capabilities of many operator interfaces supplied with microprocessor based control systems (DCS). This paper will discuss the design, implementation and experiences of using a PC/PLC system for plant wide control on a 118 MWe utility plant.

Introduction

As Utility Power Plants age, many Utilities are extending the life of their older power plants and retrofitting new control systems equipped with a Graphical User Interface (GUI). Most often, the GUI selected is supplied as part of a Microprocessor Based Control and Monitoring System which has commonly become known as a Distributed Control System (DCS). The reason for choosing the DCS as the backbone of the control system has been the boiler controls require complicated interlocking control loops and often incorporate Data Acquisition and Alarm features. Traditionally, costs and functionality have dictated the DCS be used for the boiler and plant modulating controls while Programable Logic Controllers (PLC), are used for the digital and sequential controls. To provide a common operator interface platform, the PLC's must be incorporated into the DCS via a Gateway or Datalink. Gateways, which translate files and/or code from one system to another can be a potential bottleneck which may affect the overall system performance.

The DCS firmware has been changing to include more robust I/O and often incorporating ladder logic capabilities, which offers the option of using a DCS as the plant wide control system thus eliminating the need for the slow Gateways. In addition to the changing DCS functionality, the processing power and I/O capabilities of PLC's has been increasing, opening the option of using PLC's and a PC based GUI for plant wide control.

This paper discusses the design and application of a boiler and plant control system consisting of PLC's and a Personal Computer (PC) based Operator Interface. To form the basis of discussion, the concepts and limitations of the project will be reviewed.

PLANT DESCRIPTION

The power plant control design strategies and systems described in this paper are based upon the Nova Scotia Power Incorporated (NSPI) Glace Bay Generating Station located in Cape Breton County, Nova Scotia, Canada. The plant is a former co-generation facility which in earlier years supplied steam to a local heavy water plant. The power plant consists of two 550,000 lb/hr cyclone coal/oil fired drum type boilers, one 38 MWe topping turbine (1800 psi) and four 17.5 MWe and one 20 MWe low pressure (600 psi) turbine/generators connected on a common header. The controls upgrade was part of a larger improvements project which included the addition of ID Fans and electrostatic - precipitators and the 20 MWe condensing unit. The vintage of the controls equipment being replaced was 1960's and installed in 1966. The final feed water pressure is provided by steam driven turbine pumps. Figure 1, shown below, depicts the present day steam and gas side of the plant configuration.

FIGURE 1

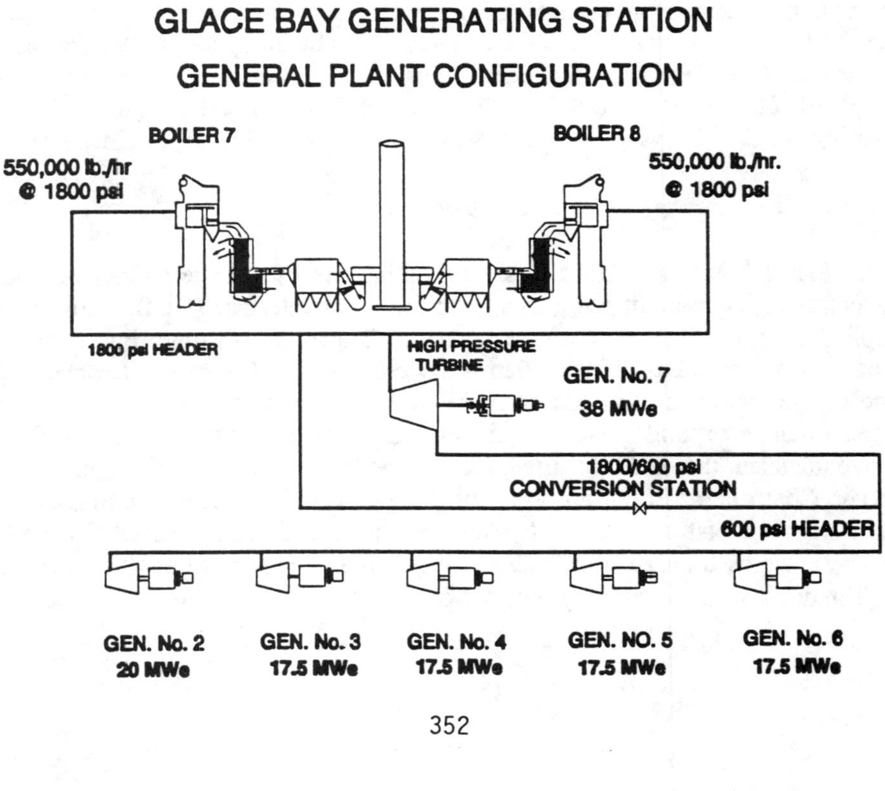

OLD CONTROL SYSTEMS

The existing controls in the plant at the project conception consisted of many self contained pneumatic controllers and interfaces, hard wired relay systems and hard wired switches mainly mounted in a remote control desk located in a central control room. The desk was partitioned into two sections, one desk contained all of the boiler controls (analog and digital) and the other section contained all of the turbine/generator and sub station switching controls. Since improved plant performance from both a reliability and environment perspective was desired, replacement of the major boiler controls was performed in conjunction with the performance and environmental improvements to the plant. Space in the existing control room was limited so the existing boiler desk had to be removed. Since the entire boiler controls were to be replaced, it was decided at this time to replace the boiler control desk with a GUI based control system. Bids were solicited and a contract awarded to Fossil Power Systems (1988) Inc. of Dartmouth, Nova Scotia, based upon price and proven past experience.

PLANT OPERATION

The normal plant operation is to be available for peak load and emergency load requirements during the winter heating period. The plant is capable of idling at about 6 - 8 MWe gross output with two cyclones burning fuel oil and steam being supplied to two low pressure turbines through the steam conversion station. Station net full load burning coal is 118 MWe and 110 MWe burning Fuel Oil with the high pressure turbine on and 90 MWe when the high pressure turbine is bypassed through the steam conversion station. Normal operating period for the plant is from the start of November to the middle/end of March.

DISTRIBUTED CONTROL AND MONITORING SYSTEM (DCMS)

DCMS ARCHITECTURE AND PARTITIONING

The DCMS is a PLC/PC based distributed control system which acts as the main control system for the plant and performs the majority of data acquisition functions. The PLC's include all plant logical functions, control algorithms and I/O. The PC's handle all trends, logging and operator interface functions. Communications between the two mediums is provided at each PC by a highway communications card which allows the PC direct access to the Data Highway. All main control PLC's are fully redundant and are basically partitioned into a common control system for plant motor control and modulating loops (MCS), two fuel control systems (BAS 7 & 8) and a data acquisition system (DAS). Each low pressure turbine has a local PLC and a local display unit for turbine monitoring. A stand alone ash system also incorporates a PLC control as does the electrostatic precipitator rapper control. Neither of these systems were deemed appropriate to be accessed by the DCMS. The operator controls for the substation and the turbine/generator controls were not part of the retrofit. The functionality of each main PLC's pair is as follows:

The MCS performs the following:

- Plant wide analog control (two 1800 psi, 550,000 lb/hr boilers and feedwater system)
- Motor control for all common plant equipment (feedwater pumps and valves)
- Some data acquisition I/O and alarms

The BAS (one each boiler) performs the following:

- Motor control for boiler fans (ID and FD)
- Logical protection per NFPA Guidelines (flame failure, purge etc.)
- Soot blower motor control
- Cyclone control
- Ignitor control

The DAS performs the following:

- Analog inputs for monitoring and alarming (thermocouples,RTD's and transmitters)
- Digital alarm inputs
- Topping turbine vibration signal multiplexing and supervisory logging
- Provide highway access for low pressure turbine PLC's

Figure 2 shows a simplified configuration of the PLC Data Highway for the system.

The DCMS is provided with ten (10) 486 personal computers (PC's) which communicate with the PLC's via the Data Highway. Eight of the PC's are located in a 9 section control console from which the operators may interface with the PLC's by use of a bus mouse. Each of the operator PC's is capable of trending, displaying graphics and displaying alarms. The desk layout is such that four PC's are used for control of each boiler but common controls may be done from any PC. Each PC has 16 Mbytes of RAM which is sufficient to maintain all real time data and the graphics which allows exceptional screen update times (typically less than 1 second). The system was also provided with two alarm printers and a color ink-jet plotter for screen dumps. The screen dumps are handled through the server so the operator station is not tied up during a screen dump.

The remaining PC's function as the file server, which handles all historical trending and archiving to the mass storage medium and the engineering work station which is used for all graphic development and PLC programming. Any of the other PC's may be converted to a engineering work station by adding a key board (a valuable feature during the commissioning process) and also any PC may function as a file server for updating any or all hard disks on the other PC's.

Figure 3 shows the LAN connections between the various PC's.

FIGURE 2

SIMPLIFIED DATA HIGHWAY STRUCTURE

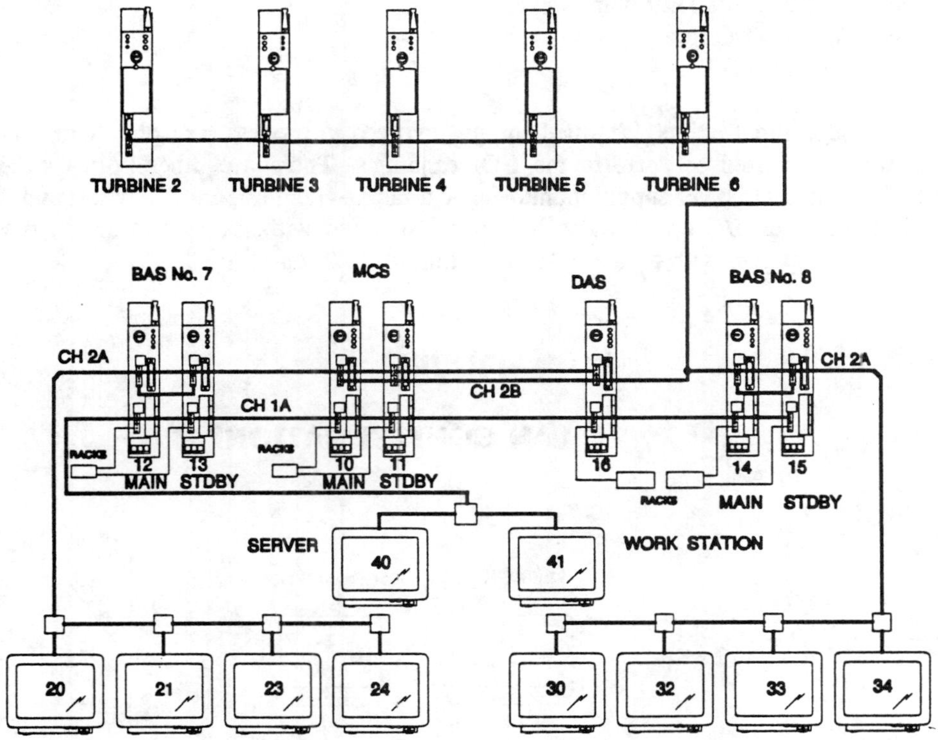

SYSTEM I/O

For terms of reference, the control system consists of the following type and quantity of I/O:

Digital:

643	24 VDC inputs
132	120 VAC/VDC isolated inputs
66	120 VAC outputs
42	24 VDC (Interposing Relay Outputs - 10 Amps @ 250 VAC)
144	24 VDC (Interposing Relay Outputs - 1 Amp @ 250 VAC)

Analog:

287	4 - 20 mAmp inputs
36	4 - 20 mAmp outputs
144	thermocouple inputs (J, K and T)
48	RTD inputs (120 ohm Nickel)
16	0 - 5 VDC inputs
6	± 100 mVDC

Of the 501 analog input points, 208 of them are involved in process control. The 1564 points given above is what was used to perform the I/O scanning. There are other points which do internal housekeeping (internal power supply monitoring and rack switching), and many internal tags which will not be reflected in an I/O count. The I/O points supplied with the system are in the area of 1800 points, including spares, with a data base consisting of 2076 tags.

FIGURE 3

PC LAN CONFIGURATION

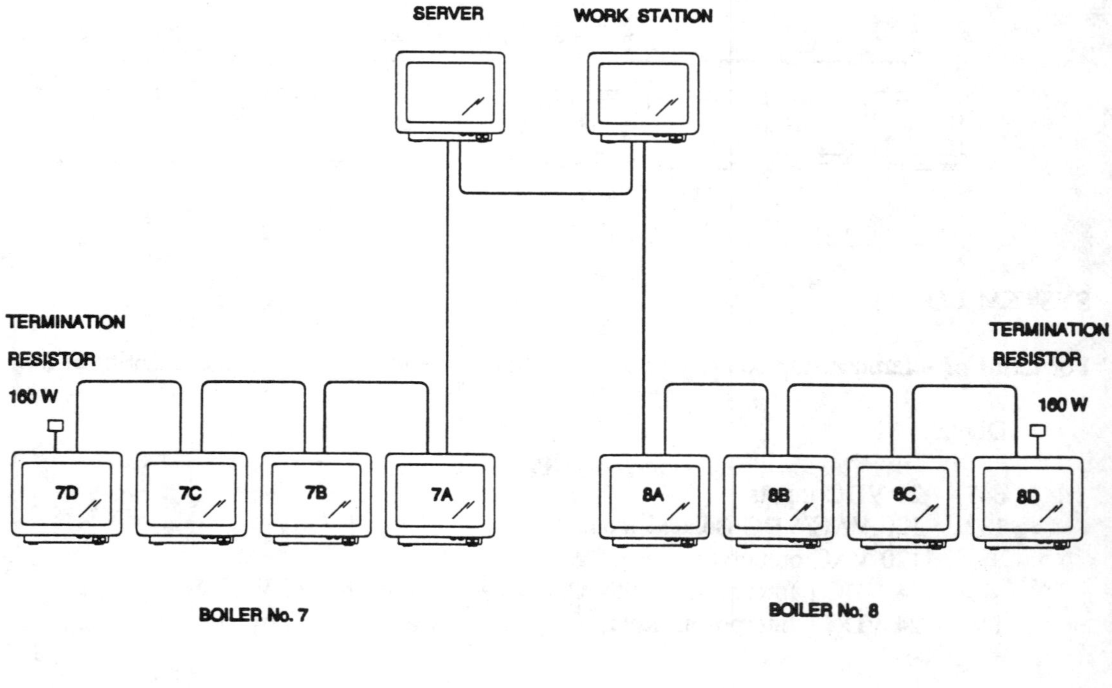

BOILER MODULATING CONTROLS

Advanced analog control capabilities of PLC's has been available for several years but are not perceived as sophisticated when compared to other options for performing complex digital based analog schemes found in utility boiler application. A new generation of larger memory, multiple highway PLC's were introduced which offered a cost benefit, low risk opportunity to experiment with PLC's for use in complex analog control schemes. The size of the Glace Bay modulating control system is relatively small when compared to most utility boilers, the application however, is as complex as most of the larger units. The modulating control system is contained completely in the Main Control System (MCS) redundant PLC's. The controls are partitioned into three groups; common, boiler 7 and boiler 8. Operator interface partitioning is obtained by dedicating a different highway to each set of PC's. The system design philosophy is on a top down approach and contains the following 39 control loops:

- Plant Master (+ Bias)
- Boiler Master (2 + Bias)
- BTU Flow {Coal BTU Compensator}(2)
- Cyclone Coal Flow { Feeder Speed }(4)
- Cyclone Fuel Oil Flow (4)
- Combustion Air Flow { Secondary Air } (4)
- Coal Transport Air Flow { Primary/Tertiary Air } (4)
- Furnace Pressure { ID Fan } (2)
- Cyclone Differential Pressure { FD Fans } (2 + Bias)
- Main Steam Temperature { Attemperator Spray } (2)
- Drum Level { Feed Water Flow } (2)
- Back End Temperature { Steam Coil Air Heaters } (2)
- Feed Water Pressure { Steam Turbine Boiler Feed Pumps } (1)
- Deareator Level (1)
- Deareator Pressure { Pegging Steam } (1)
- Auxiliary Steam Pressure { 600/250 psi } (1)
- Auxiliary Steam Pressure { 250/100 psi } (1)
- Fuel Oil Pressure (1)
- Fuel Oil Temperature (1)
- Ignition Oil Pressure (1)

This PLC pair is the most heavily loaded from a programming and I/O sense and is able to average a 250 millisecond scan time. This scan time enables screen updates on the operator interface to typically be under 1 second. As many of the normally *canned* functions of other type of control systems, such as function generators, were not available for the PLC's, subroutines and look up tables were created to fulfil these needs. The mathematical logic associated for these functions is included on the suppliers logic drawings and explained in English in the program comments.

Since the salient feature of this PLC system is the boiler modulating controls, a description of the combustion control design is warranted.

COMBUSTION CONTROLS

The plant configuration as previously given consists of two identical boilers supplying a common header. This is different from the majority of utility applications where the steam demand from a single boiler is supplied to a single turbo/generator set. The Glace Bay configuration is similar to many industrial applications where multiple boiler feed onto a common header.

The basis of the design philosophy for the plant and the individual boiler masters is to maintain a calorific energy balance between energy input to the boilers (fuel, air and feedwater) and the energy output (steam flow) from the boilers. With both boilers in service and the Plant Master in Auto, the firing rate demand is essentially a feed forward signal in MBTU's. This signal is derived from the calculated energy input required to maintain the sum of the steam flows from both boilers at the set point conditions (steam pressure and temperature). The feed forward bias signal is modified to compensate for changes in:

boiler stored energy (pressure setpoint ÷ actual pressure)
deviations of the feedwater and/or combustion air temperature from design criteria
energy input requirement changes due to differences between steam and feedwater flows

Steam Header Pressure, the balance point in the boiler energy input/output relationship, is maintained via a proportional-plus-integral controller. Adaptive tuning techniques are employed to calculate the optimal proportional (K_p) and integral (K_i) gains for the pressure controller versus boiler load (steam flow demand).

The Plant Master Firing Rate Demand is divided between the two boilers using a simple ratio calculation based on the total MBTU capability of the plant. The ratio split between the two boilers of the total plant firing demand is determined by the operator using a setpoint adjustment on the Plant Master Hand Auto Station. For example, if the total plant MBTU demand as determined by the Plant Master Pressure Control loop was 1200 and the Ratio entered by the operator was 50%, then the plant master total demand would be divided equally between the two boilers and all four (4) cyclones (ie. a boiler master demand of 300 MBTU to each cyclone). In the event that a cyclone is tripped or shut down by the operator to clean a plugged feeder, the boiler ratio calculation logic automatically recalculates the required ratio based on the calculated fuel flow and capability of the individual boilers to deliver the requested load.

Expanding on the example above, if the individual boiler demand of 600 MBTU (half of the plant demand) is equally shared among the four cyclones available (two each boiler) and a feeder clutch is disengaged to clean a plugged coal feeder, the Plant Master Firing rate ratio will be recalculated to reflect the fact that one boiler is now only capable of delivering half of its rated load. The ratio will be recalculated to be 33% of the plant demand to the boiler with one cyclone running and 67% to the other boiler. This recalculation of the firing rate demand to both boilers enables the load originally carried by the cyclone with the plugged feeder to be distributed equally amongst the three (3) remaining cyclones.

The firing rate demand to each boiler is passed on to a fully metered cross limited control system consisting of individual cyclone fuel (coal/oil) and secondary air controllers. Each boiler is fitted with two in-situ O_2 analyzers which are used for trimming excess air during the combustion process. A simplified overall view of the combustion controls can be seen in Figure 4.

FIGURE 4

SIMPLIFIED COMBUSTION CONTROL

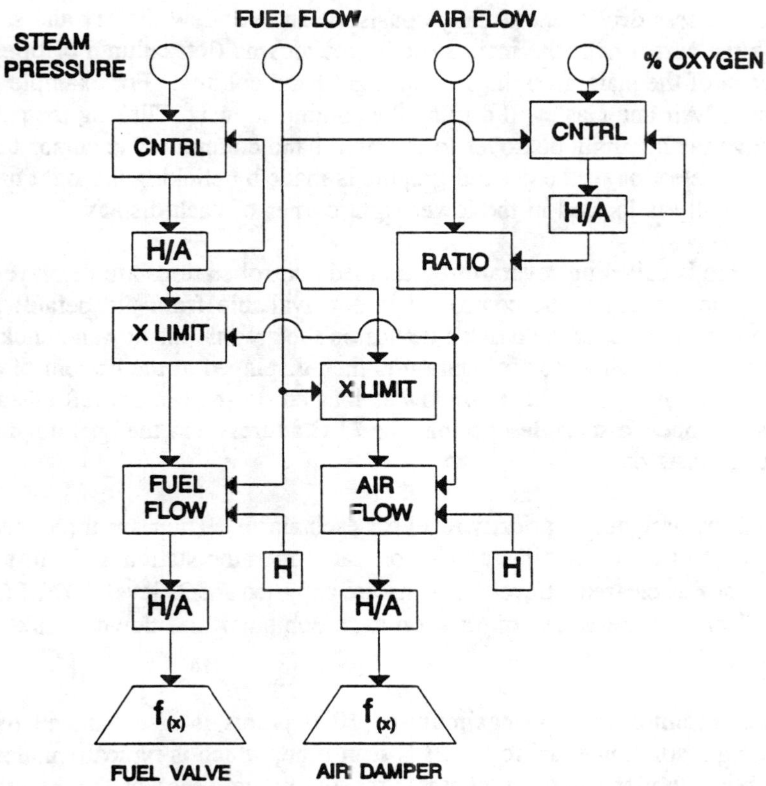

DATA ACQUISITION AND OPERATOR INTERFACE

The Data Acquisition System is handled in a three tier approach. All I/O passes through the PLC's, thus, all digital and analog input scaling, conditioning and alarming is performed in the PLC's. Each operator station handles and processes trend data and current alarms. The file server sorts logged alarms and trend data for archiving on the mass storage medium (Tape Drive).

The Operator Interface is provided by eight personal computers each using a run time version of the operator interface software and each with its own PLC highway interface card. This allows each station to be fully independent from the other operator stations. The data highways and operator software are partitioned such that four operator stations are attached to only the particular boiler fuel control system (ie. Boiler 7 BAS) and both the modulating control and data acquisition PLC's. This functionally eliminates the operator from making incorrect changes to the wrong boiler. Graphics are quite detailed and are accessed by use of a bus mouse and a menu system. By use of the menu system, any graphic display may be accessed from the current display. Further, each display has a five line alarm window box which is always displayed in the upper right corner of each graphic. Graphics that contain modulating controls incorporate a 10 minute mini trend window which may be displayed in the lower right of each display.

Graphic groups are process driven and usually consist of an overview display and several detailed sub displays. Running the cursor over the main system breakdowns (left column of titles) will display the group listing for each of the main groupings in the right hand column. For example placing the cursor on the main grouping "Air and Gas" will display the column shown. Clicking the right hand button on air and gas will move the highlight box over to the right hand column. The cursor then may be run up and down the list and selection of the desired graphic is made by clicking the right hand mouse button. This mini menu normally is located in the lower right corner of each display.

Once a desired graphic is called up several programmed control stations are displayed at the bottom of the screen. If a device that is to be controlled is not available from the default popup, it may be selected by passing the cursor over the desired device on the graphic display and clicking the right hand button on the mouse. The desired control station is then displayed at the bottom of the display screen in a predetermined location. The help window is located on the bottom left of each screen, it will display a device tag number, a short description, the PLC address and the operator interface ID of any device that the cursor rests on.

Alarming is handled by assigning a priority level for each alarm. Their are three levels of alarms with level 1 being the most critical. Alarms appear on each operator station and must be acknowledged before the highlight box is cleared. Level 1 alarms are assigned *RED*, level 2 *YELLOW* and the lowest level is *BLUE*. All alarms are stored on disk on each computer and down loaded once a day to the mass storage medium.

Sequence of Events Monitoring to a maximum of 1000 points is also handled by each BAS PLC (Internal SEM) with a resolution equal to the PLC scan time, which is typically under 20 milliseconds. Trigger events such as a boiler trip or an operator manual request enables the monitor report and data is down loaded to the server for printing. This resolution is slow when compared to most DCS or packaged SEM's, but is sufficient for most typical relay trips.

Screen dumps are available from each station. Initiation of a screen dump does not tie up the operator station as all pertinent data is down loaded to the server which creates the dump file to be plotted at the color printer.

Graphic displays were a project team effort. The plant operations group and the head office engineering staff constructed pictorially what was desired to be displayed on the screen while the supplier was responsible for making the graphics functional as well as the development of all the control stations and screen standards.

DRAWINGS AND PROCEDURES

As with any major retrofit, documentation is of prime importance. Documentation for this project fell under two different responsibilities although the entire team cross checked one another. Drawings from the control system supplier included detailed logic drawing and I/O layout drawings. Drawings by the owner included cable schedules, schematics, P & ID's and loop drawings. As no intelligent CAD type program was available for the PLC configuration, it was of extreme importance that the detailed logic diagrams contained relevant programming information for trouble shooting and commissioning purposes.

The data base for the project was also used by the control system. The operator interface software allows the use of a relational data base which allowed a multipurpose data base to be constructed. The owner remained largely responsible for the data base as the bulk of the information was derived by the owners engineers and technicians.

OBSERVATIONS AND COMMENTS

This control system has been in use for two heating seasons or 1.5 years. A system based upon PLC's and Personal Computers is certainly viable and has been proven to be effective. There are a few drawbacks with such a system. Future development by PLC suppliers should incorporate the higher level functions required for complex modulating controls into the processors and should include a CAD based programming package to enable the modulating programs to be written both faster and easier. These obstacles were overcome by writing subroutines and implementing look up tables, which were very effective, are often not intuitive to the non programmer. This method requires precise documentation to enable the end user to easily follow the programmers intent. A summary of the salient pros and cons of the system as compared to a combined DCS and PLC system are offered as follows:

Key features:

- All logic functions are contained in the same type controllers - thus less training required
- All operator interface stations contain the same operating and application software
- The system is responsive
- The system is flexible
- The system as a whole consumes less power as compared to conventional DCS suppliers
- The multiple highway capability
- Reduced spare part types in inventory

Areas of Improvement Required

- CAD type programming package for analog controls
- Expanded Instruction Set to include popular instructions for analog controls
- Data Base should be relational to PLC's as well as operator interface
- Integrated system diagnostic tools for trouble shooting
- Closed loop simulator option

The quoted price of this system compared favorably with the next lowest bidder, a DCS, by about 20% in 1991. The system is not configured as a typical DCS one would purchase for use in a utility plant, but offers the majority of the features expected of a system of this size. As many of the componets are available from several vendors, ie. computers, the life expentency and support for this system appears to be achievable and economic.

© 0074-056X/95/363-370/$0 + .50pp

EXTENDING THE CALIBRATION INTERVALS OF PRESSURE TRANSMITTERS IN NUCLEAR POWER PLANTS

H.M. Hashemian
D.W. Mitchell

Analysis and Measurement Services Corporation
AMS 9111 Cross Park Drive
Knoxville, Tennessee 37923 USA
Phone: (615) 691-1756
Fax: (615) 691-9344

KEYWORDS

Nuclear Power Plant, Instrument Channel, Calibration Drift, On-Line Monitoring, Calibration Reduction, Signal Validation

ABSTRACT

Process instrumentation channels in nuclear power plants are calibrated at every refueling outage and surveillance tested once every month or once every quarter depending on the plant. These activities usually involve extensive manpower and personnel radiation exposure, and have the potential of producing maintenance-induced errors, reactor trips, and wear and tear of the plant equipment. A search of the Licensee Event Report (LER) database, which documents instrumentation problems in nuclear plants, has shown that a notable number of instrumentation problems and plant trips have been caused by hands-on maintenance and testing activities. Furthermore, a review of historical calibration and surveillance data from nuclear power plants has revealed that less than ten percent of instrument channels are normally found to drift out of tolerance over a typical fuel cycle of eighteen months. That is, a majority of the effort currently spent on calibration of instrument channels can be saved if a means is established to accurately identify those instrument channels which have drifted, and must therefore be calibrated. Recognizing the side-effects of manual calibrations, a method for automatic testing of the calibration of pressure transmitters was developed and successfully implemented in a nuclear power plant. The method is based on continuous monitoring of the D.C. output of the transmitters to identify those that are drifting and must therefore be calibrated. This will allow one to limit the calibration effort to those transmitters that need a calibration as opposed to calibrating all transmitters, as is done now.

1. INTRODUCTION

The process instrumentation channels in nuclear power plants are calibrated periodically to ensure a safe and efficient operation. In most cases, these calibrations involve hands-on procedures which require physical access to the process sensors and involve extensive manpower and radiation exposure to the test personnel as well as maintenance-induced wear and tear of the plant equipment. Furthermore, a review of historical calibration data from nuclear power plants has shown that less than ten percent of the instrument channels in nuclear power plants are found out of calibration.

It has been proposed that an on-line monitoring system can be implemented in a nuclear power plant to record the output of the instrument channels while the plant is operating, and use simple averaging and intercomparison techniques to identify those channels that are drifting. This will help limit the hands-on calibration activity to those channels that have drifted, and thereby reduce as much as 90 percent of the effort currently spent in calibrating process instrumentation channels in nuclear power plants.

This paper presents the results of a research and development effort being conducted for the U.S. Nuclear Regulatory Commission (NRC) to evaluate the adequacy of on-line monitoring techniques for instrument calibration reduction in nuclear power plants. The project includes both laboratory and field tests. The laboratory tests are performed in a water test loop consisting of numerous sensors and signal conditioning equipment. The field tests are performed at the McGuire Nuclear Power Station. This is a four loop Westinghouse Pressurized Water Reactor (PWR) that is operated by the Duke Power Company who is serving as the host utility for the project. Approximately 170 live signals from the primary and secondary systems of the plant are being monitored at McGuire for drift. To date, the laboratory and field tests have shown that the on-line monitoring technique is a viable approach for testing the calibration stability of process instrumentation channels.

2. TECHNICAL BACKGROUND

Following the 1979 accident at the Three Mile Island Nuclear Power Station Unit 2, the NRC implemented a number of new requirements to ensure that the reactor operators are provided with accurate, timely, and reliable information about the status of the plant under normal and accident conditions. To ensure that reliable signals are used in operator aids, the Electric Power Research Institute (EPRI) initiated research and development activities in the early 1980s in an area that is now known as "signal validation."[(1)] Signal validation techniques have been used previously in the aerospace and aviation industries for flight control and space vehicle applications.

Signal validation consists of a variety of signal processing techniques implemented in nuclear power plants to ensure that sensor drift, response time degradation, bias, noise, and other sensor or system anomalies do not mislead the reactor operators. Signal validation depends on the redundancy of sensors in nuclear power plants and the physical relationships between the process parameters to check the consistency of the measurements, predict the expected values of process variables, and detect, isolate, and characterize any significant anomaly in the instrument channel.

EPRI's efforts in the signal validation area have not only produced improvements in operator aids, but also have laid the foundation for the development of on-line methods for testing the calibration of instrument channels. In addition to EPRI, a number of research and development organizations, universities, national laboratories, and utilities have worked in the signal validation area. As a result, numerous techniques have been researched, developed and documented under a variety of names. Three examples of these techniques are described below.

2.1 Like Signal Comparison (Cross Calibration)

This method involves scanning the output of a number of instrument channels that are measuring the same process parameter and determining the deviation of each channel from the average of all channels (excluding the outliers). This method is popular in PWRs for on-line testing of calibration of temperature sensors at isothermal conditions. In this application, the method is referred to as cross calibration (Figure 1).

A problem with the like signal comparison technique is that it may not account for common mode drift unless a newly calibrated sensor is included in the comparison and/or a method such as analytical redundancy, as defined below, is used to provide an independent estimate of the process parameter.

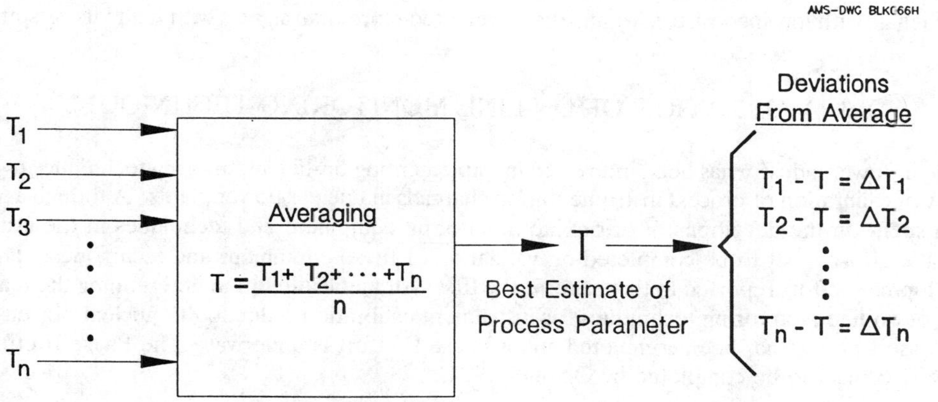

Figure 1. Illustration of Principle of Cross Calibration

2.2 Analytical Redundancy

This method is used when an adequate number of physically redundant channels is not available for intercomparison and when an independent measure of a process parameter is needed to account for common mode drift. Analytical redundancy depends on theory to produce fictitious sensors or instrument channels to increase the redundancy. More specifically, it uses a group of diverse signals as the input to a physical or empirical model, or an artificial neural network, to produce an independent process signal. Figure 2 illustrates the principle of analytical redundancy.

2.3 Parity Space

This method is used to determine the consistency between a group of redundant signals. The common components of the signals are subtracted out and the remaining components are compared, two at a time. Based on the differences between the residual components in each pair, an inconsistency index is generated and used for diagnostics. This index is also used as a weighting factor in weighted averaging of redundant signals. Signals with low inconsistency indices are weighted more than signals with high inconsistency indices.

3. VALIDATION OF ON-LINE MONITORING TECHNIQUES

The nuclear power industry has been interested in implementing on-line monitoring techniques to extend the frequency of calibration of process instrumentation channels in nuclear power plants. Although a lot of effort has been spent on the development of on-line monitoring equipment and techniques in the last ten years, a systematic effort is yet to be completed on validation of these equipment and techniques. The research and development effort reported in this paper is the first systematic attempt in determining the feasibility and accuracy of on-line monitoring techniques for instrument calibration reduction in nuclear power plants. To date, a Phase I project has been completed and a Phase II effort is underway. The Phase II effort is a two year R & D project to be completed by October 1994.

The results of the Phase I effort have been documented in the NUREG/CR-5903, which has been written by AMS under a contract with the NRC. [2] This report documents the initial efforts involving both laboratory and in-plant validation research conducted under a joint R&D effort between AMS, the NRC, and Duke Power Company.

3.1 Laboratory Test Results

The feasibility of on-line monitoring techniques are being studied through a series of experiments performed on representative temperature and pressure sensors and associated signal conditioning equipment installed in a laboratory test loop.

An empirical model of the laboratory loop pressure was developed and validated with data from the sensors in the loop. Representative results are shown in Figure 3 in terms of a time history plot for the loop pressure signal. It is apparent that the model prediction agrees reasonably well with the measured data indicating the validity of the model.

Figure 4 shows time history traces for three flow transmitters installed in the laboratory test loop, with a drift intentionally induced in FT-1-1 to show how the drift manifests itself in the deviation plot. Note that the drift in FT-1-1 causes all three signals to exhibit drift behavior in the deviation plot at the beginning of the

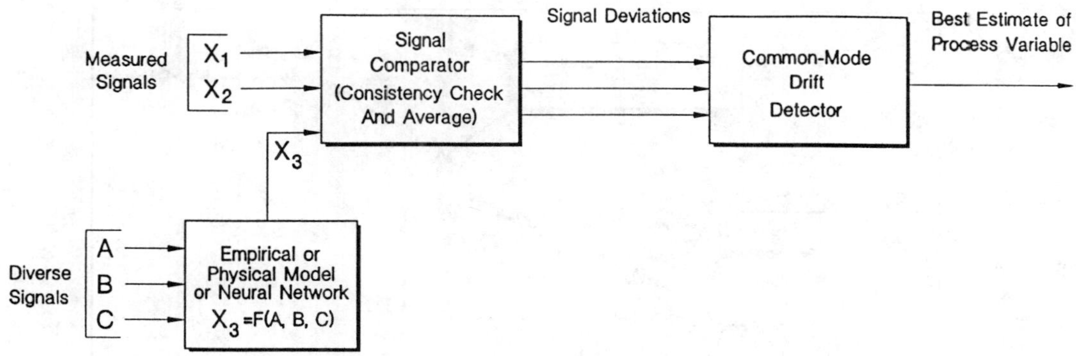

Figure 2. Principle of Analytical Redundancy

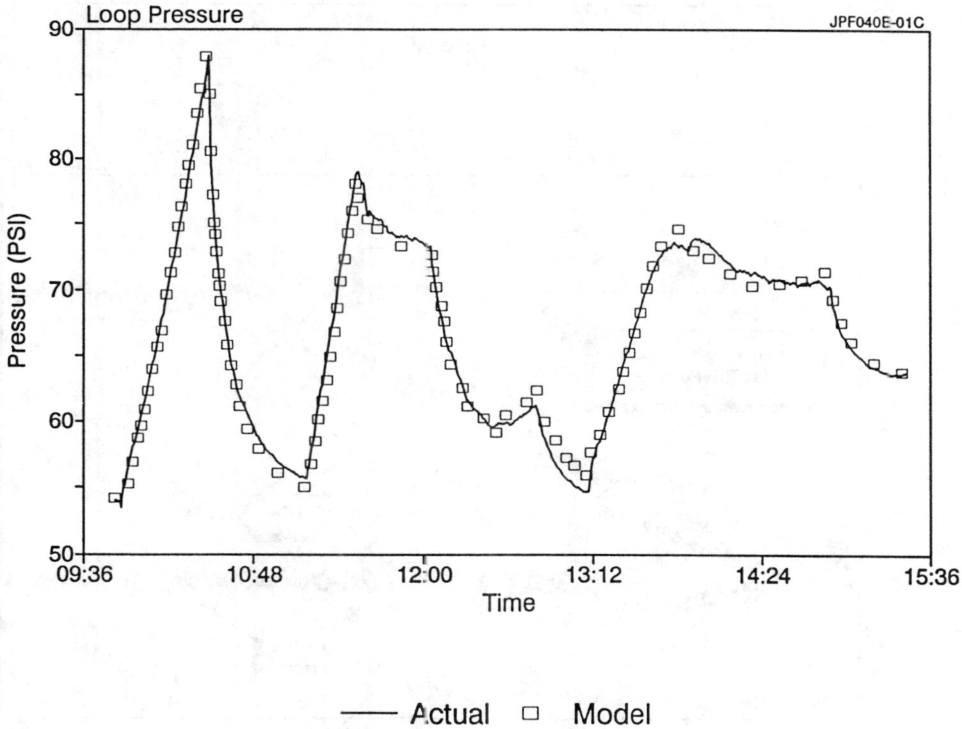

Figure 3. Agreement Between Results of Empirical Model and Actual Loop Pressure Data

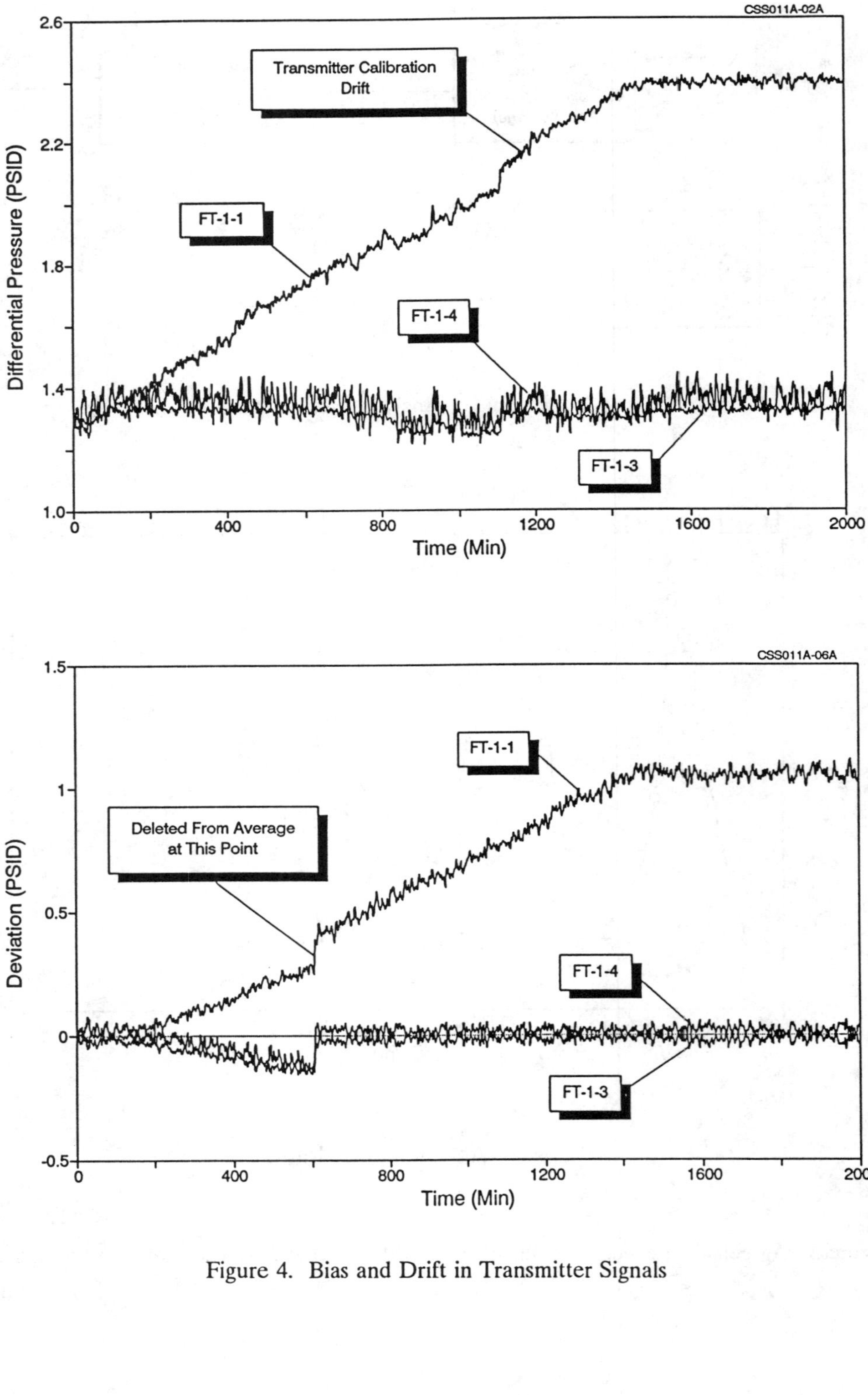

Figure 4. Bias and Drift in Transmitter Signals

data. Also, note the sudden shift in the deviation traces at about 600 minutes into the data. This occurs because FT-1-1 is excluded from the average at this point in time when its deviation exceeds a preset criteria. This causes all three signals to exhibit a step change in the deviation plot.

3.2 In-Plant Test Results

The in-plant validation of on-line monitoring techniques is being performed at the McGuire Nuclear Power Station. The monitoring began in March 1992 at the beginning of the plant's eighth fuel cycle and will continue for two complete fuel cycles. Approximately 170 signals from both safety and non-safety related instrument channels are being monitored.

The main purpose of the in-plant validation effort is to determine if the on-line monitoring system is able to identify the channel drifts that are normally identified by hands-on calibrations. This is accomplished by comparing the results from the on-line monitoring system to the actual in-plant calibrations from McGuire. An example of the results from this validation work is given in Figure 5, which shows a comparison between the hands-on calibration results and on-line monitoring results for three pressurizer level transmitters. Such results attest to the validity of the on-line monitoring method.

In addition to simple drift analysis, analytical redundancy techniques are used in the processing of the McGuire data for detection of calibration drift. Several process conditions at McGuire have been modeled using empirical and physical modeling, along with artificial neural networks. Figure 6 shows a redundant group of steam generator pressure transmitters along with the neural network results for this process group. As seen in this figure, the close agreement between the neural network prediction and the measured data from the process sensors indicates the validity of this technique.

4. CONCLUSION

A feasibility study was successfully completed on the validity of on-line monitoring techniques for remote testing of calibration of process instrumentation channels in nuclear power plants. This work involved research with temperature and pressure instrumentation in simulated reactor conditions in a laboratory, and in-plant validation work at the McGuire Nuclear Power Station, a four-loop PWR. Representative results of this work were summarized in this paper.

5. REFERENCES

1. Meijer, C.H. and Pasquenza, J.P., "On-Line Power Plant Signal Validation Technique Utilizing Parity-Space Representation and Analytical Redundancy," Electric Power Research Institute, EPRI NP-2110, Palo Alto, California (1981).

2. Hashemian, H.M., et al., "Validation of Smart Sensor Technologies for Instrument Calibration Reduction in Nuclear Power Plants," NRC Report Number NUREG/CR-5903, U.S. Nuclear Regulatory Commission, Washington, D.C. (1993).

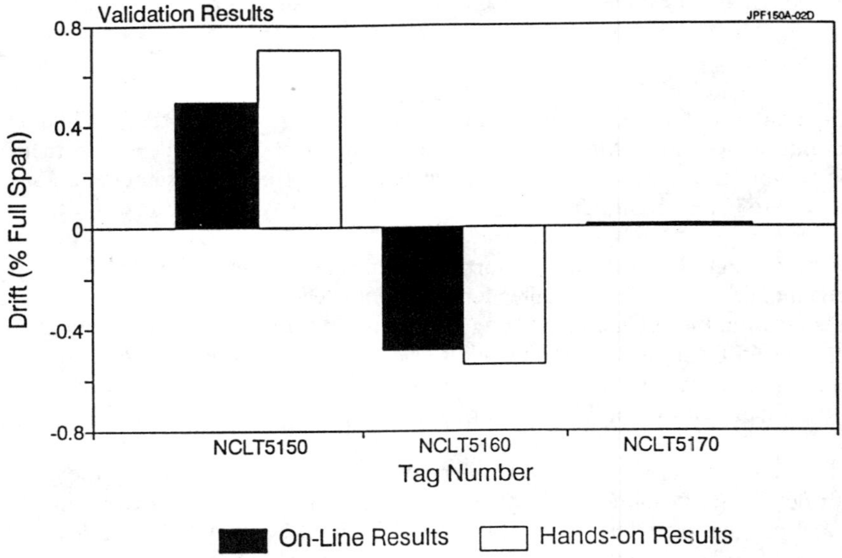

Figure 5. Hands-On Versus On-Line Calibration Testing Results for Three Pressurizer Level Transmitters at the McGuire Nuclear Plant

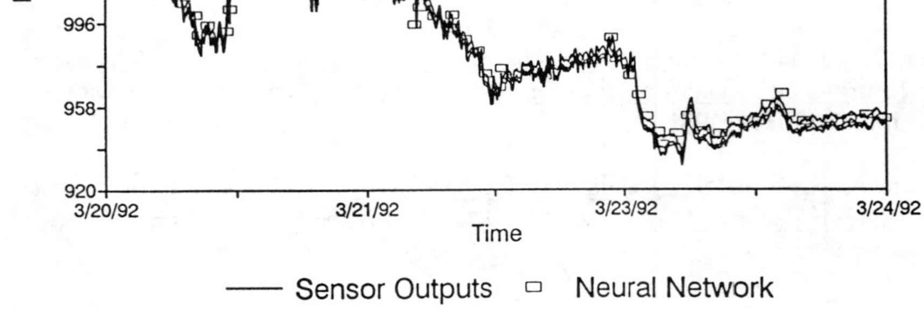

Figure 6. Comparison of Artificial Neural Network and Steam Generator Pressure Sensors

0074-056X/95/371-381/$0 + .50pp

NEW EQUIPMENT FOR ROD DROP AND CONTROL ROD DRIVE MECHANISM TIMING TESTS IN PWRs

R.E. Fain
K.M. Petersen
H.M. Hashemian

Analysis and Measurement Services Corporation
AMS 9111 Cross Park Drive
Knoxville, Tennessee 37923 USA
Phone: (615) 691-1756
Fax: (615) 691-9344

KEYWORDS

Nuclear Power Plant, Control Rod, Rod Drop Testing

ABSTRACT

Rod drop time testing is performed in pressurized water reactors (PWRs) to measure the length of time required for control and shutdown rods to travel from their fully withdrawn to their fully inserted positions. The test is performed at each refueling outage and is accomplished by interrupting the electrical circuits for one or more groups of rods causing them to drop, while at the same time monitoring and recording their positions in real time.

Rod drop time testing is currently performed manually on one rod at a time. This sometimes requires access inside containment, resulting in radiation exposure to the test personnel. To resolve the radiation exposure problem and other complexities of rod drop testing, computerized test equipment was designed, tested, and validated for automatic testing of the rod drop times in PWRs. This equipment consists of two portable test units with up to sixty-four channel data acquisition capability. This allows all of the rods to be dropped at the same time if so desired, thereby reducing the time required for the test. Utilizing high speed modems allows one test unit to be left unattended for the duration of the test and eliminates the need for personnel inside containment during testing.

Control Rod Drive Mechanism (CRDM) timing tests are performed in most PWRs at the same time as the rod drop tests. This test provides the information for determining the proper timing and sequencing of the CRDMs prior to startup. As with the rod drop time tests, the current practice is to test each CRDM individually. Test equipment was developed to aid in testing multiple CRDMs and reduce the critical path time needed for this test as well. In the process, a more precise and convenient way of evaluating the data was developed.

1. INTRODUCTION

In cooperation with Commonwealth Edison's Braidwood Nuclear Power Station, AMS began a research project in 1992 to develop new rod drop time test equipment. Rod drop time tests at Braidwood are currently performed manually which requires the tests to be performed on one rod at a time. In addition, personnel are required to be in the containment during the entire test. The new equipment was designed to allow Braidwood and other nuclear power plants to perform the tests automatically and remotely. The automatic testing will significantly reduce the amount of time required to perform the tests and will not require personnel to be in containment for the duration of the test. To date, AMS has developed equipment for rod drop testing and has performed two sets of validation tests at Braidwood and commercial tests in a three loop Westinghouse plant. As an extension of this research, control rod drive mechanism (CRDM) testing was added to further automate the plant startup activities. To date, CRDM tests have been performed in both Westinghouse and Combustion Engineering plants. The results of these developed efforts are presented in this paper.

2. BACKGROUND OF THE ROD DROP TIME TEST

Braidwood Units 1 and 2 (as well as other four-loop Westinghouse PWRs) have approximately fifty-three control and shutdown rods. When maintenance is performed on the reactor vessel head, such as during refueling, the plant technical specifications require tests be performed to verify that the control and shutdown rods will drop from their fully withdrawn to fully inserted positions within a specified length of time. These tests are called "rod drop time" tests and must be performed at normal operating temperature, pressure, and flow conditions. In most cases, the tests are performed at hot standby following an outage and are considered a critical path activity during startup.

The basic procedure for manual rod drop time tests is as follows:

1. Control room operators withdraw one bank of rods (up to eight rods).

2. Operations personnel turn power off to the RPI (Rod Position Indication) system which is located in containment, if the plant is so equipped.

3. Instrumentation and Control (I&C) personnel connect an installed test cable which extends from containment to the rod control room (located outside containment) to test points on the RPI cabinets for one of the rods. This test point provides an analog signal which is proportional to the rod's velocity as it travels from its withdrawn position to the bottom of the dashpot.

4. In the rod control room, additional I&C personnel connect the other end of the test cable to a strip chart recorder.

5. To initiate the rod drop, I&C personnel remove two fuses in the Rod Control Power Cabinets. This disconnects power from the stationary and movable gripper coils which hold the rod in place and causes the rod to drop.

6. The voltage transient generated in the RPI coils as the rod falls is recorded on the strip chart recorder and then analyzed manually to provide the rod drop time for that rod.

7. Steps 3 through 6 are then repeated to test each rod in the bank which was withdrawn in Step 1.

8. Operations personnel then restore power to the RPI system to verify that all rods are fully inserted and to provide rod position indication while the next bank of rods are being withdrawn.

9. Steps 1 to 8 are repeated for each bank of rods until the tests are completed.

It is important to note that if it were possible to sample data and to initiate the drop of all of the withdrawn rods at one time, the above steps would only need to be performed once. This was the reasoning behind the development of the system described in this paper.

3. EQUIPMENT DESCRIPTION

The equipment developed to perform this testing consists of two units. The first unit is the AMS Model ERD-1 which is set up at the RPI cabinets in containment if the plant is so equipped. The second piece of equipment is the AMS Model ERD-2 which is set up at the Rod Control Power Cabinets located outside of containment and is connected to the control computer. The equipment is connected by one twisted-pair cable currently installed in the plant for manual rod drop testing. This cable allows the equipment to communicate via modems which are installed in each unit. The features of the ERD-1 and ERD-2 are as follows.

3.1 Description of ERD-1 Unit

The ERD-1 is a microprocessor-based data acquisition system designed to be located in containment and connected to the RPI cabinets during the rod drop tests. The ERD-1 has a 64 channel analog to digital converter (A/D) to allow sampling of the analog signals from the RPI coil stacks. The A/D is capable of sampling data at rates up to 3 Kilohertz on all 64 channels simultaneously. This allows data acquisition to be performed on all rods at the same time if desired.

The ERD-1 is connected to the RPI cabinets with eight test cables. These cables connect the ERD-1 to the analog test points for all fifty-three rods. In addition, the unit can be connected to power control relays (if installed) in the RPI cabinets. This allows the RPI power to be cycled during the tests under computer control. The power is cycled due to the fact that the RPI power must be on while operations personnel are withdrawing a bank of rods so that they will have rod position information in the control room. However, before the rods are dropped, the RPI power must be off so that the voltage signals from the coil stacks can be recorded during the rod drop tests.

The RPI signals are isolated from the A/D by protection relays installed in the ERD-1. The ERD-1 also has a high speed modem which is used for communication with the equipment located outside containment. The front panel of the ERD-1 has a 2-digit status display which displays appropriate codes to indicate that the system is functioning properly. In the event of problems, the display is used to show error codes which aid in trouble shooting.

When rod drop tests are to be performed, the ERD-1 is placed in containment and the eight cables are connected from the ERD-1 to the RPI cabinets. After the ERD-1 is energized, it can be left unattended in containment for the duration of the tests. All functions of the ERD-1 are controlled remotely via the modem in the ERD-2.

3.2 Description of ERD-2 Unit

The ERD-2 test unit is positioned at the Rod Control Power Cabinets located outside of containment. The entire sequence of rod drop tests are performed from this location. The ERD-2 is connected to the control computer which is used to control the rod drop tests through its internal modem and to analyze and store the data. The control computer also contains a relay board which is connected, through the ERD-2, to the regulator circuit boards in the Rod Control Power Cabinets. The user can then shut off power to the gripper coils and allow the appropriate rods to fall at the beginning of a test. The reactor trip breaker can also be utilized to initiate the rod drop tests.

After each rod drop test is performed, a command from the control computer to the ERD-1, located in containment, sends the digitized rod drop transients to the control computer. At this time, the transients can be viewed, analyzed, printed, and stored for future reference.

4. AUTOMATIC ROD DROP TEST PROCEDURE

The automatic rod drop time test is performed using the equipment described in this paper and the following steps.

1. The ERD-1 unit is placed in containment and connected to the RPI cabinets.
2. The ERD-2 unit is connected to the Rod Control Power cabinets.
3. Operations withdraws one or more banks of rods (up to 8 rods per bank).
4. The rod drop test is initiated by the control computer of the ERD-2 unit or by opening the reactor trip breaker. This causes all the rods in the withdrawn bank(s) to drop.
5. Data is automatically acquired on all the rods by the ERD-1 unit in containment.
6. The data is transferred from the ERD-1 to the ERD-2 via modem and analyzed.
7. Steps 3 through 6 are repeated for additional bank(s) to be tested.

5. IN-PLANT TESTING

AMS has performed two sets of in-plant tests at Braidwood in cooperation with personnel in the Electrical Technical Staff group. The first tests were performed in Unit 1 in October 1992. In these tests, data was acquired using AMS equipment in parallel with the manual tests being performed by technical staff personnel. The purpose of these tests was to acquire rod drop data to use for developing the computerized analysis algorithms which were incorporated into the AMS rod drop time test equipment. Another purpose of this first set of in-plant tests was to determine the design criteria for the rod drop test equipment to be developed. During these tests, data was acquired on twenty-seven control and shutdown rods. Figure 1 shows a sample trace for one of the control rods and one of the shutdown rods which were tested. Table 1 shows the results of the analysis for all twenty-seven rods which were tested. The first column of results (Manual Calculation) are those reported by Braidwood personnel from the manual analysis of the strip chart

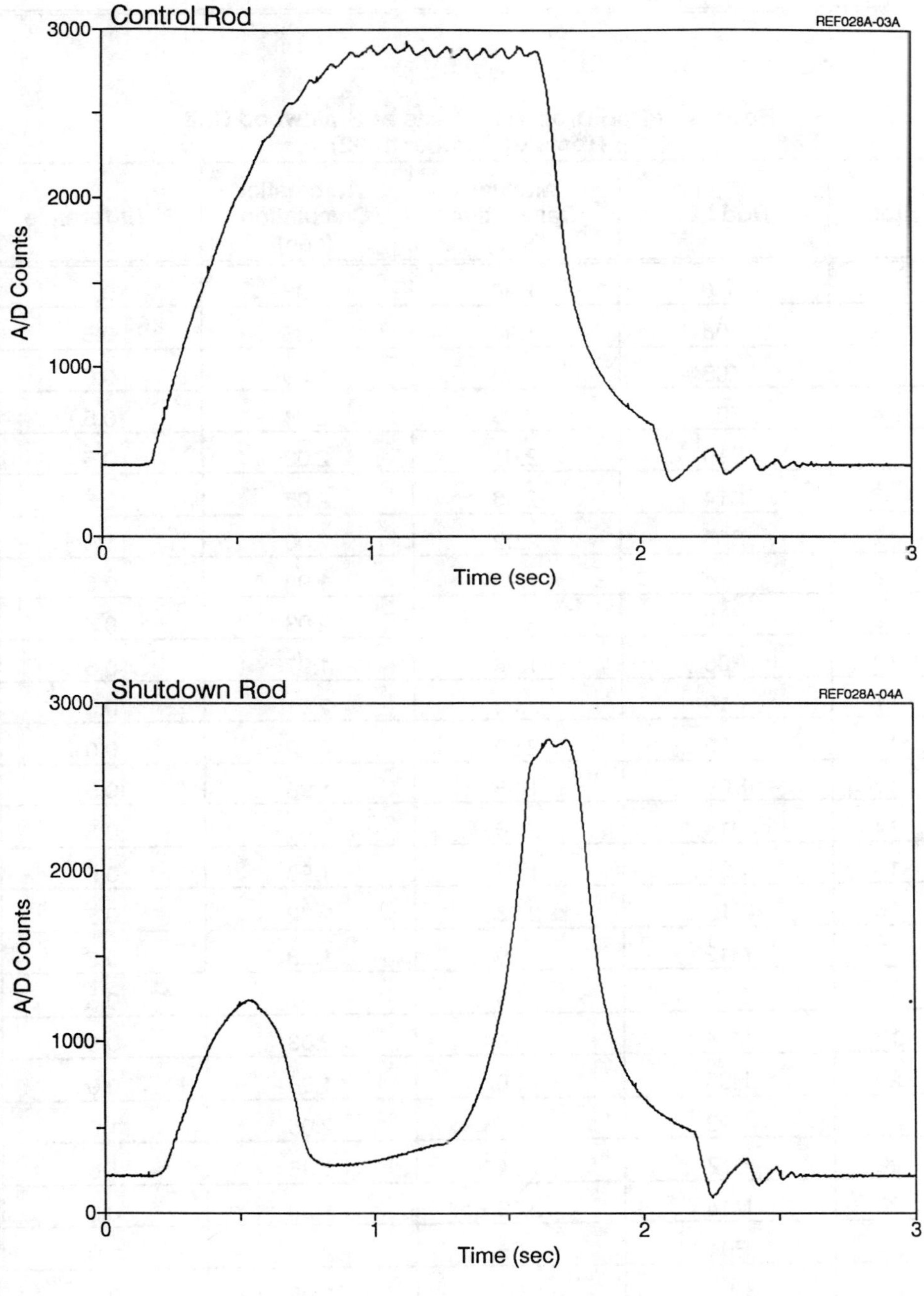

Figure 1. Sample Rod Drop Traces for Control and Shutdown Rods from Tests Performed in October 1992

TABLE 1

Results of Rod Drop Time Tests at Braidwood Unit 1 (Tests of October 1992)

Item	Rod I.D.	Manual Calculation (sec)	Automatic Calculation (sec)	% Difference
1	H10	1.96	1.96	0.0
2	F08	1.94	1.95	0.5
3	K08	1.99	1.99	0.0
4	F02	2.03	2.04	0.5
5	B10	2.01	2.02	0.5
6	K14	2.06	2.05	0.5
7	B06	2.02	2.00	1.0
8	F14	1.99	1.98	0.5
9	K02	2.02	2.03	0.5
10	P06	1.96	1.97	0.5
11	P10	2.00	2.01	0.5
12	H02	2.02	2.02	0.0
13	B08	1.98	1.99	0.5
14	H14	2.00	2.01	0.5
15	P08	1.97	1.98	0.5
16	F10	2.03	2.02	0.5
17	M12	1.99	1.98	0.5
18	D12	1.98	1.97	0.5
19	M04	2.04	2.03	0.5
20	H08	1.98	1.98	0.0
21	D02	2.07	2.05	1.0
22	B12	1.94	1.96	1.0
23	M14	2.00	1.99	0.5
24	P04	2.05	2.03	1.0
25	B04	1.99	1.99	0.0
26	D14	2.06	2.06	0.0
27	P12	1.89	1.89	0.0

recordings. The second column (Automatic Calculation) shows the results from the computer analysis of the sampled data. The average difference between the two results is 0.4 percent.

Following the first set of in-plant tests described above, AMS designed and constructed prototype equipment for performing the rod drop tests on one bank of rods at a time. This equipment was tested at Braidwood Unit 2 in April 1993. The tests were performed on two groups of rods. Figure 2 shows two composite plots of the rod drop traces from all four of the control and shutdown rods that were acquired. This type of plot is useful for visually identifying any rods which may exhibit a slow rod drop time and verifying that all of the dropped rods began moving at the same time.

A set of tests was performed in April 1994 in a three loop Westinghouse plant. This plant has 52 rods grouped in seven banks. The final production model of the rod drop test equipment was used to acquire the data on all of the rods. Table 2 lists sample results from this test.

6. CONTROL ROD DRIVE MECHANISM TESTING

In addition to the rod drop time tests described above, many plants perform control rod drive mechanism (CRDM) timing tests prior to startup. The purpose of these tests is to verify the proper timing and sequencing of the CRDM system. CRDM timing tests involve monitoring the stationary gripper coil current, movable gripper coil current, and the lift coil current simultaneously while withdrawing or inserting a bank of rods. This provides an accurate representation of the actual movement of the individual mechanisms.

Manual CRDM tests are currently performed by monitoring these signals for each rod separately using a chart recorder. Each rod is withdrawn and inserted individually while the coil currents are acquired on a chart recorder. As with the rod drop testing, if all of the coil currents from a complete bank of rods could be acquired at the same time, the duration of the test would be significantly reduced. As an extension of research performed by AMS on automating rod drop time tests, equipment was developed which allows the performance of CRDM timing tests of all rods in each bank at the same time. This results in data acquisition of a complete bank in the same amount of time it previously took to test only one rod.

Figure 3 shows three steps of a control rod for CRDM timing tests in a Westinghouse plant. The three signals shown on the figure are the stationary gripper current, movable gripper current, and the lift coil current. By acquiring the data digitally, various calculations can then be performed on the data. In these tests multiple CRDMs were tested at the same time eliminating redundant movement of the control rod banks and saving valuable critical path time.

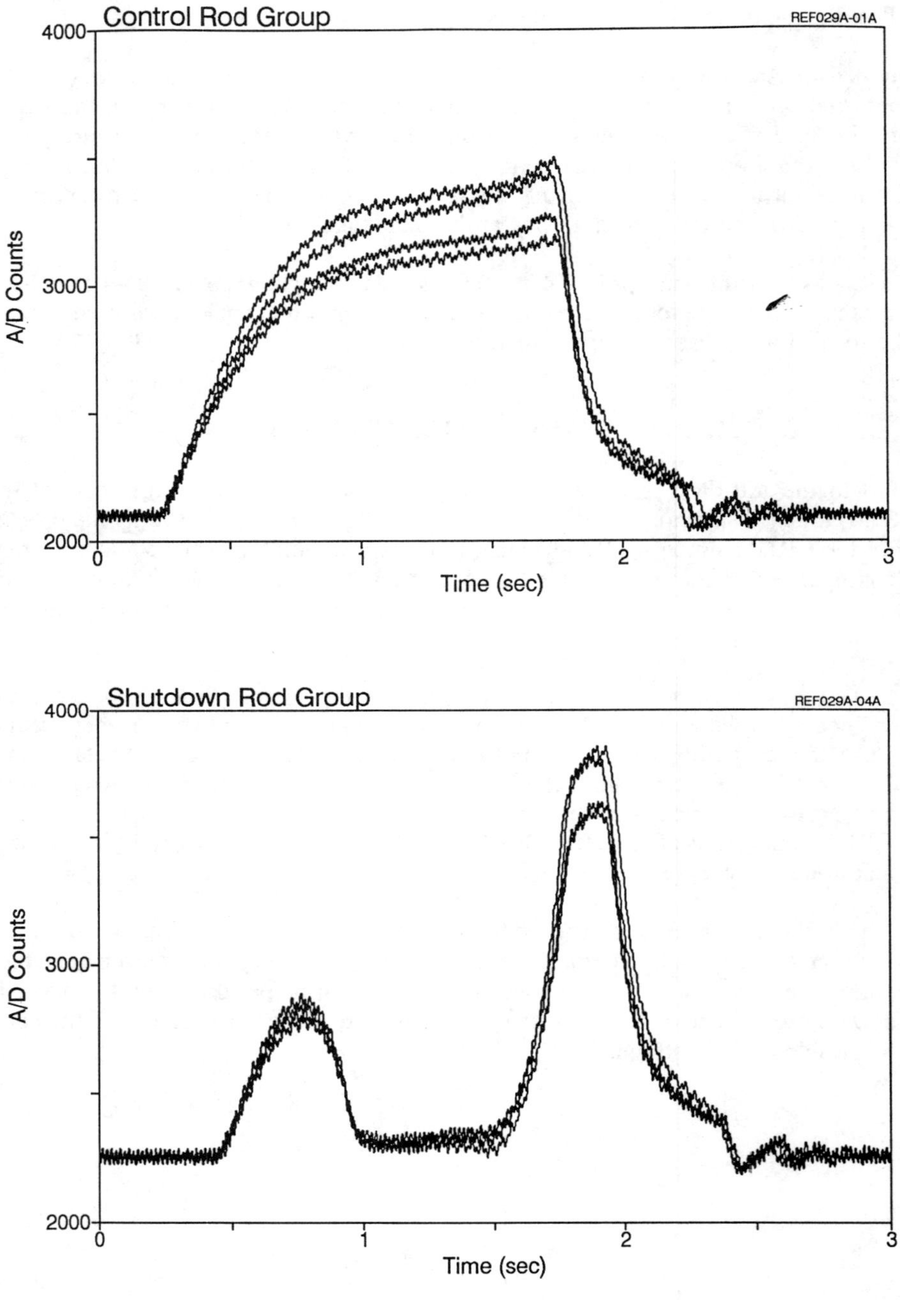

Figure 2. Sample Group Rod Drop Traces for Control and Shutdown Rods

TABLE 2 Sample Results of Rod Drop Time Tests at a Three Loop Westinghouse Plant (Tests of April 1994)		
Item	Rod I.D.	Rod Drop Time (sec)
1	F02	1.96
2	B10	1.94
3	K14	1.99
4	P06	2.03
5	K02	2.01
6	B06	2.06
7	F14	2.02
8	P10	1.99
9	D04	2.02
10	D12	1.96
11	M12	2.00
12	M04	2.02
13	H06	1.98
14	F08	2.00
15	H10	1.97
16	K08	2.03
17	G03	1.99
18	C09	1.98
19	J13	2.04
20	N07	1.98
21	J03	2.07
22	C07	1.94
23	G13	2.00
24	N09	2.05
25	E03	1.99

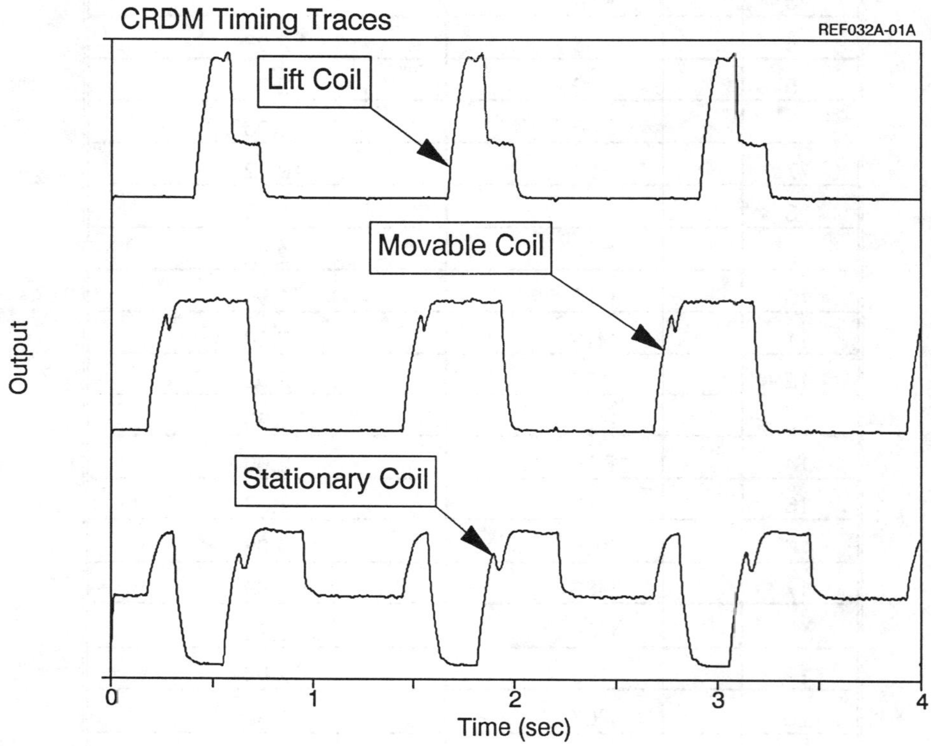

Figure 3. Three CRDM Timing Steps in a Westinghouse Plant

7. CONCLUSIONS

Automatic rod drop time and CRDM timing test equipment has been developed and validated for use in Pressurized Water Reactors. This equipment provides the following advantages over the current practice of manual tests:

1. The equipment can be used to drop any number of rods at once instead of testing the rods one at a time. This capability will reduce the critical path time required for the tests by as much as a factor of ten.

2. Use of this equipment eliminates the need to pull fuses to initiate the rod drops. The rod drops are initiated automatically under computer control or by opening the reactor trip breaker.

3. The analysis of the rod drop data is performed by computer and the results are available immediately after the test is performed.

4. Connections are made to the RPI cabinets once at the beginning of the test and personnel are not required in containment during the test. Power to the RPI cabinets can be controlled remotely by the computer. This avoids the need for operations personnel to be available to cycle the RPI power on and off during the test.

5. CRDM testing using an automated data acquisition system to acquire data on a complete bank at a time instead of only one rod at a time also decreases the time required during startup for control rod testing. Compared to using a chart recorder, digitally acquiring the data allows for easy calculations of various timing information and proper manipulation and storage of the data.

0074-056X/95/383-396/$0 + .50pp

PREPARING FOR ADVANCED CONTROLS

by

Richard C. Anderson
ARCO Products Co.

Presented

at the

Fourth International
Joint ISA POWID / EPRI

Controls and Instrumentation
Conference

on

June 6th, 1994

Preparing for Advanced Controls

ABSTRACT

This paper will discuss the implementation of advanced control(AC) systems in power generation facilities by extracting parallels to implementation experience within the hydrocarbon processing industry. The primary focus will be the foundation needed for AC systems. This would include topics such as; organization, benefit studies, proposal specifications, personnel requirements, training, hardware & software selection, vendor selection, documentation, etc. An overview of the benefits realized by the hydrocarbon processing industry will be presented for those unfamiliar with AC systems.

The improvements in unit profitability and process stability made possible by advanced control projects have been well documented over the past several years. Most of the publicly available material tends to emphasize the technical aspects of the project, usually centered on the success achieved by using a particular control technology. However, the success of an advanced control project usually depends just as much, if not more, on project execution. Unfortunately, the importance of project execution to advanced control is usually found out only through the experience obtained on the first few projects at a plant site.

There are many issues that must be resolved before advanced control implementation can take place. These include:

- Project Organization
- Personnel Requirements
- Cost / Benefit Studies
- Proposal Specifications
- Instrumentation
- Hardware Platform
- Software/Technology Selection
- Vendor Selection
- Documentation
- Training

These issues are common to all advanced control projects. Early attention to each of these will pay dividends as an advanced control project proceeds by providing: faster implementation and acceptance, eliminating rework, and reducing long term maintenance costs.

There is a common theme, ***communication***, that runs throughout each of the issues. Much of what is presented in this paper involves providing clear communication between the interested parties, whether internally within the user's organization or externally to and from an advanced control vendor. Any aspect of the advanced control project that is unclear will only cause unnecessary and potentially costly delays if left until the project is underway.

I. Project Organization

Most companies have their own "standard" way of organizing project work. Many different styles of organization will work. An AC project, like any other will run smoother if a project team is assembled as early as possible.

Communications are an important aspect of the AC project. Methodology should be established that will keep all project team members informed on current issues. Likewise, procedures for filing and retrieving information in a central location should be set. This includes electronic as well as paper-based information.

The improvements in unit profitability and process stability made possible by advanced control projects have been well documented over the past several years. Most of the publicly available material tends to emphasize the technical aspects of the project, usually centered on the success achieved by using a particular control technology. However, the success of an advanced control project usually depends just as much, if not more, on project execution. Unfortunately, the importance of project execution to advanced control is usually found out only through the experience obtained on the first few projects at a plant site.

There are many issues that must be resolved before advanced control implementation can take place. These include:

- Project Organization
- Personnel Requirements
- Cost / Benefit Studies
- Proposal Specifications
- Instrumentation
- Hardware Platform
- Software/Technology Selection
- Vendor Selection
- Documentation
- Training

These issues are common to all advanced control projects. Early attention to each of these will pay dividends as an advanced control project proceeds by providing: faster implementation and acceptance, eliminating rework, and reducing long term maintenance costs.

There is a common theme, ***communication***, that runs throughout each of the issues. Much of what is presented in this paper involves providing clear communication between the interested parties, whether internally within the user's organization or externally to and from an advanced control vendor. Any aspect of the advanced control project that is unclear will only cause unnecessary and potentially costly delays if left until the project is underway.

I. Project Organization

Most companies have their own "standard" way of organizing project work. Many different styles of organization will work. An AC project, like any other, will run smoother if a project team is assembled as early as possible.

Communications are an important aspect of the AC project. Methodology should be established that will keep all project team members informed on current issues. Likewise, procedures for filing and retrieving information in a central location should be set. This includes electronic as well as paper-based information.

There are two primary fields of process control expertise that may need to be covered by Process Control personnel, depending on the application being considered. Most plants are now instrumented with a Distributed Control System, or DCS. However, many AC packages, due to size and complexity, require execution on a Host computer such as a VAX/VMS system, UNIX system, or even a PC. While process control engineers are usually familiar with both types of system, most tend to be stronger technically on one of the two aspects. Therefore, on AC systems that will work with a combination of DCS and Host computers, two process control engineers with the appropriate specialization may be required.

Since most process control engineers have chemical or electrical engineering backgrounds, it is usually necessary to have someone with real-time computing systems experience associated with the project. Real-time process computing systems, while physically similar, are very different from business computers in the way that they are used. Business machines are normally required to supply information to a wide number of users. Whether the user gets the requested information now or 10 minutes from now is usually not very important. Process computing systems have only a few users, of which the most important is always the process control application and it's associated inputs and outputs. If information is expected once a minute, then it had better be ready once a minute, as failure can have very undesirable effects on the plant operation. Another aspect of real-time computing is that system software and operating changes have to be scheduled carefully. Unlike business machines, system maintenance can't be shifted to "off-peak" hours. Real-time systems have no "off-peak" hours, so all maintenance must be coordinated to minimize system downtime.

The final basic field of expertise that should be covered is instrumentation. Many, if not most, AC projects uncover a few locations in the plant where additional measurements are needed. While process control engineers can identify the instrumentation that is desired, an instrumentation engineer is usually needed to specify the particular type of instrument for the service and how it will be installed.

III. Cost / Benefit Studies

Cost / benefit studies are used to decide if a proposed AC project is feasible. Basically this involves performing a statistical analysis of current and past plant performance, then determining the value of improving the standard deviation of each controlled variable. A cost / benefit study is best performed without a preconceived idea of how the advanced controls will work. Rather, the cost / benefit study should pinpoint the most important controlled and controlling variables, guiding the project team to the objectives that will maximize benefits.

The cost / benefit study is usually performed by a process control engineer. This is because the determination of the amount of reduction in setpoint standard deviation is often a judgement based on past experience. For this reason, it is often advisable to hire an outside vendor to perform the study. A vendor has the advantage in that the costs and benefits can be based on projects with similar units at other plant sites. Also, the cost of a study is usually low, as the vendor hopes to entice the plant to use their services to install the AC system.

However, care should be taken when requesting a study from a vendor. The degree of certainty on the costs and benefits can vary widely depending on the amount of effort expended by the vendor. Beware of very quick and cheap studies. Such studies often assume that your plant is exactly like the last plant that was studied, and will assign benefits without really examining the history of your plant. Make sure that the study reports the statistical analysis of data from your plant. Otherwise, the user must accept a larger degree of uncertainty in the projected benefits. While most power plant processes are relatively straight-forward, simulation studies may be required for processes that show extreme non-linear behavior.

Payback periods for AC systems vary widely, depending on the process to be controlled and the state of the existing instrumentation and computer hardware. In general, most AC projects will pay out within a year or two. However, it is not uncommon to find projects that achieve payback in only a few months, especially if all hardware required for the project is already in place. Often, the return from an AC project can be used to justify the upgrading of a DCS or host computer system.

Since most power plants do not have extensive experience with developing AC systems, a brief description of how project benefits are determined is necessary. Most AC benefits are the result of stabilizing the control of one or more process quality parameters. In the power industry this would most likely be the MW of electricity or steam production rates. A common benefits calculation method is described below. This method has been proven through project post-audits for many different refining and chemical processes.

Calculating Benefits

The first step for calculating benefits is to determine how much variability is currently encountered in the controlled variables, including constraints, for the process to be evaluated. This is done by statistically analyzing history data from the plant. It is important to use data on as quick a time frame as possible to get a true indication of the amount of swing in the control variables. Averaged historical data should not be used as it tends to smooth the variability. From the analysis the user should be able to determine the standard deviation of all controlled variables.

The next step is deciding how much of the variability that can be reduced by AC system implementation. If a vendor is performing the study and has prior experience with the unit in question, then this experience will dictate the expected reduction. However, if there is no prior work on which to base the estimate, or if the study is being performed by the user, a conservative rule of thumb is that a 50% reduction in standard deviation for a controlled variable can normally be achieved.

Translating the reduction in standard deviation into throughput requires an additional assumption. Prior to advanced control, a process unit is typically run such that certain constraints are violated at a particular frequency. For instance, a maximum temperature constraint is violated maybe once a week. The usual assumption in these cases is that continuing the once a week constraint violation is acceptable performance for the AC system. Therefore, the AC system can increase throughput (or other quality parameter) such that the average value is moved closer to constrained operation. Refer to figure 1. for a graphical representation of this logic.

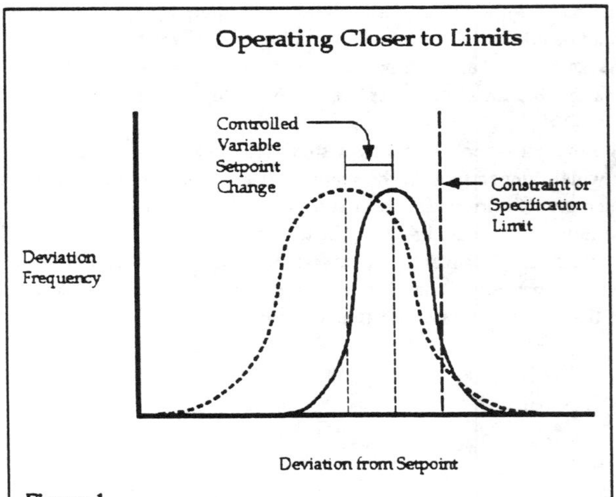

Figure 1

Once the expected improvement in the controlled variable has been determined, it is usually fairly easy to calculate the estimated benefits. For profit generated by an increase in throughput or output, don't forget to factor in the load related process cost increases such as additional energy requirements.

In some cases, reducing the number constraint violations or quality specification excursions are more important than increasing production. The number of excursions are reduced by reducing standard deviation of the controlled variable while holding the same average value. A cost will need to be determined for the "average" excursion. The benefits are calculated by multiplying the number of excursions prevented and the average excursion cost. This method is illustrated in figure 2.

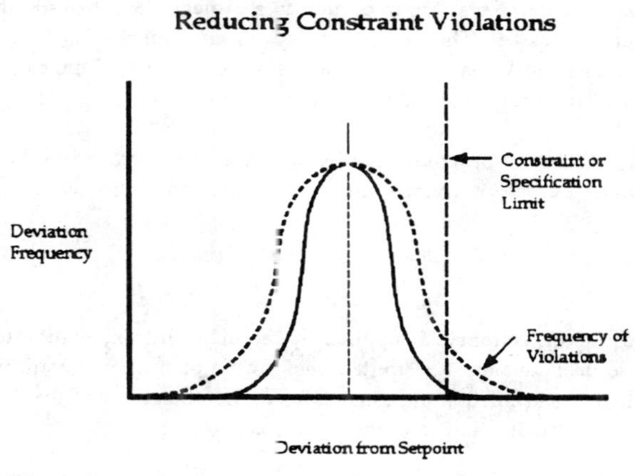

Figure 2

IV. <u>Proposal Specifications</u>

A well written proposal specification is a critical part of a successful AC project. In essence, an AC proposal specification is a formal way of communicating, "What do we really want?", to prospective vendors.

The act of preparing a proposal specification provides a very valuable service to the user. Often an AC project is started based on the belief that installation will improve process operation. Writing the proposal specification forces the project team to focus on exactly what improvements are expected. Otherwise, prospective vendors must make assumptions as to the project objectives. The result in competitive bid situations are widely varying bids that run the gamut from a bare-bones to a leading edge type AC system. It is almost impossible to make an objective judgement in these cases.

Another benefit of writing the proposal specification is that it usually brings out the "hidden" objectives of the project team. These might include:

- Control technology preferences
- Hardware preferences
- System interface preferences
- Documentation requirements
- Expected division of labor between client & vendor
- Training requirements
- Graphical interface requirements
- Ease of operation objectives

There are many additional little items that are important to a project team that are otherwise verbally communicated as a project progresses. This leads to delays in implementation, cost overruns, frustrated team members, and in the worst case an AC system that doesn't work properly. This can be prevented by taking the time to prepare a comprehensive proposal specification.

Once written, any variances from the proposal specification must be approved by the project manager and documented in written form. Verbal exceptions must not be allowed in any case.

V. Instrumentation

An AC system works with the measurements supplied by the plant instrumentation to make decisions on how to improve operations. Those decisions are transmitted back to the plant through still more instrumentation in the form of control valves. If the instrumentation is inadequate, the objectives of the AC project are doomed. Problems with instrumentation usually fall into one of two categories:

- Missing Instrumentation
- Unreliable or Broken Instrumentation

A cost / benefit study will normally highlight missing instrumentation needed to implement advanced controls. The missing instrumentation should be divided into two groups: those required and those that enhance advanced control. Installing new required instrumentation usually requires a lengthy lead time, so early identification is the key to avoiding project delays. In some cases, a unit turn-around is needed, and the entire project must wait, sometimes for years. Instrumentation that enhances advanced control performance requires a judgement call. The incremental benefit expected from enhancing the instrumentation system must be weighed against the cost for installing and maintaining the instrumentation.

Unreliable and/or broken instrumentation can be a much more subtle problem. Identifying such problems is one of the important reasons for having someone with extensive process experience on the project team. Operations personnel can usually provide the team with information such as: instrument readings that are very noisy, control valves that stick, control loops that are poorly tuned or in manual, analyzer sampling systems that plug frequently, etc. Such problems must be investigated as soon as possible. If the problems cannot be repaired, then the AC system design must be adjusted accordingly.

VI. Hardware Platform

Most issues concerning the hardware platform for the AC system can be condensed into a single question, "Do you have the computing resources to reliably execute the planned system?" AC systems can be executed on a wide variety of hardware platforms, in both the DCS and host computer environments. Most companies will have a preference as to which brand of computing hardware they will use. For the purposes of installing an AC system, the most common issues are: memory, CPU, disk storage, data interface capability, and number of users (including applications). In most cases, the expected benefits will easily justify any upgrades that are necessary for the AC system.

Once the resource requirements have been determined, and the current computing environment (if any) is found to be deficient, there are several options available for meeting the resource requirements. Among these options are: retro-fitting the existing system with upgrades, replacing the current system with a completely new system, or building a network of computers and workstations. The most reasonable option will vary depending on the current computing environment and amount of additional resources needed.

Try not to use a computer set up for business and administration needs as an AC platform. The differences in execution priorities between a business machine and a real-time process control machines are great. Business machines are usually concerned with crunching data for a large number of users simultaneously. The response time that it takes to process a user request is usually not critical (although easily annoying). Real-time process control systems must always give first priority to data input/output and execution of a few AC applications to the exclusion of all other users. The response time to application requests is critical to the reliability of the AC system. It is very difficult to make both business and process control users happy when using a common platform.

Another issue, already discussed in the Personnel Requirements section, is the need for someone knowledgeable in real-time computing to take care of the system. This usually requires someone familiar with both the DCS and host computers, as data interfaces are critical to the success of an AC system.

VII. Software/Technology Selection

The selection of the control technology is usually the most interesting aspect of an AC project. Usually there are several different technologies that can be used to achieve a given control objective. Each technology will have it's own advantages and disadvantages. While cost is always a major issue, many other factors must be taken into consideration.

For instance, some control software can only be used on certain types of operating systems. If the existing computer system does not provide the correct operating system, then the cost of converting the software to a new operating system (including the schedule impact of debugging the conversion) must be taken into account. Otherwise, a new computer and operating system must be factored into the project cost. In addition, some thought must be given as to who will support the new operating system.

Another issue is the expected maintenance effort and cost expected for a given control technology. A technology with a low initial cost might possibly require greater maintenance resources long-term. In many cases it is better to go with a higher initial cost if lower maintenance is expected. A control system that requires constant care will often degrade into an unused system.

Ease of understanding a particular control technology has an impact on it's final acceptance. If the Operations personnel cannot understand what the AC system is doing at any moment in time, the system will probably decline into disuse. Remember, the AC system is being commissioned to make operations easier to control. If it isn't, only constant vigilance by supervision will keep the system on-line. This is usually not an efficient way to run a plant.

Robustness of control is normally hard to evaluate from vendor proposals. The benefits of a AC system can only be achieved if the controls stay on-line. If possible, try to obtain client references willing to comment on this issue. Of course, the vendor will give references where the technology has been a success, but generally one can find out what kind of effort it takes to achieve success.

In some situations, the only viable technology may be unproven in the field. This may be because the technology itself is new or because a trusted vendor is moving into a new business area. The project team should carefully assess all cost and time estimates for using a new technology. Unfortunately, most of us tend to be overly optimistic with these estimates. New technology implementation invariably has small problems which cause schedule delays. There is also the chance that the problems and delays will cause Operations personnel to distrust the finished AC system for an extended period of time, decreasing the expected benefits. Plant sites implementing their first AC system are advised to avoid this situation and pick a more established technology, avoiding the possible complications.

Multi-unit facilities can create an additional factor in technology selection. For maintenance reasons, it is usually desirable to limit the number of technologies implemented within a facility. Therefore, a technology flexible enough to handle all or most of the units will often be the best choice. This minimizes the amount of training, procedures, and in-house expertise that must be developed for each succeeding project. Once a technology has been chosen, subsequent projects that desire a different technology must provide justification that the expected benefits outweigh the costs of maintaining two different systems.

VIII. Vendor Selection

Choosing a vendor for advanced control is as important as the technology selection. The best technology is no guarantee that a project will be successful. If your company's business style does not mesh well with the vendor's, the AC project is liable to be a long and frustrating experience. Some of the criteria that should be used in selecting a vendor are:

- Technology
- Experience
- Personnel
- Customer Support
- Price

Often the choice of a particular control technology will mandate the use of a specific vendor. Likewise, the use of specific hardware, especially at the DCS level, can also require that specific technology be used. Be careful that the desire to use a particular technology or hardware does not blind the project team to the other aspects involved in selecting a vendor. Fortunately, there are usually several competing technologies for most AC objectives. The project team should take the time to investigate as many as possible to find which gives the best fit for a particular project.

It is best if the vendor installing and designing your AC system has prior experience with implementing advanced control with the computing hardware on your process. However, perfect fits cannot always be found. For instance, a vendor may have experience with your DCS, but not with your Host system. Another possibility is that a vendor will have a well established control technology, but may not have implemented it on your type of process. If the plant is installing an initial AC system it is probably best to select a vendor that has been in business several years and has prior experience with your particular type of process, rather than a vendor applying a new technology.

The vendor personnel who will work on your project can greatly influence the schedule and ultimate success of an AC system. A good technology can easily be poorly applied by inexperienced personnel. Ask each vendor to supply resumes of their personnel who would work on your project. Try to personally interview as many as possible. The vendor should be required to commit core personnel to a project team prior to project kick-off. Retain the right to affirm any substitutions on the core team. Once the projected has started, be very suspicious if the vendor wants to change out more than one or two core team members, as it is usually a sign of organizational problems. Do not agree to personnel switches until you are satisfied that the replacement is suitable.

Most vendors will offer some form of post-project support. At the very least, a vendor should warrant an AC system to be free of defects for at least one year from the plant acceptance of the commissioned system. Licensed control technologies will usually require the client to pay a yearly maintenance fee. The maintenance fee will entitle the client to periodic upgrades of the control technology and sometimes some engineering hours to handle unforeseen problems that may occur. Some of the larger AC vendors will provide customer support hot-lines to answer questions and/or schedule on-site support should a problem be found. Smaller AC vendors typically operate less formally and allow the client to directly contact a project team member when problems occur.

Price will always be a big factor in choosing a vendor. It is the one item that anyone external to the project team can immediately understand. However, choosing a vendor solely on price can lead to problems. If vendors know that they will be selected solely on price, then they will, of course, bid accordingly. However, this can often lead to the offering of "stripped down" AC systems with lower prices but also possibly lower benefits and higher long term maintenance costs. This can be prevented by producing a comprehensive proposal specification for the vendor's to bid against.

IX. Documentation

The most important rule in obtaining good documentation for an AC project is to provide very specific requirements. The project team needs to decide what documentation will be needed, how it will be organized, and the expected delivery time. The project team should decide the content and level of detail required for each piece of documentation. These requirements will vary widely depending on factors such as the hardware and software used, availability of in-house technical advanced control support, over-all system complexity, etc. A typical AC project is likely to require the following documentation:

- Project Correspondence File
- Cost / Benefit Study Report
- Functional Design Report
- Detailed Design Report
- On-Site Activity Logbook
- Operations Manual
- Engineering Manual
- Systems Maintenance Manual
- Post-Audit Report

Describing the usual contents of these documents is beyond the scope of this article, but a few comments are necessary.

The user should carefully review proposed documentation from the each AC vendor during the bidding process. If it does not contain the expected information or if it is in a format that is difficult to follow, ask about the cost to have in written according to your standards. This additional cost will need to be added to the total project cost when evaluating bid proposals. Do not accept the non-standard documentation.

Try to keep documentation for Operations separate from all other documentation, especially engineering documentation. There are at least two good reasons for doing so. First, it keeps the operator from having to sort through pages of information intended for a different audience while trying find what is needed. The Operations manual should have just the information required to run the AC system, such as: graphics screens, key stroke sequences, responses for error messages, manual entry requirements, etc. Leave technical details such as model development for the engineering manual. The second reason is that the Operations manual will most likely require a separate review by the Operations department. Often what an engineer (vendor or client) considers important is not what the operator feels is important, and the Operations Manual must be written accordingly. It is also likely that the Operations Manual will need to be included as a section of a unit operating manual. This is much more difficult if the operating documentation is imbedded within another document.

The On-Site Activity Logbook can be an important tool in both tracking the AC project as well as minimizing time spent on maintenance after the system is completed. Most engineers keep some sort of personal record as to their daily activities while commissioning an AC system. The On-Site Activities Logbook serves as a central location for storing these daily reports. The user should require all personnel, both in-house and vendor, to record each days activities, including problems encountered. This allows the user to determine whether the commissioning effort is proceeding according to plan, and helps to highlight problems areas that need attention. Once the commissioning is complete, the logbook helps answer questions such as, "Why did we do it this way?" and, "What did we do to solve this before?"

The user should give some consideration to the form the documentation will take. It is a good idea to ask the AC vendor to supply an electronic version of all delivered documentation. This will make it easier to revise the documentation as changes occur to the process and the AC system. Make sure that the electronic version is delivered in a format your system can read. Also, it is likely that some form of on-line help for operators and engineers who will be using the AC system will be desired. Do not use on-line help screens as the project documentation. Due to system limitations, it is usually difficult to put the degree of detail in on-line help screens that is required for good AC system documentation. Despite the desire to go to the paper-less work environment, the author has yet to see an on-line documentation system that beats flipping through the pages of a well-written manual.

X. Training

An AC project usually requires that four distinct types of personnel be trained:

- Operations
- Process and Control Engineers
- Computer Systems Engineers
- Management

The need for training Operations personnel is obvious. If the operators are not trained properly, they will not understand what the AC system is doing. As a result, the AC system will most likely be turned off at the first hint of a problem or some other situation where the operator does not understand the control action being taken. It is recommended that the operators receive training in three sessions: prior to commissioning the AC system, at the time of commissioning, and a final session within a few months after acceptance of the AC system. Also, some provision should be made for how to train new operators on the system in the years to come. One method for doing so is to videotape the original training sessions or to prepare such a tape for the new operators.

Another possibility for training the operators is the use of dynamic simulation. This method is usually very expensive, especially if an exact duplicate of your process is to be modelled. Generic process simulations can be built easily, but this usually degrades their value as a training tool for an AC system. However, many pieces of equipment in a power plant are essentially duplicates (for instance, multiple turbines with HRSGs). When duplicate equipment is used, it may be economically feasible to use dynamic simulation as a training tool by modelling just one of the equipment trains.

The Process and Control Engineers, in addition to knowing how to operate the AC system, need instruction on how to maintain the system after the vendor leaves. Their training should include the basic operator training as well as items such as:

- Engineering error message interpretation
- Debugging techniques available
- How and when to change engineering or tuning constants
- How to update economic data
- Procedures for obtaining printed reports
- Who and when to contact for Vendor support

Training of the Process and Control Engineers should be performed after installation but prior to commissioning of the AC system.

The Computer Systems Engineer needs to know the details of how the application is executed and maintained on the AC computer(s). The systems training should be provided during the installation of the software onto the AC computer(s). After commissioning is complete and the system has be cleaned-up, a final review of the system training should be performed.

Finally, the plant Management staff should receive some brief overview training about the AC system. The overview should provide general knowledge as to process objectives that the AC system can and cannot achieve. Also, they should be informed of the impact of any data that they may be asked to provide for the system. In particular, the requirement for accurate economic data so that true optimal performance can be achieved should be demonstrated.

For instance, misleading economic data can cause some very interesting effects on plant operation. In one case, an ethylene plant optimization system was given economic data that reflected the final prices for most of the components in the plant product streams. However, due to the effect of downstream processing of some components, not all of the product prices reflected the proper value at the ethylene unit boundaries. This caused the optimizer to adjust the operating conditions to make less ethylene while producing more side products. After the use of the product prices was properly explained, the relative prices were modified and the optimizer change the operating targets to reflect the true objective of maximizing profit by increasing ethylene production while also maximizing side product value.

XI. Summary

Preparing for advanced controls is an interesting challenge that requires the integration of engineering and computing disciplines. The success of the final advanced control system depends on the decisions made in the early stages of project development. This paper discussed several issues that need to be resolved during the start of an advanced control project. The importance of clear communication of requirements and expectations on each of these issues cannot be over-emphasized.

0074-056X/95/397-403/$0 + .50pp

STEAM TEMPERATURE CONTROL WITH RATE OPTIMAL CONTROL

by
Craig Backman
Kris Hanson
Jay Werre
Montana-Dakota Utilities
and
Jerry Richerson
Honeywell

Introduction

Montana-Dakota Utilities has experienced significant steam temperature control problems throughout its operating history. This has resulted in constraints on unit swings, controls instability, and significant steam temperature excursions. One of the major contributing factors to the instability was slagging. Because of the slagging, the boiler characteristics changed dramatically between outages. This caused significant tuning problems to the point that automatic control was not possible. After extensive evaluation, advanced multivariable model based control was purchased for this application. This paper describes the evaluation and initial results of model based steam temperature controls.

Background

Montana-Dakota Utilities is the operator of the Coyote Station generating plant and is one of the four owners of the plant, the other three owners being Northern Municipal Power Agency, Northwestern Public Service Company, and Otter Tail Power Company. The Coyote station is a 415 MW unit located near the center of North Dakota, in the heart of lignite country. The unit which is a B&W drum boiler design came on line in May of 1981. The unit burns lignite, which has a British Thermal Unit (BTU) rating of 6,900 BTU per pound, using 12 cyclones. Coyote Station experienced control problems due to the boiler characteristics and controls implementation using several different control systems.

From initial operation, steam temperature control was a problem. The boiler characteristics are variable, dependent mainly on the cleanliness of the boiler. The control stability followed cleanliness on a one-to one basis. As the boiler fouled, a tight coupling is observed between steam temperature control, the boiler master and the pressure control. Under heavy slagging conditions, steam temperature controls were typically set to manual.

The thermal inertia with varying slag deposits on the boiler made it very difficult to set up the tuning, even on main steam, to constants that were acceptable. The reheat controls, on both attemperator and gas recirculation loops had been re configured (hard wired) a number of times with no improvement in operation. Tuning was sometimes a day-to-day basis. Since the steam temperature controls were implemented in two different control systems, the tuning and re-configuration was difficult for the technicians. With these conditions, we were experiencing steady state super heat temperature swings up to 20 degree. F. On load swings, the superheat temperatures varied as much as 40 degrees

Obviously this had an impact on both heat rate and availability. Steam temperature control was improved somewhat through operator awareness and manual attemperator control. Gas recirculation was used for both control of furnace exit gas temperature (FEGT) and steam temperature control - a compromise between the two requirements. Measuring furnace exit temperature, especially at high loads, is extremely difficult. Thus we calculated furnace exit temperature based on data from our PMAX performance calculations. Using this technique, the operator was able to manually adjust the Gas Recirculation fans to keep the exit gas temperatures in a range that minimized the slagging. This prevented using the Gas Recirculation fans for steam temperature control as they were initially intended.

Evaluation

We began investigating better solutions to these problems two years ago. We decided we needed to replace both the reheat and superheat control systems, which were initially a mix of different types of analog control systems. Additionally, with better technologies available to measure FEGT, we added loops to control FEGT, improving slagging control.

We also evaluated using both conventional PID based control and model based control. Our decision between the two approaches was based primarily on two criteria; initial cost, and life cycle cost. Evaluating initial cost is fairly straight forward. Evaluating life cycle cost is another matter. It involves performance, maintainability, and equipment impact (in this case equipment impact involves how well it supported life of the boiler and turbine). Our control vendor informed us that implementing new controls in a DCS using PID would control main steam temperature within a range of eight to ten degrees Fahrenheit. They offered as an option, model based control which would provide additional controllability with expectations of maintaining steam temperature within three degrees Fahrenheit. There was a cost difference of $50,000 between the two systems and we had to evaluate the value of this cost difference.

The Rate Optimal Control provided several significant benefits over conventional control.

1. Greater ability to coordinate multiple loops and their interactions. For instance, the steam temperature control is de coupled from the other control loops, thus removing one of the causes of the boiler swings.
2. Tighter temperature controls with rate of change control. Temperature swings of three degrees F and low rate of change of temperature will result in lower stresses on boiler tubing, boiler headers and the main turbine. With the significant heat damage to the heavy water wall components of this boiler, this capability alone justified the additional cost of model based control.
3. More robust control. The soot blowing and slagging will have less effect on the controllability. Model based control as applied to steam temperature control will provide the benefits of continuously de coupled control which will help stabilize other control loops.

Implementation

In July of 1993 we elected to proceed with the model based control provided by Honeywell called Rate Optimal Control. With the ability to control temperature ramps, we anticipated savings from reduced outages associated with tube leaks, and we anticipated extending furnace tube life. This is very important since we will perform a major tubing replacement during the same outage we install the new controls. The second reason we selected this technology was its maintainability. We could not hope to achieve these benefits (reduced outage time and extended furnace tube life) if the system was not maintained and kept in automatic. Control loops must be tuned regularly whether conventional PID or model based control. With the high degree of variability of temperature transfer characteristics due to slagging already experienced, the new controls had to exhibit robust control and ease of tuning. The method of tuning is different between PID and control. With PID control, you set Gain, Rate, and Reset. With control you modify (update) the model. We determined that we needed a system such that any engineer or technician who would normally be expected to tune a PID loop, could also adjust (tune) the model. The control had the capability to continuously develop a new model for use by the technician in addition to automatically calculate new models at desired operating conditions. This built in modeling capability gave the plant technical staff the tools necessary to keep the models updated under changing plant conditions.

Once we placed an order, we became concerned about the model based control. We were concerned about how the model would be built, what it would look like, and how would we maintain these controls with our personnel. In fact we kept asking the vendor to supply SAMA's of the model so we could review and approve. However, since the technology is so different from the PID technology with which we are familiar, it was difficult to understand the implementation. This lead to some concern during the implementation cycle. The following is a brief description of the control technology and the steam temperature implementation to give an idea of what we faced.

The steam temperature control is based on the Honeywell designed, model-based, non-linear, discrete digital multivariable control algorithm that utilizes finite variable horizon technology. In order to understand how the steam temperature controls operate, an understanding of the controller is necessary. (See Figure #1 for the control functional structure). The controller handles setpoint changes in an open loop manner, in either a time-based or rate-based mode using a function called the reference processor. In a similar manner disturbances, such as slagging, are rejected closed loop, again in either a time-based or rate-based mode. The Disturbance Processor produces the corrective disturbance control signal required to correct for plant upsets. The disturbances are identified by comparing actual measurements to the variables forecast by the model. The rate and time may be modified and the mode may be changed during operation without causing upset. This flexibility provided us with the comfort that the model based control could handle the variability in our boiler response. The actual plant model for each of the plant functions being controlled is stored in the Model Reference. These models are automatically calculated by the system based on plant disturbances established by the technicians. The model selected for operation is stored in the Model Reference. Based on different operating conditions, new steam temperature models can be loaded on line and automatically to allow for different load and plant conditions, such as slagging.

One other key aspect of the controller is the Interference Processor. The Interference Processor is the mechanism by which each controller communicates to the others to provide anticipatory information as to a planned change in the output. This is the function which allows us to separate the interaction of superheat, reheat, gas recirculation fans, and load. Using its knowledge of the plant model and the user's rate requirement, the Interference Processors coordinates and continually modifies the activity of the various ROCs to meet the control objective specified by the user. This allows the controller to deal with the interactive nature of boiler control such as superheat, reheat, and gas recirculation.

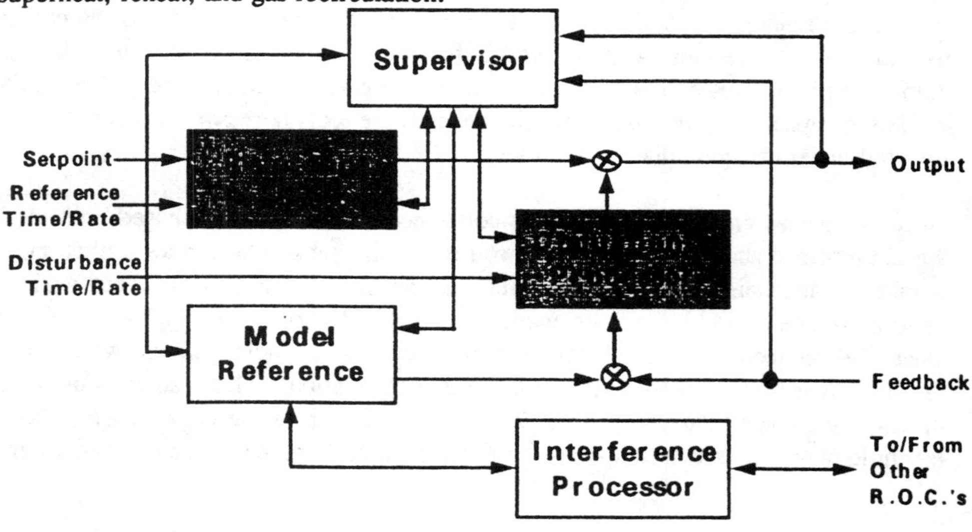

Figure 1

A simplified structure of the advance steam temperature controls is shown in Figure #2.

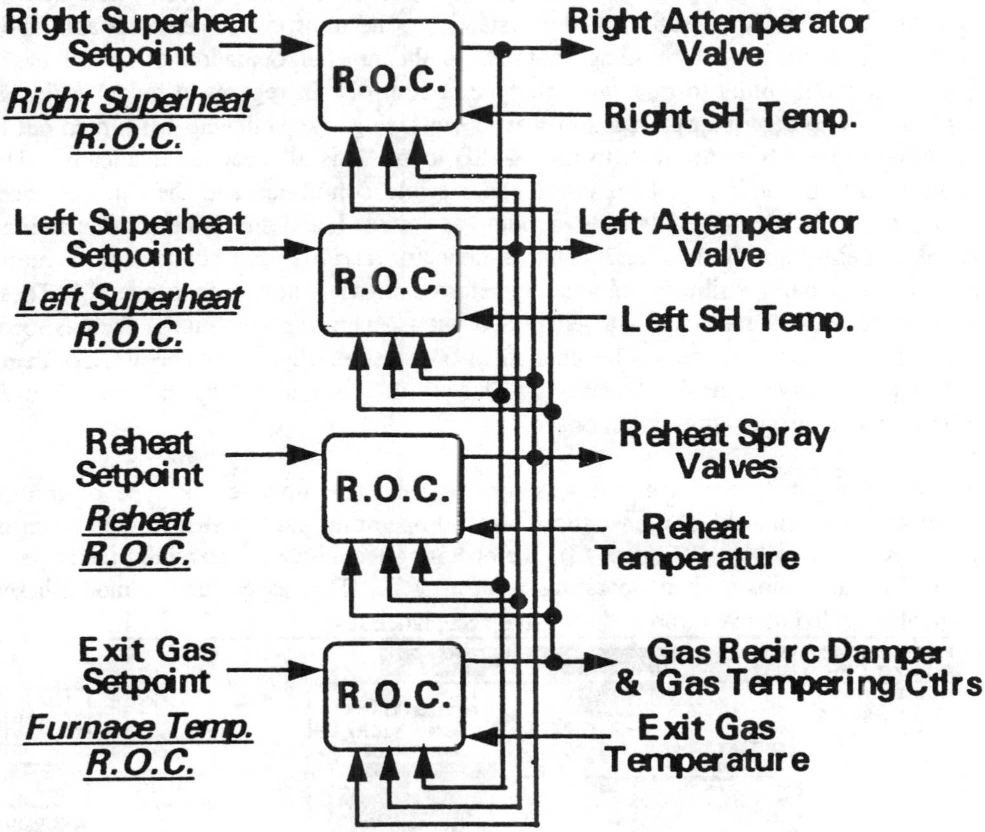

Figure #2

As can be seen four controller are used. A controller is used for each of the Superheat attemperator valves, the reheat spray valve and the Gas Recirculation damper Each controller has its own process model as well as a model of the interaction between each of the other three controller. In this manner both the process response and the interaction between each of the outputs can be modeled and controlled. By breaking the controls into a controller for each aspect of the control problem, i. e. right and left superheat, reheat, and Gas Recirculation, the set-up and modeling is greatly simplified. The base model for the right and left super heat, reheat, and exit gas controls were determined for each controller. After the base models were determined, the interaction between each controller was modeled and provided for the interference processor. This ability was very important in order to provide for the de coupling of the interaction of the controls. As the plant characteristics continues to change after the outage, new models can be loaded. This can be on a manual basis or automatically depending on the frequency of the changes.

In January of 1994 we performed the factory checkout of the system. The official agenda was to verify operation of the system, but the unofficial objective was to gain confidence in the model building method, in the normal operation of model based control, and our ability to maintain this type of control. In regards to model building, the best way to describe our reaction was to say seeing was believing. It turned out to be easier to build a model than to tune a PID loop. It is all done automatically. The operator sets the unit at a load point under stable conditions and then the engineer initiates a model build. Basically a bump is initiated and the system calculates an equation describing the process. This is done for various load points and operating conditions to build a library of what is referred to as "knowledge packets". These knowledge packets are then switched in and out as operating conditions, such as load, soot blowing, etc. change. The change over is essentially instantaneous, occurring between successive control calculations. This should allow us to stay in automatic over a wider range of operating conditions.

The other major concern we had was how to view and operate this type of control technology. During the demonstration, we had operating personnel review the control graphics. As seen below (Figure #3), the operating technique is essentially the same as using PID technology when operating from a CRT. This aspect of the model based control provided us reassurance during the acceptance test.

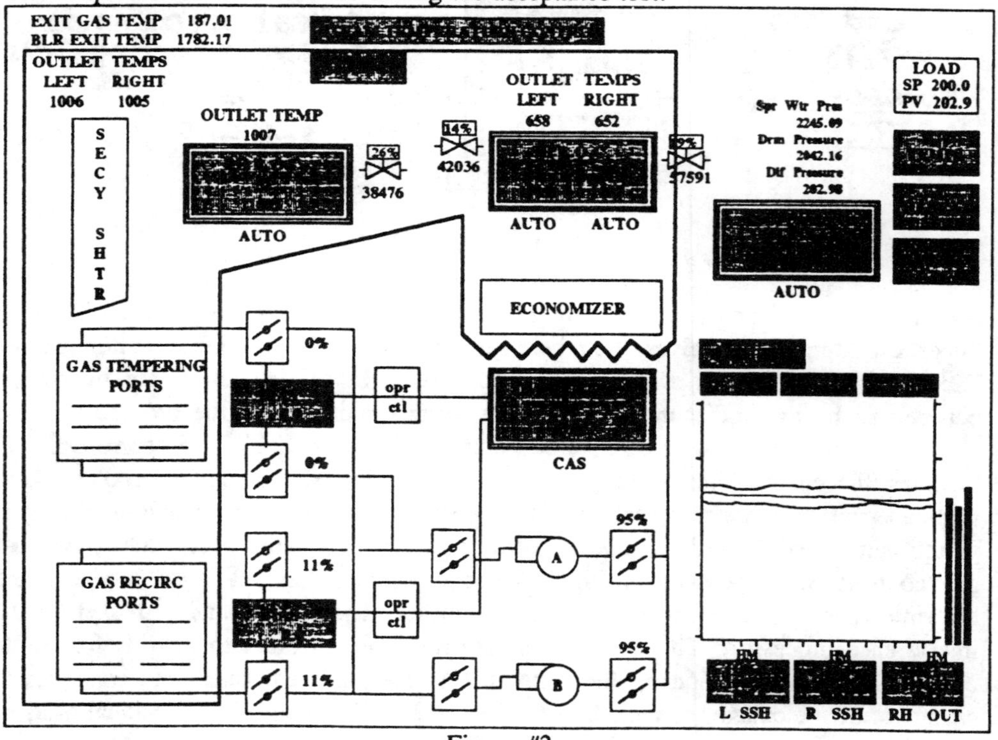

Figure #3

Results

Based on the factory acceptance test, we were confident that the controls would perform as we envisioned and that we can maintain this type of control. The outage was completed on May 15, 1994. Actual modeling did not start until May 17 due to other plant equipment problems. The initial modeling of the superheat controls was completed in one day. With the initial models and without any interference modeling completed, the superheat controls were maintained to within 5 degrees F. Under upset conditions, such as soot blowing or adding cyclones, the temperatures were maintained to within 10 degrees F. The superheat dead time was identified at 60 to 75 seconds. One surprise in the tuning of superheat attemperator flow was that the spray valves have a 5 per cent minimum movement. Obviously this results in coarser control than desired. With this limitation, super heat control within 5 degrees F is very satisfactory. Preliminary results of the reheat controls are similar. The reheat apparent dead times of up to five minutes and a time constant similar to the super heat temperature. However, there is a small change in reheat temperature for each disturbance.

At this point, we have all models completed. We are currently working with the interaction between the furnace exit gas temperature and the reheat temperature. Our original objective was to control furnace exit gas temperature at 1850 degrees F. Under clean boiler conditions, this furnace temperature was not sufficient to maintain reheat temperatures. As the boiler fouled, we have been able to maintain the reheat and exit gas temperature controls in automatic and met reheat temperature. With the ability to decouple the two actions, we now have the flexibility to adjust the operating points and evaluate the impact on slagging and reheat. This process of evaluation will take a period of several months to establish our new desired operating point.

Summary

The controls were installed and in automatic in less than four days from start up. Because of the long time delays involved with steam temperature controls, it takes several hours to establish a model, place it into operation, and observe the results. This is different from Honeywell's experience of using the Rate Optimal Controller for a boiler front end. For a boiler front end with the faster process time constants, a complete model and test for the front end could be completed in four to six hours. The time for modeling the temperature controls takes several days. Automatic control of super heat temperature, reheat temperature, and furnace exit gas temperature has been obtained. Super heat temperature is being maintained to less than 4 degrees F under typical boiler conditions. We are waiting to see the controls response under dirty boiler conditions; however, based on initial results we are confident that the results will be good under dirty boiler conditions. Overall the controls modification met all of our expectations. We are able to maintain superheat, reheat, and exit gas temperatures in automatic controls with minumum temperature excursions under all operating conditions for the first time in the boiler history.

0074-056X/95/405-414/$0 + .50pp

THE ALLOWABLE VALUE IN THE WESTINGHOUSE SETPOINT METHODOLOGY - FACT OR FICTION?

C. R. Tuley Jr.
Fellow Engineer
Westinghouse Electric Corporation
P. O. Box 355
Pittsburgh, PA 15230-0355

T. P. Williams
Senior Engineer
Westinghouse Electric Corporation
P. O. Box 355
Pittsburgh, PA 15230-0355

KEYWORDS

Setpoints, Allowable Value, Instrument Operability

ABSTRACT

The concept of a fixed magnitude Allowable Value was first utilized in November, 1974, as the determination point for process rack (and by implication, channel) operability for Westinghouse plants. It was modified in August, 1982, to include additional instrument uncertainties and two calculation paths to determine the limiting value. In June, 1991, it was proposed that in the Westinghouse setpoint methodology, a fixed magnitude Allowable Value is not a very sensitive indicator of equipment operability. It was suggested that equipment operability should be based in part on the concept of relativistic drift. Relativistic drift should be determined to be less than a specified value, e. g., within design. In June, 1992, a revised algorithm was proposed which demonstrates significantly less sensitivity to a fixed magnitude Allowable Value than the original algorithm. This paper discusses some of the misconceptions and misuses of the concept of a fixed magnitude Allowable Value as an indicator of channel operability within the Westinghouse setpoint methodology. Also discussed is an approach to determine relativistic drift and to define its use as a more sensitive indicator of process rack module or transmitter operability. Use of this approach will result in the definition of a trending program for determination of drift for the satisfaction of surveillance interval extension programs. A model Technical Specification is proposed for discussion.

INTRODUCTION

In the Custom Technical Specifications, e.g., Zion Units 1 & 2, the Trip Setpoint is a single value. An instrument channel whose as-found condition for the bistable exceeds this value is considered to be inoperable. This resulted in a large number of unnecessary reporting occurrences. Unnecessary because allowances had been made in the uncertainty calculations/judgements performed at that time, for calibration and drift errors for the process racks. To reduce the reporting of these occurrences, the concept of the Allowable Value was created, i.e., two values were established for an instrument channel, the nominal setpoint (Trip Setpoint) and an operability point (Allowable Value). This concept was first utilized in the D. C. Cook Unit 1 Technical Specifications, issued in November, 1974. Westinghouse agreed with the concept as a means of allowing the plant to set a trip setpoint at a nominal value and determine operability of the process racks on a monthly basis. The original basis for the difference between the Trip Setpoint and the Allowable Value was process rack drift. It was noted at that time that other parameters may contribute to the as-found condition of an instrument channel, but the NRC determined that the allowance for rack drift was a reasonable starting point and licensed the change in approach. This change resulted in a significant reduction in the number of channels found inoperable during performance of the required surveillance.

However, reporting occurrences were still created because of procedural and equipment differences. Plant procedures still had calibration tolerances that were two sided in nature but were forced to be one sided to insure maximum drift allowances (see Figure 1). Measurement and test equipment errors at some plants were also noted to be quite significant due to desires for flexibility in test equipment selection. In response, Westinghouse revised its methodology for determination of the Allowable Value, including explicit allowances for process rack calibration accuracy, drift, measurement and test equipment accuracy and bistable setting tolerances. Discussions with the NRC were initiated in 1980 and approval was received in August, 1982. The revised Allowable Value was first used in the Virgil C. Summer Technical Specifications. This resulted in a further reduction in the number of reporting occurrences for process rack inoperablity. Additional reductions in reporting occurrences took place when the NRC changed reporting requirements to loss of functionality as opposed to loss of a channel in the 1983 - 1984 time frame. Since that time, Westinghouse has reviewed its basic approach to instrument uncertainty calculations, particularly with the use of digital protection system designs, and has suggested that the current concept of the fixed magnitude Allowable Value may no longer be a reasonable determination of channel operability. Noted below are examples of why this is believed to be true.

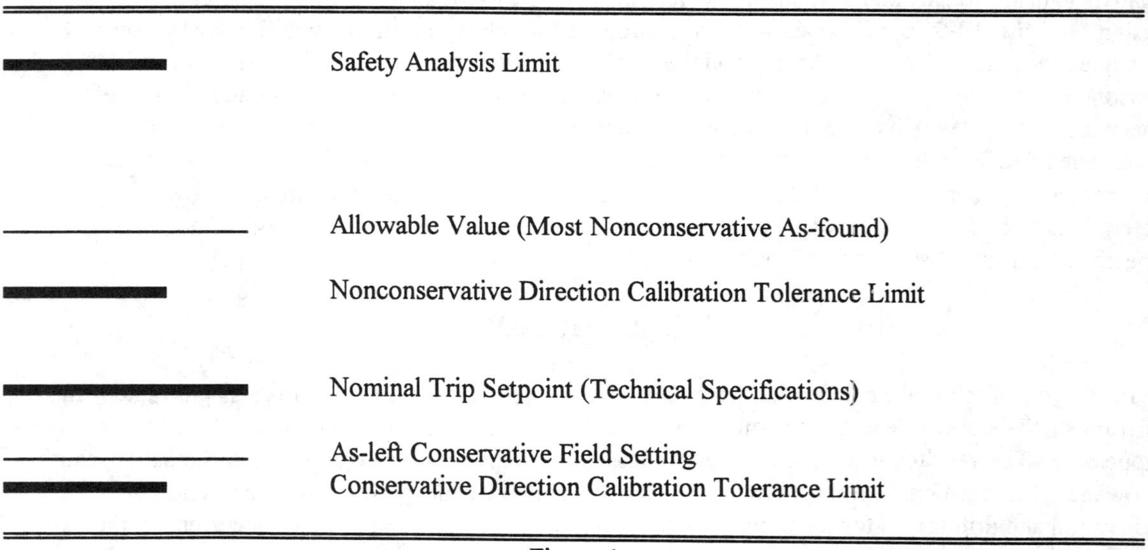

Figure 1
Typical Relative Calibration Tolerance and Allowable Value Relationship

ALLOWABLE VALUE MISCONCEPTIONS/MISUSES

Conservative Setting of Trip Setpoints

The Custom Technical Specifications encouraged the setting in the field a nominal trip setpoint conservative with respect to the Trip Setpoint defined in the document. An example of this concept for Pressurizer Pressure High Reactor Trip, is a field setting of 2380 psig for a Technical Specification Trip Setpoint of 2385 psig when the Safety Analysis Limit is 2440 psig. While this was not explicitly encouraged, it was the only way for a plant to allow for calibration accuracy and process rack drift. The Technical Specifications explicitly noted that an instrument channel's process rack/bistable found to trip at a value in excess of the defined Trip Setpoint value should be considered inoperable. Thus to insure some allowance for drift and to allow instrument technicians some margin for field setting errors, the as-left condition for a channel was typically defined in plant procedures as 0.5 to 1.0 % span conservative with respect to the Trip Setpoint in the Technical Specifications. Because the difference between the Trip Setpoint and Allowable Value in the Standard Technical Specifications (STS) was defined as process rack drift only, many plants continued to procedurally define the channel's as-left condition in the conservative direction with respect to the Trip Setpoint. Since many plant procedures allow calibration tolerances significantly larger than what can be achieved in the field, e.g., a calibration tolerance of ±0.5 % span when ±0.2 % span can be easily achieved for single input protection functions, a significant, implicit increase in the process rack drift allowance can occur. Thus a plant could perform surveillance, determine that a channel was operable, because the as-found condition was within the Allowable Value, but in actual fact, the channel operability is suspect because the drift for the channel was 50 to 100 % greater than that allowed in the channel uncertainty calculation and possibly greater than that specified by design.

Increased Measurement and Test Equipment Accuracy

The original assumption for measurement and test equipment accuracy (M&TE) in the Westinghouse setpoint methodology is that SAMA PMC 20.1-1973[1] is satisfied. PMC 20.1-1973 has since been replaced by ANSI/ISA S51.1-1979[2] but the requirements for M&TE are essentially the same, i.e., M&TE should have accuracy ratios of at least 4:1 (calibration tolerance to M&TE) and preferably 10:1 or greater. For some devices, e.g., pressure transmitters with turndown ratios of 3 to 3.75:1, it can be difficult to achieve field M&TE accuracy ratios of better than 1.25:1, even with the use of very accurate digital readout devices. But for process racks, there is a considerable quantity of various digital readout devices that can achieve accuracy ratios of 4:1 and even 10:1 for digital process rack designs. Thus achieving an M&TE accuracy ratio of 10:1 for 30 year old analog process rack designs is not difficult. However, there can be an insistence by plants on continuing the use of M&TE accuracy ratios of 1:1 in uncertainty calculations. It is recognized that this results in a conservative assumption for a channel error, but the assumption results in a nonconservative magnitude for the Allowable Value. If the plant uses an M&TE accuracy ratio of 10:1 in the field, but uses an accuracy ratio of 1:1 in the calculation, the pre-1994 Westinghouse algorithm would effectively increase the Allowable Value by 25 %. This results in an increased allowance for process rack drift of up to 100 % larger than that utilized in the uncertainty calculation and would be at least four times the amount of drift typically seen for a process rack channel. This can be significantly in excess of design. For digital process rack designs with self check/self calibration capability, the M&TE accuracy ratio should be well within the process racks self calibration accuracy to insure the proper functioning of that aspect of the design. Use of an M&TE accuracy ratio of 1:1 for this capability is defeating the purpose of incorporating this feature into the design.

Overstatement of Drift

When performing an uncertainty calculation there are several competing effects, e.g., utilizing the maximum possible drift to insure a bounding total uncertainty, bounding the expected drift to decrease the probability of exceeding the drift value, decreasing the drift value to decrease the total uncertainty to reduce the operational impact of a Trip Setpoint. However, none of these approaches are the most appropriate when trying to set a channel operability limit. A weakness of past Westinghouse approaches has been utilizing only the Safety Analysis Limit viewpoint, i.e., values that are conservative in the determination of satisfying the Safety Analysis Limit are acceptable for use in channel operability determinations. This is inappropriate because what is a conservative value for determination of the difference between the Safety Analysis Limit and the Trip Setpoint may be a significant overstatement of the expected drift for the process racks. For example, the expected drift value for Westinghouse analog process racks is a factor of two to four smaller than the value used in the uncertainty calculations. Thus for channel operability determination, it would be better to compare the actual drift with the expected drift for the channel. It would be best to compare the actual drift to past drift over the last several surveillance intervals for that specific channel, i.e. channel specific trending. For digital process racks with self check/self calibration capabilities, any drift noted in excess of the self calibration accuracy should be considered indicative of failure in an operability determination. Digital process racks with this capability will have failed if they are unable to perform calibration corrections, i.e., the channel will have shifted such that the instrument span limits can no longer be satisfied or the correct analog to digital conversion can no longer be performed.

Inclusion of Calibration plus Drift

The original Allowable Value was an allowance for process rack drift only. The 1982 revision of the Allowable Value included the terms calibration accuracy, rack drift, M&TE and bistable setting accuracy. If a plant used an M&TE accuracy ratio of 10:1 and calibrated the channel in a string very accurately, i.e., did a very good job because the as-left condition of the channel was very close to the desired nominal value, then a very significant quantity of margin to the Allowable Value was built in for drift. For the case where the as-left condition is 0.1 to 0.2 % span from nominal, an allowance for drift of 400 to 800 % of expected is created. Satisfying the current fixed magnitude Allowable Value in this instance would not result in an appropriate evaluation of channel operability or provide the important hint of a channel initiating a failure mode. While it is inappropriate to penalize a plant for doing an excellent job of calibration, it is also inappropriate to perform operability evaluations based on a single fixed magnitude Allowable Value that has no relationship to the performance of the channel.

Inclusion of Process Measurement Accuracy Terms

Several functions in the Westinghouse protection system design are subject to quite large Process Measurement Accuracy (PMA) terms. PMA terms are non-instrument related effects, e.g., streaming in the Hot Leg or Cold Leg, reference leg heatup effects in level channels, effects of changes in process pressure in level channels, axial and radial power redistribution effects on neutron leakage fluence for Nuclear Instrumentation System (NIS) channels and secondary side power calorimetric effects on NIS channels. The variance of these PMA terms can cause significant difficulty in determining channel operability, particularly during startup. However, it should be recognized that if these terms are accounted for, additional variance in indicated levels or NIS power or current can be expected. It is inappropriate to utilize an Allowable Value as the determination of channel operability for a function that exhibits these effects. For example, the Allowable Value for any NIS function of Westinghouse design will account for process rack drift only or the sum of process rack calibration accuracy, M&TE, process rack drift and bistable setting accuracy only. All of these parameters can be evaluated by test and measurement in the process racks with a known input source and a digital voltmeter. At no time has Westinghouse ever included a PMA term in the determination of an Allowable Value. PMA terms can and do vary as functions of thermal power level, axial and radial power distribution, cold leg temperature, core life, level and ambient temperature. To include the most limiting nonconservative PMA terms in the determination of an Allowable Value would result in the definition of an extremely large magnitude value that has no relationship to the operability of the process rack modules.

It is straight forward to demonstrate that the process rack channel modules are operable, i.e., are accurately calibrated, have not drifted significantly, thus will satisfy the Allowable Value and will trip in an appropriate manner (less than the Safety Analysis assumed thermal power level). Yet the channel indicates nonconservatively, e.g., 5 to 10 % Rated Thermal Power lower than actual, due to redistribution effects or temperature. This condition can occur at low thermal power levels with deep control rod insertion when the NIS channel process rack modules have just been calibrated using a known current source and a very accurate DVM. This is acceptable as long as the errors due to redistribution or temperature have been accounted for in the overall channel uncertainty calculation. However, plants have declared NIS channels performing in this manner inoperable because the bistable would not trip at an indicated thermal power less than the Allowable Value noted in the Technical Specifications. Yet there is not a thing wrong with the channel (it is reporting the plant condition as it sees it), the effects noted are accounted for in the uncertainty calculations but the effects noted were not considered in the determination of the affected Allowable Value.

Conversely, the process rack channel modules can respond appropriately to a known current input, but because the variances in PMA terms have not been accounted for, the channel will not trip appropriately and **should** be considered inoperable. In this second instance the NIS channel process rack modules are responding appropriately to the input signal. However, due to not accounting for the redistribution or temperature effects in the overall channel uncertainty calculation, the channel is inappropriately scaled, i.e., the assumed input currents are too high to cause a trip at the required thermal power level, and thus the channel is inoperable even though the calibration and drift procedures were satisfied. It is therefore important to determine the conditions under which an action is required and the differences that may be introduced by the plant not being in that configuration. Those differences should then be evaluated and accounted for in the overall uncertainty calculation if necessary. But accounting for those same differences would not be appropriate in the determination of an Allowable Value or in the determination of acceptable operation with respect to an Allowable Value if they are not specific to the definition of an Allowable Value.

Using the Allowable Value as an As-left Tolerance

A troubling use of the Allowable Value by some plants has been as an as-left tolerance. The plant calibration procedure requires that a channel be calibrated to within a specified tolerance, as-left$_{cal}$. The channel functional test procedure requires that a channel be found to be less than the Allowable Value. This same procedure allows a channel found to be less than the Allowable Value to be left at the same point, even if the as-found condition is outside the calibration procedure as-left tolerance. This effectively defines a second as-left condition, as-left$_{AV}$, where as-left$_{AV}$ is farther from the nominal Trip Setpoint than as-left$_{cal}$. The problem that arises from this practice is that the initial condition for the next surveillance interval is outside the assumptions of the uncertainty calculations, i.e., the effective calibration accuracy based on as-left$_{AV}$ results in a larger calibration tolerance than the calibration accuracy assumed in the uncertainty calculations. Thus it allows continued use of equipment that has not been verified to meet the as-left$_{cal}$ tolerance, a basic requirement for operability. This can result in one of two conditions, 1) an as-left condition closer to the Allowable Value, where further drift in the nonconservative direction may result in exceeding the Allowable Value prior to the end of the surveillance interval, or 2) an as-left condition significantly farther away from the Allowable Value, i.e., drift in the conservative direction with respect to the nominal Trip Setpoint. This second condition results in the equivalent of the case described in the section Conservative Setting of Trip Setpoints. An initial condition assumption made in all Westinghouse uncertainty calculations is that the channel as-left condition is within that specified by the calibration accuracy term. This is treated as a two sided error, thus any time a channel is found outside the specified calibration accuracy, it must be returned to within that accuracy for the start of the next surveillance interval, regardless of the direction of the drift (conservative or nonconservative).

For the Westinghouse setpoint methodology, the concept of a fixed magnitude Allowable Value being the sole determinant of channel operability is insufficient. Each of the examples presented above identifies ways in which the plant may do an excellent job in setting up a channel, but due to a limited viewpoint will not perform an initial operability evaluation based on the actual drift of the channel. It has also been noted that the plant can intentionally introduce safety margin, which even decreases the magnitude of the channel operational margin, yet hides the true magnitude of the channel drift when an as-found value is compared against the current fixed magnitude Allowable Value. Finally it is possible for the plant to introduce large M&TE errors or the vendor to overstate the magnitude of the process rack drift, both of which are conservative in the determination of the acceptability of a Trip Setpoint or Safety Analysis Limit, but are nonconservative practices with respect to determination of an Allowable Value that is used for operability determination.

RECOMMENDATIONS

Calibration

To reduce some of the operability determination nonconservatism that can be introduced by overly conservative calibration practices or inappropriate use of the Allowable Value as an upper limit for a process rack as-left value, Westinghouse has maintained for several years the advisability of use of a two sided calibration tolerance about the nominal Trip Setpoint[3]. In this approach, the nominal Trip Setpoint is defined as a nominal equality with both a high and a low limit (limits in both the conservative and nonconservative directions) on the as-left calibration condition. This is in contrast with the current Technical Specification approach of a one sided inequality (limit only in the nonconservative direction). The proposed limitation on calibration tolerance limits the magnitude of undetected drift allowed in an operability determination as compared to the old method which uses a comparison of as-found to the Allowable Value.

Measurement & Test Equipment Accuracy

To reduce the nonconservatism introduced into an operability determination by M&TE, Westinghouse maintains that equipment with the best reasonable accuracy be utilized for the performance of instrumentation calibration and for the determination of the as-found condition. For process racks, an M&TE accuracy ratio of 10:1 is easily achievable. For transmitters, achieving an equivalent ratio is very difficult, particularly without the use of deadweight testers. Based on their sensitive nature, Westinghouse does not recommend the use of deadweight testers in an uncontrolled environment, i.e., they are not recommended for in-plant field use. However, even though there are practical limitations on the M&TE accuracy ratios on the front end of transmitter calibration, this should not preclude the use of the most accurate, reasonably available digital multimeter on transmitters.

Limiting Drift

Based on the operability determination limitations inherent in the use of a fixed magnitude Allowable Value, it is suggested that the drift magnitude utilized in Allowable Value determination calculations be less than that assumed in the channel uncertainty calculations. For example, if an instrument channel process rack module(s) drift value of X % span is used to determine the acceptability of the difference between the Safety Analysis Limit and the nominal Trip Setpoint, a drift value of (X/2) % span may be used to determine the Allowable Value. Under most circumstances, a limitation of this magnitude would assist in reducing the effects of a calibration as-left on the conservative side of the nominal Trip Setpoint when comparing an as-found value with a fixed magnitude Allowable Value on the nonconservative side of the nominal Trip Setpoint. For example, assume for a decreasing protection function a channel with a nominal Trip Setpoint of 10 % span, a process rack calibration tolerance of ±0.5 % span, an Allowable Value of 9 % span (based on process rack drift only), a process rack as-left value of 10.5 % span, a process rack as-found value of 9.25 % span and an M&TE accuracy ratio of 10:1 (so it can be ignored). The as-left is within the calibration tolerance, thus it is acceptable. The as-found is greater than the Allowable Value, thus under the current Technical Specifications the channel is considered operable. However, the original assumption for drift in the determination of the Allowable Value is 1 % span (nominal Trip Setpoint minus Allowable Value). It is obvious that the difference between the as-left and as-found conditions is greater than 1 % span. Therefore the process rack has experienced drift greater than that assumed in the uncertainty calculation and potentially greater than that specified in design. Based on this second determination, the channel may not be operable. If a fixed magnitude Allowable Value must be determined, utilization of the drift limitation of (X/2) precludes allowing for drift in excess of that assumed in the uncertainty calculations or that specified in design. It is recognized that if the magnitude of the calibration tolerance or the M&TE accuracy approximates the magnitude of the limited drift, the actual effect of this process would be to eliminate measurable drift. This approach results in some drift allowance only if the as-left condition is close to the nominal value and the M&TE accuracy ratio is 10:1 or better.

Relative Drift - Trending

The most effective way to minimize the effects of comparison of an as-found value with a fixed magnitude Allowable Value is to perform a different type of comparison. It has been noted that operability determination based on comparison to a fixed magnitude Allowable Value is subject to several weaknesses. These weaknesses are eliminated when operability is determined by a comparison of relative drift. Relative drift is determined by the difference between as-left and as-found recorded values. Under relatively controlled conditions, i.e., measurement ambient temperatures within 10 °F, power supply voltages within 3 %, process pressure at atmospheric (for Δp measurement systems), input values ramped up slowly or in small step increments (essentially static conditions) and five time constants waited prior to measurement, the determination of relative drift is the most accurate means of determining module response characteristics over a period of time. Recording of as-left and as-found data (especially by entry into a computerized data base at the time of test performance) then allows the use of trending relative drift for each specific process rack module and transmitter. Significant variation from past performance can then function as a tripwire to perform additional tests or evaluations, e.g., calibration/reference accuracy characteristics, response time characteristics and short term drift characteristics. These evaluations, both calibration and drift, will acquire additional significance by placing more and more reliance on less and less data as the surveillance interval increases.

Proposed Technical Specification

The current Technical Specifications for protection systems for Westinghouse NSSS plants can be described in one of three ways, 1) "one column" where the Trip Setpoint is based on an inequality with a one sided calibration tolerance, 2) "two column" where the Trip Setpoint and Allowable Value are based on inequalities with a one sided calibration tolerance and 3)"five column" where the Trip Setpoint and Allowable Value are based on inequalities and three additional uncertainty calculation parameters are used in an operability determination equation in the event the Allowable Value is exceeded. As currently utilized, each has areas of nonconservatism when used for operability determinations. To address these areas of nonconservatism, Westinghouse has utilized aspects of each type to form a hybrid Technical Specification.

The revised specification would be a single column nominal Trip Setpoint. In the Bases of this specification the following would be noted:

1) The Trip Setpoint is a nominal value with acceptable calibration to the nominal value being defined as nominal value ± calibration tolerance.
2) The calibration tolerance is defined as that value necessary to confirm the Reference Accuracy as defined by SAMA PMC 20.1-1973[4].
3) M&TE accuracy for the process racks should be defined and controlled (it is recommended that equipment meeting an accuracy ratio of 10:1 or better be defined in procedures or administrative control documents).
4) M&TE accuracy for the transmitter should be the most accurate practicable for the measurement of pressure or Δp (it is recommended that digital multimeters meeting an accuracy ratio of 10:1 or better be defined in procedures or administrative control documents).
5) Initial determination of process rack operability should be defined as rack drift (as-left minus as-found) less than the required value for the surveillance interval. In the event excessive rack drift is

noted, operability of the device may be confirmed by evaluation of other device characteristics, e.g., trended drift data, calibration characteristics, response time characteristics and short term drift characteristics.

6) Initial determination of transmitter operability should be defined as transmitter drift (as-left minus as-found) less than the required value for the surveillance interval. In the event excessive transmitter drift is noted, operability of the device may be confirmed by evaluation of other device characteristics, e.g., trended drift data, calibration characteristics, response time characteristics and short term drift characteristics.

7) A device (RTD, relay, etc.), transmitter or process rack module, whose as-found value is in excess of the calibration tolerance but within the operability criteria is considered operable but must be recalibrated such that the as-left value is within the two sided calibration tolerance.

The requirements contained in the Bases of the Technical Specifications effectively set up administrative control of the calibration, M&TE accuracy ratios and drift of the transmitter and process racks. They allow operability to be defined for both transmitter and process racks and, while setting minimum standards, allow the use of other characteristics to confirm operability. The Technical Specification limit is for the parameter of primary interest in the Westinghouse setpoint methodology, i.e., the nominal Trip Setpoint. The Bases allow administrative control of those parameters which vary from channel to channel and device to device. As a side benefit, this revised Technical Specification should not change just because a piece of hardware is changed in the plant. With the "two column" and "five column" Technical Specifications changing a transmitter, the process racks (for example, analog to digital), a DVM or even a Process Measurement Accuracy term magnitude can result in a change in one of the values noted in the Technical Specifications. Westinghouse rarely changes the nominal Trip Setpoint, thus the revised Technical Specification allows the flexibility of change to the plant hardware or procedures without explicit changes to the Technical Specifications.

CONCLUSIONS

Based on evaluation of the various mechanisms noted, it has been concluded that for the Westinghouse setpoint methodology use of a single process rack as-found value for comparison with a fixed magnitude Allowable Value is insufficient to determine channel operability. In such comparisons, it is quite easy to conclude that process rack drift is consistent with the assumptions of the setpoint methodology yet the actual drift experienced by the channel or device is 200 to 400 % of drift allowed by design or expected. Such determinations are misleading with regards to actual device performance. It is suggested that a more appropriate comparison is relative drift (as-left minus as-found) versus design drift. It is also more appropriate to control the as-left condition of the device more closely, i.e , use of the most accurate M&TE practicable and use of a two sided calibration tolerance about a nominal equality setpoint as opposed to a one sided tolerance with a limiting inequality setpoint. It is also suggested that operability can be defined not only by relative drift, but also by conformance with other design performance characteristics, e.g., calibration/reference accuracy, response time and short term drift. The Technical Specification proposed is a hybrid of the three different approaches currently in use in Westinghouse NSSS Technical Specifications. The hybrid utilizes a single nominal Trip Setpoint with bases that define conditions placed under administrative control. This results in control of the one parameter believed to be the most important in the Westinghouse setpoint methodology and flexibility to reflect actual plant hardware and control practices. Based on the above, it is concluded that the use of the current fixed magnitude Allowable Value can be more fiction than fact when it comes to device operability determination.

DEFINITIONS

As-found The condition in which a transmitter, process rack module or process instrument loop is found after a period of operation. For example, after a period of operation, a transmitter is found to deviate from the ideal condition by -0.5 % span. This would be the as-found condition.

As-left The condition in which a transmitter, process rack module or process instrument loop is left after calibration or bistable trip setpoint verification. This condition could be better than the calibration accuracy for that piece of equipment. For example, the permitted calibration accuracy for a transmitter is ±0.5 % span, while the worst measured deviation from the ideal condition after calibration is +0.1 % span. In this instance, if the calibration was stopped at this point, i.e., no additional efforts were made to decrease the deviation, the as-left error would be +0.1 % span.

REFERENCES

[1] Scientific Apparatus Makers Association Standard PMC 20.1-1973, "Process Measurement & Control Terminology," New York, NY, 1973, p 36.

[2] Instrument Society of America Standard 51.1-1979, "Process Instrumentation Terminology," Research Triangle Park, NC, 1979, p 32.

[3] Tuley, C. R., Miller, R. B., "The Significance of the Nominal Trip Setpoint in the Westinghouse Setpoint Methodology," Instrumentation, Controls, and Automation in the Power Industry, Vol. 34, Proceedings of the Thirty-Fourth Power Instrumentation Symposium (1st Annual ISA/EPRI Joint Controls and Automation Conference), St. Petersburg Beach, FL, June, 1991, p 134.

[4] Scientific Apparatus Makers Association Standard PMC 20.1-1973, "Process Measurement & Control Terminology," New York, NY, 1973, p 4.

© 0074-056X/95/415-430/$0 + .50pp

UPGRADE TO 1970s VINTAGE TURBINE CONTROLLERS FOR IMPROVED PERFORMANCE INFORMATION AND CONNECTION TO PLANT DISTRIBUTED CONTROL SYSTEM AT CONESVILLE UNITS 5 AND 6

Drake B. Bosler
Principal Engineer
Turbine Control Upgrades
Westinghouse Electric Corp
Process Control Division
200 Beta Drive
Pittsburgh, PA 15238

Richard L. Gail
Performance Supervising Engineer, I&C
American Electric Power Service Corp
Columbus Southern Power Co.
Conesville Power Plant
47201 County Road 273
Conesville, OH 43811

KEYWORDS

Turbine Control, DCS, Modernization, Plant Networks, System Integration

ABSTRACT

This paper describes the decision process which led to an upgrade of 1970s vintage turbine control systems and the replacement of a similar vintage Data Acquisition System (DAS) with a new Distributed Control System (DCS). It also discusses details of the installation and start-up of the turbine control systems upgrade. This upgrade made it possible to connect the turbine control systems to the DCS data highway, and from there to a plant information network. A review of the turbine information now available to operators via the DCS operator stations is presented, along with the resulting improvements that have been realized in plant operations and performance.

BACKGROUND

The Conesville Power Plant is located in Conesville, Ohio, 50 miles to the east of Columbus. Units 5 and 6 are coal-fired, Combustion Engineering tangential, 2500 psi drum type controlled circulation boilers with Westinghouse turbine/generators; Combustion Engineering flame supervisory safety systems; and United Oil Products flue gas desulfurization systems. The units are rated at 400MW, and were placed on-line in 1976 and 1978, respectively.

The original DAS was manufactured by Foxboro. This mid-70s vintage system was based on DEC minicomputers and offered limited CRT functions (alarm and character values; no graphics; black and white display). System reload was done with paper tape, which was cumbersome and time consuming even under ideal conditions. Historical information was not available except for data logged periodically to hard copy printers. One system serviced both units.

The same vintage as the DAS equipment, the Westinghouse Digital Electro-Hydraulic (DEH) MOD I turbine control system is based on the Westinghouse P2000 CPU. Each DEH features an analog manual backup subsystem and an operator interface with back-lighted pushbuttons, analog meters, and two five-digit digital displays. There is no rotating memory, and the original paper tape loader had been replaced with a PROM loader. Prior to the upgrade, historical data recording was not available. One DEH is installed on each unit as a stand-alone package. Originally, each DEH had an RS232 data link to the DAS for Automatic Turbine Startup (ATS) message display on the DAS CRTs.

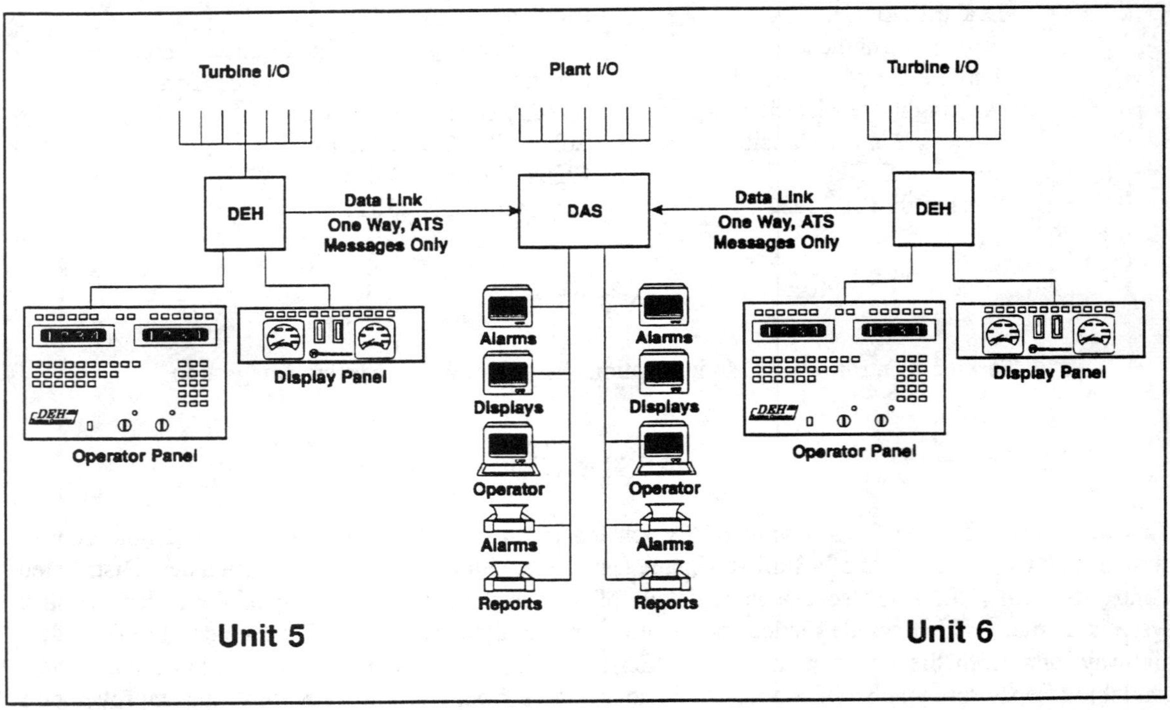

FIGURE 1

Block Diagram of Original DAS and Turbine Control Systems

FACTORS BEHIND DECISION TO UPGRADE

Vendor support was no longer available for the DAS, and there were problems with parts and service availability. Plant personnel were not capable of trouble-shooting the system in the event of a failure, and therefore depended on a small, third-party firm for this support. System failures had been increasing over the years, and there was a concern that loss of the DAS might result in unit shutdowns to prevent potential damage due to insufficient data for operations. This situation was further complicated by the fact that a single DAS monitored both units.

The utility decided to proceed with a replacement of the single DAS with two separate DCS systems. This replacement will not be discussed further in this paper, except for the connection between the DEH and the DCS. The system chosen for the DAS replacement was the Westinghouse Distributed Processing Family (WDPF II).

For the DEH, technical support, parts, service and upgrades were still available from the original vendor and another third party. The utility decided to upgrade the existing systems rather than replace them with new systems. This decision was based on the high availability record of the systems, the existence of support, and the type of upgrades that were available.

The utility first decided to proceed with a stand-alone upgrade for each DEH. This upgrade would provide graphic display capabilities for operations and maintenance, and historical storage and retrieval for performance monitoring and maintenance. Westinghouse was chosen as the vendor for these upgrades because they had expertise with the existing systems and turbines. They could provide an upgrade which was based upon current product and the upgrade could be easily connected to the DCS system to integrate the turbine information on to the data highway. By following this upgrade approach, capital expenditures for replacing the turbine control systems could be deferred while the plant could immediately reap many of the benefits of a new system.

DESCRIPTION OF TURBINE CONTROL SYSTEMS UPGRADE

The existing DEH core memory boards (quantity 3, total 40K 16 bit words of memory) were replaced by a single Static RAM board (64K 16 bit words of memory) with the same form factor as the original boards. The new RAM board has dual-ported memory and an on-board microprocessor which can interface to external computers via serial ports using a ModBus(tm)-type protocol. Two ports are RS422 and able to drive long distances at high speed (up to 38.4Kbaud); one is RS232 for remote modem communications; and one is RS232 to accommodate a dumb terminal which serves as a built-in datascope.

In this upgrade, one RS422 port is connected to an IBM PS/2 PC running OS/2 and the vendor's current, standard PC software for graphics, historical storage and logging. A serial communications driver interfaces with the RAM board to acquire data which is converted into point values in the PC. The converted point values then appear to the standard software as if they had come from the data highway. Point values are polled from the DEH at various rates, depending on how often that data is updated by the existing DEH application software. These rates vary from 1 second to 70 seconds.

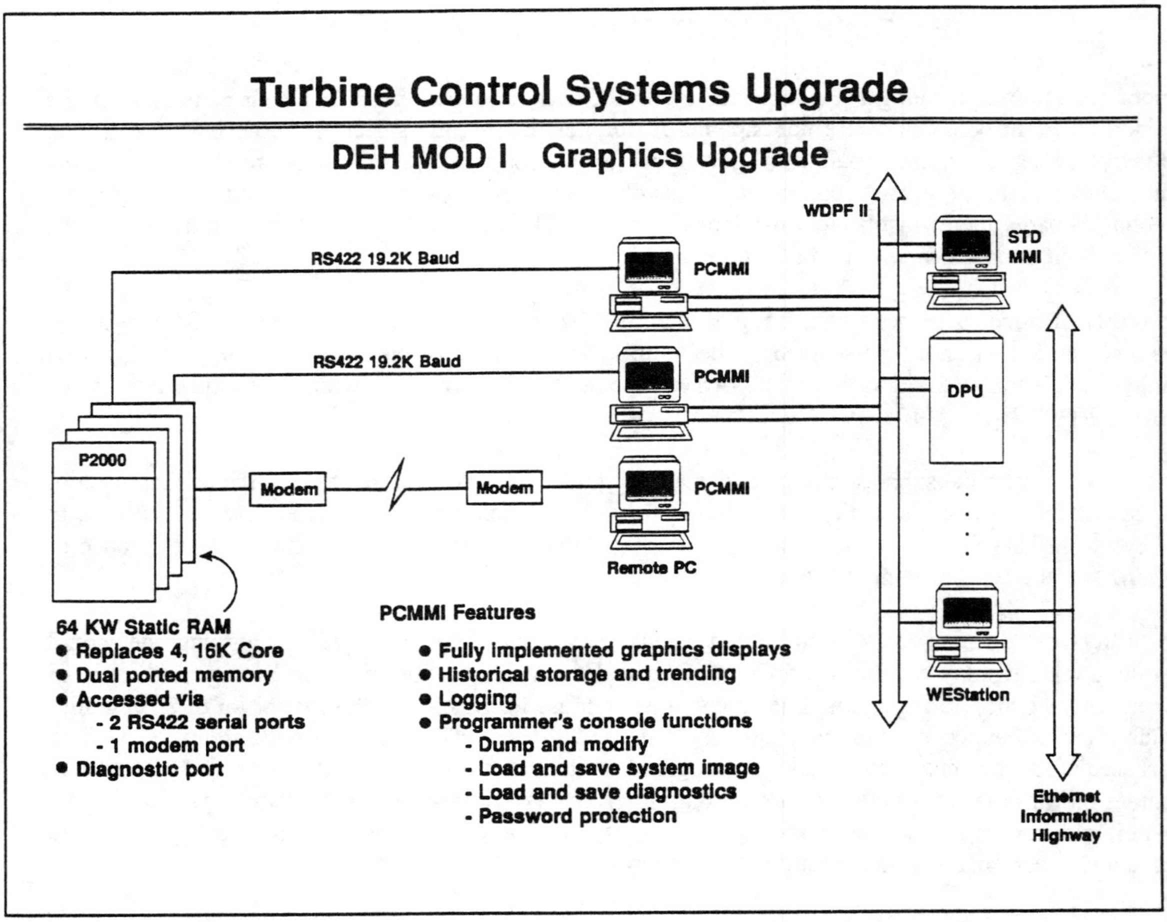

FIGURE 2
Block Diagram of the Turbine Control Upgrade

A set of custom graphics which display all pertinent data in a form similar to the graphic displays available with the vendor's current DCS-based turbine controller was also included. These displays were adapted to fit the type of data and operation of the existing DEH. The graphics make it possible for operators to view data in a fashion which allows them to make better decisions. See Appendix A for a complete list of the graphic displays.

In addition, all of the ATS function operator guide messages were grouped together into four Operator Guide graphics which display start-up information as the DEH software sees and interprets it. These displays are in a checklist format, with text displayed as blue (OK) or red (NOT OK). This vastly simplifies an operator's job compared to the previous method of watching messages scroll through a small window on the black and white DAS CRT. There are many messages in the ATS function, and prior to this upgrade, important messages could be easily overlooked.

An additional 25 graphics related to the ATS function were also included with the upgrade. These graphics feature a message status overview from which an operator can poke to another screen containing the actual text of each message. Active messages are highlighted. Due to memory constraints in the original systems, many of the original messages were abbreviated and thus cryptic. With the graphic upgrade, they have been expanded into a more understandable form.

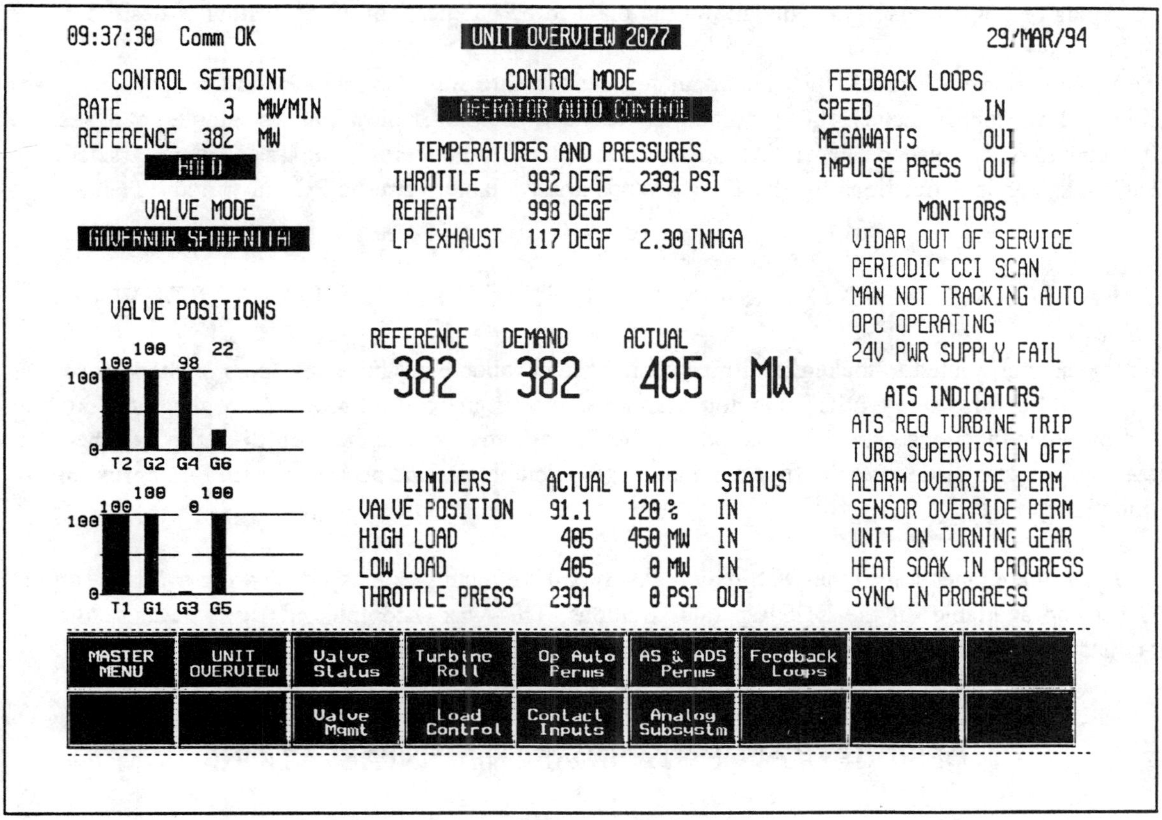

This display is the most commonly used graphic during normal unit operation. The operator can quickly glance at the display and see the most critical information about the turbine control operating status, control information, and supervisory status. Items of interest include the information displayed in reverse-video (control mode, valve mode and control setpoint status), analog data (values displayed in different colors depending on conditions), and digital data (text displayed in different colors depending on conditions).

FIGURE 3
Unit Overview Display

For engineering personnel, a custom program was provided to allow direct access to the P2000 CPU memory for system loading, programming changes, and problem debugging. This replaced the original typewriter-based "Programmer's Console" and paper tape loading devices with a windowed function featuring pull-down menus. Loading of the system memory can be accomplished in less than 1 minute, as compared with 1 hour for paper tape and 3 to 10 minutes with PROM. If required, the system memory can also be saved to document a change of tuning constants; something which could not be done with either paper tape or the PROM loader.

Source code and tools are provided with the upgrade so that engineering personnel can easily modify or add points or graphic displays. This allows the plant to be independent of the vendor if desired.

The vendor's current, standard PC historical storage software was also included in the upgrade. It was initialized for approximately 100 important I/O and computed variables. Data is sampled at 1 second, 10 seconds, or 60 seconds depending on type, and stored after being compared against a selectable deadband. Standard functions for displaying trends are available from the PC via window menus.

FACTORS BEHIND DECISION TO CONNECT TO DCS DATA HIGHWAY

Plant personnel wanted to make the turbine control data available on the DCS Operator Stations so that there would be only one type of operator interface device in the control room. Also, the data could be used by other functions once available on the DCS data highway. The decision to connect to the DCS data highway was made after the first stand-alone graphic upgrade had been in service for approximately 6 months.

Following the connection to the DCS data highway, all data and graphics which were available on the PC became available on the DCS Operator Stations. This was accomplished easily because the PC graphics are in the same standard format as the rest of the DCS system graphics.

DESCRIPTION OF STAND-ALONE INSTALLATION

The Unit 5 system was upgraded first, while the boiler was down for a 3-day tube repair outage. First the P2000 CPU was checked, then core boards were removed, and the RAM board was installed. The cable from the CPU cabinet to the PC in the control room had already been pulled. The plant-specific DEH application software had already been prepared and placed on a PC floppy disk at the vendor's lab facility using an identical operational DEH. After the RAM board was installed, diagnostics were run to check operation of the CPU. The DEH software was then loaded from the PC and the system was restarted.

In the control room, each point value displayed on the PC was compared to a value accessed from the pushbutton panel to verify correctness of the information. Training on the operation of the graphic display was then conducted for operators and technicians.

The operators immediately recognized the value of the larger picture which could be seen with the graphical representation of the turbine data, and were very pleased. Navigating through the screens is very simple, by "double-clicking" the mouse on graphic "buttons" which contain the title of each available screen.

During the installation on Unit 5, a problem with the CPU power supply was noted, and the supply was changed. As was discovered later, however, the real problem was with the I/O driver board which was causing distortions in the power supply waveform. This problem was not even noticed until the day after the installation, when the random stops began to occur. After further diagnosis, the I/O board was replaced and the problem was corrected.

One lesson which was learned on the Unit 5 installation is that problems can occur when older systems are powered down and restarted, and when boards are removed and reseated. There may also be problems which exist, but are not known. It is prudent to allow enough time for contingencies and to be sure to have ready access to "known good" spare parts before beginning any task such as this upgrade.

The complete upgrade, excluding training, took only one-half day. The Unit 6 system was similarly upgraded 6 months later during an outage.

DCS DATA HIGHWAY INTEGRATION

The process of adding the PC with the DEH information to the DCS data highway was done in several small steps. The description below applies to a single unit integration. The total process of integrating both systems was accomplished with approximately 40 labor-hours per unit over the course of a three week period.

First, the PC was moved from the control room to the Instrument and Control Shop. There the PC would fulfill a new dual purpose. It would serve as the DEH interface device as well as an engineering station on the DCS for the I&C department. The cable between the PC and the turbine control system was moved at the same time, reconnected, and the PC was restarted in the new location.

Next, the data highway cables to connect the PC were pulled into place. The highway layout had been planned such that the PC would be at the end of the physical highway routing. After the cables were pulled, each of the redundant highway cables was connected and reterminated at the PC location. All of this work was done in advance of the arrival of the highway interface hardware, without disrupting the operation of the DCS.

Finally, the data highway interface hardware was added to the PC and attached to the coax connectors on the previously pulled cables. The PC configuration file was modified to select the device driver for the highway interface, and the PC was rebooted onto the data highway. At this time, additional software for the engineering station functions was loaded, initialized, and tested. In addition, appropriate window selection menus were created. While this software was unrelated to the turbine controls upgrade, it does allow plant personnel to derive additional value from the purchase of the PC.

To integrate the turbine data with the DCS data, the system point directory had to be updated with the new DEH points. The DCS and DEH shape libraries were merged to provide all the necessary shapes used in the turbine graphics. The unedited DEH graphics source code was then recompiled against the new point directory to provide the proper links. All of this work was done on the PC while the DCS was fully operational.

Following this work, the new point directory, shape library and graphics were downloaded via the data highway into each DCS Operator Station. To facilitate display selection, a custom key on each operator station keyboard was assigned to display the Turbine Unit Overview graphic from which operators could select any of the other turbine displays.

All of the integration was done while the units were on-line with no disruption to unit or DCS operation, except for the few minutes required to download to each operator station.

Now that all of the turbine data is available on the DCS Operator Station CRTs, the operators use the original pushbutton panel only for control operations such as mode change, setpoint change, or valve testing.

RESULTS AND SAVINGS

With the recent completion of a plant-wide information network, data for units 4, 5, and 6, including critical turbine information passed from the existing DEH systems, is now available to plant management personnel at their desks.

With the highway connection completed, the PC historical storage capability is no longer required because the operators are using the DCS historical storage functions which are accessible from all operator stations. Operators can add turbine control points to the trend list at any station, and display data at up to 1-second resolution.

For four years, there had been a nasty bump in the load ramp at around 260MW, but the root cause could not be determined. The resulting boiler control instability had caused an occasional unit trip. This situation had required slowing the load ramp to get through the range, and had made it impossible to put the unit on Automatic Generation Control (AGC).

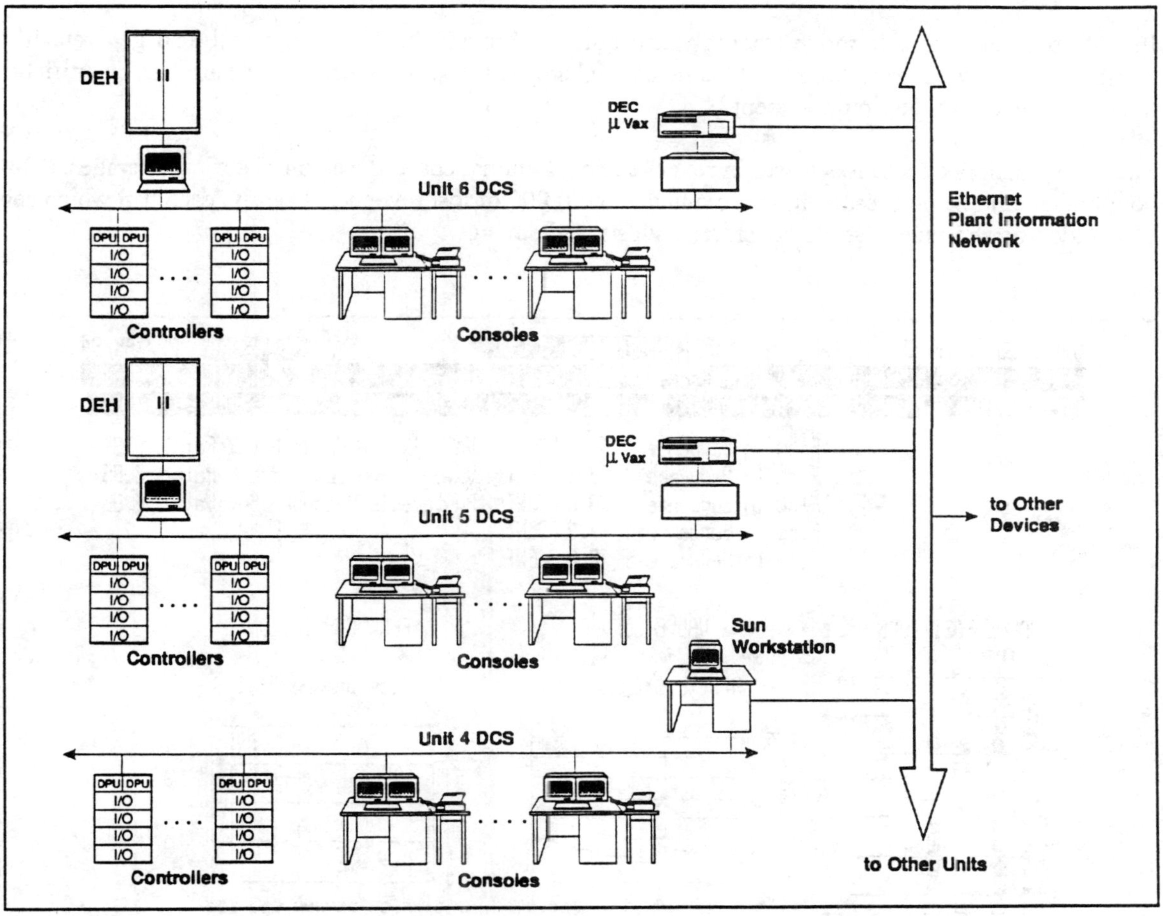

FIGURE 4
Block Diagram of Final Interconnected Systems

By utilizing the graphics and the trending features of the upgrade (the stand-alone PC, before highway connection), plant performance engineers were able to watch the operation of all of the valves during the load ramp and determine that one valve was coming open too early in the sequence. Using the trending, they were then able to more finely tune and adjust the valve curves to eliminate this problem.

The Plant Performance Section has projected a heat rate improvement of approximately 20 BTU/KWH as a result of understanding this problem and using the upgrade to optimize the turbine valve curves to expend the maximum amount of energy per valve sequence. This equates to roughly $38,400 savings of fuel per unit per year, based on a three-year historical average fuel cost for these units.

In addition, being able to move smoothly and quickly through the load range will also generate fuel savings and allow better response to load demand changes. However, these savings are less quantifiable than the direct heat rate improvement.

Each trip which occurs causes losses in revenue and additional costs. Based on historical averages, a 24-hour outage on one unit results in approximately $100,000 of lost revenues. Clearly, each trip which can be avoided or shortened can be considered substantial savings.

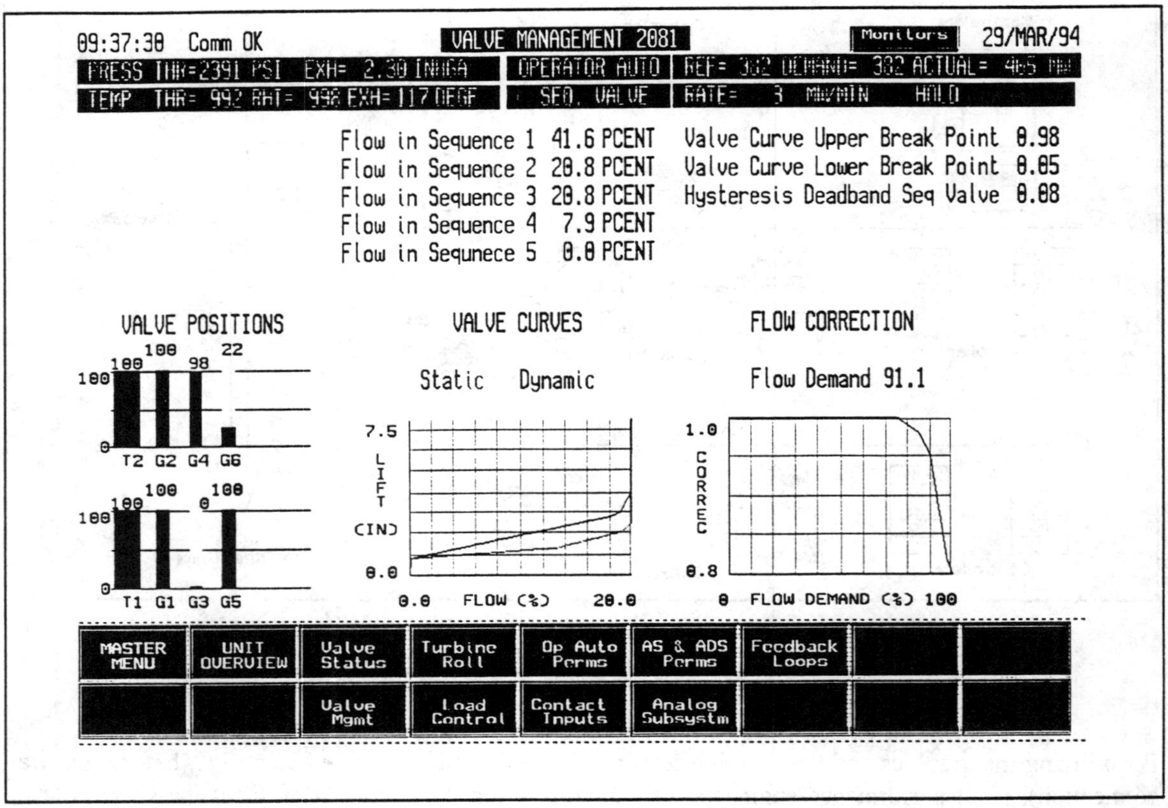

Engineer and operators interested in the flow rates through each governor valve during operation in sequential valve mode can see the breakdown through each sequence (GV 1 and 2, GV 4, GV 5, GV 6, GV 3)

FIGURE 5
Valve Management Display

With the turbine data available in graphic display format, the operators can make better decisions faster, thus reducing the time it takes to get the unit on-line. In particular, the ATS checklist screens make this process faster and more consistent. However, the results have been difficult to quantify, and are based on the general opinions of the operators.

One unit trip occurred after the upgrade which could not be explained. Operator procedures were suspect, but with the additional historical data provided by the upgrade, it was determined conclusively that a procedural error was the cause of the trip. The resulting explanation to management was therefore concrete, and procedures were reviewed to eliminate this error in the future. As stated previously, each trip which can be avoided or shortened can be considered as substantial savings.

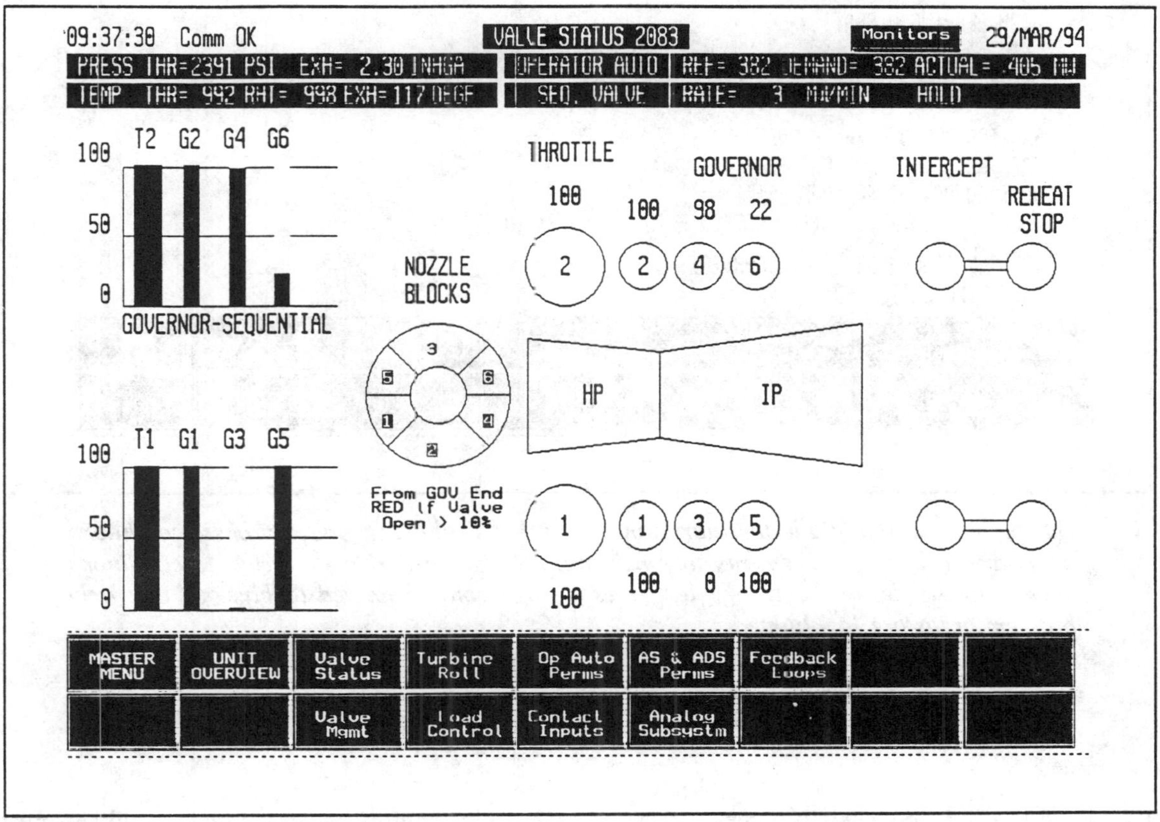

The valve status display gives a physical representation of the governor and throttle valves relative to the HP/IP section and steam entrance into the nozzle block. With the bar graphs and digital data, an operator can quickly learn the concept of governor valve control and it's effect on unit operation.

FIGURE 6
Valve Status Display

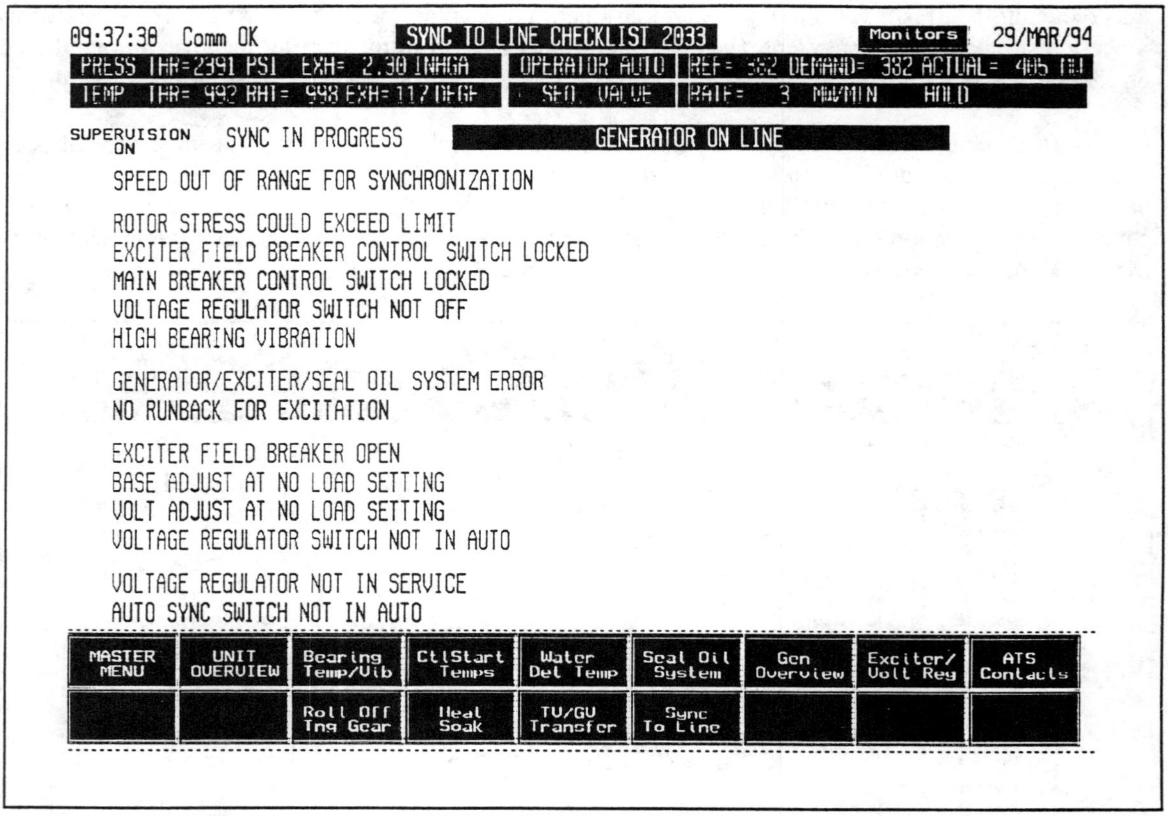

During unit startup, plant operators can follow this display to monitor process conditions necessary to meet requirements to synchronize the generator to the grid. As conditions change to the desired state, the text lines change color from red to blue, so that only problem areas are highlighted.

FIGURE 7
Sync-to-Line Checklist Display

By adding this upgrade, capital monies for an eventual turbine control system upgrade could be spread over several years. The estimated cost of a replacement system versus the upgrades is considered to be greater than 10 to 1.

The fact that all turbine data is now available on the highway and accessible by the plant performance calculations may make it possible to do turbine efficiency testing in the future at any time and any load, without adding extra temporary equipment. This would represent a tremendous savings in labor hours.

CONCLUSION

In the present climate of tight financial constraints, when vendor support for original systems is still available and the systems themselves are still performing at high levels of availability, certain upgrades can be very beneficial and provide a high Return-on-Investment (ROI). Capital investment for eventual replacement of the systems can be deferred into the future, and better planned. This is especially true of upgrades which can provide integration of information from the existing control systems into modern DCS systems and plant information networks. Substantial savings, both in dollars and productivity, can result when using this approach.

APPENDIX A. List of Graphic Displays

MASTER MENU
List of all graphic titles, grouped by Base system or ATS system
UNIT OVERVIEW
Overview of all critical information about the operating status of the controller

BASE SYSTEM

VALVE STATUS
Overview of turbine valve layout for valve testing, including nozzle block
TURBINE ROLL
Overview of all shaft information related to speed roll, including resonance ranges, critical speeds, vibration, expansion and eccentricity
OPERATOR AUTOMATIC PERMISSIVES
List of permissive conditions for both speed and load control, including valve tracking information
AUTO SYNC AND DISPATCH PERMISSIVES
List of permissive conditions for automatic synchronizing and automatic dispatch control
FEEDBACK LOOPS AND SPEED SELECTION
List of permissive conditions for feedback loop operation, and all speed inputs
VALVE MANAGEMENT
Valve positions, flows through each valve sequence, and a plot the flow vs lift curve
LOAD CONTROL FUNCTIONAL
SAMA functional diagram of the load control operation, including constants and dynamic values
CONTACT INPUT STATUS
Text description of each contact input, with hardware location and dynamic status
ANALOG SUBSYSTEM
Dynamic offset and span correction factors used by the A/D subsystem, including RTD CJB temperatures

ATS SYSTEM

TSI AND BEARING METAL TEMPS
Graphic representation of the shaft with vibrations, bearing metal temperatures, differential expansion, eccentricity, bearing
oil pressure and temperature
CONTROL START TEMPERATURES
Overview of steam paths and major metal temperatures
WATER DETECTION TEMPERATURES
Side section of the HP/IP case with cover and base metal temperatures

APPENDIX A. List of Graphic Displays (Cont'd)

SEAL OIL SYSTEM
Graphic representation of the seal oil system and dynamic data

GENERATOR
Generator information, including stator RTDs, hydrogen information, and stator current

EXCITER AND VOLTAGE REGULATOR
Side section of exciter with air cooler information, and dynamic text for digital conditions

ATS CONTACT INPUT STATUS
Text description of each contact input, with hardware location and dynamic status

ROLL OFF TURNING GEAR CHECKLIST
Dynamic operator guide information for roll off turning gear

HEATSOAK
Dynamic operator guide information for heatsoak, including trend for IP blade ring and IP bore temperatures

THROTTLE/GOVERNOR VALVE TRANSFER
Dynamic operator guide information for TV/GV transfer, including valve positions and trend of steam chest metal and saturated steam temperatures

SYNC TO LINE CHECKLIST
Dynamic operator guide information for synchronizing generator to line

ATS MESSAGES
A total of 25 screens, one overview of all messages, from where an operator can poke down to another 24 screens with individual message text. Active messages are highlighted.

NOTE

All displays include a header with critical data on operating status and status of the communication link to the turbine control system

APPENDIX A. List of Graphic Displays (Cont'd)

SEAL OIL SYSTEM
Graphic representation of the seal oil system and dynamic data

GENERATOR
Generator information, including stator RTDs, hydrogen information, and stator current

EXCITER AND VOLTAGE REGULATOR
Side section of exciter with air cooler information, and dynamic text for digital conditions

ATS CONTACT INPUT STATUS
Text description of each contact input, with hardware location and dynamic status

ROLL OFF TURNING GEAR CHECKLIST
Dynamic operator guide information for roll off turning gear

HEATSOAK
Dynamic operator guide information for heatsoak, including trend for IP blade ring and IP bore temperatures

THROTTLE/GOVERNOR VALVE TRANSFER
Dynamic operator guide information for TV/GV transfer, including valve positions and trend of steam chest metal and saturated steam temperatures

SYNC TO LINE CHECKLIST
Dynamic operator guide information for synchronizing generator to line

ATS MESSAGES
A total of 25 screens, one overview of all messages, from where an operator can poke down to another 24 screens with individual message text. Active messages are highlighted.

NOTE

All displays include a header with critical data on operating status and status of the communication link to the turbine control system

1994 POWID/EPRI AUTHOR INDEX

REINFORCE YOUR CREDIBILITY AND EXPERTISE...WITH ISA REPRINTS

Order quantities of your important ISA article!

If you have published an article in *INTECH, Industrial Computing, Motion Control*, or ISA Conference Proceedings, you can use reprints of that article to enhance your sales presentation or other promotional material.

Article reprints are a great way to highlight your company, product, or industry. And showing your by-line featured in a respected trade publication convinces your audience that **your message is more than just hype.**

Your ISA Reprint can include any of the following[1]:

- A cover headlining the article or paper title, author(s), plus your company's name and/or logo
- A reference or part number
- 3-hole punched copies
- Shrinkwrapping sets of 25, 50, 100, etc.

[1]Some options carry a small additional cost.

ISA Reprints are a proven tool for reinforcing your message.... Order today!

BLACK AND WHITE PRICE CHART[2]

(Other quantities, full color reprints, and additional customized options available. Call ISA for quotes.)

Quantity	Number of Reprint Pages (Count both sides of article. Cover is 1 additional page.)								
	1-2	3-4	5-8	9-12	13-16	17-20	21-24	25-28	29-32
100	$115	$198	$259	$331	$416	$489	$562	$654	$660
200	127	232	293	354	438	522	605	735	786
300	138	267	328	389	484	580	674	815	840
400	150	301	362	423	530	637	743	896	919
500	159	330	391	452	570	689	806	971	997
1000	205	474	535	596	772	948	1122	1344	1374
2000	366	773	995	1171	1519	1868	2157	2494	2903
5000	515	1037	1420	1734	2238	2730	3164	3616	4435
10000	860	1612	2398	3114	3963	4743	5607	6347	8263

[2]Prices are subject to change. Shipping is additional.